T0323450

GLOBAL ENVIRONMENTAL
SUSTAINABILITY

GLOBAL ENVIRONMENTAL SUSTAINABILITY

Case Studies and Analysis of the United Nations' Journey toward Sustainable Development

DR. CHOY YEE KEONG

Faculty of Economics, Keio University, Tokyo, Japan

ELSEVIER

Elsevier
Radarweg 29, PO Box 211, 1000 AE Amsterdam, Netherlands
The Boulevard, Langford Lane, Kidlington, Oxford OX5 1GB, United Kingdom
50 Hampshire Street, 5th Floor, Cambridge, MA 02139, United States

Notices
Knowledge and best practice in this field are constantly changing. As new research and experience broaden our understanding, changes in research methods, professional practices, or medical treatment may become necessary.

Practitioners and researchers must always rely on their own experience and knowledge in evaluating and using any information, methods, compounds, or experiments described herein. In using such information or methods they should be mindful of their own safety and the safety of others, including parties for whom they have a professional responsibility.

To the fullest extent of the law, neither the Publisher nor the authors, contributors, or editors, assume any liability for any injury and/or damage to persons or property as a matter of products liability, negligence or otherwise, or from any use or operation of any methods, products, instructions, or ideas contained in the material herein.

British Library Cataloguing-in-Publication Data
A catalogue record for this book is available from the British Library

Library of Congress Cataloging-in-Publication Data
A catalog record for this book is available from the Library of Congress

ISBN: 978-0-12-822419-9

For Information on all Elsevier publications visit
our website at https://www.elsevier.com/books-and-journals

Publisher: Candice Janco
Acquisitions Editor: Peter J. Llewellyn
Editorial Project Manager: Chiara Giglio
Production Project Manager: Kumar Anbazhagan
Cover Designer: Matthew Limbert

Typeset by MPS Limited, Chennai, India

Dedication

This book is dedicated to the memory of my beloved father Choy Tuck,
my beloved mother Lee Yong Kan, and my beloved sister Chui Soh Lan
who have been my constant source of great encouragement
and inspiration.

Contents

4. Greening for a sustainable future: The ethical connection 213

List of Figures

List of Tables

List of Boxes

List of Appendixes

Preface

Looking back on the world's first international environmental conference, the Stockholm Conference held by the United Nations in 1972, which brought leaders across the globe together to facilitate global consensus on the urgency to address the challenges of global environmental degradation, we wonder what has been achieved so far. We know that the United Nations, committed to addressing this challenge, has made continuous and unrelenting efforts at convening a host of international environmental meetings and conferences to advance its core mandate of bringing about a wiser use of our natural environment. Hundreds of international declarations, agreements, regulations, protocols, agendas, and action plans were adopted at these conferences to facilitate and guide the global community to efficaciously address the challenges of balancing its three core values of sustainable development, namely, environmental protection, economic growth, and social equity. Thus a new world order was presented on the horizon held together by these three pillars of sustainable development. Since then, this has been the *raison d'être* for framing development discourse across the globe. Virtually all the countries in the world have endeavored to streamline their development path for sustainable use of the natural environment deemed critical to the wellbeing and livelihood of not only the present but also the future generations.

Yet, more than 45 years later, it has turned out that the United Nations' canonical formula for a sustainable world has unfortunately resulted in chaotic development. It may well be that in a neoliberal economic world order dominated by the imperative of keeping up with progress, unrestrained expropriation of environmental resources spurred by ever-increasing scales of production and consumption has become the norm. Anthropogenic transformation of our Earth system such as irreversible habitat destruction, unprecedented scale of ecosystem disintegration, and persistent level of greenhouse gas emissions are some of the notable examples. The stark truth is that humans are exerting heavy pressure on our Earth system at an unprecedented rate.

Today the threat of environmental doom driven by the increasingly widening gap between economic growth and environmental protection, and unsustainable resource use is making itself felt. Not too long ago in 2017, a perfect storm of life-threatening environmental disasters hit many countries affecting millions worldwide. Extreme hurricanes such as Hurricanes Harvey, Irma, and Maria swept across the United States with devastating consequences. The year 2017 was also ranked second hottest worldwide on record by NASA, just behind a sizzling 2016, and the hottest year on record without the short-term warming influence from El Niño. While temperature soared to 53.7°C (128.7°F) in the Southwestern Iranian city of Ahvaz, one of the Earth's hottest temperatures ever recorded, Shanghai experienced its hottest day in 145 years with temperatures soaring to a record 40.9°C

(105.62°F). Drought in Eastern and Southern Africa has put 38 million people at risk of food shortage and malnutrition. Heavy snow in Northern Afghanistan resulted in a deadly avalanche which buried towns in up to 10 feet of snow, crushing, or freezing hundreds of people to death. In addition, sea ice both in the Arctic and Antarctic continue its declining trend due to rapid ice melt caused by global warming which could trigger polar "tipping points" and uncontrollable climate change at global level.

In the year 2018, even a stronger risk of environmental catastrophe surfaced on the global sphere of "sustainable development" as a rude awakening. More precisely, a series of extreme weather conditions was wreaking havoc around the globe in the same year, from prolonged heat waves in the northern hemisphere with very extreme conditions near the Artic, unprecedented wildfires in Sweden, to life-threatening floods and landslides in China, India, and Japan. Super Typhoon Mangkhut, the strongest storm observed on Earth in 2018 battered across the Philippines, Hong Kong, Macau, and China, causing massive flooding and devastation in the regions. Japan witnessed its exceptionally brutal year in 2018 with a cascade of extreme events hitting the country. These included the historic rainfall pattern with a 72-hour precipitation record, deadly temperature spikes, and super typhoons. Similarly the record-breaking hurricanes, Hurricanes Michael and Florence in the Atlantic, and Typhoons Jebi and Trami in the Pacific were wreaking havoc around the world in the same year.

These summary statements are by no means exhaustive but they suffice for the present purpose of emphasizing that things are getting worse and harder to deal with now than they have been in the past. They also underscore the fact that the United Nations' concept of sustainable development embraced by the global community has ended in nature exacting revenge on us in cruel and unusual ways. This leads us to conclude that after more than 45 years of promoting environmental sustainability, the United Nations has apparently failed to prepare the global community to contain worldwide environmental threats. The perilous environmental status quo also points to the direction that humans are increasingly losing touch with nature.

All these are clearly indicative of the emergence of a new economic order, cemented by the anthropocentric worldview of development in which humans regard themselves as separate from, independent of, and superior to the natural world, and that resources may be rightly and justifiably exploited for human benefit. In this dominant anthropocentric paradigm, nature is a horn of plenty and a bottomless sink, and there is no sense of interconnection between humans and nature. This sounds the death knell for the United Nations' overarching environmental mandate which has turned into a false hope of a sustainable future. Equally obvious is the fact that sustainable development as embraced by the global community firmly holds to the dictum that sustainable development is economic growth and material progress ad infinitum.

Thus the iron rule of the biophysical limits to economic expansion was hard to exhibit as a model of sustainable development across the anthropocentrically dictated global economic system. Developed and developing countries alike continue their self-interested focus on tapping excessively into the patrimony of nature without constraint. Conceivably, sustainable development has ended in a plunder of our planetary system. Against the backdrop of this anthropocentric paradigmatic change, the environmental controlling and management systems across the globe are simply not functioning effectively as they have expected to. In many instances they are broken. The consequence is that our planet is at breaking point more than four decades later and sustainable development has moved into an impasse. This raises the most pertinent question: With the present dire environmental conditions, can our Planet still be saved? This will be the gist of my analysis.

My intention here is to allow readers to conceptualize the concrete reality of the United Nations environmental protection efforts and their implications for a sustainable future in the interests of all humanity. While this United Nations environmental conundrum has been considerably and extensively discussed among academics and scholars from various perspectives and in a host of disciplines, each advocate mostly confine his/her argument within a particular discipline or concept, or isolated for separate study in accordance to its specific objective. This makes it difficult for researchers, environmentalists, NGOs, policy makers, or students alike to gain a holistic view of the global environmental sustainability issues in their entirety. This, in turn, hampers the articulation of proper modes of control and policymaking. In an attempt to make headway in the urgent task of containing this critical problem, this book seeks to develop more adequate perspectives and concepts for an analysis of the complex process which has led to the present environmental conundrum, and to offer a way out of it.

At the outset, I present a critical analysis of the efforts undertaken by the United Nations to decisively break the vicious circle of environmental degradation for the past 45 years. I further show how the

United Nations' grand schemes of environmental protection were too much in dissonance with the dynamics of the ethical link between the economy and the environment. To enable the readers to understand clearly the economy−environment dilemmas, I shall present the United Nations' environmental protection discourse over the past few decades alongside a rich and complex account of contemporary global environmental trends.

This urgent reality calls for solutions to circumvent the economy−environment predicament. The upshot is that the prevailing looming environmental crisis is a crisis of the human heart: only humans can save our planet. For a start, we have to shed our anthropocentric skin to view ourselves as custodians of nature. To be sure, even the most cursory examination of the prevailing environmental trends in the human-centered world reflects that this is an extraordinarily formidable task. Nonetheless, it is not impossible if we frame our sustainable development discourse from a different angle.

We must embark on a bold new direction guided by the prism of the ecumenical disciplines of environmental philosophy. We have to take the leap to view sustainable development as a never-ending process of environmental improvement which underpins human long-term existence. Taking cognizance of the legitimate view that the real looming crisis lies not in our Earth system but in our heart, we must reconstruct our mental representations of the natural world based on the moral beliefs on the rightness and wrongness of our actions toward it. The quintessence of the argument is that humans are the major destructive force of Mother Nature, so it is humans themselves who can turn the tide of environmental impoverishment around. Clearly how we think about and how we view our Earth system matter; it bespeaks what we value and shapes what we do. Simply put, our values serve as a strong guiding force in our dealings with nature. I believe that the classical foundations of philosophical studies contribute immensely to understanding this important debate. More specifically, Aldo Leopold, Albert Schweitzer, Jeremy Bentham, Immanuel Kant, Émile Durkheim, René Descartes, and Francis Bacon's conceptual insights are a remarkable *tour de force* in that they collectively provide a persuasive scholarly line of argument in this complex discussion.

In undertaking this ambitious and arduous academic task, I acknowledge that it is unrealistic for any single concept to fully capture the complexity of sustainable development. Hence, this book takes a multidisciplinary study and interdisciplinary approach, embracing such branches as science, environment, ecology, economics, politics, philosophy, anthropology, and empirical studies. I certainly do not claim to be a specialist in all these fields, but years of research have enabled me to glean from them the perspectives of a hopeful sustainable future.

This study is considered to be authoritative because its findings are supported by unusually rich concepts with semantic precision, empirical analysis, and actual field research. The book also serves to supplement many books and articles in sustainable development, environmental ethics, and environmental value systems written from philosophical perspectives. It also reinforces many of their armchair theoretical analyses with empirical verification based on solid evidence gathered from an extensive field study.

I hope this book will serve as a wake-up call to the United Nations, the world leaders, the global community, environmentalists, and the public at large of the urgency to recast the hitherto all-encompassing, politically versatile, and anthropocentric concept of sustainable development and reorient our way forward on a genuine course of sustainable future. Those who staunchly believe that the international environmental negotiations, treaties, declarations, and agreements of the past few decades have enabled us to rescue our plundered planet need to realize that the goals of sustainability are not attainable by those means alone. Sustainability will not be achieved without putting environmental ethics and moral philosophy at the forefront of development discourse. This book is intended to serve as a therapeutic solution to morally heal the ethical wound of sustainable development.

It is also my sincere hope that the book will fuel human collective moral consciousness of and ethical engagement with nature and induce responsive ways to reattune and transform mere human rhetoric of environmentalism on paper to one of the effective policies in fine print and real action. A world driven by an anthropocentric view of nature is an ecologically destabilized world. It is time to reverse our anthropocentric drive toward destroying our planetary system in the name of sustainable development. Frederick Engels, almost one-and-a-half centuries ago in *The Part Played by Labour in the Transition from Ape to Man* (1876), said: "Let us not, however, flatter ourselves overmuch on account of our human victories over nature. For each such victory, nature takes its revenge on us." This cogently and partly explains the root cause of the 2017 and 2018 years of disaster: human beings were the cause, and it is only human beings who can heal our planet.

Dr. Choy Yee Keong
Keio University, Tokyo, Japan

August 2020

Acknowledgments

I am deeply indebted to many people and institutions whose support has been instrumental in bringing this work to fruition. First of all, I wish to express my gratitude to Prof. Ayumi Onuma (Keio University), Prof. Takashi Iida (Keio University), Prof. Eiji Hosoda (Chubu University), Prof. Kitakawa Hideki (Ryukoku University), Prof. Tadayoshi Murata (Yokohama National University), Prof. Skoko Sakai (Kyoto University), Prof. Masahiro Ishikawa (Koichi University), Prof. Andrew Alex Tuen (University Sarawak Malaysia), Prof. Mazlin Mokhtar (National University of Malaysia), Prof. Lee Khai Ern (National University of Malaysia), Prof. Wang Chen-Hao (National Taiwan University of Science and Technology, Taiwan), Dr. Wang Chin-Tsan (Ministry of Science and Technology, Taiwan), Dr. He Yanmin (Otemon Gakuin University), Dr. Yamaguchi Rintaro (National Institute for Environmental Studies), Tomonori Ishida (National Institute for Defense Studies, Ministry of Defense, Japan), Dr. Goh Chun Sheng (Harvard University), Dr. Kim Woo Jin, Dr. Wei Hongbiao, Dr. Lee King Siong, Lai Choong Hon, Max Choo Ming Hang, Xu Yiran, Usat Ibut and Sano Tatsuhiko for their support in the course of writing this book. I also wish to extend my thanks to Candice Janco (Publisher), Marisa LaFleur (Acquisition Editor), Peter Llewellyn (Acquisition Editor), Michelle Fisher (Editorial Project Manager), Chiara Giglio (Editorial Project Manager), Matthew Limbert (Cover designer), Indhumathi Mani (Copyrights Coordinator), and Kumar Anbazhagan (Production Manager) from Elsevier for their advice and valuable suggestions. I am also grateful to four anonymous reviewers for their helpful comments on my earlier proposed draft of this work and an anonymous expert reviewer for his careful reading of my final manuscript and his many insightful comments and suggestions for improvement. Above all, my heartfelt gratitude goes to Choy Yee Hong, Chooi Yee Kuan, Chooi Yee Wah, Chooi Yee Nam, Chee Sow Lin, Chooy Sam Mooi, Chye Siew Fong, Teng Foh Mooi, Choy Khai Luen, Choy Sook Fan, Choy Sook Yee, Choy Sook Wai, Choy Sook Theng, Choy Sook Yan, Yee Phooi See, Yee Phooi Phooi, Yee Kit Hoe, Chin Yee Meow, Goh Choon Keat, Yong Yoke Heng, Leaw Siew Mooi, Chin Chee Keong, Chin Chee Han, Jasmine Chin Yong Shya, Chin Mee Kuen, Chin Yuen She, Chooi Hui San, Chooi Zhan Hoong, and all my family members, whose relentless support and encouragement over the years leaves a debt that I can hardly pay.

I am grateful to the National Institutes for Humanity, Research Institute for Humanity and Nature (RIHN), Kyoto, Japan (Project No. D-04) for funding the larger part of my indigenous field research in the state of Sarawak in Malaysia. I also wish to acknowledge the financial support from the Ministry of

Education, Culture, Sports, Science and Technology (MEXT) Supported Program for the Strategic Research Foundation at Private Universities, 2014–18 (MKS1401), and the Grant-in-Aid "Scientific Research (C)," 2016–19 (MKK349J), MEXT, Japan, in the course of writing this book. Last but not least, any flaws that remain in the manuscript are solely the responsibility of the author.

Dr. Choy Yee Keong

Keio University, Tokyo, Japan

August 2020

1

Introduction: Sustainable development—a preliminary reflection

1.1 The rise and fall of sustainable development: A historical perspective

The concept of sustainable development is not new. Historically it may be traced back to the traditional agricultural practices among the ancient tribes in Sri Lanka, the Sonjo and Chagga tribes in Eastern Africa; and in the American and European continents which embraced sustainability ethics by integrating environmental concerns with economic activity, some of which may date back to as far as 2000 years ago (Marong, 2003). The concept of intergenerational equity is also not a recent invention. Its pedigree appeared more than 500 years ago in the *Codex Leicester*, which may have been the first book to unveil the ethics of sustainable development by emphasizing the need to preserve the environment in the interests of future generations while pursuing economic progress (May, 1998).

However, this environmentally sustainable and morally justifiable development ideology gradually eroded with the advent of the industrial revolution in the 1800s in Western Europe, characterized by massive capitalist modes of production and consumption. In 1860, the three leading industrial countries, namely, Great Britain, Germany, and the United States, were producing over a third of total global output, further increasing to a little under two-thirds of a much larger total by 1913 (WTO, 2014). In 1820, economic progress measured in terms of per capita gross domestic product of the richest countries was about three times that of the poorest, and by 1913 the ratio had increased from 1 to 10 (WTO, 2014; Boon and Eyong, 2009). The drive for capitalist expansion and industrial advancement inevitably ignited a mad rush for imperialist acquisition and accumulation of sources of raw material in developing countries such as those in Africa and in Latin America (Boon and Eyong, 2009).

1.2 Reemergence of sustainable development: The United Nations environmental protection initiatives

The dominant western industrial culture and the neo-liberal capitalist expansion come at the cost of global natural resource depletion and environmental impoverishment. Out of

1

a deep concern over the scale and persistent trends of the earth's degradation, the United Nations convened the first world conference on the environment, the Stockholm Conference, in 1972 with the view of returning to the mode of development that was environmentally sustainable. The Stockholm Conference contributed to triggering the rise of environmentalism and fostering the gradual emergence of an environmental movement across the world in the 1970s. Many countries across the developed region responded positively to the Stockholm environmental protection initiatives to arrest any further global environmental decline. However, many developing countries in the South, especially in Africa and Latin America, were less enthusiastic. Despite the Stockholm recognition of the need for development in the South, they still viewed its environmental inspiration suspiciously as a new form of colonialism aiming to sustain the developed countries' continued access to raw commodities and deprive them of their rights to develop.

The tensions that emerged from the incongruent North—South environmental and development agendas were, to a great extent, diffused with the release of the Founex Report on Development and Environment prior to the convention of the Stockholm Conference and the adoption of the United Nations Cocoyoc Declaration in 1974. Both documents explicitly acknowledged the legitimate rights of the developing countries of the South to development and helped to convince them that environmental issues were equally relevant in their development discourse (Marong, 2003). The sense of distrust between the North and the South was further mitigated with the publication of the Brundtland Report in 1987 (WCED, 1987). The Report viewed environmental protection and economic development as not necessarily incompatible. To the developing countries, the Brundtland concept of development is more palatable because it allows development to proceed while protecting their environment. More specifically it provides a basis for them to locate the link between environment and development as the foundation for a just and equitable economic order which is socially and environmentally sustainable.

Following Stockholm and Brundtland, a multitude of environmental summits and conferences were convened, and hundreds of regional and international environmental treaties were made to reinforce global commitment to promote environmentally sustainable development. In response to the high aspirations and ideals of the United Nations' environmental initiatives, the international community has put in place comprehensive environmental institutions and legal systems to promote sustainable resource use and environmental conservation.

But how much has the planet changed since then? To answer this question, I embarked on an extensive empirical research to assess the nature and extent of environmental degradation across the world, including the world's largest economy, the United States, and the world's fastest developing country, China, which is also the world's second largest economy after the United States. The result of the assessment reveals that environmental degradation poses a serious threat to sustainable development in the world today and achieving environmentally sustainable development remains one of the greatest challenges of human civilization. The root of the problem lies in how we perceive the environment and our place in it. Among the most influential moral mental representations of our planet are ethical beliefs about the rightness and wrongness of our actions toward our Earth system. Clearly the philosophical principles of environmentalism provide a strong foundation for this line of enquiry. In digression, the Earth system is itself an integrated system which may be subdivided into four interconnected "spheres," namely, the geosphere, atmosphere, hydrosphere, and biosphere.

1.3 The ethics of sustainable development

Rightly the ethics of sustainable development constitutes one of the most valuable pointers in what direction our scientific investigation should stretch to reach deterministic and persuasive results. Fundamentally the ethical content of sustainable development matters because environmental sustainability issues are about the relationship between humans and the natural world. Although global environmental consensus created through the adoption or ratification of international agreements is a vital aspect of an effective global response to environmental degradation, they cannot be relied on exclusively as impetus for action. No number of international environmental conferences or agreements will have any enduring or lasting effect if most member countries fail to undertake a real and genuine commitment for effective actions. However, a real and genuine commitment to environmental preservation is impossible without any indication of "appropriate concern for, values in, and duties to the natural world" (Rolston, 1999: 407).

This book attempts to conceptually and theoretically assess the ethical constraints to sustainable development by considering human responsibility to the natural world with the aim of providing a philosophical framework for rethinking our relationships with nature. The philosophical implications of proenvironmental behaviors are empirically reified based on evidence drawn from field research with the forest dwelling indigenous people in Malaysia. It is demonstrated that the philosophical concepts of environmentalism which are holistically mirrored in the culture of the indigenous communities, have an important role to play in promoting global environmental sustainability.

It is instructive to note that moral principles per se may end up in a rhetorical or symbolic gesture of environmentalism if it is not activated in the minds of humanity. In other words, human moral attunement to nature requires a means of activation that contributes to its application in a real-world system. It is argued that the means of activation are premised on the effective implementation of environmental and moral education which serves to bring to the forefront the practical implications of human moral views of nature on proenvironmental behaviors. In plain language, environmental and moral education constitutes a *conditio sine qua non* for transforming changes in human environmental attitudes and behavior toward environmental sustainability. The rest of the book will explore how this can be systematically achieved.

1.4 Structure of the book

Most of the chapters to follow have the environmental worldview thesis as their common feature. Chapter 2, The United Nations' Journey to Global Environmental Sustainability Since Stockholm: An Assessment, explores comprehensively the evolution of the concept of sustainable development since the 1972 Stockholm Conference as a main framework for understanding the relationship between the economic, social, and environmental problems confronting us today. The UN Scientific Conference on the Conservation and Utilization of Resource held in 1949 to examine sustainable management of natural resources provides a good starting point for this analysis. The analysis is then divided into three distinct periods: 1970–1980, 1990–2000, and 2001 onwards. Each period is embedded in one or two path-breaking conferences which produced important environmental documents that spurred the development and rapid expansion of international environmental rules and principles, and reinforced the high aspirations of sustainable development as the fundamental guide for our common environmental future.

Chapter 3, The United Nations' Journey to Global Environmental Sustainability Since Stockholm: The Paradox, takes this theme further by assessing the principles of sustainable development in addressing the precarious state of our environment. Drawing from global environmental evidence on the state of our planet, it demonstrates that the implementation of sustainable development has hitherto proven to be daunting. The assessment reveals that the roots of the many massive environmental problems facing us today are the lack of an ethical vision of resource use in the utilitarian pursuit of economic growth and the lack of an awareness of the moral causes of our environmentally destructive practices in our dealings with nature.

By way of illustration, the book, drawing from the environmental worldview of Francis Bacon, also examines the philosophical concepts underscoring the United States government led by Donald Trump in articulating the global direction on key environmental issues of common concerns such as climate change and sustainable resource use patterns. It shows how the dominant anthropocentric view of nature and the "American first policy" as embraced by the current administration has led to what we call here the "Donald Trump environmental protection rollback" which threatens to roil back global efforts in arresting the increasing scale of human impact on our fragile Earth system.

The unprecedented scale of environmental problems confronting the world today call for serious philosophical and ethical soul searching aimed at addressing unsustainable development practices. This will be the main focus of Chapter 4, Greening for a Sustainable Future: The Ethical Connection. The main task is to explore the philosophical and ethical concepts of environmentalism so as to broaden and expand our environmental perspectives in challenging and fruitful ways. With this end in view, this chapter examines what contribution environmental philosophy and ethics can make toward achieving environmental sustainability. It attempts to examine the moral principles based on theoretical analyses of contemporary environmental ethics ranging from anthropocentrism to biocentrism and ecocentrism.

Taking the theoretical base as a vantage point, this chapter ventures deeper into exploring the philosophical rationale for environmental conservation based on Aldo Leopold's land ethic, Albert Schweitzer's reverence for life ethic, and Immanuel Kant's deontological concepts. These philosophical thoughts provide very rich practical clues and holistic inspirations for a profound understanding of the theoretical and conceptual underpinnings of human proenvironmental and moral behaviors. They demonstrate clearly that environmental philosophy constitutes an indispensable element in harnessing a more expansive human appreciation of the complexity and beauty of the Earth system. They also represent an important driving force for sustainable environmental attitudes and ethical behavior to guide our actions toward fostering a harmonious relationship with the natural world, with "love and respect," in Aldo Leopold's words.

However, acknowledging that the theoretical argument and the practical importance of environmental ethics in promoting environmental sustainability may be less than convincing in the absence of empirical verification. Chapter 5, The Nexus of Environmental Ethics and Environmental Sustainability: An Empirical Assessment, presents the extensive field research embarked on to investigate into the indigenous worldviews of the moral relationship between humanity and the natural world, conducted in the state of Sarawak in Malaysia.

The field research is related to the theoretical concepts developed in the previous chapter, and contributes to deepening our understanding of the crucial role of environmental ethics in turning the tide of environmentally unsustainable human behavior, while also reinforcing the theoretical argument as discussed in Chapter 4, Greening for a Sustainable Future: The Ethical Connection. The theoretically discussed and empirically substantiated ethical frame of environmental sustainability provides a moral wake-up call on the urgent need for our ethical engagement with nature in addressing global environmental problems.

However, whether this wake-up call resonates well with humanity hinges on environmental and moral education to increase public awareness on environmental problems and their inevitable consequences which impact on human civilization. This theme which constitutes the last section of this chapter will be discussed in relation to the theoretical and empirical studies carried out earlier. It argues that environmental education plays a crucial role in raising public environmental awareness which is a precondition for inducing changes in environmental attitudes and moral behavior. This would probably translate into a real commitment to environmental conservation against the throes of the current environmental catastrophe.

The emphasis is on promoting ways of thinking about the aspects of environmental sustainability through environmental education. This will be the focus of Chapter 6, The United Nations Environmental Education Initiatives: The Green Education Failure and the Way Forward. Tracing the beginnings of the environmental education programmes launched by the United Nations/UNESCO since Stockholm, it critically examines to what extent they have succeeded in promoting environmental awareness and proenvironmental behaviors. As the United Nations/UNESCO environmental education agendas have largely fallen short, we turn to Émile Durkheim and René Descartes' theoretical insights to understand their flaws, and also to gain valuable advice on how they could be improved.

By now, the repertoire of the analytical frames developed can be applied to critically examine how and to what extent global efforts in promoting environmental sustainability may be enhanced.

The last chapter, Chapter 7, Summary and Conclusion, drawing from the preceding chapters, provides an overall assessment of the United Nations' efforts at addressing global environmental problems. It reveals succinctly that, for more than 45 years since Stockholm, the United Nations has developed progressively through its continuous efforts in creating and developing international agreements, treaties, and guidelines in addressing global environmental problems. However, evidence shows that all this change and progress have failed to halt the escalating rate of decline in the state of the global environment. It may well be that global debate on environmental issues has fundamentally concentrated on scientific and economic concerns while the more important ethical issues of environmentalism have often been ignored in global discussion. Thus it has been impossible to trigger a dramatic shift in how nations consider their ethical concerns about and moral responsibilities for the environment when optimizing its economic use.

It is concluded that in contrast to the ecological history of naturally induced environmental disasters over the past billion years that were beyond human control, the environmental problems we face today are fundamentally caused by humans and are practically within our ability to control or prevent. International environmental conferences or agreements alone, no matter how sophisticated or extensive, will never be sufficient to contain our persistent and increasingly destructive impact on the global environment. Mapping the way forward necessarily calls for ethical and moral change in public attitudes, and policies toward more environmentally sustainable modes of resource use and management. This could be enhanced through the promotion of environmental and moral education. It is also demonstrated that how the Coronavirus pandemic, currently wreaking havoc around the world, has much to teach us about how we should deal with the global environmental crisis confronting humanity today.

The United Nations' journey to global environmental sustainability since Stockholm: An assessment

2.1 Introduction

Conservation and sustainable management of the natural environment, including its biodiversity, is critical for sustaining long-term human existence. Acknowledging this, and in response to the increasing threat of environmental degradation in the west, the United Nations convened its first and paradigm-breaking global environmental conference, the Stockholm Conference in 1972 to launch "a new liberation movement to free men from the threat of their own thralldom to environmental perils of their own making" (United Nations, 1972: 45). The Conference also sought "to inspire and guide the peoples of the world in the preservation and enhancement of the human environment" (United Nations, 1972: 3). Since then, a multitude of international environmental conferences and summits have been held and hundreds of multilateral agreements, treaties, and declarations adopted, generating the momentum needed to promote sustainable use and management of natural resources and environmental protection globally.

As a result, the environment has become a major issue on the international development agenda, and developed and developing countries alike have also been galvanized into action. Countries across the global divide have set up their ministries or departments of environment and produced their national development agenda with great emphasis on environmental protection. To reinforce environmental protection efforts, countless environmental laws and regulations have been enacted by all member states of the United Nations. This has been instrumental in shaping the international environmental legal regimes and sustainable policy agenda to reverse environmental degradation on our planet. The main purpose of this chapter is to assess the United Nations' impact on the evolution of global environmental regimes in promoting environmental sustainability.

The term environmental sustainability is associated with responsible human decisions and actions in interactions with the natural environment. It may well be that many natural systems can withstand human disturbances or external pressure only up to a certain threshold (or "tipping point") beyond which ecological discontinuities and possibly irreversible consequences may occur (Srebotnjak et al., 2010). Thus within the present context, environmental sustainability may be defined as sustaining the ecological integrity of the natural system while optimizing its economic use (Choy, 2015a). Conceptually, ecological integrity may be explained using Holling's concept of sustainability which is expressed in terms of ecological stability and resilience (Holling, 1986).

Stability and resilience are descriptive concepts which posit the dynamic properties of a system. Stability refers to the ability of a natural system to return to its equilibrium state or viable level of regeneration after a temporary ecological disturbance has resulted from, for example, an act of economic exploitation (Holling, 1973; Choy, 2015a). The criterion is that the more rapidly the system restores itself and the less it fluctuates, the more stable it would be (Holling, 1973: 14). The term "resilience" is a measure of the persistence of systems and of their ability to absorb change and disturbance in the face of human perturbation while maintaining its organizational structure (Holling, 1973: 14). It may be technically defined as "the magnitude of disturbance that can be absorbed before the system changes its structure by changing the variables and processes that control behavior" (Holling and Gunderson, 2002: 4). Thus a system may said to be Holling sustainable or ecologically sustainable if and only if it is able to revert to its stable or resilient position when exposed to external disturbances.

However, it must be noted that in a real-world system, determining the critical ecological thresholds requires an in-depth understanding of the ecosystem dynamics underlying uncertainties around multiple and stochastic disturbances (Sasaki et al., 2015). Granted, given our very limited scientific understanding of the complex system dynamics of natural systems and the ecological surprises arising from the interaction between ecological and the social systems, it would not be easy, and may be practically unfeasible to decide when, how and to what extent to intervene to enable the ecosystem to absorb various shocks and disturbances and reorganize itself into a Holling resilient state. Despite being confronted with uncertainty, we must apply the precautionary measures based on the philosophical principle of environmental conservation as discussed in Section 4.41 to avoid irreversible anthropogenic disturbances of our global ecosystem, that is, the resilience-based precautionary strategy.

2.2 The United Nations' journey to global environmental sustainability: The evolution of the Stockholm green era

Environmental issues were not a major concern of the United Nations in the period following its establishment in 1945. In 1949, the United Nations convened its first conference on the environment, the UN Scientific Conference on the Conservation and Utilization of Resources, to address the urgent issue of the improvident use of the earth's dwindling resources against the background of dramatic population growth and increasing resource demand (United Nations, 1950). The Conference focused primarily on six major themes— land, water, forests, wildlife and fish, fuels, energy, and minerals. However, the main thrust of the Conference was on ways to manage these natural resources to sustain economic and social development rather than from a conservation perspective (Jackson, 2007) (Timeline 2.1).

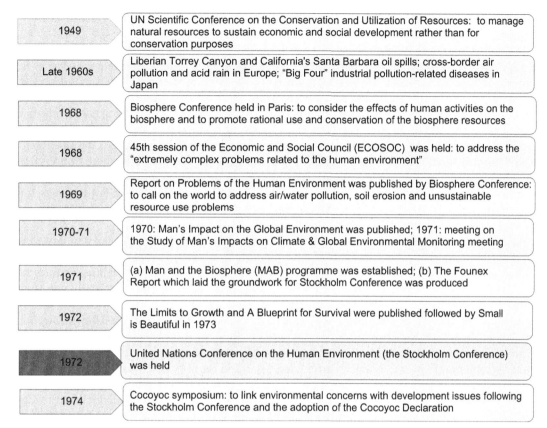

1949	UN Scientific Conference on the Conservation and Utilization of Resources: to manage natural resources to sustain economic and social development rather than for conservation purposes
Late 1960s	Liberian Torrey Canyon and California's Santa Barbara oil spills; cross-border air pollution and acid rain in Europe; "Big Four" industrial pollution-related diseases in Japan
1968	Biosphere Conference held in Paris: to consider the effects of human activities on the biosphere and to promote rational use and conservation of the biosphere resources
1968	45th session of the Economic and Social Council (ECOSOC) was held: to address the "extremely complex problems related to the human environment"
1969	Report on Problems of the Human Environment was published by Biosphere Conference: to call on the world to address air/water pollution, soil erosion and unsustainable resource use problems
1970-71	1970: Man's Impact on the Global Environment was published; 1971: meeting on the Study of Man's Impacts on Climate & Global Environmental Monitoring meeting
1971	(a) Man and the Biosphere (MAB) programme was established; (b) The Founex Report which laid the groundwork for Stockholm Conference was produced
1972	The Limits to Growth and A Blueprint for Survival were published followed by Small is Beautiful in 1973
1972	United Nations Conference on the Human Environment (the Stockholm Conference) was held
1974	Cocoyoc symposium: to link environmental concerns with development issues following the Stockholm Conference and the adoption of the Cocoyoc Declaration

TIMELINE 2.1   The evolution of the Stockholm Conference.

Acknowledgment of international environmental problems as distinct from the conventional resource-use issues arose in the late 1960s spurred by several widely publicized events. These included the environmental and ecological damage caused by massive oil spills such as that caused by the grounded Liberian Torrey Canyon supertanker which poured 120,000 tonnes of heavy crude oil onto hundreds of miles of British and French coastlines in 1967. Two years later in 1969, another tragic oil spill occurred from an off-shore well in California's Santa Barbara Channel (M'Gonigle and Zacher, 1981).

Other environmental issues were the cross-border air pollution and acid rain in Europe in the late 1960s, the widespread use of pesticides and fertilizers and other sources of pollution such as water pollution. Severe industrial pollution in Japan in the 1960s also fueled growing concerns about human unsustainable activities (Kapur, 2015). The "Big Four" industrial pollution-related diseases in Japan were "Minamata Disease" and "Niigata Disease," both caused by mercury poisoning, "Itai–Itai Disease" caused by cadmium poisoning, and "Yokkaichi Asthma" caused by air pollution. The publication of the paradigm-breaking books and articles such as the *Silent Spring* on the use of potentially harmful pesticides (Carson, 1962) and "The Tragedy of the Commons" on widespread ecological degradation and unsustainable resource use has also contributed toward shaping the perception of critical environmental issues (Hardin, 1968).

The first major sign of global environmental concern was reflected in the convention of an intergovernmental world conference, the Biosphere Conference, held in Paris in September 1968 to consider the effects of human activities on the biosphere, including the effects of air and water pollution and deforestation and to promote rational use and conservation of the biosphere resources (Sands et al., 2012). A report on *Problems of the Human Environment*, published in the year after the Conference, suggested that the world should undertake serious action to address air and water pollution, soil erosion, and the profligate use of natural resources (ECOSOC, 1969; Haq and Paul, 2012). The biosphere conference led to the establishment of the Man and the Biosphere (MAB) program in 1971. The MAB is an intergovernmental scientific program which aimed to establish a scientific basis for the improvement of relationships between man and the environment and to curb biodiversity loss (Haq and Paul, 2012).

Another major breakthrough in the late 1960s was the formal inclusion of the environment for the first time into the list of international concerns at the 45th session of the Economic and Social Council (ECOSOC) held in 1968 as a result of the relentless efforts of Sweden to push for international policy action to urgently address the "extremely complex problems related to the human environment" (Howe, 2014: 70). The proposal to hold a conference on the human environment was raised by Sweden at this 45th session through a convincing memorandum outlining the purpose of the Conference aimed at raising public awareness among governments and society around the world on the seriousness of environmental problems (Ivanova, 2005). Sweden also extended its offer to convene the conference in its capital city and to make a significant financial contribution (Howe, 2014). As a result, ECOSOC passed a resolution [resolution 1346 (XLV) of July 30, 1968] to recommend to the General Assembly to convene a United Nations conference on the problems of the human environment. In digression, it may be noted in light of the above that in 1967, Inga Thorsson, a Swedish diplomat, and Sverker Äström, Sweden's permanent representative to the United Nations, objected to the United Nations plans to convene a conference on the peaceful use of atomic energy, the fourth in a row, on the grounds that it would benefit mostly the limited number of nuclear industries in the North (Ivanova, 2005; Howe, 2014).

The General Assembly, noting "the continuing and accelerating impairment of the quality of the human environment caused by such factors as air and water pollution, erosion and other forms of soil erosion, secondary effects of biocides, waste and noise" (United Nations, 1968: 2), decided to pass a resolution, Resolution 2398 (XXXIII) in response to the need to convene a global conference, called the United Nations Conference on the Human Environment in Stockholm (the Stockholm Conference). The main purpose of the Conference was to encourage, and to provide guidelines to governments and international organizations to design policy actions to protect and improve the human environment and to remedy and prevent its impairment by means of international cooperation (United Nations, 1969).

In the meantime, to lend force to the urgent need for the Conference, four major collaborative and overlapping interdisciplinary studies of the global climatic and ecological effects of human activities were produced more or less concurrently between 1969 and 1972. They are: (1) Man's Impact on the Global Environment which studied a wide range of environmental issues including desertification, pollution of the air and oceans, and other harms, and warned of the impending risk of global warming (SCEP, 1970); (2) Following up SCEP 1970, a meeting on the Study of Man's Impacts on Climate was held focusing on the analysis of climate change and warned further of the dangers of anthropogenic emissions of greenhouse gases (GHGs) and particle pollutants (SCEP, 1970; Matthews et al., 1971); (3) Global Environmental Monitoring focused on the accumulation, review and assessment of available information on human-induced environmental changes and their impacts, and also sought to assess methodologies of measurement of environmental parameters, among others (SCOPE, 1971); and (4) The Limits to Growth, produced by the Club of Rome (Meadows et al., 1972). Except for The Limits to Growth, each of these studies was undertaken explicitly for the 1972 United Nations Conference on the Human Environment (Sohn, 1973; Howe, 2014).

2.3 The Stockholm impediment and the Founex Report

The developing countries were, however, wary about the North's environmental movement. They feared that "the humanitarian concern for environment can far too easily become a selfish argument for greater protectionism" (United Nations, 1971: para. 54, p. 27). They also feared that "the concern for environment may become a priority unto itself in the developed countries" and that this would hinder or slow down their economic development (United Nations, 1971: para. 58, p. 29). To the developing region, the North environmental protection initiatives were no more than a neo-imperialist ploy to keep former colonies in a poor state of development (Gaines, 1997). Many developing countries distrusted Stockholm as an attempt to "ratify and even enhance existing unequal economic relations and technical dependence, miring them in poverty forever" (Hecht and Cockburn, 1992: 849). They therefore threatened to boycott the Conference (Selin and Linnér, 2005). Several developing countries such as Brazil charged that the global conference was "a rich man's show to divert attention from the real needs of developing countries" (Engfeldt, 2009: 41).

The developing countries also asserted that major environmental problems identified for the Stockholm Conference such as industrial pollution, uncontrolled urban development, and wilderness degradation, were a consequence of unrestrained industrialization in the United States, Europe, and the Soviet Union. While these nations were concerned about the impacts of environmental degradation on the quality of life of their people in relation to, for example, recreational open spaces, clean air and water, and endangered species, the developing countries were more concerned about the problems of poverty and related issues such as access to clean water, malnutrition, and proper sanitation.

The developing countries, especially in Latin America and Africa, were worried that the environmental movement in the West would aggravate their poverty problems by stifling development (Howe, 2014). On a more radical note, political ecologist, Professor Héctor Alimonda from Universidade Federal Rural do Rio de Janeiro in Brazil lamented that environmental degradation was caused by persistent colonialism and he further claimed that "over the five centuries, entire ecosystems were destroyed by the implementation of monoculture export crops" (Alimonda, 2011: 22, quoted in Martinez-Alier et al., 2016: 37).

Wade Rowland summarized the above North–South environmental and development dilemmas succinctly: "Opinions among the developing nations ranged from an assumption that problems relating to the environment were a concern for the highly developed nations alone [...] to a belief that the developed nations were using environmental doomsday predictions as a racist device to keep the non-white third world at a relatively low level of development. Environmental concerns were a neat excuse for the industrialized nations to pull the ladder up behind them" (Rowland, 1973: 47). In a nutshell, the preStockholm era was marred by a politics of contestation and reluctant participation by the South in the northern global environmental movement (Najam, 2005).

On the other hand, the industrialized countries such as Canada, the Netherlands, Sweden, and the United States, in their attempt to secure political commitment to environment protection, argued that the origins of environmental problems were fundamentally the same in all countries and it was indeed the poor who would benefit most from environmental protection in addressing issues such as soil deterioration, water pollution, urbanization, negative effects of dam construction, and loss of wildlife. They further contended that in the process, they would have possibly avoided many costly mistakes made earlier by the industrialized countries in their quest for economic development (Engfeldt, 2009).

In addressing the North−South environmental and development conflicts and in an attempt to secure the support of the developing countries for the Conference, the Secretary-General for the Stockholm Conference, Maurice Strong, established a panel made up of 27 experts from developing and developed countries (the Founex Panel) which met in Founex between June 4 and 12, 1971 as part of the preparations for the Conference to discuss the issues surrounding the relationship between development and environment (United Nations, 1971). The panel affirmed that "the current concern with environmental issues has emerged out of the problems experienced by the industrially advanced countries" and that "these problems are themselves very largely the outcome of a high level of economic development," and "the creation of large productive capacities in industry and agriculture" (United Nations, 1971: para. 2, p. 3). The panel further revealed that the scale of environmental disruptions constituted serious hazards to human health and wellbeing in many communities.

In recognition of the developing countries' concerns, the panel acknowledged the distinction between "development as cause of environmental problems" as had been experienced primarily experienced by the industrialized countries (United Nations, 1971: para. 5, p. 4) and "environmental problems that reflect the poverty and the very lack of development" as experienced predominantly in the developing countries (United Nations, 1971: para. 4, p. 23). The panel recognized that addressing environmental problems in the developing countries called for the need to address the problems of under development and dire poverty. This is reflected explicitly in its argument that "the kind of environmental problems that are of importance in developing countries are those that can be overcome by the process of development itself" (United Nations, 1971: para. 5, p. 4). More specifically, the panel recognized that "development becomes essentially a cure for their major environmental problems" (United Nations, 1971: para. 5, p. 4).

The recognition of environmental issues in developing countries is an aspect of widening the development concept beyond the conventional and more narrowly conceived objective of economic growth measured in terms of increase in gross national product. To secure support for environmental protection from the developing countries, the panel further recommended that "the increased cost burden arising from greater attention to environmental problems should be accompanied by a greater willingness to provide additional assistance" by the developed countries to developing the South (United Nations, 1971: para. 13, p. 7). The panel produced the Founex Report which played a critical role in laying the ground work for the 1972 Stockholm Conference. This is examined in the next section.

2.4 The Stockholm Conference and the North–South greening conflicts

The Founex Report played a critical role in bridging the policy and conceptual differences in the overriding objectives of environmental protection and economic development between the north and the South. The Report stimulated developing countries' participation at Stockholm based on the understanding that environmental commitments reached at the Conference would not be used to restrain their development efforts (Marong, 2003). However, the Soviet Union and most of the East Bloc countries boycotted the Conference over the Western exclusion of the German Democratic Republic, a nonmember country, from the Conference on the basis that it was neither a member of the United Nations nor of any specialized United Nations agency (van Dever, 2006).

The boycott affected, but did not seriously frustrate, the Conference as it was finally held between June 5 and 16, 1972 in Stockholm after lengthy negotiations and preparations. The Stockholm Conference formally known as the United Nations Conference on the Human Environment, with its theme *Only One Earth*, is the first global conference on human environmental impact in United Nations history which formally placed the environment on the global political agenda. Here, it may be reiterated that as noted above, the 1949 United Nations Scientific Conference on Conservation and Utilization of Natural Resources was the first global United Nations environmental conference which was held to discuss sustainable use of natural resources. As far as the Stockholm Conference is concerned, it is often regarded as the first United Nations global environmental conference that formally placed the environment on the global political agenda.

The Stockholm Conference was attended by 113 member states of the United Nations calling "for a common outlook and for common principles to inspire and guide the peoples of the world in the preservation and enhancement of the human environment" (United Nations, 1972: 1). The Stockholm Conference gave birth to the concept of sustainable development, commonly known as "eco-development" in those days, as a way of reconciling the competing claims of economic development and environmental protection (Dresner, 2008).

However, the Conference did not have a cordial feel from the beginning with the developing countries sharply criticizing the industrialized countries as the main cause of the present global environmental decline because of their unrestrained economic exploitation of the earth's system. The developing region also lamented that environmental concerns postulated by the affluent north, associated with achieving a better quality of life with the enjoyment of a clean environment, was an attempt to stifle developmental aspirations in the poverty-stricken South. The Ivory Coast, for example, showed reservation about placing environmental issues as a global priority. At the Conference, it announced that it would prefer more pollution problems to poverty problems "insofar as they are evidence of industrialization" (Rowland, 1973: 50).

More telling is the statement by the Indian Prime Minister, Indira Gandhi, who claimed that poverty was the worst form of pollution: "How can we speak to those who live in villages and in slums about keeping the oceans, the rivers and the air clean when their own lives are contaminated at the source? Are not poverty and need the greatest polluters?" (quoted in Tinker, 1975: 481). She further pointed out that "the inherent conflict is not between environment and development but between environment and reckless exploitation of earth's resources in the name of development" (quoted in Kumar and Kumar, 2015: 11). Peter Walker, Britain's Secretary of State for the Environment put it bluntly, "many developed countries have turned rivers of beauty into open sewers, the very air we breathe into poisonous gases" (quoted in Tinker, 1975: 481).

In Latin America, the governments rejected the idea of the "Limits to Growth" and opted for distinct styles of development in accordance with ecological and social reality. In consonance with the argument contained in the Founex Report, the Latin American governments further emphasized that in their regions which had been plagued by an uneven distribution of income for centuries, the solution to environmental problems was not to halt economic progress in the developing region but in changing unequal distribution of power and wealth in the world (Pintasilgo, 1992; Martinez-Alier et al., 2016). The Brazilian delegation, for example, made it clear that it needed to develop first to be able to clean up later and "no growth" was unacceptable (Edwards and Roberts, 2015; Roberts and Newell, 2017). The World Bank President, Robert McNamara also echoed the same sentiments at the Conference, confessing that continued economic growth in both the North and the South necessarily constituted a precondition for improved global environmental protection and management. Indeed, the developing South had persistently argued that the only way to circumscribe local environmental problems was through rapid industrial progress and economic growth (Rowland, 1973).

Of great concern to the developing South, especially Brazil, which was an important player at the Conference, was the fear that the industrialized North might use ecological concepts such as "our Spaceship Earth" and the "common heritage of mankind" embodied in international environmental treaties to restrict their sovereign rights to exploit their own natural resources, to alter their development paths, to reduce aid, or to slow down investments (Edwards and Roberts, 2015; Roberts and Newell, 2017). The Brazilian delegation also lamented that no one should claim that all nations have a share of the earth's resources that forms a "common pool" or the "World Trust." The delegation viewed that this "beautiful assumption" requires sharing also of economic and political power, industry, and financial control, which the wealthy nations found quite unthinkable (Roberts and Newell, 2017). Brazil's noncompromising environmental position struck a chord with most other developing nations, while the Sri Lankan ambassador warned that: "we must not, generally speaking, allow our concern for the environment to develop into a hysteria" (Edwards and Roberts, 2015: 43). The developing countries also argued for "additionality" of financial aid for environmental improvement. The "additionality" refers to additional international assistance apart from those resources already identified for international official development aid (ODA) target of 1% of the world's rich countries Gross National Income (GNP) for the Second Development Decade in the 1970s. The 1% target is the same as the target set for the First Development Decade in the 1960s (Stokke, 2009).

Meanwhile, the economically secure developed countries which approached environmental issues as a quality of life issue were of the view that overpopulation and the Western-style of industrialization in the developing countries would accelerate the rate of global environmental decline (Ellison, 2014). Worried about the world shrinking resource base and spreading pollution, they stressed the importance of environmental protection and argued that it was an obligation of all countries in the world to take necessary domestic environmental measures to protect their environment and to refrain from causing harm beyond their borders in the process of pursuing economic growth.

2.5 Outcomes of the Stockholm Conference

The Conference put the industrialized rich on a collision course with the poor developing South over the priorities of economic growth and environmental protection. These seemingly endless North–South conflicts reflected the need for a re-thinking and this led to the adoption of the following three major sets of nonbinding decisions and recommendations which bear the marks of compromise between the North and the South:

a. Declaration of the United Nations Conference on the Human Environment (Stockholm Declaration) which aimed to guide the international community on the preservation and enhancement of the human environment. The declaration consists of 26 principles covering a multitude of issues ranging from environmental education, science and technology, nuclear weapons, institution, transfer of financial and technological assistance, the need for social and economic development in the developing region, and the sovereign right of resource exploitation, among others. Principles 1 to 4 of the declaration explicitly acknowledge the need to exercise prudent use of natural resources in line with ecological balance of the biosphere or the carrying capacity of our planet for the benefit of present and future generations.
b. Resolution on Financial and Institutional Arrangements for International Environmental Cooperation which sought to promote effective implementation of measures by the international community to safeguard and enhance the environment.
c. The Stockholm Action Plan on the Human Environment which entailed concrete and specific recommendations for national and international actions. It includes 109 recommendations on international measures to arrest global environmental degradation (Sohn, 1973). This document is intended to supplement the Stockholm Declaration.

The Conference led to the establishment of the United Nations Environmental Program (UNEP) within the United Nations system. UNEP serves as an institutional catalytic agent for the promotion and coordination of global environmental activities (Appendix 2.1). It also laid the foundation for the creation and adoption of a series of important documents at and after the Conference to enhance environmental sustainability. These include:

i. the 1972 World Heritage Convention which calls on the Member States to protect and preserve not only World Heritage sites but also national heritage situated in their territories. The Convention has had impact in raising greater awareness of the importance of protecting these sites for the benefit of both present and future generations;
ii. the Convention on International Trade in Endangered Species (CITES) of Wild Fauna and Flora, an international treaty drawn up in 1973 which aims to promote wildlife conservation, to prevent overexploitation of wild fauna and flora, and to regulate international trade in species threatened with extinction including 5000 animals and 29,000 plant species (CITES Secretariat, 2010); and

iii. the 1979 Convention on the Conservation of Migratory Species (CMS) of Wild Animals, or the Bonn Convention, was concluded in recognition of Recommendation 32 of the Action Plan adopted at the Stockholm Conference to protect migratory species of wild animals of global importance. The Convention recognizes that wild animals including terrestrial, aquatic, and avian migratory species in their innumerable forms are an irreplaceable part of the earth's natural system and each generation owes a responsibility to conserve these biological resources for the benefit of future generations. The Convention pays specific attention to protecting and conserving wild animal species that regularly cross-national borders or that migrate in international waters. The convention entered into force only in November 1, 1983.

It may further be added that in 1980, the International Union for Conservation of Nature and Natural Resources (IUCN), in collaboration with UNEP, published an important document on biodiversity conservation, the World Conservation Strategy, which aimed to spearhead eco-development by identifying priority conservation issues and key policy options (IUCN-UNEP-WWF, 1980; Drexhage and Murphy, 2010). Three major concerns of the Strategy are (1) the maintenance of essential ecological process, (2) the preservation of genetic diversity, and (3) sustainable use of species and ecosystems (for a brief summary, see Appendix 2.1).

APPENDIX 2.1   United Nations engagement with environmental sustainability in the postStockholm era (1974−2000 period).

1974	UNEP/UNCTAD symposium on. "Patterns of Resource Use, Environment and Development Strategies" (or the Cocoyoc Symposium) held at Cocoyoc, Mexico	The main purpose of the symposium was to identify the economic and social factors leading to environmental deterioration and to discuss patterns of resource use, environment, and development strategies. The symposium led to the adoption of the Cocoyoc Declaration which called on "political leaders, governments, international organizations and the scientific community to use their imagination and resources to elaborate and implement programs aimed at satisfying the basic needs of the poorest people throughout the world" (World Development, 1975: 144).
1979	First World Climate Conference held in Geneva	Established the WCP in 1980 to promote global research on important climate issues including ozone depletion and global warming.
1980	IUCN (currently the World Conservation Union) in collaboration with the UNEP, and the WWF	Published the World Conservation Strategy to promote sustainable resource-use within the carrying capacities of the ecosystems (ecological sustainability). Particularly, its main objectives are (a) to maintain essential ecological processes and life-support systems, (b) to preserve genetic diversity, and (c) to ensure the sustainable utilization of species and ecosystems. It defines conservation as "the management of the human use of the biosphere so that it may yield the greatest sustainable benefit to present generations while maintaining its potential to meet the needs and aspirations of future generations".

(Continued)

APPENDIX 2.1 (Continued)

1982	United Nations 48th plenary meeting	World Charter for Nature was adopted to protect the ecological integrity, diversity, and genetic viability of ecosystems. It also acknowledged the importance of protecting the intrinsic value of natural resources and called for our understanding of human dependence on nature for long-term socioeconomic sustenance.
1982	Third UNCLOS III	The United Nations Convention on the Law of the Sea, which entered into force only in 1994, aims to prevent, reduce and control pollution of the marine environment, and protect fish stocks from depletion. In particular, highly migratory species of fish and marine mammals would be accorded special protection.
1983	The WCED (the Brundtland Commission) was formed	WCED aimed to hold hearings across the globe and to unite countries to pursue sustainable development. It led to the publication of the Brundtland Report or Our Common Future which defined sustainable development as "development that meets the needs of the present without compromising the ability of future generations to meet their own needs."
1985	Vienna Convention for the Protection of the Ozone Layer	Provided a framework for efforts to protect the globe's ozone layer from the effects of human activities and led to the adoption of the Montreal Protocol on Substances that Deplete the Ozone Layer in 1987, designed to reduce the production and consumption of ozone depleting substances.
1989	The IPCC was established	To focus on scientific assessment on all aspects of climate change and their environmental and socioeconomic impacts with a view to formulating realistic response strategies.
1990	The Intergovernmental Panel on Climate Change (IPCC)	Published the first IPCC Scientific Assessment Report and warned of the impending threat of global warming and its effects on natural and human systems. The Report is made up of three volumes and an overview, namely, (1) Climate Change: The IPCC Scientific Assessment (1990), (2) Climate Change: The IPCC Impacts Assessment (1990), and (3) Climate Change: The IPCC Response Strategies (1990), and (4) First Assessment Report Overview.
1991	IUCN (currently the World Conservation Union), the UNEP, and the WWF	Published Caring for the Earth with the aims of securing a widespread and deeply held commitment to the ethic for sustainable living and to integrate conservation and development.
1992	UN Conference on Environment and Development (the Earth Summit/Rio Summit) held in Rio de Janeiro, Brazil	Documents adopted: (i) The Rio Declaration on Environment and Development—a program of action which spelled out 27 guiding principles for the

(Continued)

APPENDIX 2.1   (Continued)

		management of natural resources and environment; (ii) Agenda 21—a 40-chapter and 800-page agreement that laid down 115 specific programmes to help to achieve sustainable development; (iii) Statement of Forest Principles—a brief document containing 15 principles to guide the management, conservation and sustainable development of all types of forests; (iv) The UNFCCC—an international treaty for global cooperation to combat climate change by limiting the emission of greenhouse gases, and (v) CBD—a legally binding multilateral agreement with three main goals: conservation of biodiversity; sustainable use of biodiversity; fair and equitable sharing of the benefits arising from the use of genetic resources. Its overall aim is to protect the diversity of species and habitats in the world. The summit also paved the way for widespread discussions of sustainable development around the issue of climate change.
1995	The Intergovernmental Panel on Climate Change (IPCC)	Released the Second IPCC Assessment Report comprising (1) Working Group I: The Science of Climate Change, (2) Working Group II: Impacts, Adaptations and Mitigation of Climate Change: Scientific-Technical Analyses, (3) Working Group III: Economic and Social Dimensions of Climate Change, and (4) IPCC Second Assessment. The essential message of the Report is that carbon dioxide remains the most important contributor to climate change and the warning that uncontrolled human activities threatened to alter the Earth's climate to an extent unprecedented in human history with destructive impacts such as sea level rise.
1997	Rio +5 Summit (Earth Summit II) held in New York	Special session to review and appraise the implementation of Agenda 21and to deepen the commitments made at Rio (1992).
2000	Millennium Summit held in New York	Adopted the Millennium Declaration which contained eight Millennium Development Goals: (1) eradicate extreme poverty and hunger; (2) achieve universal primary education; (3) promote gender equality and empower women; (4) reduce child mortality; (5) improve maternal health; (6) combat HIV/AIDS, malaria and other diseases; (7) ensure environmental sustainability; and (8) develop a global partnership for development. On environmental the goal of environmental sustainability, the Declaration stated that efforts must be taken to counter the threat of the planet being irredeemably spoiled by human activities. It also called for a new ethic of conservation and stewardship to protect the environment. The Summit committed member nations to achieve these goals by 2015.

(Continued)

APPENDIX 2.1 (Continued)

1990	The Intergovernmental Panel on Climate Change (IPCC)	Published the first IPCC Scientific Assessment Report and warned of the impending threat of global warming and its effects on natural and human systems. The Report is made up of three volumes and an overview, namely, (1) Climate Change: The IPCC Scientific Assessment (1990), (2) Climate Change: The IPCC Impacts Assessment (1990), and (3) Climate Change: The IPCC Response Strategies (1990), and (4) First Assessment Report Overview.
1991	IUCN (currently the World Conservation Union), the UNEP, and the WWF	Published Caring for the Earth with the aims of securing a widespread and deeply held commitment to the ethic for sustainable living and to integrate conservation and development.
1992	UN Conference on Environment and Development (the Earth Summit/Rio Summit) held in Rio de Janeiro, Brazil	Documents adopted: (i) The Rio Declaration on Environment and Development—a program of action which spelt out 27 guiding principles for the management of natural resources and environment; (ii) Agenda 21—a 40-chapter and 800-page agreement that laid down 115 specific programmes to help to achieve sustainable development; (iii) Statement of Forest Principles—a brief document containing 15 principles to guide the management, conservation and sustainable development of all types of forests; (iv) The UNFCCC—an international treaty for global cooperation to combat climate change by limiting the emission of greenhouse gases, and (v) CBD—a legally binding multilateral agreement with three main goals: conservation of biodiversity; sustainable use of biodiversity; fair and equitable sharing of the benefits arising from the use of genetic resources. Its overall aim is to protect the diversity of species and habitats in the world. The summit also paved the way for widespread discussions of sustainable development around the issue of climate change.
1995	The Intergovernmental Panel on Climate Change (IPCC)	Released the Second IPCC Assessment Report comprised (1) Working Group I: The Science of Climate Change, (2) Working Group II: Impacts, Adaptations and Mitigation of Climate Change: Scientific-Technical Analyses, (3) Working Group III: Economic and Social Dimensions of Climate Change, and (4) IPCC Second Assessment. The essential message of the Reports is carbon dioxide remains the most important contributor to climate change and warned that uncontrolled human activities threatened to alter the Earth's climate to an extent unprecedented in human history with destructive impacts such as sea level rise.
1997	Rio +5 Summit (Earth Summit II) held in New York	Special session to review and appraise the implementation of Agenda 21 and to deepen the commitments made at Rio (1992).

(Continued)

APPENDIX 2.1   (Continued)

2000 Millennium Summit held in New York	Adopted the Millennium Declaration which contained eight Millennium Development Goals: to (1) eradicate extreme poverty and hunger; (2) achieve universal primary education; (3) promote gender equality and empower women; (4) reduce child mortality; (5) improve maternal health; (6) combat HIV/AIDS, malaria and other diseases; (7) ensure environmental sustainability; and (8) develop a global partnership for development. On the goal of environmental sustainability, the Declaration stated that efforts must be taken to counter the threat of the planet being irredeemably spoiled by human activities. It also called for a new ethic of conservation and stewardship to protect the environment. The Summit committed member nations to achieve these goals by 2015.

CBD, Convention on Biological Diversity; *IPCC*, Intergovernmental Panel on Climate Change; *IUCN*, International Union for Conservation of Nature; *UNCLOS*, United Nations Conference on the Law of the Sea; *UNCTAD*, United Nations Commission on Trade and Development; *UNEP*, United Nations Environment Program; *UNFCCC*, United Nations Framework Convention on Climate Change; *WCED*, World Commission on Environment and Development; *WCP*, World Climate Program; *WWF*, World Wide Fund for Nature.

2.6 The Cocoyoc Symposium

Since Stockholm, various conferences have been convened and a plethora of multilateral and bilateral international environmental documents have been produced and adopted aiming to reconcile the potential conflicts between economic development and environmental sustainability. One of the more significant events was the Cocoyoc Symposium held at Cocoyoc, Mexico in 1974. The symposium, which was jointly organized by the United Nations Environment Program (UNEP) and the United Nations Commission on Trade and Development (UNCTAD), was the first major attempt to link environmental concerns with development issues following the Stockholm Conference. The Cocoyoc Declaration adopted at the symposium called for the need to reform international economic order in the postcolonial world, placing emphasis on the promotion of eco-development. The Cocoyoc Symposium fundamentally dealt with the synthesis of environmental sustainability and economic development especially viewed from the developing countries' perspective.

The Cocoyoc highlighted the bio-capacity limits of the Earth's system denoted as "the outer limits" in fulfilling the "inner limits" of human needs, namely, food, shelter, clothing, education, and health amidst explosive population growth. It placed poverty eradication as the primary goal of all development. While the Stockholm Conference aimed to address global environmental issues in the context of economic development, the Cocoyoc Symposium fundamentally dealt with the synthesis of environmental sustainability and economic development especially viewed from the developing countries' perspective.

More explicit than the Stockholm Declaration, the Cocoyoc Declaration focused on issues of environmental justice, particularly on unequal North–South economic relationship, inequitable distribution of resources, misuse of cheap imported materials, and over and super-consumption by the industrialized North, which directly affected the physical integrity of the earth's system. It bluntly stated that "A growth process that benefits only the wealthiest minority...is not development. It is exploitation" (UNEP and UNCTAD, 1974: 896).

In addressing the above issues, the Cocoyoc Declaration called for the reform of the existing economic order and supported UNEP's efforts to design strategies and assist projects for ecologically sound socioeconomic development (eco-development). The call for reform was closely connected to the United Nations Declaration on the Establishment of a New International Economic Order adopted at the Sixth Special Session of the General Assembly in May 1974. The new order was based on a host of principles, one of which was the "full and effective participation on the basis of equality of all countries in the solving of world economic problems in the common interest of all countries, bearing in mind the necessity to ensure the accelerated development of all the developing countries" (United Nations, 1974).

The Stockholm initiatives reinforced by the Cocoyoc Declaration seemed to provide a new journey of hope both in terms of inspiration and guidelines for the governments and people worldwide to preserve the human environment. Particularly, at Stockholm, it was widely acknowledged that extensive international cooperation among nations and global organizations is significantly important for achieving environmental sustainability as environmental problems do not recognize any political boundaries. This set the stage for the creation of various environmental protection agencies and ministries, and the creation of a host of environmental laws and policies by countries across the world (Drexhage and Murphy, 2010).

2.7 The Stockholm Conference and the emergence of global environmental regimes

It would appear that the Stockholm Conference, as pointed out by Maurice Strong, "marked the first time that nations of the world collectively acknowledged that something has gone wrong with the way in which man had been managing...his relationship with the natural world on which his own survival depends" (quoted in Jacobsen, 1973: 35). Before Stockholm, national environmental policy efforts, especially those targeting genuine environmental protection, were rare, but became common after the Conference. Also, most countries, despite the existence of environmental protection laws even before Stockholm, were fundamentally concerned about sectoral or specific aspects of environmentalism and failed to consider the interconnection of the multiple factors affecting the natural environment (Caponera, 1972). Besides, in the developing countries, many of their environmental legislations were not adequately enforced, rendering them environmentally ineffective.

As a case in point, during the preStockholm period, various environmental laws were passed in the United States—one of the few countries which had developed environmental laws before the Stockholm Conference (Hironaka, 2014). These included the Clean Air Act (1963), the Motor Vehicle Air Pollution Control Act (1966), Air Quality Act (1967), and Clean Air Act (1970). However, at the time of the enactment, these legislations were perceived as necessary for expanding federal control over urban planning and public health rather than for environmental protection. Only with the environmental awareness and understanding developed since the Stockholm Conference have these laws been reconceptualized to embrace environmental protection (Hironaka, 2014).

It may well be that before Stockholm, only a few countries, primarily in the West, had environmental ministries. However, in the postStockholm period, more than 100 countries set up environmental ministries and agencies which developed policies and legislations to deal with environmental issues. In Europe, for example, in the wake of the 1972 Stockholm Conference and growing public awareness and concerns about the biophysical limits to growth, the European Commission established its first Environmental Actions Program (EAP) in 1973 with some of the most important objectives: (1) to prevent, reduce, and contain environmental damage, (2) to conserve an ecological equilibrium, and (3) to promote rational use of natural resources, among others. The second EAP, established in 1977, was essentially a follow-up to the first EAP with greater emphasis placed on nature protection (Hey, 2005). The United Kingdom established its environmental ministry for the first time in 1971, while Canada, Austria, Denmark, East Germany, Netherlands, Australia, and New Zealand established their respective ministries in 1972 (Selin and Linnér, 2005).

In Asia, even before the Stockholm Conference, Japan had enacted 14 environmental protection laws in one session at the so-called "Pollution Diet" after decades of corporate cover-ups of toxic pollution (Kapur, 2015). Japan established its environmental ministry in 1972 (Selin and Linnér, 2005). In the early 1970s, the Chinese authorities in Maoist China began to consider its environmental problems fairly seriously (Taneja, 1998). They became concerned about the environmental issues facing the country and acknowledged the need to take environmental protection seriously in national planning in the face of rapid economic expansion (Cai and Voigts, 1993). The Stockholm Conference seemed to provide China with the opportunity to revisit its unprecedented environmentally destructive industrialization policies implemented under Mao Zedong's "Great Leap Forward" campaign. To note in passing, The Great Leap Forward (1958—60) was Mao Zedong's plan to transform China's economy from a predominantly agrarian society into a modern, industrial communist society through rapid industrialization (Section 3.18).

Indeed, it is often claimed that the Stockholm Conference was the turning point in the development of environmental protection in China (US Department of Commerce, 2002). After Stockholm, the term "environmental protection" began to appear in the Chinese media and in official documents (Taneja, 1998). Since then, China enacted a series of environmental laws and regulations to reinforce its environmental governance for the promotion of sustainable development (Mol and Carter, 2007, see Sections 3.13.1 and 3.17 for a detailed discussion). Similarly, in India, Indira Gandhi upon her return from Stockholm, began to pass various environmental laws such as the Water (Prevention and Control of Pollution) Act in 1974 and to amend its constitution to include environmental protection among the principles governing State policy (Dembowski, 2001).

In the case of Southeast Asia, environmental issues were not explicitly recognized as a concern when the geo-political region, the Association of Southeast Asian Nations (ASEAN) comprising Indonesia, Malaysia, Singapore, Thailand, and the Philippines, was established in 1967. However, following Stockholm, the region began to integrate environmental concerns into its regional development planning (Koh and Robinson, 2002; see Section 3.12.1). To avoid confusion, Southeast Asia is a subregion of Asia which lies from the South of India to the East of China while ASEAN is a regional intergovernmental organization (IGO) which brings together the 11 Southeast Asian countries comprising Brunei Darussalam, Cambodia, Indonesia, Laos, Malaysia, Myanmar, the Philippines, Singapore, Thailand, Vietnam, and East Timor or Timor-Leste. In 1967, ASEAN comprised only five countries as mentioned above.

Also, during the 1970s, all national governments in Latin America created legal and administrative structures for natural resource management. Indeed, the creation of UNEP in 1972 stimulated Latin American active engagement in natural resource and environmental management especially from 1975 onwards. For example, with the support of UNEP, the Spanish Iniciativa de Copenhague para Centroamérica y México was created and various courses and seminars on environmentalism were also organized (Martinez-Alier et al., 2016).

2.8 The Stockholm environmental impacts: Some remarks

However, enlightened ecological awareness and mere acknowledgment of looming environmental disasters do not necessarily reflect real and genuine aspirations for ecologically sound socioeconomic development. It may well be that immediately after the Stockholm Conference, the emerging environmental initiatives or policies, especially in the developing South as briefly elucidated above, were typically symbolic in nature.

In particular, the severe economic recession and energy crisis in the 1970s had undermined states' willingness to implement environmental controlling measures which would strangle the economy further. Thus the differences between the approaches of the affluent North and the developing South toward environment and development were increasingly acute in the 1970s; and responses to the Stockholm environmental initiatives from the South were not particularly promising. As a matter of fact, the Stockholm Conference did not directly lead to a sustainable development movement (Egelston, 2013). Indeed, in reality, it hardly made a ripple in halting unsustainable levels of environmental exploitation and ecological abuse across the globe, especially in the developing South where environmental awareness of the disturbing reality did not lead to actions commensurate to the environmental problems occurred in the same period (PSCD, 2012).

As it turned out, the creation of laws or regulations to promote environmental protection in the South was generally conceived as an unjustified limitation on economic activities, and existing legal and administrative structures often avoided wholly embracing eco-development principles that guide and promote the rational use of resources (Caponera, 1972). That said, the proliferation of environmental controlling and resource management instruments in the 1970s did not automatically translate into genuine environmental protection efforts in the South as the region was more concerned about economic development and poverty issues.

For example, for most of the 1970s, environmental efforts in China were overshadowed by Mao's environmentally destructive development philosophy which was still widely embraced by the Chinese government. This was evidenced by the increasingly acute environmental pollution, including air and water pollution, that prevailed in the 1970s amidst rapid industrial development (Wang, 1989; Mu et al., 2014; Delang, 2016; see also, Sections 3.13.2 and 3.18). It may well be that Mao's Great Leap Forward in the 1950s saw an era of extreme human interference and transformation of nature. The traditional Chinese environmental ideal of "harmony between heaven and humans" was abrogated in favor of the slogan "Humans Must Conquer Nature" which exerted immense influence on human environmental attitudes and the magnitude and impact economic activities during the Maoist period (Shapiro, 2001). Thus no significant environmental regulations were enacted following Stockholm until the reformist Deng Xiaoping firmly took the political helm in 1978 (Kapur, 2015, see Sections 3.13.1 and 3.17).

Deforestation in Southeast Asia in the same period had also become increasingly environmentally destructive despite having various environmental protection and resource management regulations in place (Sodhi et al., 2004; Choy 2015a, 2015b, 2018, 2019; see also, Section 3.12). In India, the environmental legislation enacted after Stockholm were, to a great extent, mere rhetoric and ineffective in reality because the government was reluctant to enforce laws which would inhibit industrial growth (Chandra, 2015). Indeed, in India, even basic laws concerning urban planning and fundamental hygiene and public health were hardly implemented (Dembowski, 2001). In Brazil, it would seem that the Stockholm Conference had failed to make an impact on the alarming rates of deforestation in the Amazon forest in the 1970s caused by uncontrolled commercial ranching activities (Sawyer, 1998; see also, Sections 3.10.3 and 3.10.4).

Despite this, it is however fair to say that the Stockholm Conference did successfully raise global environmental awareness. It has put the environment permanently on the international political agenda, marking the coming of age of the green evolution of global environmental regimes in the latter part of the United Nations sustainable development history. More importantly, the Stockholm Declaration adopted at the Conference constitutes the first body of rules with goals to protect the essential components of the environment which are considered as a "global public good" (Chechi, 2016: 217).

It may be noted that we need to assume a response lag to the Stockholm impacts. Between the time a participant country enacts environmental policies or regulations and the point at which these are implemented may be short or long depending on the efficiency of political institutions, the socioeconomic conditions, and the speed of development in the country concerned. Also, proper implementation of the environment polices may take place gradually, rather than abruptly as there is a learning period to adhere to. Thus it is no surprise that the Stockholm Conference could hardly be said to have made waves in driving real movement for halting unsustainable development practices across the globe, especially in the South in the 1970s, as shown earlier.

Our analysis will now proceed with determining whether the United Nations' efforts in the postStockholm era have made any advance in greening the global economic system in harmony with the natural world.

2.9 The postStockholm era: The second wave of United Nations engagement with environmentally sustainable development

As examined below, in the follow-up to the Stockholm Conference, a series of summits and conferences were held and hundreds of environmental documents were adopted. These were held to sustain the momentum of the Stockholm call for eco-development, and to raise international awareness of the looming environmental threats such as climate change, as well as to review environmental progress, and to reinforce global commitment to environmental sustainability (Timeline 2.2).

1979	First World Climate Conference (WCC-1) was held in Geneva, Switzerland: raised massive waves of global concern over the threat of atmospheric greenhouse gas concentration and global warming
1982	Stockholm+10: to review the progress of the Stockholm Action Plan and to forward recommendations for UNEP activities for the coming decades
1983	World Commission on Environment and Development (WCED) or Brundtland Commission was established
1987	Brundtland Report or Our Common Future was published: popularized the concept of sustainable development
1990	Second World Climate Conference (WCC-2) held in Geneva: to lay the foundation for the negotiations of global climate treaties followed by the WCC-3 held in 2009 to establish "an international framework for climate services," etc.
1992	United Nations Conference on Environment and Development or the Earth Summit was held
1992	Conference for the Adoption of the Agreed Text of the Convention on Biological Diversity (CBD) was held in Nairobi, Kenya for the adoption of the CBD agreed text
1994	First Conference of the Parties to the Convention on Biological Diversity (COP1) was held in Nassau, Bahamas (see Appendix 2.3 for the Timeline events)
1995	The United Nations Framework Convention on Climate Change (UNFCCC) first Conference of Parties (COP1) held in Berlin, Germany (see Appendix 2.2 for the Timeline events)
1997	The Earth Summit II was held in New York: to review and appraise the implementation of Agenda 21

TIMELINE 2.2   The second United Nations wave of environmental sustainability.

2.9.1 The first World Climate Conference (1979)

The first World Climate Conference (WCC-1), convened by the World Meteorological Organization (WMO), in collaboration with the United Nations Educational, Scientific and Cultural Organization(UNESCO), the Food and Agriculture Organization of the United Nations (FAO), the World Health Organization, the UNEP, International Council for Science (ICSU) and other scientific partners, in 1979 in Geneva, Switzerland, for example, raised massive waves of global concern over the threat of atmospheric greenhouse gas (GHG) concentration and global warming caused by uncontrolled human economic activities (Zillman, 2009). The Conference created the urgency for deep reform and mitigating policies to address the problem (Appendix 2.1). Subsequent to this, the World Climate Program (WCP) was created to provide a framework for international cooperation in research on climate issues including ozone depletion. The WCP was followed by the Vienna Convention for the Protection of the Ozone Layer (1985) and the establishment of the Intergovernmental Panel on Climate Change (IPCC) in 1989 (Appendix 2.1). This was followed by the publication of the IPCC First and Second Assessment Reports in 1990 and 1995 (Appendix 2.1).

Following the WCC-1, the WCC-2 was also held in Geneva in 1990 under the sponsorship of WMO, UNESCO, UNEP, FAO, and ICSU. The Conference adopted the WCC-2 Ministerial Declaration—an important document that set the essential parameters for negotiation and establishment of the United Nations Framework Convention on Climate Change (UNFCCC) (Zillman, 2009). This laid the foundation for the negotiations of global climate treaties as discussed in the subsequent section.

The WCC-3 was held in Geneva under the theme, "Climate Prediction and Information for Decision Making" in 2009. Its focus was to establish "an international framework for climate services that links science-based climate predictions and information with the management of climate-related risks and opportunities in support of adaptation to climate variability and change in both developed and developing countries" (GFCS, 2009a: 3) The WCC-3 High Level Declaration was adopted at the Conference which aimed to establish a Global Framework for Climate Services to strengthen production, availability, delivery, and application of science-based climate prediction and services, among others (GFCS, 2009b: 3)

2.9.2 Stockholm +10 (1982) and the Brundtland Report (1987)

Ten years after Stockholm, the UN Environment Program convened a special Stockholm +10 Conference in 1982 in Nairobi to review the progress of the Stockholm Action Plan and to forward recommendations for UNEP activities for the coming decades (Selin and Linnér, 2005). The Conference which expressed "serious concern about the present state of the environment" revealed that progress made by governments in addressing major environmental issues had been slow and that the Stockholm Action Plan had only been partially implemented. The Nairobi Declaration adopted at the Conference reaffirmed the principles of the Stockholm Declaration and called on the international community to intensify environmental protection efforts. It also proposed the establishment of the World Commission on Environment and Development (WCED), also known as the Brundtland Commission, in 1983 as an independent commission under the Norwegian Prime Minister Gro Harlem Brundtland, to propose long-term environmental strategies for "the achievement of common and mutually supportive objectives which take account of the interrelationships between people, resources, environment and development" (United Nations, 1983). This paved the way for the publication of the Brundtland Report or *Our Common Future* in 1987.

The Brundtland Report may be regarded as a green effort to crystallize the conservationist thinking in early environmental sustainability literature such as the three landmark publications, namely, *Limits to Growth* (Meadows et al., 1972), *A Blueprint for Survival* (Goldsmith et al., 1972) and *Small is Beautiful* (Schumacher, 1973), and the idea of "Only One Earth" that was promoted at the 1972 Stockholm Conference, the Founex Report, and the Cocoyoc Declaration. In the sphere of environmental conservation policy, the Report is considered an ideological document that summarized and conceptualized major ecological strategies emerging during the green decades of the 1960s and 1970s. These include the Common Heritage of All Mankind first introduced in the 1960s, Man and the Biosphere Program (1971), the World Heritage (1972), World Conservation Strategy (1980), the World Charter for Nature (1982), (Larsson, 1999), and "eco-development," a major theme of discussion at the United Nations Conference on the Human Environment (the Stockholm Conference) held in 1972.

The Brundtland Report popularized the concept of sustainable development which was formally introduced in the World Conservation Strategy. Defining sustainable development as "development that meets the needs of the present without compromising the ability of future generations to meet their own needs," the Report placed great emphasis on the concept of intergenerational equity as well as the need to protect the natural resource base and the natural environment while promoting economic growth (WCED, 1987: 43). This is to ensure that future generations will not be worse off than the present generation in terms of resource endowment. To achieve this, the Report calls for the need to observe the physical limits and biophysical constraints to development.

More specifically, it states that development would be guided by the "ability of the biosphere to absorb the effects of human activities" (WCED, 1987: 8). That is to say, humans in the present generation owe a moral responsibility to the future generations to exercise prudent use of the natural environment within its environmental sustainability thresholds (the Holling concepts of sustainable resource-use) to fulfil the principle of intergenerational equity. The Report placed the environment in the political context and generated a second wave of United Nations engagement for a rapid global transition to a more environmentally sustainable world. The Report recommended convening another UN Conference that would bring together both the environment and development agendas under one roof. Here, it may be noted that while sustainable development as used in the World Conservation Strategy is more concerned about ecological sustainability and less concerned about economic growth, the Brundtland concept of sustainable development places its focus on reconciling both while promoting social equity.

The Brundtland concept of sustainable development may be regarded as a global refinement and expansion of eco-development as used in the 1970s such the Cocoyoc Declaration which defined eco-development as "ecologically sound socioeconomic development (eco-development) at the local and regional level" in the developing South (UNEP and UNCTAD, 1974: 900). In acknowledging that "environment" and "development" are inseparable, the concept of sustainable development under Brundtland's perspective entailed the integration of development and environment in policies in all countries, industrialized and developing alike. While the concept of eco-development had failed to enter into the mainstream of the larger policy community, the Commission deliberately defined and embedded the notion of sustainable development within such a wide scope that it has been able to gain recognition by economists, policy-makers, politicians, social groups, and the public at large. As a result, the Brundtland concept of sustainable development became a central theme of the environment and development discourse in the 1980s (Selin and Linnér, 2005).

2.9.3 United Nations Conference on Environment and Development (The Earth Summit, 1992)

In response to the recommendation of the Brundtland Commission, the United Nations convened the Earth Summit in Rio de Janeiro, Brazil, in 1992 "to help Governments rethink economic development and find ways to halt the destruction of irreplaceable natural resources and pollution of the planet" (United Nations, 1997a). The summit was unprecedented both in terms of its size and scope of concern. It was attended by 172 heads of state and government and some 2400 representatives of nongovernmental organizations (NGOs), while 17,000 people attended the parallel NGO Forum (United Nations, 1997a). The summit laid the foundation for the global institutionalization of sustainable development through the adoption of the following crucial documents which commit the world to promoting sustainable development:

i. The Rio Declaration and Development—a program of action spelling out 27 guiding principles for the management of natural resources and the environment.
ii. Agenda 21—a 40-chapter and 800-page agreement laying down 115 specific programs to help achieve sustainable development.
iii. The Forest Principles or "Non-legally binding authoritative statement of principles for a global consensus on the management, conservation and sustainable development of all types of forests"—a brief document containing 15 principles to guide the management, conservation, and sustainable development of all types of forests.
iv. The United Nations Framework Convention on Climate Change (UNFCCC)—an international treaty for global cooperation to combat climate change by limiting the emission of GHGs.
v. Convention on Biological Diversity (CBD)—a legally binding multilateral agreement with three main goals: conservation of biodiversity; sustainable use of biodiversity; fair and equitable sharing of the benefits arising from the use of genetic resources. Its overall aim is to protect the diversity of species and habitats in the world.
vi. The United Nations Convention to Combat Desertification (UNCCD), adopted on June 17, 1994 and entered into force in 1996, led to a global and legally binding coalition between developed and developing countries to combat desertification (Kutter, 2009). The principal objective of UNCCD (objective 1) is to "combat desertification and mitigate the effects of drought, particularly in Africa," thus contributing to long-term sustainable development.
vii. Following (vi) above, in recognizing land degradation/desertification as a threat to environment and the very base of sustainable livelihood in many countries especially in the arid, semi-arid and dry sub-humid areas around the world, the United Nations

Conference on Desertification held in Nairobi adopted a Plan of Action to Combat Desertification in 1977. Despite this, land degradation in the regions especially in Africa had intensified, prompting the United Nations General Assembly at the Earth Summit to establish an Intergovernmental Negotiating Committee on Desertification to prepare the Convention to Combat Desertification which was concluded in 1994 (Kutter, 2009).

Agenda 21, and in particular the Forest Principle, acknowledged for the first time the important role of forests and expressed the need for their sustainable management (Dine, 2012). More importantly, the Agenda provides a blueprint for achieving sustainable development worldwide. Indeed, following the summit, most countries created new coordinating environmental and development mechanisms and drew up their Local Agenda 21 documents and action plans for promoting sustainable development. Another important feature of Agenda 21 is its recognition of the importance of NGOs or civil society as "one of the fundamental prerequisites for the achievement of sustainable development is broad public participation in decision-making." It further acknowledges the "commitment and genuine involvement of all social groups" (Agenda 21, Chapter 23: Strengthening the Role of Major Groups). The Agenda also stressed the importance of the "integration among national and local government, industry, science, environmental groups and the public in the process of developing effective approaches to environment and development" (Agenda 21, Chapter 8: Integrating Environment and Development in Decision-Making).

The Rio Declaration, building upon the Stockholm Declaration and Agenda 21, laid a solid foundation to guide national action and international cooperation in promoting environmentally sustainable development through balancing the integration of the three pillars of sustainability, namely, social, economic, and environmental progress (Appendix 2.1). Principle 15 of the Rio Declaration is particularly noteworthy for explicitly providing an environmental controlling principle through its precautionary principle: "Where there are threats of serious or irreversible damage, a lack of full scientific certainty shall not excuse States from taking cost-effective measures to prevent environmental degradation" (United Nations, 1992a: 3–4).

The status of NGOs is further reinforced under Principle 10 of the Rio Declaration which broadly states that "environmental issues are best handled with the participation of all concerned citizens at the relevant level." Also, Principle 27 postulates a new social partnership between the state and its people "in the further development of international law in the field of sustainable development" while Principle 10 mandates appropriate access to information, and encouragement of public participation in the decision-making process. Particularly, Principle 10 seeks to commit national governments to an inclusive process of public participation involving all concerned citizens at the relevant levels in environmental decision-making, and to improve conditions such as accountability and effective access to judicial and administrative proceedings for good environmental governance at the national level.

Two of the more important outcomes which are directly associated with some of the most pressing environmental problems confronting the world today, namely, global warming and biodiversity loss, are UNFCCC and CBD. The following section discusses why the initiatives built under these two conventions are crucially important to protect the resilience of the Earth's system underpinning humans' long-term existence.

2.9.4 The Earth Summit II: Review of Agenda 21

In the latter part of the United Nations' journey to global environmental sustainability, in 1997, a special session known as the United Nations General Assembly Special Session was held in New York to review and appraise the implementation of Agenda 21, that is, progress achieved over the previous 5 years since the Earth Summit held in Rio de Janeiro in 1992, and also to reinforce global joint environmental protection efforts. The event is also known as the Earth Summit + 5 or Earth Summit II. Briefly, the main objectives of the summit were:

- To call on governments and the international community to commit to sustainable development and to build up momentum for its implementation at the international, national, and local levels.
- To identify reasons for failure to achieve goals set in Rio and to suggest corrective action.
- To recognize Rio achievements and identify actions and means, including innovative approaches, cooperation or financial assistance that will boost them.
- To define priorities for the post97 period.
- To raise the profile of issues addressed insufficiently by Rio (NILOS, 1998).

There was no major breakthrough at the meetings as discussions became bogged down in North–South differences over the provision of financial resources and the transfer of environmentally friendly technologies in promoting sustainable development globally (Osborn and Bigg, 1998; Baker, 2006). For instance, many countries in the South saw the decline in levels of official development assistance (ODA) from an average of 0.34% of the donor country's gross national product in 1991 to 0.27% in 1995 for development agreed at the Rio Earth Summit 5 years earlier as an impediment to global partnership (Osborn and Bigg, 1998; Baker, 2006).

The Program for the Further Implementation of Agenda 21 was adopted at the summit. The document generally painted a dismal picture of global environmental progress. It acknowledged that "the state of the global environment has continued to deteriorate... Some progress has been made in terms of institutional development, international consensus-building, public participation, and private sector actions and, as a result, a number of countries have succeeded in curbing pollution and slowing the rate of resource degradation. Overall, however, trends are worsening..." (United Nations, 1997b: para. 9).

Despite this, some important outcomes of the summit are worth mentioning:

- A detailed assessment of progress since Rio and a program for further implementation of Agenda 21 was agreed upon.
- Political commitment to the promotion of sustainable development was confirmed.
- UNCED targets and commitment for ODA were confirmed.
- A more focused program of work for the Commission on Sustainable Development (CSD) for the next 5 years was developed (Osborn and Bigg, 1998).

The summit also served as an important platform for delegates to discuss and exchange information on sustainable development problems and solutions throughout the world. This indirectly influenced the official negotiations on various issues at the summit (Osborn and Bigg, 1998). The summit was also significant in that, unlike the summit preparatory process where the NGOs had limited access to delegates and negotiations, they were able, for the first time, to participate at the plenary session or allowed in at ministerial-level consultations (Baker, 2006).

Viewed from this perspective, the summit was a milestone event in that it brought NGOs on board to keep up the pressure and help mobilize the public in readiness for sustainable development and environmental protection. The key role of NGOs in global environmental sustainability especially in relation to climate change was also acknowledged by the former British Prime Minister, Tony Blair and the former United States President Bill Clinton (Carpenter et al., 1998).

2.9.5 United Nations Framework Convention on Climate Change

The UNFCC which entered into force on March 21, 1994, followed by the first Conference of Parties (COP1) in Berlin, Germany in 1995, brought the international community to a roundtable discussion to strengthen global commitment to mitigate climate change. It paved the way for a solid international climate governance to spearhead progressive global, national and local action to address climate change by stabilizing "greenhouse gas concentrations in the atmosphere at a level that would prevent dangerous anthropogenic interference with the climate system" (UNFCCC, 2006: 21). This led to the declaration of the Berlin Mandate which committed industrialized nations through legally binding obligations to reduce GHG emissions, while exempting developing countries from the same commitment. The Mandate led to the adoption of the Kyoto Protocol at COP3 held in 1997. The Protocol legally binds developed country parties to stabilize GHG emission standards and the detailed rules for the implementation of the protocol were adopted at COP7 held in Marrakesh in 2001, known as the Marrakesh Accord.

The Kyoto Protocol, a legally binding treaty was adopted in Kyoto, Japan, on December 11, 1997 and entered into force on February 16, 2005. The Protocol commits industrialized countries to limit and reduce GHG emissions by 5.2% below the 1990s base year level by 2008–12 in aggregate (the first commitment period). In 1990, the global output of carbon dioxide was 22.7 billion tonnes (Rosen, 2015). The adoption of the Kyoto Protocol represents a landmark decision that shaped international climate politics over the past two decades as indicated in Appendix 2.2. It provides the following three important mechanisms in addressing GHG emission problems:

i. Emission Trading known as the "carbon market";
ii. Clean Development Mechanism (CDM) which involves investment in sustainable development projects; and
iii. Joint Implementation (JI), a mechanism enabling industrialized countries to implement joint-projects with developing countries.

APPENDIX 2.2   United Nations Climate Conference (COP1–COP25).

Year		Venue	Remark
1995	The first Conference of the Parties (COP1)	Berlin, Germany	To review the effectiveness of the agreements contained in the Convention in combating climate change. The findings of the review pointed to the necessity of creating a legally binding protocol instead of voluntary commitments under the Convention and with new, national emissions reduction targets, and clear time frames.
1996	COP2	Geneva, Switzerland	Adoption of Intergovernmental Panel on Climate Change (IPCC) second Assessment Report which warned of a discernible human influence on global climate change. The meeting highlighted the urgent need for a binding protocol in addressing greenhouse gas emission problems.
1997	COP3	Kyoto, Japan	Adoption of the Kyoto Protocol on Climate Change, a legally binding agreement which commits industrialized countries to limit and reduce green house gas emissions by 5.2% below the emission level of 1990 (22.7 billion tonnes) by 2012 (first commitment phase: 2008–2012).
1998	COP4	Buenos Aires, Argentina	Agreement on the completion of the detailed structure of the Kyoto Protocol by the 6th meeting of the Conference of the Parties at the latest. The meeting was characterized by informal discussions on the need for developing countries to commit to reducing green house gas emissions.
1999	COP5	Bonn, Germany	Issues on monitoring commitments and the design of the Kyoto mechanisms, especially the (Clean Development Mechanism, CDM) were discussed and guidelines were drawn up for industrialized countries' national emissions reports.
2000	COP6-1	The Hague, Netherlands	Aimed to clarify the details of the Kyoto Protocol but was unable to reach an agreement between the umbrella group, which included US, Australia, Canada, Japan, and Russia, the developing countries and the European Union. The main issues of contention were temporary carbon storage function of natural forests and its inclusion in the Clean Development Mechanism as well as the question on the need of binding rules for reduction commitments.
2001	COP6-2	Bonn, Germany	Reached an agreement on the main unresolved issues of the Kyoto Protocol and led to the adoption of the Bonn Agreements on international climate policy which established the conditions needed to ratify and implement the Kyoto Protocol.
2001	COP7	Marrakech, Morocco	Adoption of Marrakesh Accords which consisted of a package of 15 decisions on structuring and implementing the Kyoto Protocol. This included a system for monitoring compliance, using the Kyoto Mechanisms of carbon sinks credit, and the promotion of climate action in developing countries. The adoption of the Accords speraheaded the way for the Kyoto Protocol's entry into force.

(Continued)

APPENDIX 2.2 (Continued)

Year		Venue	Remark
2002	COP8	New Delhi, India	Negotiations on the details of the Kyoto Protocol were essentially completed. Decisions were made on the design of the Clean Development Mechanism and the use of funds provided by industrialized countries for climate action in developing countries. Work program aiming at raising awareness of climate issues and mainstreaming them into the Parties' educational programmes was initiated.
2003	COP9	Milan, Italy	Successful conclusion of the 2-year negotiations on the rules for afforestation and reforestation projects in developing countries, closing the last gap in the Kyoto Protocol's rules of implementation.
2004	COP10	Buenos Aires, Argentina	Issues pertaining to funding, institutional reform, capacity building, and technology transfer to promote the implementation of Framework Convention on Climate Change in developing countries were considered at the meeting.
2005	COP11	Montreal, Canada	Adoption of Montreal Action Plan, a roadmap to a post2012 international climate regime. The Kyoto Protocol was fully implemented with its organizational structure strengthened with a robust review regime and increased funding.
2006	COP12	Nairobi, Kenya	The meeting concentrated on the discussion of the African issues including capacity building and assistance in the development of concrete projects and participation in the Clean Development Mechanism (CDM).
2007	COP13	Bali, Indonesia	Adoption of the Bali Action Plan which involved parties' negotiation on concrete commitments and contributions to emissions reductions (including a reduction of deforestation), adaptation, technology, and financing up to and beyond 2012.
2008	COP14	Poznań, Poland	Key elements of a new climate regime, focusing primarily on the necessary national greenhouse gas reduction targets and financial support for climate action in developing countries were discussed.
2009	COP15	Copenhagen, Denmark	The Copenhagen Accord which defined some central components of future international climate policy was drawn up and a "1.5 or 2.0 degrees C" limit emission standard was agreed upon by a large group of industrialized and developing countries. Also, the industrialized countries pledged up to US $30 billion for climate action in developing countries from 2010 to 2012. A Technology Mechanism and a REDD + Mechanism aimed at supporting developing countries in technology programmes and reducing emissions from deforestation and forest degradation were considered.
2010	COP16	Cancún, Mexico	With the adoption of the Cancun Agreements, the agreements reached in Copenhagen were turned into official decisions, developed further and operationalized. These include the

(Continued)

APPENDIX 2.2   (Continued)

Year		Venue	Remark
			official recognition of the two-degree target, forest conservation (REDD + Mechaism), technological cooperation and capacity-building in developing countries. Industrialized countries pledged US $100 billion dollars per year by 2020 to support climate action efforts in developing countries.
2011	COP17	Durban, South Africa	Agreed to continue the Kyoto Protocol for a second period from the beginning of 2013. The summit also adopted the Green Climate Fund, aimed at providing financial support for industrializing and developing countries in their climate action efforts. The ADP established a timeline and set out the motivation, guidance and process for developing a new international treaty to replace the Kyoto Protocol upon its expiration in 2012.
2012	COP18	Doha, Qatar	The summit spearheaded more immediate climate initiatives needed to meet the two-degree target as agreed under the Copenhagen Accord. Doha Amendment to the Kyoto Protocol to reduce green house gases emissions by at least 18% below 1990s emission levels by 2020 (second commitment phase: 2013–20).
2013	COP19	Warsaw, Poland	Entered into negotiation in advance of the Climate Change Conference in Paris in 2015 for a worldwide climate agreement on issues pertaining to regulations on mitigation, adaptation, financing, technology, transparency, and capacity-building.
2014	COP20	Lima, Peru	Laid the foundations for negotiations on the new global climate agreement to be agreed at the Climate Change Conference in Paris in 2015.
2015	COP21	Paris, France	Adoption of the Paris Agreement which aims to strengthen the global response to the threat of climate change by keeping a global temperature rise this century well below the 2°C and parties to the Conference agreed to accelerate and intensify actions and investments needed for a sustainable low-carbon future.
2016	COP22	Marrakech, Morocco	Main aim: "to maintain momentum on climate action and continue strengthening the global response to the threat of climate change."
2017	COP23	Bonn, Germany	Main aim: "to accelerate climate action toward the completion of the work program under the Paris Climate Change Agreement."
2018	COP24	Katowic, Poland	Main aim: to finalize the guidelines known as the Paris Rulebook for the full implementation of the Paris Agreement among the 197 parties to the Convention to limit global warming to well below 2°C and ideally 1.5°C by the end of the century.

(Continued)

APPENDIX 2.2 (Continued)

Year		Venue	Remark
2019	COP25	Madrid, Spain	Main aim: to formalize the rules by which the Paris accords would be implemented, and to accelerate actions by which the decarbonization pledges made under the Accord could be achieved.

ADP, Durban Platform for Enhanced Action; *CDM*, clean development mechanism; *COP*, Conference of the Parties; *IPCC*, Intergovernmental Panel on Climate Change.

The concept of "carbon market" represents a new mechanism for transferring money and technology to the developing countries in their GHG reduction efforts such as carbon sink protection through forest conservation. UNFCCC also regularly provides comprehensive scientific and technical assessment reports on climate change under the IPCC (Appendix 2.2).

Undeniably, as reflected in Appendix 2.2, the UNFCCC summits from COP1 to COP24 provide an important global platform for international negotiation in addressing global emission standards. At COP15 held in Copenhagen in 2009, for example, China, Brazil, South Africa, and other major developing countries had agreed to fulfill the treaty obligation for the first time to reduce emission standards (Tollefson and Gilbert, 2012). The developed countries also agreed to a collective commitment under the Copenhagen Accord adopted at COP15 to scale up additional financial resources to US $30 billion for the period 2010–12 to reduce emissions from deforestation and forest degradation, to enhance capacity-building and to adapt to climate change. They also committed to jointly mobilize US $100 billion a year by 2020 to address the needs of developing countries (UNFCCC, 2010).

As noted in Appendix 2.2, agreements reached at COP15 were made official on the adoption of the Cancun Agreement at COP16 and subsequent meetings eventually led to the adoption of the Doha Amendment to the Kyoto Protocol at COP18 held in 2012 which committed developed countries for much larger emission cuts by at least 18% below 1990 levels by 2020. Global emission reduction initiatives were further reinforced with the adoption of the "Lima Call for Climate Action" at COP20 held in Lima where all parties to the Conventions (developed and developing countries alike) agreed to submit their proposed measures, the "intended nationally determined contributions" to combat climate change (UNFCCC, 2014). This constitutes the basis for the establishment of a universal climate change agreement to be decided at the 2015 Paris Convention (COP21) and scheduled to be implemented by 2020.

The Lima agreement represents a significant breakthrough from the conventional Kyoto Protocol which had primarily and legally committed a small subset of developed countries (Annex I Parties) to the exclusion of developing countries (Non-Annex I Parties) in GHG emission reduction. This accomplished the aim of the 2011 Durban Platform for Enhanced Action which sought to include all parties including the poorest and the most vulnerable countries to the Convention under a common legal framework.

At COP21 held in Paris in 2015, the parties to the Convention overwhelmingly adopted a new first-ever universal, legally binding international climate agreement, the Paris Agreement.

The Agreement which was also ratified by the world's two biggest polluters, China and the United States, sets out a global action plan for emission-cutting pledges known as "intended nationally determined contributions" (INDCs), to limit global warming to below 2°C or, if possible, below 1.5°C as agreed under the Copenhagen Accord (UNFCCC, 2015). At the time of writing, 186 countries including the United States are parties to the Paris Agreement. Countries that have signed but stop short of ratifying the Agreement include Angola, Eritrea, Iran, Iraq, Kyrgyzstan, Lebanon, Libya, South Sudan, Turkey, and Yemen. However, it is relevant to note that the current President of the United States, Donald Trump, has formally notified the United Nations in November 2019 that he would withdraw the country from the Paris Agreement.

In the following year, just 3 days after the entry into force of the Paris Agreement, COP22 was held in Marrakech, Morocco from November 7 to 18, 2016 "to maintain momentum on climate action and continue strengthening the global response to the threat of climate change" (United Nations, 2016a). The Conference was claimed by Minister of Foreign Affairs and Cooperation of Morocco, Salaheddine Mezouar, to be "within a climate of hope and of legitimate aspirations for all of humanity" (United Nations, 2016b). The Marrakech Action Proclamation for our Climate and Sustainable Development and the Marrakech Partnership for Global Climate Action were adopted at the Conference to address climate change problems.

Following Marrakech, COP23 was held from November 6 to 17, 2017 in Bonn, Germany to accelerate climate action toward the completion of the work program under the Paris Climate Change Agreement (UNFCCC, 2018a). A year later, COP24 was held in Katowic, Poland from December 2 to 14. One of the main aims of summit is to finalize the guidelines known as the Paris Rulebook for the full implementation of the Paris Agreement among the 197 parties to the Convention to limit global warming to well below 2°C and ideally 1.5°C by the end of the century (UNFCCC, 2018b). Almost 200 countries signed up to the 156-page "Rulebook" for implementing the Paris Climate Change Agreement. Following Katowic, COP25 was held in Madrid, Spain from December 2 to 13, 2019, to reinforce the political will for more climate ambition and action as pledged under the Paris Accord.

However, the summit, the longest ever climate talks on record in United Nations history, only ended with unambitious and watered-down commitments to the drastic cuts in GHG emissions as pledged under the Paris Agreement. For example, nations including Brazil, Australia, and Saudi Arabia opposed commitments to enhance climate action while the United States under Donald Trump has yet to show any serious commitment to tackle climate crisis. Also, all delegates have failed to finalize the rules of the Paris Agreement and the key decision on global carbon markets were left for the next summit to be held in Glasgow, Scotland in 2021—pushed back by a year because of the outbreak of coronavirus (Farand, 2019; Economist, 2019; Ambrose, 2020).

2.9.6 Convention on biological diversity and biodiversity conservation

The CBD served as an important international legally binding agreement to help address all aspects of biological diversity including very complex global ecological degradation issues.

It focuses on human impacts on global biodiversity degradation, and seeks to conserve biodiversity at all levels—genetic, population, species, habitat, and ecosystem, which underpins the integrity of the life-support system of the biosphere.

The Convention has near universal participation among countries (CBD Secretariat, 2016a). There are currently 196 parties to the Convention with 168 signatures attested (United Nations, 2018; Tsioumani et al., 2019). All the 196 parties have ratified, acceded, approved, or accepted the Convention with the United States remains the only industrialized country in the world resisted from ratifying the legally binding document. The following are the three main goals of the Convention:

1. conservation of biological diversity (or biodiversity);
2. sustainable use of its components;
3. fair and equitable sharing of benefits arising from genetic resources.

As reflected in the above pivotal goals, the Convention linked for the first time ecological conservation and sustainable exploitation. The Convention calls on the parties to implement protected areas and biodiversity conservation measures and to regulate activities that are detrimental to the ecological integrity of biological diversity. More specifically, Article 8 of the CBD encourages parties to the Convention to:

- Establish a system of protected areas or areas where special measures need to be taken to conserve biological diversity;
- Develop, where necessary, guidelines for the selection, establishment, and management of protected areas or areas where special measures need to be taken to conserve biological diversity;
- Regulate or manage biological resources important for the conservation of biological diversity whether within or outside protected areas, with a view to ensuring their conservation and sustainable use;
- Promote environmentally sound and sustainable development in areas adjacent to protected areas with a view to furthering protection of these areas;
- Cooperate in providing financial and other support for in situ conservation, particularly to developing countries.

The Conference for the Adoption of the Agreed Text of the CBD was held in Nairobi, Kenya, in 1992 by the UNEP. This was followed by the First Conference of the Parties (COP1) held in 1994 in Nassau, Bahamas. The Convention led to the adoption of 13 decisions in relation to the rules of procedure for the COP, financial resources and mechanism and, clearing-House mechanism for technical and scientific cooperation, among others. In the following year, COP2 was held in Jakarta, Indonesia. The Conference adopted 23 decisions. Some of the most important decisions were matters pertaining to forests and biological diversity; conservation and sustainable use of marine and coastal biological diversity and, cooperation with other biodiversity-related conventions, among others.

Following COP2, COP3 was held in Buenos Aires, Argentina, in 1996. The Convention led to the adoption of 27 decisions (Appendix 2.3 for some of the important decisions reached at the meetings). Since 1996, the COP meetings are now held every 2 years. The meetings lead to the adoption of more decisions in relation to biodiversity conservation and sustainable resource-use. In 2000, for example, COP5 held in Nairobi, Kenya led to the adoption of a more targeted treaty—the Cartagena Protocol on Biosafety which entered into force in 2003. The treaty aimed to protect biodiversity from the potential risks posed by living modified organisms (LMOs) resulting from modern biotechnology. In general, each party to the Convention is obliged to sustainably manage and conserve its own biological diversity (Appendix 2.3).

APPENDIX 2.3   United Nations Convention on Biological Diversity (COP1–COP14).

Goal 1	Conservation of biodiversity
Goal 2	Sustainable use of the biological components of biodiversity
Goal 3	Fair and equitable sharing of benefits arising from genetic resources

Year	Name of meeting	Venue	Some major outcomes
1994	COP1	Nassau, Bahamas,	13 decisions including matters pertaining to (i) setting up rules of procedure for the Conference of the Parties, and (ii) the adoption of policy, strategy, program priorities and eligibility criteria for access to and utilization of financial resources, and (iii) preparation of the participation of the Convention on Biological Diversity in the third session of the Commission on Sustainable Development.
1995	COP2	Jakarta, Indonesia	23 decisions including matters pertaining to (i) calling upon the international community to make contributions for the preparation, (ii) conservation and sustainable use of marine and coastal biological diversity, and (iii) publication and distribution of scientific and technical information.
1996	COP3	Buenos Aires, Argentina	27 decisions including matters pertaining to (i) the adoption of the Memorandum of Understanding between the Conference of the Parties to the Convention on Biological Diversity and the Council of the Global Environment Facility and (ii) affirmation of the program of work for terrestrial biological diversity/forest biological diversity, and (iii) reaffirmation of mutually supportive activities under the Convention on Biological Diversity and activities under other conventions, processes and institutions relevant to the achievement of the objectives of the Convention.
1998	COP4	Bratislava, Slovakia	19 decisions including matters pertaining to (i) the review of the status and trends of the biological diversity of inland water ecosystems and options for conservation and sustainable use of biological resources, (ii) conservation and sustainable use of marine and coastal biological diversity, and (iii) issues related to biosafety.
1999	First Extraordinary Meeting of COP	Cartagena, Colombia	3 decisions including matters pertaining to the adoption of the Cartagena Protocol and interim arrangements.
2000		Montreal, Canada	
2000	COP5	Nairobi, Kenya	29 decisions including matters pertaining to (i) the adoption of the Cartagena Protocol on Biosafety which aims to ensure the safe handling, transport and use of living genetically

(Continued)

APPENDIX 2.3   (Continued)

Goal 1	Conservation of biodiversity
Goal 2	Sustainable use of the biological components of biodiversity
Goal 3	Fair and equitable sharing of benefits arising from genetic resources

Year	Name of meeting	Venue	Some major outcomes
			modified organisms (LMOs), (ii) progress report on the implementation of the program of work on the biological diversity of inland water ecosystems, marine and coastal biological diversity and forest biological diversity, and (iii) endorsement of ecosystem approach and operational guidance to promote conservation and sustainable use of biological resources in an equitable way.
2002	COP6	The Hague, Netherlands	32 decisions including matters pertaining to (i) the ratification, acceptance, approval or accession of the Intergovernmental Committee for the Cartagena Protocol on Biosafety (ICCP), (ii) agricultural biological diversity, global strategy for plant conservation, and (iii) recognition of the necessity for the application of ecosystem approach in national policies and legislation, and to integrate the approach in thematic and cross-sectoral programmes of the Convention at the local, national and regional level.
2004	COP7	Kuala Lumpur, Malaysia	36 decisions including matters pertaining to (i) implementation of the expanded program of work on forest biological diversity, (ii) global strategy including the establishment of the global partnership for plant conservation, and (iii) enhancement of the implementation of the ecosystem approach for addressing the three objectives of the Convention in a balanced way.
2006	COP8	Curitiba, Brazil	34 decisions including matters pertaining to (i) the adoption of program of work on island biodiversity, (ii) adoption of second edition of Global Biodiversity Outlook, and (iii) the implementation of the Convention and its Strategic Plan, and the acknowledgment of the reports of the Millennium Ecosystem Assessment, in particular the Synthesis Report.
2008	COP9	Bonn, Germany	36 decisions including matters pertaining to (i) in-depth review of ongoing work on alien species that threaten ecosystems, habitats or species, (ii) review of implementation of goals 2 and 3 of the Strategic Plan, and (iii) preparation of the Third edition of the Global Biodiversity Outlook.
2010	COP10	Nagoya, Aichi Prefecture, Japan	47 decisions including matters pertaining to (i) the adoption of the Nagoya Protocol on Access to Genetic Resources and the Fair and Equitable Sharing of Benefits Arising from their Utilization to the Convention on Biological Diversity

(Continued)

APPENDIX 2.3   (Continued)

Goal 1	Conservation of biodiversity
Goal 2	Sustainable use of the biological components of biodiversity
Goal 3	Fair and equitable sharing of benefits arising from genetic resources

Year	Name of meeting	Venue	Some major outcomes
			(the Protocol), (ii) updating of national biodiversity strategies and action plans to promote the implementation of the Strategic Plan (led to the adoption of revised and updated Strategic Plan for Biodiversity, including the Aichi Biodiversity Targets, for the 2011−20 period, and (iii) adoption of the Third edition of the Global Biodiversity Outlook.
2012	COP11	Hyderabad, India	33 decisions including matters pertaining to (i) review of progress in implementation of national biodiversity strategies and action plans and related capacity-building support to Parties, (ii) progress monitoring in relation to the implementation of the Strategic Plan for Biodiversity 2011−2020 and the Aichi Biodiversity Targets, and (iii) review of implementation of the strategy for resource mobilization, including the establishment of targets.
2014	COP12	Pyeongchang, Republic of Korea	35 decisions including matters pertaining to (i) mid-term review of progress in implementation of the Strategic Plan for Biodiversity 2011-2020 including the Fourth edition of the Global Biodiversity Outlook, and actions to enhance implementation, and (ii) review of progress in providing support in implementing the objectives of the Convention and the Strategic Plan for Biodiversity 2011-2020, and enhancement of capacity-building, technical and scientific cooperation and other initiatives to assist implementation, and (iii) integration of biodiversity into the post-2015 United Nations development agenda and the sustainable development goals.
2016	COP13	Cancun, Mexico,	33 decisions including matters pertaining to (i) Progress in the implementation of the Convention and the Strategic Plan for Biodiversity 2011−20 and toward the achievement of the Aichi Biodiversity Targets, (ii) Progress toward the achievement of Aichi Biodiversity Targets 11 and 12, (iii) Strategic actions to enhance the implementation of the Strategic Plan for Biodiversity 2011-2020 and the achievement of the Aichi Biodiversity Targets.
2018	COP14	Sharm El-Sheikh, Egypt	The Conference adopted a number of decisions on various strategic, administrative, financial, and ecosystem-related issues. These included 37 decisions under the CBD COP; 16 decisions under the Cartagena Protocol COP/MOP; and 16 decisions under the Nagoya Protocol COP/MOP.

CBD, Convention on Biological Diversity; *COP*, Conference of the Parties; *ICCP*, Intergovernmental Committee for the Cartagena Protocol on Biosafety; *LMOs*, living genetically modified organisms, *MOP*, Meeting of the Parties.

A supplementary agreement to CBD, the Nagoya Protocol on Genetic Resources and the Fair and Equitable Sharing of Benefits Arising from their Utilization (ABS), was adopted together with the Aichi Targets at the CBD held in 2010 (CBD Secretariat, 2011a). The Nagoya Protocol provides a legal framework for the effective implementation of CBD's objectives of "fair and equitable profit sharing." The Aichi Targets sought to reduce or halt the loss of biodiversity worldwide (CBD Secretariat, 2011b). Parties to CBD, including those from Southeast Asia, Africa, or Latin America are legally obliged to create their own national biodiversity strategies and action plans to translate the Aichi targets into action.

All in all, since the inception of CBD in 1992, the United Nations has convened 15 meetings including two extraordinary meetings to review, revise or reinforce environmental conservation plans and efforts (Appendix 2.3). Following Target seven of Millennium Development Goals (MDGs) on environmental sustainability, the aim of CBD, as explicitly stated in its Strategic Plan for the Convention adopted in 2002, is to significantly reduce "the current rate of biodiversity loss at the global, regional, and national level" by 2010 (CBD Secretariat, 2010a: 15). Some of the most important targets of CBD are: (1) reducing the rate of loss of biodiversity components including biomes, habitats, ecosystems, and genetic diversity, (2) promoting sustainable use of biodiversity, (3) addressing the major threats to biodiversity, including those arising from invasive alien species, climate change, pollution, and habitat change, and (4) maintaining ecosystem integrity, and the provision of goods and services provided by biodiversity in ecosystems, in support of human wellbeing, among others (CBD Secretariat, 2005).

This "2010 target" which was endorsed at the World Summit on Sustainable Development (WSSD) held in 2002 and reaffirmed in the Johannesburg Plan of Implementation (JPOI) adopted at WSSD has become the driving force behind many of the biodiversity conservation efforts at the national, local, and international levels. Contracting Parties to the Convention are required to establish and enforce national strategies and action plans to conserve, protect, and enhance biological diversity. In response to this condition, member countries across the world have drawn up their National Biodiversity Strategy and Action Plans (NBSAPs), the principal instruments for implementing the Convention at the national level. To date, 190–196 (97%) of the Parties to the Convention have developed their NBSAPs in line with Article 6 of the Convention (CBD Secretariat, 2018; CPSG, 2019).

At the 10th conference held in 2010, the Convention adopted a new and updated plan, the Strategic Plan for Biodiversity 2011–20, along with its 20 Aichi Targets. The Aichi Targets, comprising 20 targets under 5 strategic goals, aimed to strengthen global commitment to biodiversity protection through "effective and urgent action to halt the loss of biodiversity in order to ensure that by 2020, ecosystems are resilient and continue to provide essential services, thereby securing the planet's variety of life, and contributing to human wellbeing, and poverty eradication...." (CBD Secretariat, 2010b: 8). Two of the most important targets are halving the rate of natural habitat loss and protecting 17% of the world's land area in nature reserves by 2020. Contracting Parties agreed to translate the Aichi Targets' overarching framework of biodiversity management into revised and updated national biodiversity strategies and action plans by 2015 (UNEP-WCMC and IUCN, 2016).

At the 13th meeting (COP13) held in Cancun, Mexico, in December 2016, the Convention adopted 33 decisions which focus on engaging actors in support of the implementation of the Strategic Plan for Biodiversity 2011–20, and on traditional knowledge, access and benefit-sharing, among others (CBD Secretariat, 2016b). Two years later in 2018, the UN Biodiversity Conference was held in Sharm El-Sheikh, Egypt under the theme "Investing in biodiversity for people and planet." The Conference covered various meetings including the fourteenth meeting of the COP14. At COP14, the Convention adopted 37 decisions in relation the strategic, administrative, financial, and ecosystem-related issues that serve to enhance the implementation of the Convention and its Protocols (Tsioumani et al., 2018).

The CBD's 15th major conference (COP15) was scheduled to be held in Kunming, Yunnan Province, China in October 2020 with the aim of adopting a new post2020 global framework for biodiversity over the next 10 years. The theme of the Conference is "Ecological Civilization: Building a Shared Future for All Life on Earth" The post 2020 global biodiversity framework is expected to serve as a stepping stone toward the 2050 Vision of "Living in harmony with nature" through the effective implementation of a long-term strategic approach "for achieving the objectives of the Convention, the Strategic Plan for Biodiversity 2011–2020, its Aichi Biodiversity Targets and the 2050 Vision for Biodiversity" (CBD Secretariat, 2019: 7). However, the dates for the COP15 Conference originally scheduled on October 15–28 have been postponed due to the Covid-19 pandemic. The new dates for the meeting are tentatively scheduled to take place during the second quarter of 2021

2.9.7 The United Nations Framework Convention on Climate Change and Convention on Biological Diversity: Some remarks

While climate change has been recognized by the UN General Assembly as the "common concern of mankind" (United Nations, 1988), the UNFCCC specifically acknowledges this "common concern" and placed it at the center of the international climate change negotiation regime. It promotes a precautionary approach to addressing climate change, and articulating the principle of intergenerational equity as emphasized in the Brundtland Report, paying special attention to protect the global climate system in the best interests of future generations. UNFCCC also incorporates Principle 7 of the Rio Declaration on the Environment and Development on the principle of "common but differentiated responsibilities" in addressing climate change. The principle states that:

> States shall cooperate in a spirit of global partnership to conserve, protect and restore the health and integrity of the Earth's ecosystem. In view of the different contributions to global environmental degradation, States have common but differentiated responsibilities. The developed countries acknowledge the responsibility that they bear in the international pursuit of sustainable development in view of the pressures their societies place on the global environment and of the technologies and financial resources they command. *United Nations (1992a)*

The principle suggests that the international community share a common responsibility for protecting the global atmosphere, but in view of the different contributions to the problem and the financial and technological constraints facing the developing countries, the responsibility for addressing climate change should be differentiated among States (Rajamani, 2000). Accordingly, Article 3.1 of UNFCCC specifically states that: "The parties should protect the climate system for the benefit of present and future generations of humankind on the basis of equity and in accordance with their common but differentiated responsibilities and respective capabilities. Accordingly, the developed country parties should take the lead in combating climate change and the adverse impact thereof" (United Nations, 1992b).

It may be remarked in the above light that identifying an issue as "common concern of mankind" implies the involvement of the international community in matters that would otherwise fall solely within the State jurisdiction (Bowman, 2010). In other words, it implies "a common responsibility to the issue based on its paramount importance to the international community as a whole" (Glowka et al., 1994: 3). A necessary implication of the common concern of humankind is that States have a duty to cooperate to ensure that an issue of paramount importance to the international community is addressed for the benefit of present and future generations (Horn, 2004).

The UNFCCC framework built on the above principles constitutes an international climate framework for all parties to undertake a common responsibility to reduce GHG emissions and to work collectively to avoid "dangerous anthropogenic interference with the climate system." The UNFCCC thus provides important agreed rules and principles for world leaders beyond boundaries of States to make progress on global treaty and effective solutions to address climate change—a common concern which will potentially affect all humanity, including present and future generations.

The CBD explicitly recognized the World Charter of Nature's proclamation that "Every form of life is unique, warranting respect regardless of its worth to man" (United Nations, 1982) in its first preambular paragraph which focuses on the intrinsic value of the global environment and the interdependence of its components which constitutes our life-support system. The interdependence of its components and the severity of current environmental problems call for global solutions, and hence, "the conservation of biological diversity" is affirmed as a "common concern of mankind" in CBD (IUCN, 2010; French, 2016).

The CBD provides the first international biodiversity laws of its kind to recognize that the conservation of biological diversity is a common concern for all of humanity. As noted earlier on, the CBD also has developed a range of principles to promote collective international action to protect global biological resources especially in the biodiversity-rich countries in Asia, including China, Africa, Latin America, and Southeast Asia (Choy, 2015a,b, 2016, 2018, 2019).

Recognizing biodiversity conservation as being a common concern reflected a general awakening of the urgency of the global biological and environmental sustainability issues especially in the biological resource-rich developing South. Also, as Duncan French argued succinctly, considering local biological resources as being a common concern has the effect of generating global interest, and thus "removes the conceit of exclusive domestic domain" (French, 2016: 343−344). This, arguably, tends to promote global environmental responsibility for sustainable use and management of biological resources such as tropical rainforests which are related to the common concern of climate change, that is, its conservation is critical for "the evolution and for maintaining life sustaining systems of the biosphere" in terms of carbon sequestrations and sinks −a critically important ecosystem function which underpins human long-term existence (United Nations, 1992c: preamble ii).

It is progressively clear that the UNFCCC and the CBD create the necessity for international cooperation and collective responsibility on various critical and interrelated environmental problems which transcend national boundaries and that cannot be resolved solely by any one country acting alone. Thus together, these two conventions provide an indispensable global platform for world leaders to make progress on global treaty and effective solutions to global environmental issues, in particular, global warming and global biodiversity depletion which are common themes for all nations.

It is also worth mentioning that the 1992 Rio Conference constitutes a marked shift from the 1972 Stockholm Conference in that the scope of its environmental debates, environmental management and control principles, global environmental governance agenda, and international treaties, among others, are far more sophisticated and comprehensive. It is claimed that the 1992 Rio Conference provided "the vision and important pieces of the multilateral machinery to achieve a sustainable future" (UNEP, 2011: 7). The following section discusses how the momentum of the Rio environmental initiatives was carried forward to the Earth Summit II held in 1997.

2.10 The United Nations' journey to environmental sustainability 2001 and beyond: The third wave

2.10.1 The United Nations Millennium Summit

Despite many setbacks, the United Nations continued to advance progressively and aggressively to energize global political momentum in promoting sustainable development by further stimulating environmental interest and awareness, especially the threat of climate change and biodiversity loss. The importance of the environment was further reaffirmed at the Millennium Summit hosted by the then Secretary-General Kofi Annan in New York in 2000. Eight MDGs with specific targets were set to be achieved by 2015 one of which was MDG7—Ensure Environmental Sustainability (United Nations, 2000a; see Appendix 2.4) (Timeline 2.3).

APPENDIX 2.4   United Nations engagement with environmental sustainability (2001 onwards).

2001	IPCC	Released Third IPPC Assessment Report comprising the full scientific and technical assessment of climate change in three volumes: (1) Climate Change 2001: The Scientific Basis, (2) Climate Change 2001: Impacts, Adaptation and Vulnerability, (3) Climate Change 2001: Mitigation, and (4) Climate Change 2001: Synthesis Report.
2002	World Summit on Sustainable Development held in Johannesburg	Adopted the Johannesburg Declaration on Sustainable Development and the Plan of Implementation of the World Summit on Sustainable Development (Johannesburg Plan of Implementation). The meeting primarily involved reviews of progress of the aims set out in Agenda 21 and to agree on a new global deal on sustainable

(Continued)

APPENDIX 2.4   (Continued)

		development. The meeting was basically concerned with practical issues of implementation rather than creating new treaties and targets. Partly influenced by the Millennium Development Goals and recognizing the need for development in the developing countries, the meeting made a fundamental shift from environmental issues to social and economic development.
2005	World Summit held in New York	To reaffirm the commitment by all governments of the member countries to achieve the Millennium Development Goals by 2015. The summit also recognized the serious challenge posed by climate change and made a commitment to take action through the UN Framework Convention on Climate Change to address the issues concerned.
2007	IPCC	Released the Fourth IPPC Assessment Report comprising three volumes and a synthesis report: (1) Contribution of Working Group I to the Fourth Assessment Report of the Intergovernmental Panel on Climate Change, (2) Contribution of Working Group II to the Fourth Assessment Report of the Intergovernmental Panel on Climate Change, (3) Contribution of Working Group III to the Fourth Assessment Report of the Intergovernmental Panel on Climate Change, and (4) Contribution of Working Groups I, II, and III to the Fourth Assessment Report of the Intergovernmental Panel on Climate Change. The last Report summarizes the key findings of the three Working Groups. It also highlights the inextricable relationship between climate change and sustainable development. The major findings of the Report are: (i) There is strong certainty that most of the observed warming of the past half-century is due to human influences, and there is a clear relationship between the growth in manmade greenhouse gas emissions and the observed impacts of climate change, (ii) The climate system is more vulnerable to abrupt or irreversible changes than previously thought, (iii) Avoiding the most serious impacts of climate change—including irreversible changes—will require significant reductions in greenhouse gas emissions, and (iv) Mitigation efforts must also be combined with adaptation measures to minimize the risks of climate change.
2010	UN Summit on the Millennium Development Goals held in New York	The summit adopted the global action plan, "Keeping the Promise: United to Achieve the Millennium Development Goals" which reaffirmed

(Continued)

APPENDIX 2.4   (Continued)

		world leaders' commitment to the MDGs and sets out a concrete action agenda for achieving the Goals by 2015s.
2012	The United Nations Conference on Sustainable Development (the Earth Summit or the Rio +20 Summit) held in Rio de Janeiro, Brazil	The conference reaffirmed global commitment to achieve internationally agreed development goals, including the Millennium Development Goals by 2015. It has also developed a set of SDGs by expanding the Millennium Development Goals to serve as the post 2015 development agenda. The Conference also discussed green economy in the context of sustainable development and poverty eradication and adopted various guidelines for the promotion of a green path of development. The Conference adopted The Future We Want"—the declaration on sustainable development and a green economy adopted at the Conference.
2013/ 2014	The Intergovernmental Panel on Climate Change (IPCC)	Released the Fifth IPPC Assessment Report made up of three volumes and one synthesis, namely, (1) Working Group I Report "Climate Change 2013: The Physical Science Basis," (2) Working Group II Report "Climate Change 2014: Impacts, Adaptation, and Vulnerability," (3) Working Group III Report "Climate Change 2014: Mitigation of Climate Change," and (3) "Climate Change 2014: Synthesis Report". Some of the major findings are (i) Human influence on the climate system is clear, and recent anthropogenic emissions of greenhouse gases are the highest in history. Recent climate changes have had widespread impacts on human and natural systems, (ii) Warming of the climate system is unequivocal, and since the 1950s, many of the observed changes are unprecedented over decades to millennia. The atmosphere and ocean have warmed, the amounts of snow and ice have diminished, and sea level has risen, (iii) Anthropogenic greenhouse gas emissions have increased since the preindustrial era, driven largely by economic and population growth, and are now higher than ever. This has led to atmospheric concentrations of carbon dioxide, methane, and nitrous oxide that are unprecedented in at least the last 800,000 years. Their effects, together with those of other anthropogenic drivers, have been detected throughout the climate system and are extremely likely to have been the dominant cause of the observed warming since the mid-20th century.
2015	United Nations Sustainable Development Summit held in New York	Adoption of the post2015 development agenda with 17 SDGs. The deadline for the SDGS is 2030.

(Continued)

APPENDIX 2.4 (Continued)

| 2019 | SDG Summit held in New York | Aims: (i) to follow up and review progress achieved so far since the adoption of the 2030 Agenda for Sustainable Development in September 2015 and, (ii) to provide leadership and guidance on the way forward that would help accelerate implementation of the 2030 Agenda and SDGs. The Summit led to the adoption of the Political Declaration, "Gearing up for a decade of action and delivery for sustainable development". |

IPCC, Intergovernmental Panel on Climate Change; *MDGs*, Millennium Development Goals; *SDGs*, Sustainable Development Goals.

2000	Millennium Summit held in New York with one of the main targets to achieve MDG7 (Ensure Environmental Sustainability) by 2015
2002	The United Nations World Summit on Sustainable Development (WSSD) or Rio+10 was held in Johannesburg, South Africa: to revive political commitment in environmental protection
2005	The United Nations World Summit was held in New York: to review the progress of MDGs including MDG7
2010	The Millennium Development Goals Summit was held in New York: to further review the progress of MDGs
2012	The United Nations Conference on Sustainable Development or the Rio+20 was held in Rio de Janeiro, Brazil: to secure renewed political commitment for sustainable development, among others
2013	Global Millennium Development Goals (MDGs) Conference (GMC) was held in held in Bogota (Colombia): To maintain the momentum of progress and political commitment to achieving the MDGs
2013	The United Nations World Summit was held in New York: to review progress of MDGs including MDG7
2015	The United Nations Sustainable Development Summit was held in New York: to adopt the Sustainable Development Goals (SDGs) or the Global Goals
2019	The SDG Summit was held in New York: to review progress in the implementation of the 2030 Agenda for Sustainable Development and its 17 Sustainable Development Goals (SDGs)

TIMELINE 2.3 United Nations third wave of environmental sustainability.

The then Secretary-General, declaring that "we have been plundering our children's future heritage to pay for environmentally unsustainable practices in the present" by "degrading, and in some cases destroying, the ability of the environment to continue providing …life supporting services for us," called on the community, including the Heads of States and Governments to adopt a new ethic of conservation and stewardship

(United Nations, 2000a: 55). The first step in building a new ethic of global stewardship was "to adopt and ratify the Kyoto Protocol," and "to ensure that its goals are met" (United Nations, 2000a: 79).

Also, environmental issues must be integrated into the policy-making process to reflect the true environmental costs and accounting based on the application of green accounting. The Millennium Summit environmental agenda was also concerned about the provision of the necessary financial support for the Millennium Ecosystem Assessment and encouraged active engagement in it based on "major international collaborative effort to map the health of our planet" (United Nations, 2000a: 65). The summit also aimed to revitalize sustainability debate and to prepare the ground for the adoption of concrete and meaningful actions by the heads of state or government in the Earth Summit to be held in 2002 (United Nations, 2000a: 56).

The Millennium Declaration adopted at the summit reinforced the urgency of the global endorsement of the ethics of resource-use and management by stating that "no effort must be spared to counter the threat of the planet being irredeemably spoiled by human activities" (United Nations, 2000b: 6). It also reaffirmed the support for Agenda 21, and recognized the need "to intensify collective efforts for the management, conservation and sustainable development of all types of forests" and pressed for the "full implementation of the Convention on Biological Diversity, among others (United Nations, 2000b: 6).

However, despite the above environmental considerations, environmental sustainability and biodiversity conservation were not a prominent part but an add-on in the MDGs as goal 7 (MDG7) (Dalal-Clayton and Sadler, 2014; ICSU and ISSC, 2015). The environmental foundation of MDGs was also overshadowed by the inclusion of the targets for human development, namely Target 10 which aimed to halve, by 2015, the proportion of people without sustainable access to safe drinking water and basic sanitation, and Target 11 which sought to improve "the lives of at least 100 million slum dwellers". These targets would have been more appropriately included under MDG1 on the eradication of extreme poverty and hunger. Also, the specific goal on environmental issue, that is, to "integrate the principles of sustainable development into country policies and programmes and reverse the loss of environmental resources," hardly contained any quantifiable targets. It also failed to address the complex interaction and interconnectedness between environmental sustainability and socioeconomic development. As a result, efforts to translate MDG7 into international and domestic action have been slow and we are far from achieving the target of environmentally sustainable development (WWF, 2015).

2.10.2 The United Nations World Summit on Sustainable Development or Rio +10 (2002)

Two years after the Millennium Summit, a second review of the progress of the Rio agreements was carried out at the WSSD held in Johannesburg, South Africa in 2002. The aim of WSSD was to revive political commitment and global cooperative efforts in environmental protection out of deep concern over the continued alarming rate of environmental degradation. It may well be that despite the collective hope for a sustainable future envisaged at the 1992 UN Conference on Environment and Development (UNCED); there was still a significant gap between promise and real action in the implementation of Agenda 21.

Thus unlike its predecessor, the main purpose of the WSSD summit was primarily concerned with the review of the progress of the 1992 UNCED, focusing on "action-oriented decisions in areas where further efforts are needed to implement Agenda 21" (United Nations, 2001: 2). In other words, rather than introducing a new environmental treaty and target negotiations, the Summit, more specifically sought "to reinvigorate global commitment to sustainable development" and "to identify measures for further implementation of Agenda 21" by means of action-oriented decisions (United Nations, 2001: 2). The MDGs and other international agreements also constituted part of the subject of discussion at the summit.

The Summit brought dialog among a wide range of stakeholders outside the United Nations system to a new level. These include, apart from the state members of the United Nations, the international and regional financial institutions, NGOs, development agencies, and civil society groups, among others. More importantly, NGOs were recognized participants of the Summit by the United Nations and they were present throughout the summit. It is noteworthy that, for the civil society groups which may play an important role in influencing government environmental policies, the WSSD Summit became a truly empowering experience as the lessons they learned and networks and alliances they formed in the participatory process would have an impact on future development in their respective countries (La Viña et al., 2003).

Two official documents, the Johannesburg Plan of Implementation (JPOI) and the Johannesburg Declaration on Sustainable Development were adopted at the Summit. The JPOI called on all countries to take immediate steps to formulate National Sustainable Development Strategies (NSDS) and begin their implementation by 2005 (United Nations, 2002a). The UN Guidance document defined NSDS as "a coordinated, participatory and iterative process of thoughts and actions to achieve economic, environmental and social objectives in a balanced and integrated manner." The strategies involved long-term and continuous undertaking "in managing progress toward sustainability goals rather than producing a 'plan' as an end product" (United Nations, 2002b: 8).

The summit also urged the governments in all countries to strengthen governmental institutions, and to enact and enforce clear and effective laws that support sustainable development (para. 163). More importantly, from the environmental perspective, the plan encouraged the development of a 10-year framework of programmes to accelerate the shift toward sustainable consumption and production within the carrying capacity of ecosystems, and to delink economic growth and environmental degradation through efficient and sustainable use of resources (para. 15). The plan was so comprehensive that it prompted the Secretary-General, Nitin Desai to claim that it "provides us with everything we need to make sustainable development happen over the next several years" (United Nations, 2002c: 1).

The Johannesburg Declaration reaffirmed the governments' commitment to sustainable development, particularly in association with the implementation of the Rio Declaration and Agenda 21 (United Nations, 2002a). It also called for "a broad-based participation in policy formulation, decision-making and implementation at all levels," including the civil society groups, "to strengthen and improve governance at all levels for the effective implementation of Agenda 21, the MDGs and the Plan of Implementation of the Summit" (United Nations, 2002a: 4).

One of the important achievements of the Summit was that nations agreed to commit to the CBD and to significantly reduce the current rate of loss of biodiversity by 2010. There were commitments (1) to achieve sustainable fisheries by maintaining or restoring fishery stocks to levels that can produce maximum sustainable yield by not later than 2015 where possible, and (2) to reverse the current trend in natural resource degradation to protect the ecological integrity of ecosystems which underpins human wellbeing and economic progress.

On the issue of climate change, Ministers at the Summit indicated their support for the ratification of the Kyoto Protocol. Seemingly in response to the Summit's call to those states which had yet to ratify the Kyoto Protocol to expedite the ratification process so that it could come into force at the earliest possible date, Canada, Mexico, Russia, India, China, New Zealand, and Korea indicated their support or intention to ratify the Protocol (ABS, 2003). This provided the prospect "for the entry into force of the Kyoto Protocol" at the earliest date possible (United Nations, 2002a: 104).

It may be noted as a matter of interest that Article 25 of the Kyoto Protocol states that it shall enter into force 90 days after "not less than 55 Parties to the Convention, including the industrialized countries which accounted for at least 55% of the total CO_2 emissions for 1990 have ratified the Protocol." As of September 2002, 94 countries with 37.1% of emissions had ratified the Protocol. With the United States, the largest emitter of GHGs which accounted for 36% of the total emission, withdrawing from the Kyoto process; the future of the Protocol was brought into imminent collapse as no other nation, or combination of nations, except Russia, the third largest emitter, which accounted for 17.4% of the total emission, could bring the Protocol into force. It may be noted that although China was the second largest CO_2 emitter, it is not a party to binding reductions in emission under the Kyoto Protocol (Adams, 2002). Russia eventually deposited its instruments of ratification in November 2004, making it possible for the 55% requirement to be met. The Kyoto Protocol finally entered into force on February 16, 2005. Viewed from this perspective, the WSSD Summit stands out as a milestone in global green development.

Another remarkable feature of the Summit is the establishment of a voluntary and multi stakeholder initiative (Type-2 partnership) which consists of "a series of commitments and action-oriented coalitions focused on deliverables and would contribute to translating political commitments into action" at all levels (United Nations, 2002c: 1). It may be noted that Type-1 partnership deals with intergovernmental agreements such as conventions and declarations negotiated by states. While Type-1 is obligatory, Type-2 is voluntary. This so-called Type-2 voluntary partnership is a multistakeholder initiative which consists of "a series of commitments and action-oriented coalitions focused on deliverables and would contribute to translating political commitments into action" at all levels (United Nations, 2002e: 1). Type-2 implementation-focused partnership initiatives have been claimed as an important solution to the "crisis of implementation" (Hale, 2003). More than 220 partnerships with a promise of US $235 million in resources were identified in advance of the WSSD Summit and around 60 partnerships were announced during the summit by various countries (United Nations, 2002f: 1). Among the partners that joined the partnership initiative included Chile, Hungary, Italy, Mexico, Sweden, Uganda, Thailand, the United Kingdom, the World Bank and the European Commission (United

Nations, 2002d). However, it must be remarked that the Type-2 partnership initiatives are not intended to serve as a substitute for Type-1 agreements committed to by governments.

Ostensibly, the emergence of the new and innovative form of global environmental governance at the WSSD summit offered much hope in mitigating the "crisis of implementation" which had come home to haunt the Rio Declaration since 1992. With governments having agreed to a wide range of commitments and effective implementation of sustainable development objectives, the world seemed to be on a more sustainable footing by consolidating continued efforts to expedite the materialization of the Rio Declaration, Agenda 21 and the MDGs.

2.11 United Nations environmental efforts in the postRio +20

2.11.1 World Summit (2005)

Continuing the momentum on global environmental efforts for a sustainable planet, the United Nations convened the World Summit in New York in 2005—the largest ever gathering of heads of state and government in UN history to mark the 60th anniversary of the United Nations. The main purpose of the summit was to review progress of MDGs, and key poverty, development and environmental issues were brought up at the meeting (United Nations, 2005; UNEP, 2006). The Summit adopted the 2005 World Summit Outcome Document which dealt with the integration of environmental concerns into development policies. The Document addressed various important environmental issues including biodiversity conservation, sustainable use of natural resources and environmental management, energy, forestry, the impact of climate change and sustainable production and consumption (United Nations, 2005).

On the issue of biodiversity conservation, the summit called on the parties to the CBD and the Cartagena Protocol on Biosafety to support "the implementation of the Convention and the Protocol, as well as other biodiversity-related agreements and the Johannesburg commitment for a significant reduction in the rate of loss of biodiversity by 2010" (United Nations, 2005: 13). The summit document also emphasized the need to tackle climate change with "resolve and urgency" and to meet all the commitments and obligations the international community had committed to in the UNFCCC and other relevant international agreements, including the Kyoto Protocol (United Nations, 2005: 12).

The summit also reasserted the MDGs including MDG7 on environmental sustainability. More importantly, the summit reaffirmed resolve to strengthen international cooperation and coordination in the implementation of the sustainable development objectives agreed upon at the major United Nations conferences and summits in close cooperation with all other multilateral financial, trade and development institutions (United Nations, 2005). There was indeed strong commitment by all governments from the North and South alike to achieve the MDGs by 2015, and all developing countries undertook to adopt national plans for achieving the MDGs by 2006.

By reinforcing past commitments to achieving sustainable development, in particular Agenda 21 and MDGs, and by highlighting the urgency of addressing the widespread environmental degradation, especially biodiversity depletion and climate change, the World Summit provided further impetus to the global community toward realizing the sustainable development agenda.

2.11.2 Millennium Development Goals Summit (2010)

Following the World Summit, the United Nations convened the MDGs Summit in New York—officially called the High-Level Plenary Meeting of the General Assembly—in 2010 with the theme "We can end poverty by 2015." The main aim of the summit was to review progress, identify obstacles and gaps and agree on concrete strategies and action agenda to achieve the eight MDGs by 2015 (United Nations, 2010a). In particular, the summit aimed to boost progress on halving extreme poverty by 2015.

From the environmental perspective, a high-level meeting on biodiversity, "Ecosystems, Climate Change and The Millennium Development Goals (MDGs): Scaling Up Local Solutions," was convened to discuss the critical linkages between biodiversity conservation, healthy ecosystems, climate change, and MDG achievement (United Nations, 2010b; Hazlewood et al., 2010). This policy forum event was attended by over 500 participants, including heads of state, ministers, and diplomats from more than 60 countries.

At the forum, the Executive Secretary of CBD, Ahmed Djoghlaf called for a holistic approach to achieving environmental sustainability, stressing that climate change could not be combated without taking into consideration biodiversity or desertification. Also, the organization of economic cooperation and development Committee had mandated that development aid could not be administered effectively without "mainstreaming green" (Ahmad et al., 2010: 4). Canadian Ministry of Environment, Jim Prentice, called for the need for strong and effective enforcement of environmental law while the Ministry of Environment from Italy emphasized a full reform of international environmental governance (Ahmad et al., 2010). One of the major messages resonating at the forum was that biodiversity conservation, carbon dioxide emission and climate change, and socioeconomic development were all intertwined and involved a wide range of stakeholders. In view of this, addressing these problems called for a holistic and comprehensive approach, with coordinated efforts and participation at all levels rather than piecemeal solutions (Ahmad et al., 2010).

The major outcome of the summit was the adoption of the document entitled "Keeping the promise: united to achieve the Millennium Development Goals" (United Nations, 2010c). The document called on the international community to strengthen political commitment and action at all levels to implement sustainable use and management of all types of forest resources, and to continue efforts to implement the three objectives of CBD and to reduce the rate of biodiversity loss (Appendix 2.4). The document also reaffirmed the UNFCCC as the primary international, intergovernmental forum for negotiating the global response to climate change, and called on the heads of state to undertake urgent action to address the problem. Also, following the outcome of the summit, the United Nations established the UN System of Task Team in September 2011 to support the preparation of the post2015 development agenda (OHCHR et al., 2013).

Arguably, the MDG summit contributed to raising the profile and popular awareness of continued deterioration of the global environment, and the slow progress of the global commitments agreed on in the past summits to achieve environmentally sustainable development, which was attributed to deficits in political will. It also helped to bring about a stronger focus on the urgent need for meaningful and politically sustained commitment for environmental protection which underpins long-term socioeconomic progress. The summit has heralded a welcome North—South determination to ensure the timely and full implementation of these commitments. The continued relevance and legitimacy of MDGs and other sustainable development objectives as well as of past international environmental commitments were further stressed in the following United Nations summit—the United Nations Conference on Sustainable Development (WSSD).

2.11.3 United Nations Conference on Sustainable Development or the Rio +20: The next milestone (2012)

The next milestone of the United Nations environmental protection initiatives was the United Nations Conference on Sustainable Development or the Rio +20 held in 2012. This conference marked the 20th anniversary of the UNCED held in Rio de Janeiro, Brazil, in 1992. Attended by more than 40,000 people including the heads of states United Nations officials, civil society leaders in 500 official side events as well as over 3000 unofficial parallel events, the summit was deemed to be the largest conference in the United Nations summit history (Doran et al., 2012). It was also echoed as a conference of action. More specifically, as Sha Zukang, the Conference Secretary-General for Rio +20 put it: "We have enough papers; we have enough conferences. What we need to do now is something really different: Rio +20 should be not another conference in normal sense; it should be a conference of action, a conference of implementation of what we have agreed twenty or ten years ago." (quoted in Pisano et al., 2012: 18).

The main objectives of the summit were to secure renewed political commitment for sustainable development, to assess the past and present summits' achievements, and to address future challenges. The Conference focused on two themes: (1) green economy (GE) in the context of sustainable development and poverty eradication, and (2) institutional framework for sustainable development (UNEP, 2011; United Nations, 2012a). A GE may be defined as "one that results in improved human wellbeing and social equity, while significantly reducing environmental risks and ecological scarcities" (UNEP, 2010: 4; UNEP, 2011: 16).

More particularly, a GE is an economy that is "low-carbon, resource efficient and socially inclusive" (UNEP, 2011: 16). It is one of the strategic ways of realizing development that resonates with the implementation of Agenda 21 in terms of reduction in carbon emission and pollution, enhancement of energy, and resource efficiency and the prevention of the loss of biodiversity and ecosystem services (UNEP, 2010, 2011). Accordingly, a GE can offer a solution to the countries in the North and South for addressing climate change, energy insecurity, and ecological scarcity by providing "a development path that reduces carbon dependency, promotes resource and energy efficiency and lessens environmental degradation" (UNEP, 2011: 16).

The summit adopted the final outcome document, "The Future We Want" the document, upholding the Rio principles and past action plans as well as the two conference themes noted above, provided a firm foundation for world leaders to promote social, economic, and environmental wellbeing in the context of sustainable development. Accordingly, member states decided to launch a process to develop a set of action-oriented Sustainable Development Goals (SDGs) by 2015 to succeed and extend the MDGs "based on Agenda 21 and Johannesburg Plan of Implementation, fully respecting all the Rio Principles," and recognized that the development of these goals could be useful "for pursuing focused and coherent action on sustainable development" (United Nations, 2012b: 63). The summit also adopted ground-breaking guidelines on GE policies as elaborated under Section III, "Green economy in the context of sustainable development and poverty eradication," of the outcome document. (United Nations, 2012b: 14).

The new Partnership for Action on Green Economy (PAGE), was launched at the meeting of the Governing Council of the UNEP held from 18 to February 22, 2013. The aim of PAGE is to support the green economic transformation in 30 countries by shifting investment to clean technology and wellfunctioning ecosystems that will generate new jobs and skills and contribute to resource conservation and poverty eradication. The four United Nations agencies involved in facilitating the green transformation processes are the UNEP, the International Labour Organization, the United Nations Industrial Development Organization, and the United Nations Institute for Training and Research (Poulden, 2013).

To boost further political commitment to sustainable development, "A ten-year framework of programmes on sustainable consumption and production patterns" was also adopted to achieve the goals and objectives as agreed in the Johannesburg Plan of Implementation (JPOI) adopted at WSSD on sustainable consumption and production (United Nations, 2012b: 59). Also, many government leaders agreed to the need for institutional reform in the existing international processes and institutions on environmental matters as a means of promoting stronger coherence and coordination on sustainable development and environmental protection (Leggett and Carter, 2012).

The Rio +20 provided a platform for governments, corporate actors, and civil society participants to host side meetings to share best practices and promote partnership in GE management. A large number of private deals on renewable energy, pollution control and other commercial and development investments were concluded at various side meetings (Leggett and Carter, 2012). The summit received a total of 712 wide-ranging commitments to sustainable development from all stakeholders, including governments, UN system, and international governmental organizations (IGOs), the private sector, civil society, and NGOs. These included the unprecedented commitment by multilateral development banks including the World Bank and the Asian Development Bank to invest US $175 billion in grants and loans to support sustainable transportation systems in developing countries by 2022, which would help to mitigate air pollution and transport related climate change problems (United Nations, 2012c; ITDP, 2012; World Bank, 2015).

Other commitments on the list included planting 100 million trees by 2017, greening 10,000 Km^2 of desert, saving 1 Mw/hour of power per day, establishing a Master's program on sustainable development practice and developing an Environmental Purchasing Policy and Waste Minimization and Management strategy, among others. Collectively, these tangible commitments mobilized about US $513 billion in actions toward sustainable development. So impressive were the commitments that the Secretary-General of the United Nations, Ban Ki-moon claimed that Rio +20 "has given us a solid platform to build on. And it has given us the tools to build with. The work starts now" (quoted in House of Commons London, 2013: 19). The Secretary-General also lauded the summit as "further evolution of an undeniable global movement for change" (United Nations, 2012d).

Arguably, the Rio +20 helped to revitalize the commitments previously made in various thematic areas or sectoral topics closely related to Agenda 21 and the Johannesburg Declaration. The summit also reinforced global commitment to promote more coordinated, coherent, and effective implementation of sustainable development policies in addressing global environmental issues. Particularly, through "The Future We Want," the global leaders have reaffirmed their commitment to fully implement the 1992 Rio Declaration, the Johannesburg Declaration and Agenda 21 based on a more coherent global governance framework.

2.11.4 Global millennium development goals conference (GMC) (2013)

To maintain the momentum of progress and political commitment to achieving the MDGs, the United Nations convened the Global MDGs Conference (GMC) in 2013. The Conference, with the theme "Making the MDGs Work," was held in Bogota (Colombia) from February 27 to 28. Close to 200 participants including lead experts, academia, governments, civil society, and united nations development programme and other UN/development officers from Africa, Asia, Europe, Middle East and Latin America discussed, and exchanged views, knowledge and evidence on MDG issues (UNDP, 2013a). A series of working papers covering various country-specific case studies on MDG development issues, progress and future challenges were highlighted (UNDP, 2013b). The Conference also reaffirmed the need to integrate the three pillars of sustainable development as highlighted at the 2012 Rio conference. The following were some of the important themes stressed at the Conference:

- MDG breakthroughs and challenges for the next generation;
- Acceleration toward achieving the MDGs and sustaining progress;
- Mainstreaming the MDGs—translating a global agenda into national action (UNDP, 2013a).

On September 23, 2013, the Secretary-General convened a 1-day high-level forum at the United Nations Trustee Council in New York with the theme "MDG Success: Accelerating Action and Partnering for Impact" "to catalyze and accelerate further action to achieve the MDGs" (United Nations, 2013a) and "to chart an ever more efficient path to impact" and "progress" (United Nations, 2013b). It also sought to "exchange life-saving ideas and promote concrete solutions to the most pressing challenges of our time," as the Secretary-General stressed (United Nations, 2013c).

The forum also covered a special forward-looking panel, "The Next Generation: Leveraging Innovations and Partnerships for Scaled Impact," to share experiences in breakthrough technological and global development issues (United Nations, 2013c). An additional commitment of US $2.5 billion to boost MDG achievement ahead of the 2015 target date for the Goals was announced at the meeting. The United Kingdom Government, the Government of Norway, and Energia (the international network on gender and sustainable energy) committed to contributing US $1.6 billion, US $75 million, and US $10 million respectively to the global fund (United Nations, 2013c).

The document "Accelerating Progress Sustaining Results: the MDGs to 2015 and Beyond" was produced after the meeting (UNDP, 2013c), presenting evidence of how over 50 countries had employed the MDGs Acceleration Framework (MAF) to develop action plans and achieve MDGs despite the bottlenecks encountered. This provided immensely valuable lessons and served as a guide in addressing various sustainable development issues through the formulation and implementation of effective action plans.

2.11.5 UN Sustainable Development Summit (2015)

Following the Global MDGs Conference and the MDGs high-level forum held in 2013, the United Nations convened the United Nations Sustainable Development Summit in New York in September 2015 to adopt the SDGs, otherwise known as the Global Goals. The bold new SDGs agenda, "Transforming our World: The 2030 Agenda for Sustainable Development," with 17 goals at its core was unanimously adopted by the 193 member states of the United Nations on September 25, 2015 (United Nations, 2015a). The 17 goals, built on the success of the MDGs, are broad in scope. They embody a whole range of issues including poverty, hunger, biodiversity, climate change, peace, justice and strong institutions, industry, innovation and infrastructure, among others. Fundamentally, these post2015 development goals aim to address the three pillars of sustainable development as mentioned above.

The post2015 development agenda is deemed to serve as "a plan of action for people, planet and prosperity." More specifically, it seeks to "stimulate action over the next 15 years in areas of critical importance for humanity and the planet" (United Nations, 2015a: 1). Apart from aiming to end extreme poverty by 2030, the SDGs also aim at protecting Earth from degradation through sustainable consumption and production and sustainable management of natural resources. It also seeks to take urgent action to address climate change. In particular, the SDGs aim to:

- Conserve and sustainably use the oceans, seas and marine resources for sustainable development by (i) the prevention and reduction of marine pollution of all kinds, especially from land-based activities, including marine debris and nutrient pollution, (ii) sustainably managing and protecting the marine and coastal ecosystem, and (iii) minimizing and addressing the impacts of ocean acidification, among others (Goal 14) (United Nations, 2015a: 23−24).
- Protect, restore and promote sustainable use of terrestrial ecosystems, sustainably manage forests, combat desertification, and halt and reverse land degradation and halt biodiversity loss. These include: (i) ensuring the conservation of mountain ecosystems, including their biodiversity, (ii) taking urgent and significant action to reduce the degradation of natural habitats, halt the loss of biodiversity and, by 2020, protect and prevent the extinction of threatened species, and (iii) taking urgent action to end poaching and trafficking of protected species of flora and fauna, and trading of illegal wildlife products (Goal 15) (United Nations, 2015a: 24−25).

The new agenda which entered into force in January 2016 will guide future direction of sustainable development until 2030. It is also expected to reinforce action plans and the means of implementation and revitalizing the global partnership for promoting a sustainable future. As Secretary-General Ban Ki-moon claimed, "The new agenda is a promise by leaders to all people everywhere… an agenda for the planet, our common home." It is, as he declared further, "a universal, integrated and transformative vision for a better world" (United Nations, 2015b).

To be sure, the new ambitious sustainable development agenda provides a framework around which the North−South heads of state can design policy action plans to improve the lives of the poor and to transform the world toward a better green future. Through renewed global partnership and commitment, every country will be expected to work toward achieving the SDGs by 2030.

2.11.6 The sustainable development goal Summit (2019)

Three years later since the adoption of the 2030 Agenda in September 2015, the United Nations convened the SDG Summit in New York in September 2019 to review progress in the implementation of the 2030 Agenda for Sustainable Development and its 17 SDGs and, to identify actions to accelerate implementation (United Nations, 2019a). The Summit featured six thematic Leaders Dialogs on:

i. megatrends impacting the achievement of the SDGs which stresses that "actions are imperative for urgently addressing climate change," among others;
ii. accelerating the achievement of the SDGs with focus on critical entry points for action in areas such as human wellbeing and capabilities, energy decarbonization, universal access and global environmental commons, among others;
iii. measures to leverage progress across the SDGs to enhance the establishment of a socially, economically and ecologically sustainable society (the three pillars of sustainable development);
iv. localizing the SDGs: the main focus is to integrate and contextualize the 2020 Agenda into local development plans and budgets;
v. partnerships for sustainable development which places emphasis on multi stakeholder partnerships that engage civil society organizations, youth, the private sector and others to ensure more inclusive implementation of the SDGs; and
vi. the 2020−30 Vision which aims to put the world on track to achieve all the targets by 2030.

The summit unanimously adopted a political declaration known as "Gearing up for a decade of action and delivery for sustainable development: political declaration of the Sustainable Development Goals Summit" (United Nations, 2019b). The declaration reaffirmed the transformative vison of the 2030 Agenda and launches an ambitious and accelerated decade of action to speed up its implementation processes. It also pledged to reinforce the effectiveness of the United Nations High-level Political Forum (HLPF) on Sustainable Development.

The summit also launched the Global Sustainable Development Report, "The Future is Now: Science for Achieving Sustainable Development" (United Nations, 2019c). The Report calls on politicians and policy-makers to speed up actions on the following six entry points:

i. Strengthening human wellbeing and capabilities;
ii. Shifting toward sustainable and just economies;
iii. Building sustainable food systems and healthy nutrition patterns;
iv. Achieving energy decarbonization and universal access to energy;
v. Promoting sustainable urban and peri-urban development; and
vi. Securing the global environmental commons.

It is evidently clear that the 2019 SDG Summit provides further guidance and motivation to the global leaders to speed up the implementation of the 2010 Agenda for "The future we want." However, whether all these visions, efforts, or initiatives will be translated into real action or not remain to be ascertained.

2.12 The United Nations' journey to environmental sustainability: Some comments

The journey from Stockholm to Rio +20 continuing into the post2015 path of Global Goals for sustainable development has provided the necessary impetus to the North and South global community to rethink and to holistically restructure their growth strategies based on international environmental standards and new sustainability guidelines. For the past few decades, in relentlessly keeping up the momentum of sustainable development through the conventions of a series of global conferences and summits, the United Nations has been able to continue bringing to global attention the well-known but forgotten idea that human beings must live within the bio-capacity of the already overburdened planet.

The United Nations' journey to environmental sustainability has by no means been easy; it has had to deal with and accommodate the extraordinarily diverse interests from all member states from different political, socioeconomic, environmental, cultural, and ideological backgrounds. The long-standing and deep-rooted North—South development and environmental divide, for example, poses one of the most formidable constraints on the international negotiation desks. At the 1972 Stockholm Conference, for example, discussions and international treaty negotiations were conducted under the shadow of mutual suspicion between the North and the South as noted above. If not for the Founex Report, the North—South deadlock could be even far worse, thanks to the foresight of Maurice Strong who was, to a great extent, able to galvanize the trust and cooperation of the developing South by recognizing the need for development in the region. Maurice's initiative was indeed a strategic breakthrough.

In bridging this divide and bringing the North—South closer to a common global environmental discourse, the United Nations, or more specifically, the Rio Declaration (Principle 7) adopted at the Rio Summit (1992) has unveiled the concept of common but differentiated responsibilities. This reinforced the Founex initiative in terms of the recognition given to the legitimacy of development concerns in the South which sees economic growth as a panacea for addressing poverty problems and development issues. This perhaps partly contributed to the success of the Rio Summit which, for the first time, was able to get the member states to agree on a number of critical issues including climate change and biodiversity loss, despite significant ideological differences among them.

More importantly, the adoption of the Kyoto Protocol followed by the Paris Agreement on Climate Change and the CBD are breakthroughs. These, together with a wide range of environmental and development principles expounded in various international documents including the Brundtland Report (1987) and The Future We Want (2012) represent milestones in the development of environmental governance across countries in the world to translate sustainable development visions into reality.

In the European Union, for example, the Rio Summit, significant progress has been made in many areas of environmental policy. Since the creation of its first EAP in the wake of Stockholm Conference as noted in Section 2.7 above, the European Union has consistently built a framework of commitments and cooperation at international, regional, and national levels which were nonexistent prior to the Rio Conference. It has created the fifth Environmental Action Program, "Towards Sustainability" (1997—2000), to guide the community towards streamlining its environmental governance and to achieve sustainable development by putting Agenda 21 into action. The relationship between the fifth Action Program to the Rio Summit is clearly stated as follows:

"The fifth Environmental Action Program was produced as the Community's main response to the 1992 Rio Earth Summit which called on the international community to develop new policies as outlined in Agenda 21, to take our society towards a sustainable pattern of development" (European Communities, 1999). Apart from the fifth Action Program, the community had also established Natura 2000 program under the 1992 Habitats Directive to enhance environmental conservation and habitat protection (European Communities, 1997). Arguably, for the past few decades since Stockholm, the environmental legislative framework and policies have undergone considerable revision and the scope of application has been much extended (see e.g., Klemmensen et al., 2007 especially Chapter 2: Development of EU Environmental Regulation).

The United States, since the Rio Conference, has actively engaged in initiating and implementing "joint implementation" (JI) projects in reducing greenhouse gas emissions. In 1993, it announced the US Initiative on Joint Implementation (USIJI), a voluntary pilot program encouraging organizations in the United States and other countries to implement greenhouse-gas reduction projects (World Bank, 1998). Under Clinton's administration, the United States established the President's Council on Sustainable Development in 1993. The main objective of the Council is to develop policy recommendations for a national strategy for sustainable development and effective use of natural resources. Based on the Rio Summit recommendation, it has also established a CSD to monitor implementation of Agenda 21 (US Department of State, 1995; Brunnée, 2004).

In Asia, the Brundtland Report provided a remarkable conceptual framework and impetus for shaping policies in postMao China based on the integration of environmental concerns and biodiversity conservation into development policy (see Section 3.13.1). For example, through a series of UNFCCC summits, China increasingly recognizes its role in the global climate problematique and has advanced a number of targets related to energy and carbon efficiency and renewable energy development in response to the global call for GHG reduction (Schunz and Belis, 2011, see, Section 3.17).

India, as a party to the CBD, has undertaken a wide range of environmental initiatives toward developing policies, enacting legislations, and implementing programmes to promote ecological conservation and sustainable use of its biological resources. These include the National Conservation Strategy and Policy Statement for Environment and Sustainable Development (1992), the National Wildlife Action Plan (2002–16), and the National Biodiversity Action Plan (2008), among others. It has also enacted the Biological Diversity Act 2002, which primarily aims at giving effect to the provisions of the CBD (UNDP, 2009). Under these environmental frameworks, various schemes such as Project Tiger, Integrated Development of Wildlife Habitats, and Project Elephant have been created to promote biodiversity conservation. In response to the UNFCCC provision, India has developed good institutional structure involving the National CDM Authority to implement CDMs to address the issue of climate change (UNDP, 2009). India has also created national and regional versions of the IPCC called the Indian Network for Comprehensive Climate Change Assessment (INCCCA). The INCCCA comprises a network of over 125 national scientific institutions to undertake scientific assessments of different aspects of climate change in different sectors (Ramesh, 2012). Since the Copenhagen summit, India has taken on a very proactive role in international negotiations for GHG emissions (Ramesh, 2012).

In Southeast Asia, with the publication of the Brundtland Report, the regional leaders also ratified various international agreements as a sign of commitment to sustainable development. In an attempt to ensure stronger effort in biodiversity conservation and environmental quality protection in line with the United Nations CBD, Malaysia, Indonesia, Thailand, the Philippines, and Vietnam, for example, have strengthened their environmental controlling framework based on the establishment of ministerial environmental departments (see Section 3.12). To demonstrate further commitment to promoting sustainable development, each nation has also created its local Agenda 21 to execute full integration of the sustainability principles and environmental concerns of Agenda 21. Environmental protection is also being reinforced based on the enactment of a wide range of environmental laws (see Section 3.12).

The above examples demonstrate that, for the past few decades, through relentless efforts from the watershed Stockholm Conference to the world's grandiose Earth Summit, the United Nations has raised an unprecedented show of collective global awareness and concern for environmental issues. At each global summit, the United Nations also served as an instrument for articulating waves of ecological thought in a world divided between the wealthy North and the poorer South, authoritatively nurtured and shaped a world environmental culture, galvanized and reaffirmed political commitment for action programmes and created global partnerships to address various pressing environmental problems confronting us today.

The United Nations conferences also provided valuable inputs for decision-making and inspired the creation of a wide range of policies that guide government in environmental management. The consensus reached at all the conferences or summits as reflected in the adoptions of various perceived significant international treaties, conventions, declarations and agreements by the Heads of State imply the integrated consideration of the three pillars of sustainable development, namely, economic growth, social development and environmental sustainability in decision-making. This marks a new dominant global sustainable development discourse which guides our planet toward a greener future.

2.13 Concluding remarks

In the wake of severe global environmental deterioration caused by uncontrolled human economic activities, especially in the industrialized North, the United Nations convened the Stockholm Conference in 1972 with the aim of calling upon the international community to reverse environmental decline by implementing environmental protection measures. The Stockholm Conference was a critical juncture in the United Nations' journey to global environmental sustainability. For the first time in UN environmental history, the environment was topmost on the global political agenda, and environmental issues which had previously not received the necessary attention they deserved became a global concern. More importantly, the United Nations has revolutionized the ecological thought of development in an era dominated by neoliberalism. The Stockholm environmental impetus was reinforced with the publication of the Brundtland Report in 1987 which provided a unifying philosophy of sustainable development—a concept of paramount influence to guide the reorientation of the neoliberal ideology of economic growth toward environmentally sustainable development.

Following Brundtland, a series of high-profile summits or conferences including the 1992 Rio +10 Conference and the 2012 Rio +20 Conference were convened to focus on raising global environmental awareness, setting global priorities in environmental issues, reinforcing global environmental commitments, supporting establishment of multi stakeholder partnerships, and guiding the design and implementation of international and local environmental action plans, among others. The United Nations' environmental protection initiatives have dramatically increased momentum to efforts at making the global community adopt a common outlook to translate commitments into practice.

At this stage of analysis, however, it is still too early to gauge whether global political commitments for a sustainable world have been translated into real actions in saving our planet from destructive growth. This will be the subject of discussion in the following chapter.

The United Nations' journey to global environmental sustainability since Stockholm: The paradox

3.1 Introduction

Since Stockholm, the United Nations, through its series of high-level summits and conferences followed by waves of ecological thought and environmental information, has not only provided a continuous means of raising global environmental awareness but also added momentum to the process. This is reflected by an unprecedented show of collective global concern over the confluence of threats arising from growing environmental degradation, especially deforestation and habitat fragmentation, biodiversity decline, and climate change.

As a center for harmonizing the deep-seated and long-standing North-South environmental-development disputes, the United Nations plays a central role in pushing for a North-South consensus by bridging differences in perspectives and interests across the divide, thus galvanizing worldwide commitment to global environmental conservation and management efforts. This is evidenced by the establishment of environmental ministries across the world with the aim of formulating concrete and realistic action plans and regulations to reverse or halt environmental decline and to address complex and interrelated development and environmental problems.

Continuous streams of all-encompassing environmental strategies, norms and measures, scientific, technical, socioeconomic, and environmental information flowing from its broad conference mandates continue to bring forth new ideas for shaping meaningful and effective global environmental governance and regimes worldwide. Serving as a political avenue for global environmental gathering, the United Nations has, for the past few decades, provided opportunities for participants, including state and nonstate actors, to exchange and share visions, and knowledge and experience on domestic environmental issues, policy implementation gaps, and national sustainable development strategies, thus helping to streamline policy-making. Debates and discussions as well as conference outcomes continue to serve as a basis and impetus for the development of international as well as local environmental law and regulations in environmental control and state obligations and responsibilities.

The United Nations has proven to be a significant fulcrum in consolidating international environmental and development cooperation in addressing a host of multilateral environmental issues. Although it is still too early to gauge the impact of the United Nations' Millennium journey which extends up to 2030, what may be claimed in light of the above is that, since Stockholm, the United Nations has been the catalyst in propelling the global community to establish comprehensive institutional frameworks and multilateral environmental regimes to address pressing environmental issues and to guide our global commons toward a green path of development.

The most pertinent question therefore is: after more than four decades of global environment and development efforts and cooperation, how has green transformation on-the-ground progressed? The discussion on this will be the focus of this chapter.

3.2 Global environmental status since Stockholm: An overall view

In answering the above question, it may be remarked that enlightened global environmental awareness and institutional reforms do not necessarily result in real or decisive actions to forestall the ultimate destruction of our planetary system. Despite the United Nations' environmental protection initiatives and shared and common consensus among the UN member states on the importance of integrating environmental protection into development policy, the evidence points to the fact that they have been less successful in preventing or significantly slowing down the accelerating trend of global environmental decline.

While the international community has implemented various measures in response to a number of bilateral and multilateral agreements negotiated at the United Nations summits, such as the Convention on Biological Diversity (CBD) and United Nations Framework Convention on Climate Change (UNFCCC), to address perceived urgent environmental threats, those measures have been woefully inadequate to reverse the environmental impasse or to cope with the looming environmental crisis. Our Earth system is still under considerable stress fundamentally caused by the strain of uncontrolled human activities on the environment and its ecological integrity is at stake.

Indeed, as the 2002 Global Environmental Outlook (GEO3) pointed out, "the level of awareness and action has not been commensurate with the state of global environment today: it continues to deteriorate" (UNEP, 2002: 1). Taking this as a point of departure, what follows is a comprehensive attempt to empirically examine to what extent humans have contributed to the decimation of our Earth system with specific reference to biodiversity loss which is largely caused by deforestation and habitat destruction and, greenhouse gas emissions which are fundamentally caused by unrestrained combustion of fossil fuels especially coal. The assessment allows us to logically observe and explicitly gauge how far the United Nations' near-5 decade journey to global environmental sustainability has taken it toward its destination.

3.3 Forest conservation and deforestation: An empirical assessment

3.3.1 Forest conservation: The *raison d'être*

Since the global forests especially tropical rainforests are not only home to the largest percentage of the terrestrial biodiversity but also play an important role in carbon sequestration, their irreversible destruction can have far reaching impacts on biodiversity loss and climate change. Taking this as a vantage point, this section first examines the states of global forest especially in the biodiversity-rich tropical rainforest regions before ventures into discussing the biodiversity and climate change issue—the core subjects of interest in this chapter.

To begin with, global forests which cover roughly 30% of the Earth's surface, nearly 4 billion ha are also the most diverse ecosystems on land, housing more than 80% of the world's terrestrial species of animals, plants, and insects (UNFCCC, 2011; IUCN, 2012). It is worth highlighting that the tropical forests which cover less than 7% of the Earth's surface are home to an estimated one-half to three-quarters of all terrestrial plant and animal species (Schochet, 2018). Also forests are home to indigenous people worldwide and more than 1.6 billion people around the world depend to varying degrees on forests for their livelihoods (IUCN, 2012; Arnold et al., 2011). More than 25% of modern medicines, worth an estimated $US108 billion a year, originate from tropical forest plants (IUCN, 2012).

Deforestation is also closely associated with climate change. To note briefly, rainforests, including peat swamp forests, store carbon through biosequestration—the process of capturing and storing carbon dioxide (CO_2) from the atmosphere by the biological process of photosynthesis. This helps to regulate global warming and climate change. It is worth noting that forests and forest soils store more than 1 trillion tonnes of carbon (IUCN, 2012). Global tropical deforestation contributes to some 10% of global greenhouse gas emissions annually, that is, about 3 billion tonnes/year between 2000 and 2005 (Union of Concerned Scientists, 2013; Mongabay, 2012). This is equivalent to the CO_2 emission of 600 million cars in the United States at 5 tonnes of CO_2/year/car (Union of Concerned Scientists, 2013). It is relevant to note that this figure is estimated based on gross deforestation, not net deforestation. That is, it does not subtract the carbon sequestration that is taken up by forest growth. Forests also help to regulate surface albedo through the absorption of heat. Albedo is a measure of the amount of solar radiation that is reflected by the earth surface into space.

3.3.2 The United Nations forest protection initiatives

Thus in view of the vital role play by forests in sustaining life on Earth as elucidated above and as noted in the previous chapter, forest conservation has emerged as a predominant issue for world discussion at the 1992 Earth Summit which saw the adoption of Forest Principles to promote sustainable management and conservation of all types of forests. Chapter 11 of Agenda 21 on "Combating Deforestation" also provides comprehensive guidelines to enhance the protection, sustainable management, and conservation of all forests. It is also relevant to note that all three Rio Conventions, namely, the Convention on

Biological Diversity (CBD), the United Nations Convention to Combat Desertification (UNCCD), and the UNFCCC adopted at the 1992 Earth Summit, also acknowledge the important contribution of forests to achieving their respective goals and objectives (CBD Secretariat, 2012).

A post-UNCED initiative, namely, the World Commission on Forests and Sustainable Development was established in 1994. The aims of the Commission were to: (1) increase the level of awareness of the world forests in preserving the natural environment and contributing to sustainable socioeconomic development; (2) widen the consensus on the data, science, and policy aspects of forest conservation and management; and (3) build confidence between North and South on forest matters with emphasis on international cooperation (Salim and Ullsten, 1999).

In 1995 the United Nations Commission on Sustainable Development established the open-ended ad hoc Intergovernmental Panel on Forests (IPF) to pursue consensus and coordinate proposals for action to support sustainable forest management (SFM) and to implement the Forest Principles and Chapter 11 of Agenda 21 (United Nations, 1995; Fry et al., 1999; Nilsson, 2001). The IPF reached consensus on around 130 proposals for action during the 2-year mandate from 1995 to 1997 (Nilsson, 2001). The IPF focused on the following issues:

1. Implementation of United Nations Conference on Environment and Development (UNCED) decisions related to forests at national and international levels.
2. International cooperation in financial assistance and technology transfer.
3. Scientific research, forest assessment, and development of criteria and indicators for sustainable forest management.
4. Trade and environment relating to forest products and services.
5. International organizations and multilateral institutions and instruments including appropriate legal mechanisms (United Nations, 1995).

The IPF reached consensus on around 130 proposals for action during the 2-year mandate from 1995 to 1997 (Nilsson, 2001). IPF was succeeded by the Intergovernmental Forum on Forests (IFF) (1997–2000) which focused on the following issues:

1. Promoting and facilitating the implementation of the proposals for action of the IPF, and reviewing, monitoring and reporting on progress in the management, conservation, and sustainable development of all types of forests.
2. Considering matters left pending and other issues arising from the programme elements of the IPF process, including the need for financial resources and matters left pending on trade and environment and the transfer of environmentally sound technologies to support SFM.
3. Other issues including the underlying causes of deforestation and forest degradation, traditional forest-related knowledge, forest conservation and protected areas (PAs), and forest-related work of international and regional organizations and under existing instruments, among others (United Nations, 2000).

The IFF had reached consensus on a further 120 proposals for action toward achieving SFM. It may be noted that the IPF/IFF proposals for action are not legally binding and they are meant for providing guidelines for the development of national forest programmes

(NFPs) which include a wide range of approaches for SFM, and to stimulate action for their implementations (Sepp and Mansur, 2006; Maguire, 2013). The NFP is an open-ended, country-driven, and adaptive process which incorporates various approaches that can contribute toward the development of sustainable forestry planning and practices at national and subnational levels and improve forest governance and foster SFM (FAO, 2012a, 2017).

The NFP is said to be applicable to all countries and to all types of forests (FAO, 2017). Indeed, the NFP concept has been widely embraced by international forest communities and today the NFP processes are applied in more than 130 countries which have national forest policies to guide toward the planning and implementation of SFM practices (Maguire, 2013; FAO, 2017). The IPF/IFF processes also enable a holistic understanding of forests as living ecosystems which provide diverse environmental benefits and other services and hence call for a more holistic approach to SFM and conservation (Chaytor, 2001).

The momentum of the United Nations' SFM initiatives continued with the United Nations Economic and Social Council through resolution 2000/35 establishing the United Nations Forum on Forests (UNFF) to promote "the management, conservation and sustainable development of all types of forests" (United Nations, 2000: 64). More specifically, the UNFF, which is the only intergovernmental body that addresses all forest issues, is created to enhance "the implementation of internationally agreed actions on forests, at the national, regional and global levels" (United Nations, 2000: 64). In more explicit terms, it seeks to carry out:

1. the principal functions as contained in the Rio Declaration, Forest Principles, Chapter 11 of Agenda 21, and IPF/IFF proposals for action;
2. to facilitate and promote the implementation of IPF/IFF proposals for action through NFPs or any other relevant integrated programmes relevant to forests;
3. to serve as a forum for continued policy development and dialogue among governments;
4. to enhance cooperation and policy/programme coordination among relevant international and regional organizations or institutions; and
5. to foster international cooperation (United Nations, 2000: 64).

The resolution also recommends the establishment of Collaborative Partnership on Forests (CPF) to support the work of the UNFF and to enhance cooperation and coordination among participants on forests (United Nations, 2000: 64–65). The CPF which was formally formed in April 2000 is an innovative intraagency partnership on forests which aims "to support the work of the UNFF in promoting sustainable management of all types of forests and to strengthen long-term political commitment to this end" (United Nations, 2002a: 1). It may be noted that Membership to the UNFF is open to all State Members of the United Nations and/or members of the specialized agencies with full and equal participation voting rights (United Nations, 2000: para. 4). The community-based forestry (CBF) comprises 14 international organizations, bodies, and convention secretariats as listed in Table 3.1. The CPF Secretariat collaborates closely with these international organizations working on forest issues to strengthen the capacities of countries in combating forest biodiversity loss, desertification, deforestation and forest degradation, and among others (CPF Secretariat, 2013).

TABLE 3.1   Collaborative Partnership on Forests member list.

Collaborative Partnership on Forests (CPF) members
1   Center for International Forestry Research (CIFOR)
2   Convention on Biological Diversity (CBD) Secretariat
3   Food and Agriculture Organization of the United Nations (FAO)
4   Global Environmental Facility (GEF) Secretariat
5   International Tropical Timber Organization (ITTO)
6   International Union for Conservation of Nature (IUCN)
7   International Union of Forest Research Organizations (IUFRO)
8   United Nations Convention to Combat Desertification (UNCCD) Secretariat
9   United Nations Development Programme (UNDP)
10   United Nations Environment Programme (UNEP)
11   United Nations Forum on Forests (UNFF) Secretariat
12   United Nations Framework Convention on Climate Change (UNFCCC) Secretariat
13   World Agroforestry Centre (ICRAF)
14   World Bank

Based on CPF Secretariat, 2013. Promoting the Sustainable Management of All Types of Forests. The United Nations Forum on Forests Secretariat, New York.

From 2001 to 2005, the UNFF convened five sessions (UNFF-1 to UNFF-5) based on its Multi-Year Programme of Work (MYPOW) as adopted at the first UNFF session held at the UN headquarters in New York in 2001 to pave the way for reaching a consensus on an agreed international forest controlling framework based on the work initiated by the IPF and IFF. Tellingly the UNFF constitutes an important forum to continue the United Nations' effort to achieve a legally binding international arrangement on forests (IAF)—an ambitious task which it failed to accomplish at the Earth Summit held in 1992.

However, the journey to this challenge has never been a straightforward or smooth process. As shown in Table 3.2, although various agreements were reached to decide on the future of IAF through the adoption of a range of resolutions and decisions, consensus remained elusive especially regarding the global forest convention on all types of forests. This was apparent in UNFF-5 where the whole process of negotiation was mired in disagreement among member countries in trying to reach a global consensus on all types of forests. The session ended without reaching any agreement on strengthening the IAF and negotiations were put on hold until the UNFF-6 in February 2006 when the member countries met to address the unfinished UNFF-5 issues (Baldwin et al., 2005; United Nations, 2006a).

Nonetheless, a major breakthrough in negotiations was achieved when delegates, in recognizing the importance of SFM accommodating transformative change and addressing development challenges, arrived at a final decision to adopt the Resolution on the

TABLE 3.2 Brief summary of UNFF-1 to UNFF-14.

2001	UNFF-1	Adoption of decisions on: (1) the establishment of a MultiYear Programme of Work for 2001−05; (2) the development of a plan of action for the implementation of the IPF/IFF proposals for action; (3) the initiation of the UNFF's work with the Collaborative Partnership on Forests; and (4) accreditation of intergovernmental organizations. The UNFF-1 also recommended the establishment of three ad hoc expert groups to support the work of the Forum (United Nations, 2001).
2002	UNFF-2	Adoption of the following major decisions: (1) combating deforestation and forest degradation; (2) forest conservation and protection of unique types of forests and fragile ecosystems; (3) rehabilitation and conservation strategies for countries with low forest cover; (4) rehabilitation and restoration of degraded lands and the promotion of natural and planted forests; (5) concepts, terminology, and definitions (for the purpose of comparability and compatibility of forest data); (6) specific criteria for the review of the effectiveness of the international arrangement on forests; and (7) proposed revisions to the medium-term plan for the period 2002−05 (United Nations, 2002b).
2003	UNFF-3	Establishment of policy decisions on the implementation of the IPF/IFF proposals for action in relation to: (1) the economic aspects of forests; (2) forest health and productivity; and (3) maintaining forest cover to meet present and future needs. The Forum also adopted the resolutions on: (1) enhanced cooperation and policy and programme coordination and (2) voluntary reporting format; among others (United Nations, 2003).
2004	UNFF-4	Adoption of resolutions on: (1) forest-related scientific knowledge; (2) social and cultural aspects of forests; (3) forest-related monitoring, assessment and reporting; (4) finance and transfer of environmentally sound technologies; and (5) review of the effectiveness of the international arrangement on forests (United Nations, 2004).
2005	UNFF-5	A referendum agreement to four global goals to: (1) reverse the loss of forest cover worldwide, (2) enhance forest-based economic, social, and environmental benefits and the contribution of forests to the achievement of internationally agreed development goals, (3) increase significantly the area of protected forests worldwide; and (4) reverse the decline in official development assistance for SFM and mobilize significantly increased new and additional financial resources for the implementation of SFM (United Nations, 2005: 4).
2006	UNFF-6	Reached consensus on the above four UNFF-5 global goals whose achievement is to be evaluated in 2015 (United Nations, 2006b).
2007	UNFF-7	Adoption of the Non-legally Binding Instrument on All of Forests which aims: (1) to strengthen political commitment and action at all levels to implement effectively sustainable management of all types of forests and to achieve the shared global objectives on forests; (2) to enhance the contribution of forests to the achievement of the internationally agreed development goals; and (3) to provide a framework for national action and international cooperation. The Forum also set up a multiyear programme for a period of 8 years until 2015 (United Nations, 2008).
2009	UNFF-8	Adoption of a resolution on "Forests in a changing environment, enhanced cooperation and cross-sectorial policy and programme coordination, regional and subregional inputs which aims to encourage Member States to strengthen the implementation of sustainable forest management in addressing the challenges of forests in a changing environment, including climate change, loss of forest cover, forest degradation, desertification and biodiversity loss, in the context of sustainable development, and to encourage Member States to use national forest programmes or other strategies for sustainable forest management to address the above challenges" (United Nations, 2009: 3−4).

(Continued)

TABLE 3.2   (Continued)

2009	UNFF-9 (special session)	Adoption of a resolution on the means of implementation for SFM which was unresolved at UNFF-8 through which the Forum established: (1) an open-ended intergovernmental AHEG on forest financing to strengthen and improve access to funds and (2) the FP on forest financing to further develop ways to mobilize resources from all sources and to mainstream the global objectives on forests in their programmes; to facilitate transfer of environmentally sound technologies and capacity-building to developing countries, among others (United Nations, 2011: 12). The Forum also adopted the resolution on "Forests for people, livelihoods and poverty eradication" which comprises a range of recommendations some of which include the exploration, development and improvement of an enabling policy environment to strengthen forest law enforcement and governance (United Nations, 2011: 5–7).
2013	UNFF-10	Adoption of resolutions on: (1) progress in the implementation of the nonlegally binding instrument on all types of forests, regional and subregional inputs, forests and economic development, and enhanced cooperation; (2) emerging issues, means of implementation and the United Nations Forum on Forests Trust Fund (United Nations, 2013a).
2015	UNFF-11	Adoption of two key resolutions on (1) international arrangement on forests beyond 2015 and (2) ministerial declaration of the high-level segment of the 11th session of the United Nations Forum on Forests on the international arrangement on "The forests we want: beyond 2015" (United Nations, 2015a).
2017	UNFF-12	Major outcome: adoption of the provisional agenda (E/CN.18/2017/1) on "Implementation of the United Nations strategic plan for forests 2017–30" (United Nations, 2017a,b).
2018	UNFF-13	A communication and outreach strategy was adopted: (1) to raise awareness of forests and trees as being vital to life on Earth and human well-being; (2) to promote SFM, (3) to raise awareness of the global forest goals and encourage implementation of the strategic plan, and (4) to amplify communications from the forest community. The Forum also adopted an omnibus resolution which addresses (1) implementation of the UNSPF; (2) monitoring, assessment, and reporting; (3) means of implementation; (4) UN system-wide contributions to the implementation of the UNSPF; (5) the contribution of the Forum to the HLPF review of the SDGs in 2018; (6) preparations for the HLPF review in 2019; and (7) information on the UN Department of Economic and Social Affairs pertaining to UNFF (United Nations, 2018).
2019	UNFF-14	The Forum adopted the UNFF-14 report and took note of the UNFF-14 Chair's Summary, which contains the following items: (1) summary of technical discussions aligned with the provisional agenda of the 15th Session of the Forum (UBFF 15) for transmittal to the 15th Session of the Forum in 2020; (2) input of the 14th Session of the Forum to the High-Level Political Forum on Sustainable Development (HLPF) in 2019; (3) information on the reform of the Department of Economic and Social Affairs pertaining to the Forum for transmittal to the appropriate United Nations bodies such as the Committee for Programme and Coordination, the Advisory Committee on Administrative and Budgetary Questions and the Fifth Committee, as well as for transmittal to the 15th Session of the Forum (UNFF-15), and (4) an indicative list of the intersessional activities suggested during UNFF-14 to facilitate and inform policy deliberations at UNFF-15 (United Nations, 2019: 23).

AHEG, Ad hoc expert group; *FP*, facilitative process; *HLPF*, High-level Political Forum; *IFF*, Intergovernmental Forum on Forests; *IPF*, Intergovernmental Panel on Forests; *SDG*, sustainable development goal; *UNFF*, United Nations Forum on Forests; *UNSPF*, United Nations strategic plan for forests 2017–30.

"International Arrangement on Forests beyond 2015" at UNFF-11 held in 2015. (United Nations, 2015a,b). It also adopted a Ministerial Declaration, shown in Table 3.2. The UNFF-11 session, under the theme "Forests: progress, challenges and the way forward for the International Arrangement on Forests," constitutes an unprecedented opportunity for member state countries to forge an international forest policy for the next 15 years to 2030 which is in accord with the post-2015 Development Agenda adopted in August 2015 (United Nations, 2015c).

The main objectives of the IAF are to promote implementation of SFM, and to enhance the contribution of forests to the post-2015 development agenda especially in relation to Goal 15 of the sustainable development goals (SDGs). Particularly the implementation of the UN Forest Instrument will be a key task for the IAF beyond 2015 (United Nations, 2015c). Goal 15 of SDGs aims to "protect, restore and promote sustainable use of terrestrial ecosystems, sustainably manage forests, combat desertification, and halt and reverse land degradation and halt biodiversity loss" (ICSU and ISSC, 2015: 71).

Another UNFF positive development was the adoption of the provisional agenda (E/CN.18/2017/1) in association with the United Nations strategic plan for forests 2017–30 (UNSPF) at UNFF-12 session held in 2017 (United Nations, 2017a,b). The Plan set out 6 Global Forest Goals and 26 associated targets to be accomplished by 2030 (Table 3.3). It is noteworthy that the Plan includes the targets under Global Forest Goal 1 of increasing the world's forest area by 3% worldwide and to maintain or enhance the world's forest carbon stocks (United Nations, 2017c: 6).

TABLE 3.3 United Nation Global Forest Goals.

Global Forest Goal 1	Reverse the loss of forest cover worldwide through sustainable forest management, including protection, restoration, afforestation, and reforestation, and increase efforts to prevent forest degradation and contribute to the global effort of addressing climate change.
Global Forest Goal 2	Enhance forest-based economic, social, and environmental benefits, including by improving the livelihoods of forest dependent people.
Global Forest Goal 3	Increase significantly the area of protected forests worldwide and other areas of sustainably managed forests, as well as the proportion of forest products from sustainably managed forests.
Global Forest Goal 4	Mobilize significantly increased, new and additional financial resources from all sources for the implementation of sustainable forest management and strengthen scientific and technical cooperation and partnerships.
Global Forest Goal 5	Promote governance frameworks to implement sustainable forest management, including through the UN Forest Instrument, and enhance the contribution of forests to the 2030 Agenda.
Global Forest Goal 6	Enhance cooperation, coordination, coherence, and synergies on forest-related issues at all levels, including within the UN System and across Collaborative Partnership on Forests member organizations, as well as across sectors and relevant stakeholders.

Based on United Nations, 2017c. Resolution adopted by the Economic and Social Council on 20 April 2017. United Nations Strategic Plan for Forests 2017–2030 and Quadrennial Programme of Work of the United Nations Forum on Forests for the period 2017–2020 (Document E/RES/2017/4). United Nations Economic and Social Council, New York. Available from: <https://documents-dds-ny.un.org/doc/UNDOC/GEN/N17/184/62/PDF/N1718462.pdf?OpenElemen>.

In addition, the UNFF-13 provides further momentum for the world leaders to address challenges relating to forest conservation. The Forum was the first UNFF session after the adoption of the UNSPF in 2017. The Forum emphasized that forests are crucial for biodiversity conservation and climate regulation. It further reaffirmed the message that SFM is considered vital to achieving all 17 SDGs including poverty eradication. It called on all actors especially forestry and environmental agencies as well as decision-makers to support and contribute to the implementation "United Nations strategic plan for forests 2017–30 and achieving its global forest goals and targets" (United Nations, 2018: 15).

Many countries including Indonesia, Malaysia, Madagascar, Philippines, and Paraguay announced their voluntary national contributions (VNCs) to the UNSPF "to achieving one or more of the global forest goals and targets set out in the United Nations strategic plan for forests 2017–2030" (United Nations, 2018). The VNCs, as claimed by the UNFF Director, constitute the backbone of sustainable forestry management. However, most of these VNCs focus on the development of plans and national policies without describing concrete actions to be adopted to achieve the goals (do Valle, 2018). At the Forum, Brazil also announced its intention to restore 10 million ha of forests by 2020 while Thailand unveiled its national efforts to increase forest area from 32% to 40% by 2038 (Ripley et al., 2018).

Following UNFF-13, the 14th Session of the United Nations Forum on Forests (UNFF-14) was held in May 2019 to discuss the implementation of the UNSPF. The main themes for discussion were "the thematic and operational priorities, priority actions and resource needs for the period 2019–2020" and "Forests and education" (United Nations, 2019: 24). Again, many countries announced their VNCs to the UNSPF to enhance global forest cover by 3% by 2030 (Goal 1 of the Global Forest Goals). Indonesia, for example, announced its plan to reduce deforestation from 0.92 million ha/year to 0.45 million ha by 2020 and 0.325 million ha by 2030 while Malaysia uncovered its plans to ensure that 50% of its territory will be forest-covered by 2030 (Nyingi et al., 2019). Undoubtedly the UNFF-14 has played a proactive role in engaging policy-makers to contribute to enhancing the achievement of the Global Forest Goals as indicated by the increased number of VNCs.

Conceivably the United Nations has for the past few decades galvanized international cooperation and action to develop coherent policies to promote SFM and reduce deforestation and forest degradation at all levels through its UN Forest Forum. Other international institutions which also contribute to promote SFM worldwide are the International Tropical Timber Organization (ITTO) and the International Tropical Timber Agreement, which regulate trade of tropical woods and promote conservation and sustainable management and use of tropical forest resources.

The related international agreements such as the UNFCCC and the CBD also play an important role to enhance SFM and reduce deforestation and forest degradation globally. For example, Target 5 of the Aichi Target stipulates that "By 2020, the rate of loss of all natural habitats, including forests, is at least halved and where feasible brought close to zero, and degradation and fragmentation is significantly reduced" (CBD Secretariat, 2012: 56). Similarly the New York Declaration on Forests launched at the UNFCCC summit in 2014 also places great emphasis in its Goal 1 to halt natural forest loss by 2030 (Climate Focus, 2015). Article 5 of the Paris Climate Agreement adopted at the 2015 UNFCCC Conference also focuses on forest conservation, enhancement, and sustainable management (UNFCCC, 2015: 23–24). The world is also motivated by the Bonn Challenge, a global effort to restore 150 million ha of the world's deforested and degraded land by 2020, and 350 million ha by 2030 (IUCN, 2017).

The incessant flows of all these forest protection institutions or organizations and SFM and conservation initiatives, and the systematic and consistent approach to strengthen national capacities and support national forestry programmes in member state countries reflect the determination of the United Nations to uphold the Holling resilience of the global forest ecosystems. To gauge the extent these novel efforts have been able to accomplish this aim through halting deforestation or reversing the trends of forest degradation globally, various selected case studies across Asia that were part of an empirical investigation of the global forest status are presented below.

3.4 Global forest tracking: Food and Agriculture Organization versus Global Forest Watch

3.4.1 Forest: Some basic concepts and definitions

Before attempting to evaluate United Nations' accomplishment in halting global deforestation, it is important to be clear at the outset about some of the forest concepts, definitions and methodologies that are used in the analysis. This is because estimates can differ markedly among sources of studies depending on the definition of forest and the assessment methodology used. For example, due to the change of Food and Agricultural Organization (FAO)'s global definition of forest by reducing the biophysical requirement of minimum height of tree stand from 7 to 5 meter (m), and the minimum crown cover or canopy cover from 20% to 10%, the estimate of global forest area increased by 300 million ha between 1990 and 2000 (Chazdon et al., 2016).

There is currently a lack of globally agreed definition of forest. For instance, it is widely quoted that there are more than 800 definitions of forest or wooded area throughout the world in forest assessment which are used to suit the different aims and purposes of the governments, organizations, scholars, and citizens groups (Lang, 2001; Harris et al., 2016; Clark, 2017). However, in 2013, H. Gyde Lund found 1597 definitions of forest in use for administrative, land cover, land use and land capability (biome, ecosystem, or ecological) purposes, and from various scopes (Lund, 2014). Even the definitions of forest by two of the most important international organizations, namely, UNFCCC and the FAO are not exactly the same (Harris et al., 2016; Yew, 2012).

While the former defined forest to include oil palm plantation, which is in fact one of the major causes of deforestation by replacing natural forests, the latter defined forest as land officially or legally designated for a "forest use," such as conservation or harvest. Thus land predominantly marked for agricultural practices, even it contains trees or land designated for urban uses such as city parks, even if it meets the biophysical criteria of forest, will be excluded from assessment (FAO, 2000, 2012b; Harris et al., 2016). In comparison, FAO defines forest as land that has an area of more than 0.5 ha and with tree crown cover of more than 10%. Furthermore the trees should be able to reach a minimum height of 5 m at maturity in situ. It also requires that land be officially or legally designated for a "forest use," such as conservation or harvest (FAO, 2000, 2012b).

In more specific terms, FAO's definition of forest excludes tree stands in agricultural production systems, such as fruit tree plantations, oil palm plantations, olive orchards, and agroforestry systems but includes rubber plantations, cork oak, and Christmas tree plantations (FAO, 2000, 2012b). However, logged-over or firebreak areas or areas temporarily devoid of trees but legally designated as areas that will be allowed to regenerate in the future are classified as forest (FAO, 2000, 2012b; Harris et al., 2016). It thus follows that harvesting or clear cutting of all trees from a tract of legally designated forest land does not constitute deforestation if "the forest is expected to regenerate naturally or with the aid of silvicultural measures within the long-term" (FAO, 2001: 25).

As Meine Van Noordwijk and colleagues revealed in their case study on carbon stocks in the Usambara Mountains of Tanzania, even though 88% of the forest had been removed from the East Usambaras area, it is still considered as forest. The reason given was that the area was classified as "a temporarily unstocked" forest because it is legally gazetted as forest land that has the potential to regenerate. This is against the fact that forest recovery from the cleared area is unlikely to materialize due to continued grazing and burning activities (van Noordwijk et al., 2009). Similarly following the FAO definition of forest, there will be no deforestation in Indonesia as land is under the control of forest institutions and is temporarily "unstocked" (van Noordwijk and Minang, 2009). Using the same argument, the conversion of rubber plantation into oil palm plantation will not result in the loss of forest plantation area and hence in forest area (Keenan et al., 2015).

3.4.2 Food and Agricultural Organization forest tracking methodology: Some basic facts

FAO which systematically publishes its statistical report on the total area of forest land worldwide every 5 years relies heavily on responses to surveys by individual countries using a common reporting framework, agreed definitions, and reporting standards. However, the self-reported official deforestation statistics as released by the national governments across the world may not reflect the true picture of their forest depletion trends. It may well be that in practice, many countries tend to use their own definitions and methods instead of following the FAO guidelines (Harris et al., 2016). In addition, many countries still use outdated data for their forest reporting. It was reported in the 2015 FAO forest report that 87 countries were using data more than 10 years old while 58, 38, and 14 countries were using data as old as 15, 20, and 25 years, respectively (Harris et al., 2016).

Thus the self-report information can be misleading. Also at times, statistics compiled by the developing countries and fed into the global data bank or international bodies may have been manipulated to hide unsustainable land and forest use practices. For example, in the case of Malaysia, National Aeronautics and Space Administration (NASA) detected massive deforestation in the country during the first quarter of 2014 and the degazettement of the legally protected forests for economic use. These data were not furnished to FAO to avoid revealing unsustainable development practices. The environmentally destructive IndoMet Coal Project (ICP) in Indonesia as discussed below is another case in point. Thus the claim of a country's success in forest conservation is contentious in that different reports based on different sources of data or definitions of forests may reveal contradictory results.

3.4.3 Global Forest Watch forest tracking methodology: Some basic facts

Considering the above, the present analysis uses a more reliable method carried out by Global Forest Watch (GFW), a dynamic online forest monitoring and alert system conducted by a team led by Matthew Hansen of the University of Maryland. The system mobilizes satellite remote sensing technology and NASA satellite data with almost 700,000 Landsat images fed through Google's computers (Hansen et al., 2013a; WRI, 2015). The data shows deforestation, protected areas (PAs), biodiversity hotspots, mining, logging, oil palm and wood-fiber plantation. Unlike FAO, GFW does not adopt a specific definition of forest. Instead it uses the term "tree cover" instead of "forest cover" to monitor and assess global forest gain or loss.

Tree cover refers to all forms of vegetation greater than 5 m in height, the height at which they can be reliably detected by satellites. These include natural forest, tree plantations such as Acacia tree plantations, oil palm, rubber, or other plantations (Weisse, 2016). GFW detects and reports all instances of tree cover loss, regardless of whether the loss is temporary due to, for example, timber harvesting followed by replanting, or permanent, for example, due to clear cutting followed by land use conversion for agricultural development (Harris et al., 2016). In contrast to the FAO's self-report system of forest assessment, GFW uses satellite-based data which has grown beyond annual updating to providing monthly and weekly deforestation alerts, thus furnishing an independent, transparent and consistent set of data that can help people around the world to better assess and manage forest changes over time (WRI, 2015; Petersen, 2016; Harris et al., 2016; see also, CIFOR, 2013). Two years after its launch, GFW revamped its system to create FORMA (FORest Monitoring for Action) 250—a daily updated and more accurate, higher resolution offspring of the former FORMA 500 dataset (Guzder-Williams, 2017).

However, GFW is also by no means flawless. For example, its wide definition of tree cover which includes oil palm, rubber, eucalyptus, and other managed stands taller than 5 m, can also be misleading. As a case in point, in association with the 2000−12 annual tree cover change report compiled by Matthew C. Hansen and others based on this definition (Hansen et al., 2013a), Tropek and others questioned the appropriateness of including ecologically poor monoculture plantations such as oil palm, rubber, or commercial tree plantations such as Eucalyptus plantations which constitute some of the biggest threats to the biologically rich tropical or natural forests (Tropek et al., 2014). They have rightly pointed out that by failing to distinguish tropical forests from plantations, it can lead to a substantial underestimate of real forest loss.

Nonetheless, Hansen et al. argued that many national forest agencies employ a land use criterion that is not tied to forest cover and its change. Definitional issue aside, seemingly in response to the criticism that FRW's forest data set has been unable to distinguish the loss or gain of natural forests from the less biologically diverse monoculture plantations, it has recently attempted to map the extent and location of plantations including oil palm plantations, pulp plantations, and other types of plantations (Weisse, 2016; Petersen et al., 2016). The study mapped 45.8 million ha of tree plantations across seven countries based on 2013 and 2014 data. These countries included seven key tropical countries, namely, Brazil, Cambodia, Colombia, Indonesia, Liberia, Malaysia, and Peru.

This newly improved platform with plantation maps allows us to assess whether recent forest clearing in some countries has occurred in natural forests or in plantations (Chazdon et al., 2016). However, it must be cautioned that these new additional plantation maps cover only 2013 and 2014 period and hence they do not reveal the land use pattern prior to 2013 (Weisse, 2016). Nonetheless, given further improvement and additional efforts at mapping plantations worldwide, the FRW' s data set can be used to distinguish natural forest from plantations and hence provide an important platform to detect deforestation (TNC, 2014).

3.4.4 Food and Agricultural Organization and Global Forest Watch forest trackings: Some remarks

In summary, it is by no means easy or even possible to provide accurate data on global forest degradation. Even two prominent global data sources as discussed above provide sharply contradicting conclusions on global deforestation. For example, while the FAO in its Global Forest Resources Assessment (FRA) published in 2015 claimed that "world deforestation slows down," GFW revealed that "global annual tree cover loss remains high" (Harris et al., 2016). However, both constitute two of the most important sources of valuable information in revealing the state of the global forest.

Also despite that FAO has been the authoritative source of information on worldwide deforestation for decades, its assessment does not necessarily reflect the real environmental status such as the ecological damage caused to the standing trees from selective logging, road construction and other commercial land conversion activities. In particular, its land use criterion in forest definition is problematic. On the other hand, compared with FAO, GFW provides more accurate and up-to-date forest cover statistics although it may also have its shortcomings due to its wide definition of forest cover as noted above.

Nonetheless, the NASA-US Geological Survey Landsat satellite which offers the best views of land images and global forest losses and gains, including deforestation of the Amazon basin, droughts, and wildfires with high-resolution image has improved greatly since it launched its Landsat 1 in 1972 (NASA, 2013; Cole, 2015). It has started work on Landsat 9 in partnership with US Geological Survey which is expected to be launched in mid-2021. The latest Landsat Earth observation satellite system will extend the Earth-observing programme, collecting global land cover data and images for the past 50 years (Cole, 2015).

The satellite system repeatedly observes the global land surface, giving scientists a valuable means of visual image analysis to distinguish between natural and human-induced changes to the landscape. Like Landsat 8 which was launched in 2013, Landsat 9 will have a higher imaging capacity than previous Landsat satellites. This contributes significantly to enhance the Landsat global land imaging archive which is freely available to the public. The improved system in turn allows us to better monitor, understand and manage the land resources needed to sustain human life (Russell, 2017).

3.5 Global forest loss: An empirical analysis

This section will demonstrate the magnitude of global forest disturbance rather than provide accurate estimates of global deforestation trends, which is currently not feasible.

The analysis will begin with a general assessment of global tree cover loss from 2001 and 2016 based on Hansen's data set. This is followed by a specific investigation into the magnitude of tropical rainforest disturbances based on various case studies in Asia. In recognizing the shortcomings of Hansen data set which does not distinguish natural forest from plantations, the assessment will be conducted with the examination of empirical evidence and facts about what resulted in deforestation in the tropics to render the argument more convincing.

Before offering a gist of the analysis, a few points need to be clarified at the outset. To begin with, tree cover loss is not always the same as deforestation (Weisse and Goldman, 2017a). Tree cover means an area cover by crown of trees including trees in plantations and natural forests which are greater than 5 m in height. Tree cover loss is defined as "stand replacement disturbance" or the complete removal of tree canopy or crown cover at the Landsat pixel scale due to human or natural causes such as forest fire (GLAD, 2018). When a pixel registers as loss, it is only indicating the tree leaves have died—it does not reveal whether the entire tree is dead or completely removed. Thus there is a possibility that some of the trees may recover in the future (Weisse and Goldman, 2017a,b). Canopy or crown cover refers to the extent an area covered by leaves of an individual tree or a group of trees. It is commonly expressed as a percentage of total ground area. For example, at 50% canopy cover, half of the total ground area is covered by the vertical projection of tree crowns (USDA Forest Service, 2011: 421).

Deforestation refers to the permanent clearance of forests and their related ecosystems for nonforest or commercial use such as commercial agricultural development, ranching, or urban use. In these cases, trees are never replanted. Hansen et al. defined forest loss as a stand-replacement disturbance or the complete removal of tree cover canopy at the Landsat pixel scale (Hansen et al., 2013a). Deforestation may also be distinguished from forest degradation which refers to changes within the forest that negatively affect the structure, function, or species composition of the stand or site, and thereby lower the potential supply of goods and services including wood, biodiversity, and biological functions (Schoene et al., 2007). Furthermore forest degradation is characterized by a reduction of canopy cover which may be caused by deforestation, uncontrolled timber harvesting, forest fire, or other nonforest use practices (Schoene et al., 2007). Indeed, deforestation caused by agricultural expansion, timber harvesting and mining activities continues to drive global tree cover loss from year to year (Weisse and Goldman, 2017b).

The present assessment reveals three scenarios of global deforestation trend based on tree cover density. Tree cover density is measured by using the following tree cover thresholds: (1) >30% canopy cover loss: an indication of relatively low deforestation dynamics, (2) >50% canopy cover loss: a reflection of the prevalence of deforestation dynamics, and (3) >75% canopy cover loss: an exhibition of high or severe deforestation dynamics. Using different tree cover thresholds indicates different estimates of deforestation rates. Despite these differences the overall message remains the same, that is, they reflect the prevalence of deforestation in different degrees.

As shown in Fig. 3.1, on a cumulative basis, global tree cover loss for >30%, >50%, and >75% tree cover thresholds between 2001 and 2016 for various selected countries were 307, 266, and 202 million ha, respectively. Russia led globally as having the highest tree cover loss followed by Brazil in South America, the United States, and Canada (Table 3.4).

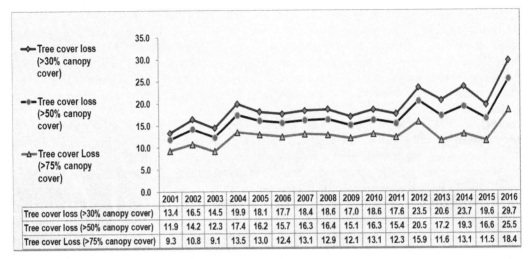

	2001	2002	2003	2004	2005	2006	2007	2008	2009	2010	2011	2012	2013	2014	2015	2016
Tree cover loss (>30% canopy cover)	13.4	16.5	14.5	19.9	18.1	17.7	18.4	18.6	17.0	18.6	17.6	23.5	20.6	23.7	19.6	29.7
Tree cover loss (>50% canopy cover)	11.9	14.2	12.3	17.4	16.2	15.7	16.3	16.4	15.1	16.3	15.4	20.5	17.2	19.3	16.6	25.5
Tree cover Loss (>75% canopy cover)	9.3	10.8	9.1	13.5	13.0	12.4	13.1	12.9	12.1	13.1	12.3	15.9	11.6	13.1	11.5	18.4

FIGURE 3.1 Tree cover loss between 2001 and 2016. *Source: Data from Hansen, M.C., et al., 2013b. Tree Cover Loss and Gain Area. University of Maryland, Google, USGS, and NASA. Accessed through Global Forest Watch on May 2018. Available from: <https://www.globalforestwatch.org/countries/overview>.*

TABLE 3.4 Total tree cover loss between 2001 and 2016.

Canopy threshold		>30%	>50%	>75%
	Russia	49,511,623	39,921,294	23,997,078
South America	Brazil	46,369,255	42,134,699	35,912,089
	United States	33,754,765	31,250,344	26,944,783
	Canada	36,012,931	29,283,429	22,306,509
South America	Argentina	5,365,943	3,517,927	1,572,703
	Bolivia	4,067,419	3,853,633	3,142,419
	Colombia	3,290,324	3,210,503	2,957,722
	Mexico	3,099,532	2,843,954	2,264,176
	Paraguay	5,099,895	3,068,031	1,616,707
	Peru	2,369,321	2,349,998	2,298,523
Southeast Asia	Indonesia	23,086,723	22,724,210	20,921,934
	Malaysia	6,804,274	6,747,818	6,426,559
Africa	DRC	10,522,673	9,839,791	7,497,861
	Madagascar	2,755,136	2,281,329	1,075,761

DRC, Democratic Republic of Congo.
Data from Hansen, M.C., et al., 2013b. Tree Cover Loss and Gain Area. University of Maryland, Google, USGS, and NASA. Accessed through Global Forest Watch on May 2018. Available from: <https://www.globalforestwatch.org/countries/overview>.

It is noteworthy that Brazil exhibited the highest tropical rainforest loss in the tropics while other South American countries including Argentina, Bolivia, Colombia, Mexico, Paraguay, and Peru also exhibited high deforestation trends. In tropical Southeast Asia, Indonesia, and Malaysia exhibited significant and persistent trends in annual tree cover loss while other countries including Vietnam, Cambodia, Laos, and Myanmar also had high loss of tree cover—between 1 and 2 million ha for all tree cover thresholds. In Africa, the Democratic Republic of Congo (DRC) showed the highest tree cover loss followed by Madagascar (Table 3.4). As shown in Fig. 3.1, tree cover loss for all threshold levels increased rapidly in 2016 from 2015, exhibiting a reference scenario for high deforestation dynamics.

It is evidently clear from the above discussion that overall, global tree cover loss, a reflection of deforestation dynamics, especially in the tropical rainforest regions, has worsened greatly since Stockholm. As shown clearly in Fig. 3.1, the global rate of tree/forest destruction is escalating year by year, especially in the tropics, reaching its peak in 2016. This is despite significant progress having been made in these countries in developing law, forest policies, or national forest management action plans to promote and guide sound forest management and sustainable forest resource use in consonance with the United Nations forest protection initiatives as discussed in Section 3.3.2.

This poses far-reaching implications for the environmental integrity of our planet. Here, it may be remarked that although the Hansen data set provides in indication of deforestation dynamics, further analysis is needed to ensure a more accurate picture of human-induced forest change. Furthermore to what extent forest policies as developed by the member state countries following the United Nations agreement and guides have been effective in addressing unsustainable forest resource use also needs to be considered.

We will now assess the effectiveness of the forestry policies and legislations in the tropical rainforest regions which were developed and subsequently streamlined to accommodate the international forestry agreement and regulations, including the UNFF guides, in promoting environmental sustainability. This assessment is based on various case studies in the tropical region. The case studies also serve to complement Hansen's data set by revealing additional and updated information that provides a more accurate reflection of the state of deforestation in the region.

3.6 Country-specific case studies

3.6.1 South America—the Big-4 (Brazil, Argentina, Paraguay, and Bolivia)

3.6.1.1 Brazil: National forest protection initiatives

Forest loss in the tropical rainforest region contributes to 32% of the global forest disappearance and nearly half of this loss has occurred in the South American rainforests (Hansen et al., 2013a). Taking Brazil as a case in point, for example, in 2005, 4 years after the inception of UNFF, it was claimed that the country had 3.6 million km^2 or 360 million ha of dense forests. This made it the world's largest rainforest region with more than three times the rainforest area of the Democratic Republic of the Congo (DRC), the second largest in global ranking (Ministry of Environment, Brazil, 2005). Brazil still remains as the world's largest rainforest region (4.6 million km^2 or 460 million ha) followed by DRC (1.8 million km^2 or 180 million ha) and Indonesia (1.3 million km^2 or 130 million ha) (Kiprop, 2017).

As early as 1934, long before the emergence of UNFF in 2001, Brazil enacted its First Forest Code to regulate logging activities and protect the forest environment (WWF, 2016a). The Code has since undergone various stages of refinement. The revised Forest Code (1965), for example, required landowners to set aside 50%−80% of their land as protected native forest (legal reserves). The Code also prohibited the clearing of vegetation in Areas of Permanent Protection such as on steep slopes and the margins of rivers and streams (da Matta, 2015). The Environmental Crime Law was also enacted to control illegal logging activities (Brito et al., 2005).

The Forest Code was further revised in 2012. Retaining the same structure and basic concepts as the old code, it introduced an innovative database and environmental management tool called the Rural Environmental Registry, which provides a way to monitor and control deforestation in private landholdings (Chiavari and Lopes, 2015).

Forest protection efforts were further enhanced when the government implemented a range of measures provided under various international instruments to which Brazil is a party, such as the UNFCCC, the CBD, Agenda 21, and the Forest Principles adopted during the 1992 Rio Conference, and the actions recommended by the UNFF (Ministry of Environment, Brazil, 2005). For example, the Brazilian government launched the Amazon Region Protected Areas Programme during the Rio + 10 Summit held in 2002. The goal of the Programme was to ensure the creation, implementation and consolidation of PAs covering at least 50 million ha in the Amazon, an area similar to that of Spain (Ministry of Environment, Brazil, 2005).

In 2004 the Brazilian government implemented the Action Plan for Prevention and Control of Legal Amazon Deforestation (PPCDAm) to arrest widespread deforestation that threatened the ecological integrity of the Amazon forest ecosystems. The Plan spread across 13 ministries and comprised 37 action groups and 214 programmes/activities (Hargrave, 2012). The government had also reformed its federal forest law in 2006 by creating new regulatory, development, and incentive policy instruments and institutions to promote sustainable forestry management and practices (Bauch et al., 2009). At the 2008 UN Climate Change Conference held in Poznàn, Brazil announced its National Climate Change Plan which included a pledge to pursue 72% cut in the rate of deforestation by 2018 (Greenpeace, 2009). In the same year, 2008, the government started a blacklist which resulted in stricter federal supervision and economic sanctions against countries to discourage deforestation (Marco Elías, 2017). The country's strong commitment to forest protection was further reflected at the 2009 UN Climate Change Conference in Copenhagen in which the Brazilian Minister of Environment announced its plan to reduce its deforestation rate by 80% by 2020 which corresponds to a 40% in greenhouse gas emissions (Kovacevic and Galindo, 2012; IBP, Inc., 2014).

Regionally Brazil also fully recognized the importance of the Amazonian Cooperation Treaty Organization (ACTO) in enhancing regional cooperation in promoting an economically, socially and ecologically sustainable policy on natural resources based on the Amazonian Cooperation Treaty signed in 1978 (United Nations, 2006b). ACTO is an intergovernmental body which comprises eight countries from the Amazon Basin: Bolivia, Brazil, Colombia, Ecuador, Guyana, Peru, Suriname, and Venezuela. The Amazonian countries also reaffirmed their commitment to strengthen the United Nations Forum on Forests "as a universal and crucial authority, aimed at implementing actions designed to increase and enhance conservation efforts and sustain the use and management of all types of forests" (United Nations, 2006b). The Amazonian members also expressed their support for enhancing global cooperation and the reversal of the decline of official development assistance allocated to forest-related activities (United Nations, 2006b).

Between 2011 and 2012 the Amazonian member countries implemented activities to monitor forest cover and to strengthen community management of forests. They also took further steps to identify additional resources for forest preservation and to promote awareness among members of the Amazonian society. More importantly they also committed to promoting international cooperation to combat illegal logging (United Nations, 2013b). Brazil is also a member of the Latin American and the Caribbean Forestry Commission (LACFC), a FAO Statutory Body, which aims to provide member countries with a forum to discuss and analyze important regional forestry issues, from monitoring of forest resources to formulating policies on sustainable management of forests and wildlife, among others (FAO, 2018).

It has often been claimed that Brazil has achieved outstanding results with its efforts to end deforestation (Kovacevic and Galindo, 2012; Hargrave, 2012; Boucher et al., 2013). The Brazilian National Institute for Space Research (INPE) revealed that 4656 km^2 (465,600 ha) of Amazon rainforest was cleared over the 12 months between August 2011 and July 2012, compared with 27,772 km^2 (2,777,200 ha) in 2004. The Brazilian government claimed that this represented a 76% reduction since 2004. This remarkable achievement was close to the country's commitment to reduce deforestation in the Amazon region by 80% by 2020, thanks to the implementation of PPCDAm action plan since 2004 (Evans, 2013; Hargrave, 2012; WWF, 2015a; Tabuchi et al., 2017). It was further revealed that deforestation rates in Brazil plummeted from 2004 to 2012 with a small increase in 2007 since the launch of PPPDAm policies in 2004 (Evans, 2013).

3.6.1.2 *The Brazilian Amazon forest: The reality*

However, as World Wildlife Fund for Nature (WWF) pointed out succinctly, merely reducing deforestation in a given territory does not necessarily mean it is heading toward a context of sustainability (WWF, 2015a). In a new report, Antonio Nobre, researcher in the government's space institute, Earth System Science Center, revealed that in the past 20 years, the Amazon has lost 763,000 km^2 (76.3 million ha), an area twice the size of Germany. In addition, another 1.2 million km^2 (120 million ha) has been estimated as degraded by cutting below the canopy and by fire (Watts, 2014). A big gap seems to exist between the stringent forestry legislation and the often nonexistent enforcement.

As the Amazon watchdog group Imazon revealed, between August 2011 and July 2012, 78% of the timber harvesting in Brazil's largest timber producer, Para state, was illegal (Schiffman, 2015; Brito et al., 2005). Furthermore, despite the revision of the Forest Code (2012) as noted above, tree cover loss remained high, at an annual average between 2 to 3 million ha for all threshold levels, indicating high deforestation dynamics (Table 3.5; see also Table 3.4 for tree cover loss between 2001 and 2016).

TABLE 3.5   Tree cover loss in Brazil between 2012 and 2016.

	2012	2013	2014	2015	2016	Grand total	Annual average
Tree cover loss (>30% canopy cover) (ha)	2,918,633	1,945,506	2,693,134	2,222,772	5,378,844	15,158,890	3,031,778
Tree cover loss (>50% canopy cover) (ha)	2,579,923	1,712,477	2,368,690	1,955,191	4,980,661	13,596,942	2,719,388
Tree cover loss (>75% canopy cover) (ha)	2,109,060	1,408,858	1,922,331	1,591,479	4,397,952	11,429,679	2,285,936

Data from Hansen, M.C., et al., 2013b. Tree Cover Loss and Gain Area. University of Maryland, Google, USGS, and NASA. Accessed through Global Forest Watch on May 2018. Available from: <https://www.globalforestwatch.org/countries/overview>.

As reported by the Brazilian NGO Imazon, one of the major causes of deforestation in the Brazilian Amazon was cattle ranging. It was further exposed that 88% of deforestation that occurred in the region between 2010 and 2015 was within the "zone of influence" or the total pasture area of the 128 slaughterhouses that process 93% of cattle raised in the region. Furthermore it is worth noting that two-thirds of the Amazon's deforested area has been turned into pastures. By 2016 the region's cattle population rose to 85 million, compared with the human population of 25 million. It is claimed that pasture is the largest source of deforestation in Brazil (Pacheco, 2017).

Another major activity that has resulted in increasing deforestation in the region is mining. It was estimated that this extractive industry resulted in roughly 10% of the deforestation in the Brazilian Amazon between 2005 and 2015. It was reported that 11,670 km^2 (1,167,000 ha) an area twice the size of the state of Delaware, of forest loss was attributable to mining activities including infrastructure built to support mineral extraction and transport during that period (Sullivan, 2017).

It does appear that with the new leader, President Jair Bolsonaro, elected into power in October 2018, the fate of the Amazon forest is at the mercy of his rigid antienvironmental stance. As a matter of fact, President Bolsonaro, a far-right individualistic utilitarian leader, has an extraordinarily strong tendency to view environmental protection or forest preservation as an obstacle to economic growth and prosperity. Hence, driven aggressively by egoistic impulses, he has vowed to dismantle the existing instrumental barriers to resource use to open up the Amazon green treasures for commercial exploitation. With the government's strong determination to scale back efforts to fight illegal logging, ranching, and mining, the Amazon forest has experienced destruction at an increasingly rapid rate (see, e.g., Londoño, 2018; Sengupta, 2018; Casado and Londoño, 2019). It has been estimated that Brazil lost about 1.35 million ha of primary rainforest in 2018, the largest loss in the world (Weisse and Goldman, 2019).

When President Jair Bolsonaro took office in January 2019, the Brazilian Amazon had lost more than 3444 km^2 or 344,400 ha of forest cover (Casado and Londoño, 2019). In early 2019, it was estimated that at least 125,000 ha of the Brazilian Amazon—the equivalent to 172,000 soccer fields—were cleared, and then burned in August (MAAP, 2019; Mendonça, 2019; see also, de Bolle, 2019).

Based on the data released by Brazil's INPE, deforestation accelerated by more than 60% in June 2019 over the same period last year. Some 769.1 km^2 (76,910 ha) were lost in that month compared with 488.4 km^2 (48,840 ha) that were vanished in June 2018. This equates to an area larger than the size of one-and-a-half soccer fields being destroyed every minute of every day (Cotovio, 2019). Worse yet, the latest INPE's data revealed that in November 2019, the rate of deforestation in Brazil surged by 104% to 563 km^2 (56,300 ha) compared with the same month in 2018—the highest number for any November since 2015 (Phys.org, 2019; RTE, 2019). For the first 11 months in 2019, a total area of 8974.3 km^2 (897,430 ha) of forest were destroyed (Phys.org, 2019; RTE, 2019). In the past 10 years, the Amazon has lost 24,000 square miles (62,160 km^2 or 6,215,972 ha) of its forest, equivalent to 8.4 million football pitches or 10.3 million American football fields. Mile upon mile of the Amazon rainforest has been cleared for a wide range of commercial developments, including cattle ranching, logging, and the palm oil industry (Vittert, 2019).

It must be admitted that the all-pervasive instrumental view of nature, that is, treating forests as an exclusive means to serve human interests, if left unsuppressed, will indubitably put the Amazon green lung in the throes of an unending series of episodes of violent destruction. This will be the subject of interest in the second-half of the book.

3.6.1.3 Argentina, Bolivia, and Paraguay: The political wills of forest protection

Other South American countries such as Argentina, Bolivia, and Paraguay have also put in place various forestry laws and policies to promote sustainable resource use and management. The three countries are also members of the FAO-LACFC which, as noted above, serves as a forum of discussion for the formulation of sustainable forestry policies to combat deforestation. All the countries have also reported on the progress made on the implementation of the Non-Legally Binding Instrument on all types Forests and achievement of the Global Objectives on Forests at the various sessions of the UNFF. These include for example, the Argentina National Report UNFF-11, the Bolivia National Report UNFF-11 and the Paraguay National Report UNFF-11 (United Nations, 2015d).

In promoting sustainable forestry management in line with UNFF initiatives or international forestry agreement, Argentina emphasized strengthening of national forest law at the Seventh Session of the United Nations Forum on Forests held in New York in 2007 (United Nations, 2007). Among the strategic steps undertaken by the country in this respect was embracing the model forest (MF) concept initiated in Canada in 1991, which involved stakeholders' participation in advancing forest culture and promoting SFM (Hall et al., 2014). The Argentine MF concept involves stakeholders from the public sector, farmers, academia, private sector, civil society, grassroots organizations, and indigenous communities. It aims at developing sustainable forest landscape planning and local forest-based development (Gabay et al., 2014).

The SFM concept was further reinforced by the enactment of Law No. 26331, Minimum Standards of Environmental Protection for Native Forests in 2007, by providing incentives and by prescribing mandatory categorization according to a set of criteria in association with the integration of the three pillars of sustainable development, namely, social, economic and environmental dimensions of resource use (Hall et al., 2014; Gabay et al., 2014). Law No. 26331, often referred to as the Forest Law, prescribes a set of Territorial Regulation of Native Forests [*Ordenamento Territorial Bosques Nativos*, (OTBN)] which requires approval by the province concerned before forest cutting may be conducted. It is explicitly stipulated that the OTBN must include zoning to designate areas into one of the following categories:

- Category I (red): high conservation value (no deforestation allowed).
- Category II (yellow): medium conservation value (sustainable use, tourism, research allowed).
- Category III (green): low conservation value (deforestation and productive activities allowed) (INCLUDEPROJECT, 2017).

Bolivia is among the countries in South America with some of the largest areas of tropical forest. It has about 57 million ha of forest in the country (FAO, 2011). The first serious but failed attempt at regulating the forestry sector took place in 1954. The Ministry of Peasantry and Agricultural Issues (currently known as the Forest Development Center) was established two decades later to revive forest management and controlling efforts. In 1974 the First Forest Law was enacted, followed by the creation of a set of regulatory controlling measures. Because of its ineffectiveness, a new Forest Law was enacted in 1996 to promote a new institutional framework for SFM. The new Forest Management Plans (FMPs) were created to regulate commercial logging activities by imposing various restrictive requirements on the extraction of timber resources (Cardona et al., 2014). The Law also established a semiautonomous administrative agency, the Superintendencia Forestal, or Forest Superintendency, to oversee that the restrictions imposed are being observed (Fredericksen et al., 2003).

Meanwhile, in Paraguay, the government has in place the Forest Law 422 (1973) and its later Resolution 11681 (1975) to regulate forest exploitation. The law stipulates that 25% of all land should remain under forest. However, the Forest Law opened the door to unrestrained forest exploitation by providing the option to transfer the remaining 25% of the forest reserves to other interested buyers who could then deforest them by 75% (Mansourian et al., 2014). Following the Rio Conference in 1992, the government launched a series of environmental reforms with support from the German Development Agency, resulting in the establishment of Sector Guidelines for a National Policy in Environment and Natural Resources in 1996. It also generated a National Strategy for Protection of Natural Resources and Environment that included a proposal for the National Environmental Policy and a proposed law for a National Environmental System (SISNAM) comprising a National Environmental Council (CONAM, a consultative policy entity) and the Ministry of Environment (regulatory institution) to promote sustainable resource management (Catterson and Fragano, 2004).

A moratorium, Forest Conversion Moratorium or "Zero Deforestation Law" was established in 2004, making it illegal to clear any forested land in eastern Paraguay (Hutchison and Aquino, 2011). The moratorium has been renewed twice until 2018 (Mansourian et al., 2014). Also the Forest Law (1973) was revised with one of the conditions to compel all rural landowners with more than 20 ha in forest areas to preserve at least 25% of their natural forest area (UNDP, 2017). However, unclear wording, alongside loose definitions of what constituted forested areas has rendered the Law being open to interpretation, even as, at the same time, it was rarely observed (UNDP, 2017). In 2008 the National Forest Institute (INFONA), replacing the former National Forest Service, was established to undertake the responsibility of implementing the Forest Law (FAO et al., 2010). Other laws, such as the Law to Promote Forestation and Reforestation, were also enacted in 1995 to strengthen conservation measures.

3.6.1.4 The state of forest in Argentina, Bolivia, and Paraguay

Despite the above environmental efforts, as reflected in Table 3.4, Argentina, Paraguay, and Bolivia exhibit high deforestation dynamics. This is basically linked to expanding agroindustry supplying global commodities markets, especially the monocultural expansion of corn, and soybean cultivation, extensive logging activities, and large-scale commercial ranching (see, e.g., Sizer et al., 2013; Muller et al., 2014; UNDP, 2017). As in Brazil, conversion to pasture is the largest cause of deforestation in Bolivia (Pacheco, 2017). Another major cause of deforestation in Bolivia is agricultural conversion of land to soy cultivation which increased more than 500% from 1991 to 3.8 million ha in 2013 (Tabuchi et al., 2017).

Against this backdrop, it may be pointed out that Argentina has been ticked off as the ninth worse country in deforestation by the United Nations. It lost 76,810 km^2 (7.68 million ha) from 1990 to 2015 (McCay, 2017). Conversion of land for soy production and intensive cattle ranging are the main causes of deforestation in the country. Even the government of Córdoba city together with the agribusiness entrepreneurs have been encroaching into the remaining 3% of the frontier forest for soy production (GJEP, 2017). Satellite images revealed a clear link between areas planted with soybean and the deforestation of native Argentinean forest (Guidi, 2016; Byrne, 2018). Between 1998 and 2006, the rate of deforestation was about 250,000 ha/year and almost 80% of this forest loss took place in the northeastern part of the country (Petras and Veltmeyer, 2014; Guidi, 2016).

Paraguay has some of the most significant forest losses compared with global averages (World Bank, 2017; Yousefi et al., 2018). Demonstrably the Eastern region of Paraguay had its original forest cover of 8.8 million ha in the 1940s reduced to 3.5 million ha, and by the 2000s, the forest cover was estimated at 2.1 million ha, or 21% that of the 1940s, due to development pressures (FAO et al., 2010). Furthermore overall, Paraguay lost almost 80% of its intact forest landscape between 2000 and 2013 (Erickson-Davis, 2018). Today, only about 7% of the original forest remains (World Bank, 2017).

In this regard, it is worth revealing the Gran Chaco, the second largest South American forest, spanning Argentina, Paraguay, Bolivia, and Brazil has turned into one of the most deforested areas on the planet. This is partly driven by agricultural expansion to the forest frontier, and hunting. It is also partly driven by a decree issued by the former Paraguayan President Horacio Cartes in 2017 which allowed the landowners to clear all the forest on their property. Shockingly in June 2018 alone, 34,000 ha of the forest were cleared—an area nearly twice the size of Buenos Aires. Between 2010 and 2018, more than 2.9 million ha of Gran Chaco forest were depleted. It was reported that 80% of Chaco deforestation during this period took place in Argentina (Chisleanschi, 2019). All in all, more than 8 million ha of this high conservation-value forest ecosystem were destroyed over just a dozen years (Yousefi et al., 2018).

In Bolivia, the rate of deforestation increased from about 148, 000 ha/year in the 1990s to some 270,000 ha/year in 2000s. On average, some 350,000 ha of land has been deforested annually for agriculture since 2011 (Tabuchi et al., 2017; Hays, 2018). To compound the problem, there are relatively few forest protections in Bolivia. Worse yet, the Forestry and Land Authority is entrusted with the potentially conflicting roles of regulating land use, forestry and agriculture, and issuing concessions for logging and farming. The government has also planned to convert some 5.6 million more ha of forest by 2025 for agriculture expansion (Tabuchi et al., 2017).

The rate of deforestation is expected to worsen in the future as the Bolivian government has planned to construct a new 190-mile road cutting through one of the most iconic and biologically diverse protected rainforests in the country. Indeed, it has been reported that the Isiboro-Sécure National Park and Indigenous Territory has been subject to alarming levels of deforestation within its borders for many years. It lost more than 46,000 ha of forest from

2000 to 2014 (Universitat Autònoma de Barcelona, 2018). Furthermore large agribusinesses such as the American soy companies, Cargill and Bunge, have been bulldozing and burning thousands of hectares of the highly diversified Gran Chac ecosystem (Yousefi et al., 2018).

It is thus clear from the above analysis corroborated by Hansen's deforestation dynamics data, forest destruction in the South American region has generally been recklessly unsustainable, destructive, and at times irreversible, fundamentally due to the human individualistic quest for economic gain or material benefit.

3.6.2 Africa: Democratic Republic of Congo

3.6.2.1 Democratic Republic of Congo forest protection initiatives

The conflict-torn Democratic Republic of Congo (DRC, formerly Zaire) is endowed with roughly 60% of the Congo rainforest, making it the second largest contiguous tract of tropical forests in the world after the Amazon (Gaworecki, 2017). Prior to 2002, forestry institutions of any kind were essentially nonexistent (Counsell, 2006). More than 20 million ha of logging titles were issued in the DRC (Greenpeace, 2007). However, because of war, violence, unrest and economic instability, it has fundamentally avoided the massive-scale of forest destruction that has taken place in other tropical rainforest countries such as Brazil and Indonesia as discussed above. Until 2002, forest management was governed by an obsolete colonial decree of April 1949 which was replaced by a technical paper called "The Logger's Guide." However, this guide which focused on the timber industry with no provision for preservation of forest had no clear legal status (Debroux et al., 2007). In view of this, the forests were under immense human pressure during and following the wars of 1996–2001 when, as reflected by the issuance of felling permits during and following the war, something in the order of 40 million ha of felling "permits" were issued (Counsell, 2006).

However, at the sixth session of the United Nations Forum on Forests, the Gabon Minister for Forests, Fisheries and National Parks, speaking on behalf of the Commission on Forests of the Congo Basin, acknowledged the importance of tropical forests in maintaining the global environmental balance. It is further revealed that the Commission was striving to develop instruments for the conservation and sustainable management of the Central African forest ecosystem, the world's second largest after the Amazon (United Nations, 2006b). With this perspective, during the past decade, the DRC has embarked on a process of legal review of old forest deeds and modernization of legal instruments, and put in place legal machinery to manage forest resources sustainably.

To ensure greater transparency, the DRC recruited an independent observer, Resource Extraction Monitoring (REM) for forest exploitation operations and a specialized company, *Société Générale de Surveillance* (SGS) to establish a strong control system to oversee timber extraction and marketing (Mpoyi et al., 2013). In 2002 it adopted the Forest Code which contained a set of new forest policies developed in the 1990s to govern forestry management in association with industrial timber production, nature conservation and community use. According to the World Bank inspection panel, the Code "introduces innovations such as traditional users' rights, including those of indigenous peoples; contributions to rural development; enhancement of the rights of local communities; and transparent allocation of future logging rights; ... [and] ... serves as a good basis for improving forest management" (World Bank Inspection Panel, 2006).

One of the most important provisions of the Code is the classification of forests according to their priority uses as contained in Articles 10−23. Accordingly forests are divided into three main categories:

1. "gazetted forests," which are primarily devoted to biodiversity conservation with the objective of conserving at least 15% of the national territory's total area (Article 14),
2. "permanent production forests" which are fundamentally for social and economic development, and
3. "protected forests" that are basically for sustainable production of timber or other forest goods/services (Debroux et al., 2007).

Other important SFM provisions are:

1. sustainable timber harvesting and biodiversity conservation: Article 100 of the Code explicitly stipulates that loggers must comply with the legal provisions pertaining to nature protection, hunting, and fishing;
2. transparency in allocation of logging rights; and
3. the Code also recognizes community-based management and traditional user rights (Debroux et al., 2007).

Some important achievements have been made since 2002 including the cancelation of 25.5 million ha of noncompliant logging concessions, and the establishment of a logging moratorium by ministerial order in May 2002 to avoid having new concessions allocated too quickly and inappropriately (World Bank Inspection Panel, 2006; Debroux et al., 2007).

Although there are various shortcomings of the Code including ambiguous terminology subject to misinterpretation, the failure to provide a specific operational framework for loggers and small-scale companies, and the absence of implementation decrees, among others (Debroux et al., 2007), it still constitutes the most important framework to date to guide sustainable forestry management in the country. This leads to the important question of the extent to which the Code has been able to promote SFM in the DRC.

3.6.2.2 Democratic Republic of Congo: The state of forests

An analysis of the trend of deforestation in DRC seems to indicate that its forest cover is fast disappearing. The Landsat time-series data show that between 2000 and 2010, the country lost 3,711,800 ha of its forest cover. The loss of forest covers in the country increased by 13.8% between the 2000−05 and 2005−10 intervals (Potapov et al., 2012; see also, Harris et al., 2017). However, the DRC's Forest Reference Emission Level (FREL) submitted to UNFCCC states that between 2000 and 2010, the DRC lost about 6.4 million ha, while between 2010 and 2014, it further lost about 7 million ha, an increase of almost 11% in the rate of deforestation. This indicates that for the period 2000−14, a total of more than 13 million ha of forest has been lost, meaning almost 1 million ha every year (CAFI, 2018). In 2018 the country lost about 481,000 ha of tropical primary forest—the second largest area loss of any country on Earth after Brazil (Weisse and Goldman, 2019).

Deforestation is mainly driven by small-scale subsistence agriculture such as slash-and-burn farming, mining operations, charcoal production, poaching, urbanization, and expansion of infrastructure. Industrial forest exploitation concentrated in a few provinces such as Equateur, Bandundu, Orientale, and Bas-Congo is increasingly becoming an important cause of rapid forest decline in DRC (Counsell, 2006).

More importantly underlying the unsustainable forestry practices in DRC is institutional failure. It appears that even as the logging moratorium was established, it was immediately violated. For example, some 15 million ha were reportedly exchanged or relocated, giving rise to a net increase of 2.4 million ha of area under concessions or permits although overall, the total area under concessions or permits has decreased from 43.5 million ha to 20.4 million ha since 2002 (Debroux et al., 2007). However, based on a "legal review" carried out by the government of all existing industrial titles, 15 illegal titles canceled in 2009 were reinstated in 2011 while cancelation was enforced only on the dormant titles (REDD-Monitor, 2016). In March 2016, the Minister of Environment, Nature Conservation and Sustainable Development indicated that "measures are underway" to lift the moratorium on the allocation of new industrial logging concessions and this is expected to result in uncontrolled timber harvesting in the name of development (REDD-Monitor, 2016; Greenpeace, 2017).

3.6.3 Indonesia: National forest policies

The United Nations' call for environmental protection and sustainable resource management has received significant attention among countries in Southeast Asia, including Indonesia, Malaysia, Thailand, the Philippines, and Vietnam. In particular, the World Conservation Strategy, the CBD, Rio Declaration on Environment, and Agenda 21 had an immense impact in these countries in streamlining their environmental protection and controlling frameworks to preserve and protect their natural environment (Choy, 2015a,b, 2016a, 2018). For example, Indonesia had established a well-defined forestry control and management framework as early as 1950, even before the Stockholm Conference, to sustainably manage its densely forested region covering some 148 million ha. This represents about 77% of its land surface area with the Kalimantan region having the largest coverage at about 47.5 million ha (FWI–GFW, 2002). In the 1950s, the "forest" category in Indonesia also included plantations of estate crops such as tea, coffee, and rubber, but these constituted only a small percentage of the total forest cover (FWI–GFW, 2002).

Starting from the Basic Forestry Law (Law No. 5 of 1967), Indonesia has been committed to managing these forest resources sustainably. The Basic Law was replaced by Forestry Law No. 41 of 1999 which includes some resource-orientation principles. The new enactment divides forests into three categories, namely, Conservation Forests, Protection Forests, and Production Forests for sustainable management purposes (Republic of Indonesia, 1999). Accordingly the Ministry of Forestry is empowered to determine and manage the national forest areas reported in 2011 to cover approximately 134 million ha (about 70% of the land surface) (Ardiansyah et al., 2015). The government also issued the Presidential Instruction (PI) No. 4 of 2005 to combat illegal logging. Other legal instruments which serve to strengthen sustainable resource management include the Environmental Protection and Management Law (2009) and Law No. 18 of 2013 on the Prevention and Eradication of Forest Degradation (WRI, 2016).

In addition, the government led by the then President Susilo Bambang Yudhoyono issued the PI No. 10/2011, Regarding the Suspension of Granting New Licenses and Improvement of Natural Primary Forest and Peatland Governance on May 20, 2011. The PI, popularly known as "the Moratorium" suspends the issuance of new harvesting licenses in primary forest and peatland forest covering around 66 million ha (Wells and Paoli, 2011; Diela, 2019). The moratorium is part of a broader $US1 billion Indonesia-Norway partnership to reduce greenhouse gas emissions from deforestation and degradation (known as REDD +). The moratorium prohibits the conversion of these Presidential Decree-protected areas for commercial uses such as for oil palm or pulpwood and logging concessions covering between 64 and 72 million ha of primary forest and peatland (Stolle and Gingold, 2011). Based on the data released by the Ministry of Environment and Forestry, the government claimed that the rate of deforestation in areas under the protection of the moratorium has decreased by 38% between 2011 and 2018 (Nicholas Jong, 2019a).

3.6.3.1 Indonesia: The state of forests

Despite this, Indonesia recorded one of the world's highest rates of increase in forest loss at 102,000 ha/year, an increase from under 1 million ha/year between 2000 and 2003 to over 2 million ha/year between 2011 and 2012 (Hansen et al., 2013a; see also, Harris et al., 2017). It is also the world's second country, after Brazil, with the highest annual net loss of forest (FAO, 2015). In the 1970s during the Stockholm period, deforestation became a real concern in Indonesia when large-scale commercial logging concessions were established for the first time which directly or indirectly led to the reduction of its forest cover by 27% to 119 million ha in 1985 (FWI−GFW, 2002). In a related note, FAO estimated that between 1985 and 1997, Indonesia lost more than 20 million ha of its forest cover, representing some 17% of the forest area in 1985. Its forest cover further decreased to 91 million ha in 2015 covering about 49.8% of its land area (FAO, 2015).

In the last half century, Indonesia lost more than 74 million ha of rainforest—an area nearly twice the size of Japan—through logging, burning, or degradation (Diela, 2019; Fig. 3.2). Monocultural land conversion especially oil palm plantation development is the biggest cause of deforestation and forest fires in Indonesia (see, e.g., Butler, 2013a; Choy, 2015a, 2016a).

Even those forests covered under the moratorium were not spared from human exploitation. For instance, in the first 7 years after the moratorium came into force in 2011, about 12,000 km^2 (1.2 million ha) of these Presidential Decree-protected areas were decimated at an average annual deforestation rate of 1370 km^2 137,000 ha. This is higher than the average of 970 km^2 or 97,000 ha prior to the enforcement of the moratorium. In addition, nearly a third of the 34,000 km^2 or 3.4 million ha of the forests and land burned between 2015 and 2018 were embedded in the "protected zones" mainly in the provinces of Central Kalimantan, Papua, South Sumatra and Riau (Nicholas Jong, 2019b). For example, Kalimantan recorded the highest forest loss within the moratorium area in 2015 (69,000 ha), followed by Sumatra (39,000 ha) and Papua (25,000 ha) (Wijaya et al., 2017).

In 2015 alone, more than 2.6 million ha of forest were burned, more than twice the size of Qatar (about 1.16 million ha). Between 2015 and 2018, some 3.4 million ha of land were burned at least once, greater than the area of the US state of Maryland (about 3.2 million ha) (Greenpeace, 2019). The burning of forests continued into the year 2019, when an estimated 857,000 ha of land were burned, of which 227,000 ha were peatland (Munthe and Nangoy, 2019; Greenpeace, 2019). This changing land use practice affects not only the ecological resilience of the regional ecosystems including dramatic global coral bleaching, but also aggravates global warming.

FIGURE 3.2   Massive burning of forest in Indonesia for monocultural land conversion.
Note: Indonesian farmers burn extensive tracts of forest and peatland every year to clear land for monocultural expansion especially oil palm plantation development, accelerating the rate of deforestation in Indonesia. *Source: Reproduced with permission from Nicholas Jong, H., 2019a. Indonesian court fines palm oil firm $18.5m over forest fires in 2015. Mongabay.com. <https://news.mongabay.com/2019/10/palm-oil-indonesia-arjuna-utama-sawit-musim-mas-forest-fires/> (accessed 09.12.19.).*

For example, the Indonesian forest fires in 2019 have emitted in at least 708 million tonnes of heat-trapping greenhouse gases (GHGs) into the atmosphere—largely as a result of burning of carbon-rich peatlands. This is almost double the emissions of 366 million tonnes from the forest fires that swept through the Brazilian Amazon between January 1 and November 30 in the same year (Nicholas Jong, 2019c; Kahfi, 2019). The scale of warming impact of the fires is more than double of the United Kingdom's total annual greenhouse gas (GHG) emissions of 370 million tonnes (see, Table 3.12 in the subsequent section). This is critically large enough to have global warming consequences.

Furthermore the yearly recurring forest fires in Indonesia have also resulted in transboundary haze pollution stretching as far as to Malaysia, Singapore, Southern Thailand and certain parts of the Philippines, adversely affecting the health of millions of people (see, e.g., Gomez and Karmini, 2019, Fig. 3.3).

It is instructive to note that the moratorium only applies to primary natural forests. Once these forests are degraded and become secondary forests, permits can be issued for commercial exploitation. Also those primary forests which are not covered by the moratorium are vulnerable to commercial exploitation especially for oil palm plantation development. Worse yet, the moratorium boundaries are regularly redrawn to remove forest or peat areas that are of interest to plantation companies (Nicholas Jong, 2019a). Indeed, it was found that 45,000 km^2 (4.5 million ha) of forests and peatlands

FIGURE 3.3 Transboundary haze pollution from Indonesian forest fires.
Note: Forest fires induced by commercial land use practices have become a recurring annual problem in Indonesia. The large-scale forest fires which send massive volumes of smoke and thick smog into the atmosphere not only affect Indonesia but also spread across its neighboring countries including Malaysia, Brunei, and Singapore, causing acute transboundary haze pollutions. *Source: Courtesy NASA Earth Observatory, 2015. Smoke Blankets Indonesia. NASA Earth Observatory, Washington, DC. Available from: <https://earthobservatory.nasa.gov/images/86681/smoke-blankets-indonesia>.*

had been removed from the map since 2011. It is also worth noting that, from 2011 to 2014, the then President Yudhoyono had given out numerous concessions covering 164,000 km^2 (16.4 million ha—an area nearly the size of Florida) out of these "protected land and forests" for various commercial uses. Under the current Joko Widodo administration, permits covering 17,000 km^2 (1.7 million ha) were issued for various commercial activities, especially oil palm plantation expansion (Nicholas Jong, 2019a; see also, Diela, 2019).

However, after claiming that the moratorium has been effective in slowing deforestation, the current Indonesian leader, President Joko Widodo, made it a permanent forest control and management instrument in August 2019 (Nicholas Jong, 2019a). The moratorium, expected to remain in force for a maximum of 3 years, aims to withhold the issuance of permits to clear a combined 16,000 km^2 (1.6 million ha) of forest areas for oil palm plantation expansion. A massive review of existing licenses is intended to be conducted to ascertain their legal status (Mongabay, 2018; Nicholas Jong, 2019d).

Although it may still be too early to judge the effectiveness of the newly reinforced palm oil moratorium in freezing the issuance of oil palm plantation permits, a preliminary examination should reflect that it is unlikely to manifest as a compelling instrument to constrain further forest destruction for land clearance for oil palm plantation expansion due to the following reasons:

1. Lack of transparency: an absence of progress reports, or data and information disclosure regarding the number of permits that have actually been issued before the freeze makes it difficult to assess its effectiveness.
2. Lack of sanctions: the moratorium, lacking the necessary obligatory force, is unlikely to be effective in curtailing developers; the individualistic utilitarian economic agents especially will not submit to the rule of sustainable land use practices.
3. Lack of lasting quality: the moratorium only covers a maximum period of 3 years. Due to its short life-span, it is unlikely to be effective in halting further forest destruction as the government will have a free hand to issue more permits upon the lapse of the moratorium to maintain and sustain its status as the global leading palm oil supplier.

Indeed, it may be remarked in light of the above that the Indonesian government has scaled back protection for some of the world's most important tropical forests because of budget cuts due to the Covid-19 pandemic. Thus, the risk for an impending episode of destructive forest fires is almost inevitable. In fact, the smog-belching forest fires have already started to break out in parts of Indonesia again as the dry season gets underway in July 2020, threatening to emit more greenhouse gases into the atmosphere.

To have an overview of the state of the forest in Indonesia, we may now turn to our discussion of the deforestation trend in Borneo Island as a whole. To start with, Borneo Island consists of Kalimantan in Indonesia, the states of Sarawak and Sabah in Malaysia and Brunei. The Kalimantan territory of Indonesia accounts for two-thirds of its land area (539,460 km^2) while the Borneo states of Sarawak and Sabah in Malaysia comprise 26.7% (197,000 km^2), and Brunei 0.6% (5570 km^2). As shown in Fig. 3.4, in 1950, the island was

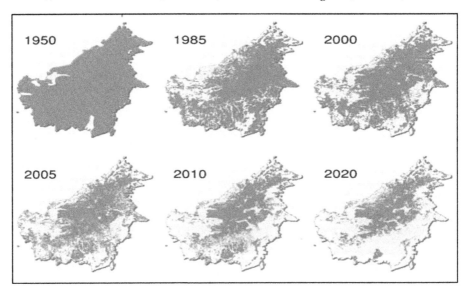

FIGURE 3.4 Deforestation in Borneo Island. *Reproduced with permission from, Ahlenius, H., 2006. Extent of deforestation in Borneo 1950–2005, and projection towards 2020. UNEP/GRID-Arendal. Available from: <http://www.grida.no/resources/8324>.*

almost completely covered with forest. Between 1985 and 2005, the island lost an average of 850,000 ha of forest annually and if left unchecked, its forest cover is expected to drop to less than a third by 2020 (WWF, 2005). The forest cover in Kalimantan decreased from 47.5 million ha in 1950 to 39.98 million ha in 1985 and further reduced to 24.35 million ha in 2005 and 20.44 million in 2010. If the prevailing rate of depletion remains unchecked, the forest cover is projected to be reduced to 12.63 million ha by 2020 (FWI—GFW, 2002; WWF, 2005). Only half of its forest cover remains today, down from 75% in the mid-1980s (UNEP, 2019a).

It is worth reiterating that the main causes of deforestation in Indonesia as a whole are: (1) uncontrolled timber harvesting and illegal logging activities; (2) extensive forest conversion into oil palm and rubber plantations (oil palm plantation development is one of the biggest threats to the remaining forests in Indonesia); and (3) forest fire. For example, from 1982 to 1983, about 3.7 million ha of forest were burned for agricultural development in Kalimantan, and in 1997/98, 6.5 million ha of forests in Kalimantan were affected by forest fires basically caused by unsustainable human practices (Butler, 2006a; see also, Choy, 2015a, 2016a, 2018b; UNEP, 2019a).

Also lack of transparency in decision-making on forest resource management has often led to the proliferation of illegal activities and corruption. Nonenforcement of the law on politically well-connected concessionaires who violated the terms of concessions also contributed to rapid deforestation in the 1990s, and even Indonesia's most celebrated national parks, the Gunung Leuser, Tanjung Puting, and the Kerinci Seblat, have not been spared from human economic invasion. It was reported that during those years, some 70% of the forest was harvested illegally and the unrestrained extraction of forest was so extensive that it wiped out at least 65 million ha of forest, which is 2.2 times the size of Italy (Contreras-Hermosilla and Fay, 2005; see also Che Yeom and Chandrasekharan, 2002).

The quest for economic growth at the expense of the environment also subjects the forest ecosystems to irreversible destruction. For example, in an attempt to realize its overarching goal of becoming a developed country by 2025, the government unveiled an ambitious plan, the Master Plan for the Acceleration and Expansion of Indonesian Economic Development (MP3EI) to direct the future course of Indonesia's development. One of the plans unveiled in this development strategy was the "Center for Production and Processing of National Mining and Energy Reserves" in Kalimantan (CMEA, 2011). One of the potential projects designated under this specific economic corridor was the indoMet coal project (ICP)

The project aims to exploit coal deposits found in East and Central Kalimantan, covering a combined area of 350,000 ha—over four-and-a-half times the size of Greater Jakarta or more than twice the size of Greater London (Denton, 2014). The coal project itself is located in the heavily forested Upper Barito Basin. The planned open-cut coal mining method as proposed under the coal project involves extensive forest clearing and habitat destruction that is environmentally destructive. The cutting down of forests involved in the construction of a coal railway infrastructure to improve extraction and production process will only aggravate the environmental quagmire (Choy, 2018). For the states of forest in Sarawak and Sabah, and Brunei, see, Sections 3.6.4.1, 3.12.12 and 3.12.13.

It is evidently clear by now that the forces of instrumentalism and materialism of sustainable resource use in Indonesia are so overwhelming that the controlling forces embodied in all the forest polices or decrees have not been able to restrain the individual utilitarian motive of uncontrolled resource exploitation.

3.6.4 Other Southeast Asian countries

3.6.4.1 Malaysia

Further to Section 3.6.3 above, in other regions in Southeast Asia, uncontrolled human economic activities also take a heavy toll on the health of their forest cover despite having in place various sustainable forestry management policies and legislation. Malaysia is an obvious case in point. Malaysia probably has one of the most advanced and sophisticated rainforest protection policies in developing Asia (Choy, 2015a, 2015b). Briefly noted, Malaysia consists of West Malaysia (with 11 states) and East Malaysia (with 2 states, namely, Sarawak and Sabah), separated from each other by the South China Sea by approximately 650 km.

Fundamentally there are three sets of forestry ordinances in Malaysia to manage forest resources sustainably. In Peninsular Malaysia, an Interim Forestry Policy was first formulated in 1952 and officially adopted as the National Forestry Policy (NFP) in 1978, and revised in 1992. The main purpose of the policy is to promote efficient use and sustainable management of forest resources for social, economic and environmental benefits. The National Forestry Act was enacted in 1984 and amended in 1993 as the National Forestry (Amendment) Act. The main purpose of the Act is to ensure effective implementation of the National Forestry Policy. Most of the amendments to the Act involve mandatory and higher fines for illegal logging practices.

To facilitate implementation of the National Forestry Act (1993), a document entitled Guidelines for the interpretation of classification of permanent forest reserves was created in 1993 which imposed stricter regulation in logging activities. The document also makes provisions for banning logging activities in soil protection forests, forest sanctuaries for wildlife, virgin jungles, reserved forests, forests for amenities, education forests and research forests (Kumari, 1995).

In Sarawak, the Statement of Forest Policy (1954) was created to permanently preserve the state's natural forests for economic and environmental reasons for the benefit of present and future generations. It also enacted the Forest Ordinance (1958, amended 1997) and Sarawak Forestry Corporation Ordinance (1995) to reinforce sustainable forestry use and management. In Sabah, the Forest Enactment 1968 (amended 1992), and Forest Rules (1969), and the Forests (Constitution of Forest Reserves and Amendment) Enactment 1984, constitute the key instruments to sustainable forestry management. Accordingly forest reserves are categorized for management purposes into seven classes, namely, Protected, Commercial, Domestic, Amenity, Mangrove, Virgin Jungle and Wildlife Reserve Forests. Other related policy instruments to strengthen or complement sustainable forestry management practices include the Sabah Forestry Policy which was adopted in 2005 to provide guidelines for the sustainable management of the state's forest resources. In addition, the Sabah Forestry Development Authority Enactment (1981) sets out the powers of the Forestry Development Authority (the Authority) to regulate forest management.

Despite the above, however, unrestrained logging activities, extensive dam infrastructure development and oil palm plantation development have resulted in extensive forest destruction (Choy, 2015a,b, 2016a). Between 2000 and 2012 alone, Malaysia cleared roughly 4.7 million ha of forests, amounting to 14.4% of its forest cover existent in 2000, an area larger than the size of Denmark. This makes Malaysia the country with the highest rate of deforestation in the world during that period compared with other countries as shown in Fig. 3.5

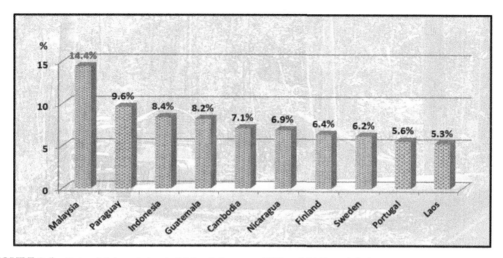

FIGURE 3.5 Rate of deforestation in Malaysia between 2000 and 2012: a global comparison.
Note: Compared with other countries such as Paraguay, Indonesia, and Nicaragua, the rate of forest depletion in Malaysia between 2000 and 2012 is the highest in the world. *Data from, Butler, R.A., 2013b. Malaysia Has the World's Highest Deforestation Rate, Reveals Google Forest Map. Mongabay.com. <https://news.mongabay.com/2013/11/malaysia-has-the-worlds-highest-deforestation-rate-reveals-google-forest-map/> (accessed 09.12.19.).*

(Butler, 2013b). Conversion of forests into oil palm plantations has been indiscriminately extensive, about 28% (318,000 ha) in Peninsular Malaysia, 48% in Sarawak (471,000 ha), and 62% in Sabah (714,000 ha) (Gunarso et al., 2013).

According to FAO, the rates of deforestation in Malaysia between the 1990−2000 period and 2000−05, have been accelerating faster than any tropical country in Southeast Asia. Based on the data from GFW, from 2001 to 2014, the total forest loss was about 5.6 million ha. However, to get into the dynamics of deforestation in the state of Sarawak is by no means easy due to government secrecy and lack of transparency in the state's commercial land and forest use practices, and the FAO statistics are unlikely to convey a real picture of its deforestation status.

Based on powerful remote sensing data, the rate of forest loss in the state appears to be increasingly rapid. According to new maps of industrial logging, oil palm plantation development, and planted forest concessions compiled and published by GFW, these concessions cover over half of Sarawak, often overlapping with sensitive intact forests that are being exploited and degraded at one of the highest rates in the world (Petersen et al., 2015) Oil palm plantation development which often precedes timber harvesting is the major cause of deforestation in Sarawak (Fig. 3.6).

For example, in the first quarter of 2014, the US NASA detected a surge in deforestation in Malaysia at an alarming rate of 150%. This is second only to that occurring in Bolivia (162%) and is significantly higher than that of Panama (123%), Ecuador (115%), Cambodia (89%), and Nigeria (63%) (Mongabay, 2014). Between 2005 and 2011, forests in Sarawak were depleted by more than 865,000 ha from 8.98 to 8.12 million ha, and during the same period, its peatland forests were reduced by 352,930 ha from 1,055,860 ha to 702,966 largely due to timber extraction and oil palm plantation development (Schrier-Uijl et al., 2013). It has been reported that total deforestation in Sarawak is 3.5 times as much as that for the

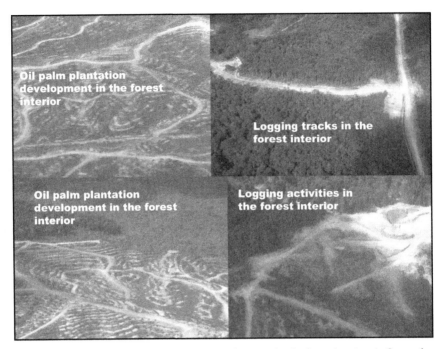

FIGURE 3.6   Deforestation and oil palm plantation development in the forest interiors in Sarawak.
Note: Photographs taken in the same geographical region by the author in 2009 on the way from Bintulu to Bario Highland by air. Based on aerial observation, extensive logging tracks have been developed in the forest interiors, facilitating the way for more forest exploitation.

whole of Asia, while deforestation of peat swamp forest is 11.7 times as much during the period between 2005 and 2010 (Wetlands International, 2011).

Furthermore one of the latest reports revealed that clear-cutting of 4400 ha of intact natural rainforest had commenced in December 2018 for a planned oil palm plantation in the vicinity of the Mulu National Park in Sarawak—a UNESCO World Heritage Site. By April 2019, it was uncovered that about a further 730 ha (16% of the total lease area of 4400 ha) had been cleared, resulting in widespread and destructive ecological impacts (Bruno Manser Fonds, 2019; Malaysiakini, 2019; Fig. 3.7).

In Peninsular Malaysia, although old growth forests are legally protected as permanent reserves, they may not be safe from human intervention. As a case in point, the Bikam Permanent Forest Reserve (400 ha) in the state of Perak was degazetted for logging and oil palm plantation development in the quest for economic gains. As a matter of fact, over a few years since 2009, over 9000 ha of the legally protected Permanent Forest Reserves in the state had been degazetted for timber production and other commercial activities (Choy, 2015a, 2016a). In the state of Kelantan, massive deforestation and illegal logging fundamentally attributed to the lack of good governance and transparency, bribery and cronyism, is wreaking havoc on the ecological integrity of its forest ecosystems (C4, 2015; Sarawak Report, 2018; Fig. 3.8).

Between 2001 and 2018, Kelantan lost around 28% of its tree cover. Tanah Merah and Gua Musang accounted for 71% of the tree cover loss during this period. The forest

FIGURE 3.7 Deforestation and planned oil palm plantation development near the UNESCP World Heritage Site in Mulu, Sarawak. *Reproduced with permission from, Bruno Manser Fonds, 2019. The Mulu Land Grab Results of a Fact-Finding Mission to Sarawak (Malaysia) on Palm Oil-Related Deforestation and a Land Conflict Near the UNESCO-Protected Gunung Mulu World Heritage Site. Bruno Manser Fonds, Basel, Switzerland. As appeared in Malaysiakini, 2019. Available from: <https://www.malaysiakini.com/news/463906>.*

destruction trend shows no sign of receding. In fact, it is worsening with plantation expansion and logging roads denuding large areas of the forest cover (Humphrey, 2019). This is despite having established various forestry polices and institutions, including the state Forestry Department, Land Office, and the Ministry of Natural Resources and Environment, to manage its forest resources. It is instructive to note seemingly in an attempt to avoid public or global scrutiny of its unsustainable land use practices, no information or statistics are available on the logging activities or the status of forest reserve on the Land Office website of Kelantan State (C4, 2015).

The irony is that, in Malaysia, many plantation development and expansion, and logging activities occur in the forest reserves, including those legally gazetted areas which, in practice, must be protected for their high conservation values. The legal or "protected" status of the forest reserves can be invalidated by the leaders in the name of "sustainable development" (read profit maximization).

FIGURE 3.8   Widespread and irreversible forest destruction caused by unrestrained logging activities and oil palm plantation development in Kelantan, Malaysia. *Reproduced with permission from, Koh Jun Lin, Malaysiakini, as appeared in Pillai, V., 2017, Orang Asli under threat as loggers 'skirt' court ruling. Malaysiakini. <https://www.malaysiaki-ni.com/news/387833> (accessed 09.12.19.).*

3.6.4.2 *The Philippines*

Triggered by decades of uncontrolled exploitation, in the two periods between 1934 and 1990 and between 1990 and 2000, the Philippines lost 10.9 million and 800,000 ha of its original forest, respectively (Rebugio et al., 2007). To put matter into perspective, in 1934, total forest cover in the Philippines was 17.18 million ha which accounts for 57.2% of the country's total area. By 2011, its forest cover had decreased substantially to 7.7 million ha, accounting for only 25.9% of the country's total area (Rebugio et al., 2007; see also, SEPO, 2015; Cabico, 2018; Fig. 3.9). It may be highlighted that the country lost about 80% of its remaining old-growth forests in the 1970s and 1980s. Rapid deforestation continued from 1990 to 1995, during which the country lost on average 3.5% of its remaining forest stand—the fastest rate in Southeast Asia during the period (Dauvergne, 2001). By 2000, the alarming rate of deforestation had left the country with only 7% of the original low land forests (Vesilind, 2002; PRB, 2006), while others put the figure at only 3% of the total land area (Bugayong, 2006; Dauvergne, 2001).

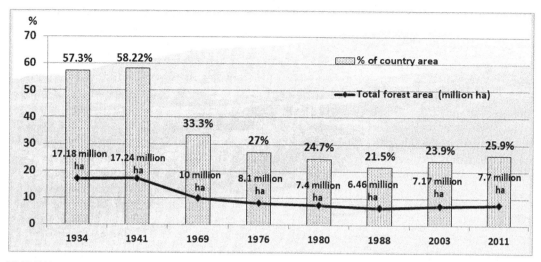

FIGURE 3.9 Forest conversion/lost in the Philippines since 1934. *Data from, Rebugio, L.L. et al., 2007. Forest restoration and rehabilitation in the Philippines. In: Don, K.L. (Ed.), Keep Asia Green. Volume 1: Southeast Asia. IUFRO World Series, Vienna, Austria.*

This is despite having in place the following sustainable resource regulatory control: (1) the enactment of the "Revised Forestry Code of the Philippines" (1975) to guide forest management, (2) the issuance of 20 policies on logging ban and moratorium imposed in over 46 provinces or nationwide to arrest indiscriminate and illegal logging practices for the past three decades, and (3) the enactment of the National Integrated Protected Areas System (NIPAS) Act of 1992 (Republic Act 7586) to conserve and to protect biological diverse areas from human exploitation (Bugayong, 2006).

However, after more than seven decades of persistent forest misappropriation, the government, to reverse the tide, implemented the National Greening Programme (NGP) in 2011. This forest rehabilitation programme aimed to increase forest cover by planting 1.5 billion trees covering 1.5 million ha of forest lands throughout the country from 2011 to 2016 (Cabico, 2018). As a result, according to FAO's 2015 Global FRA, the country experienced the fifth highest annual increase of forested area in the world from 2010 to 2015, restoring 240,000 ha/year (Manila Times, 2017; Cabico, 2018).

The programme, now called the Expanded National Greening Programme (E-NGP), has been extended to 2028 to cover an additional 7.1 million ha of unproductive, denuded, and degraded forestlands with the aim of sustaining the forest cover increase from the initial NGP (Jardeleza et al., 2019; see also, Manila Times, 2017; Cabico 2018). According to the Department of Environment and Natural Resources, over 125,000 ha of forests were rehabilitated in 2018. This brings to a total of 1.98 million ha placed under E-NGP which have been reforested since 2011 (DENR, 2018).

It needs to be pointed out that the past unsustainable land and forest use practices in the Philippines are irreversible. Although the present replanting efforts are laudable, and will, to a certain extent, help in the ecological restoration of its severely depleted forest landscape, its ecological system may not be able to revert to its former conditions (Holling sustainable conditions) and functions.

Unfortunately this inability to return to its former state is more certain given that the current greening programmes comprise the planting of a mixture of native species and, fast-growing and "biodiversity unfriendly" exotic species. It was reported that 25 million seedlings of exotic species have been planted against only 5 million seedlings of native species (Dimagiba, 2014). These include mahogany, a major invasive plant and "environmentally destructive" species capable of invading natural forests and retarding the growth of other plants. A mahogany plantation has been likened to a "green desert" in that its leaves are rarely browsed by animals (Baguinon et al., 2005). Thus it may well be that planting invasive exotic species is akin to a green cover for green assault on ecosystems.

It can be seen that the Philippine ecological quagmire serves as an important lesson to the rest of the Southeast Asian countries, in particular Indonesia and Malaysia, which have an extraordinarily strong tendency to instrumentally view nature, or land and forest resources, as a means to an end.

3.6.4.3 Thailand

Thailand has enacted the National Forest Park Act and the National Forest Reserve Act in the 1960s to promote sustainable forestry practices through the creation of protected forest reserves. It has also put in place the National Forest Conservation Act (1964) which declared 9,394,151 ha or about 59% of forestlands as national conserved forests with the aim to protect them from clearing and degradation (Ongprasert, 2011). The Forest Act (1941, amended in 1948, 1982, and 1989), National Park Act (1961, amended in 1992), and the Wild Animal Reservation and Protection Act (1960, amended in 1992) (Ongprasert, 2011), were also enacted to strengthen environmental control and sustainable resource use and practices and to prevent indiscriminate exploitation of forest resources.

Despite this, nonetheless, the country has lost nearly two-thirds of its original forest cover since 1961, falling below 23% for the first time in 1995 according to FAO estimates while the Thailand Royal Forest Department put the 1995 figure at 26% (Dauvergne, 2001; Elliott and Kuaraksa, 2008). Despite a ban on commercial logging since 1989, logging in primary forests continues and large areas of forest within existing national parks and wildlife sanctuaries have been degraded (Elliott and Kuaraksa, 2008). From 1990 to 1995, Thailand lost on average 2.6% or 329,000 ha of its remaining forests every year and only 5% of its original frontier forests remain (Dauvergne, 2001). Worst yet, between the periods 1997 and 2009, Thailand's forests declined by 43% (WWF, 2013).

Factors that contributed to the persistent deforestation trends in Thailand include (Choy, 2015a,b; Ministry of Natural Resources and Environment, Thailand, 2006; Stibig et al., 2007; Trisurat et al., 2011):

1. weak and ineffective law enforcement,
2. uncontrolled logging activities,
3. commercial or illegal exploitation of natural resources (e.g., valuable hardwood tree species, such as Siamese Rosewood, were also being harvested illegally for sale),
4. commercial agricultural expansion and promotion of cash crops on the highlands,
5. mega-dam infrastructure development,

6. promotion of the tourism industry,
7. deep-rooted corruption in forestry bureaucracy, and
8. natural resource perception of forests, and forest lands that are instrumentally valued for their economic benefit or commercial value, that is, for their timber value or agricultural potential (Leblond, 2014).

However, it is reported that the rate of deforestation has declined steadily since 2000 due to a structural shift of economic reforms away from forestry (Trisurat et al., 2019). The military junta's forest reclamation policy, launched in June 2014 via the National Council for Peace and Order (NCPO) 64/2557, had the goal of reversing the effects of past decades of deforestation in the country (ASEAN Post Team, 2018). In its 12th National Economic and Social Development Plan (2017–21), the government vowed that the proportion of land under forest should cover 40% of the country to protect and ensure the ecological integrity of the natural system (NESDB, 2017).

In spite of the above proactive measures taken during the period 2001–18, Thailand lost 1.9 million ha of tree cover, equivalent to a 9.6% decrease in tree cover since 2000, despite gaining 499,000 ha from 2001 to 2012. In 2018 alone Thailand lost 133,000 ha of natural forests (Global Forest Watch, 2019). The decrease in forest cover is to be expected due to poor enforcement of restriction policies on forest encroachment, wildlife poaching and subsistence or overexploitation, ineffective management, lack of coordination and conflict of interests among various departments, among others (see, e.g., Trisurat et al., 2019). The NCPO, for example, had utilized its powers to suppress environmental protection measures to benefit mega-project developments and industrial growth (ASEAN Post Team, 2018). Trees were also being logged from the national parks and wildlife sanctuaries due to ineffective enforcement (Tangwisutijit, 2018).

Ineffective weak policy enforcement is clearly evident in the management of the Pru Kaching National Reserve Forest, created in 1965 under the control of Thailand's Royal Forest Department. The protected forest reserve was later designated as "mangrove forest conservation area" under the responsibility of Thailand's Department of Marine and Coastal Resources (DMCR). However, a third of the 1.181 acres of the reserve had been encroached upon, cleared and replaced by oil palm plantations by smallholders. Between 2001 and 2015, about 20% of its tree cover was cleared for oil palm plantation development. Factors that contributed to this environmental quagmire are the lack of authority on the part of the Department of Marine and Coastal Resources to enforce the law against encroachers, lack of coordination among the different agencies, lack of official protection, and corruption (Stokes, 2017).

Another case in point is the Mae Wong National Park predicament. The national park is considered the largest contiguous tract of forest in all of Thailand and Southeast Asia. This nationally and globally precious protected green treasure has been facing a persistent threat of irreversible destruction. The elimination of this forest reserve is expected to result in the collapse of the forest's ecosystem, including the Huay Kha Khaeng Wildlife Sanctuary. Despite this, for the past 30 years, groups with vested interests have been persistently lobbying to construct a mega-dam, the Mae Wong Dam, within the conservation area, although this is illegal under Thai law (WWF, 2017a).

In April 2012, the Cabinet approved the project as a flagship project for flood prevention. This was met with a barrage of criticism from Thai civil society and international organizations. In 2014 the National Specialist Committee on Environmental Health Impact Assessment recommended to the National Environment Board that plans for the dam be dropped. In response, the government conceded that the dam should be redesigned and a new study would be conducted into the proposed dam's utility as a flood prevention mechanism (IUCN, 2015a).

The project returned on the agenda in 2016 after the Agricultural and Cooperatives Minister General Chatchai Sarikulya invoked special powers under Article 44 of Thailand's interim constitution to commence construction of the dam (The Nations, 2016a,b). The Article grants authorities in charge including the NCPO leaders with absolute power to give any order deemed necessary to execute key decisions for the sake of the reforms in any field. Nonetheless, the Central Administrative Court, after hearing the complaint of 151 people ruled against the project and ordered government to comply fully with the constitutions in relation to public consultations and an Environment and Health Impact Assessment before construction can be commenced (The Nations, 2016c). Despite this, the fate of the protected forest reserve is still vulnerable to human assault under the military rule of the country.

It is thus clear from the above that natural forests, although legally or officially accorded with protected status, are nevertheless vulnerable to human onslaught.

3.6.4.4 *Vietnam*

In 1943 the forest cover of Vietnam was 14.3 million ha or 44% of the total land area (de Jong et al., 2006). When Vietnam ended the war with the United States in 1975, all forests were under the State's control and managed by the Ministry of Forestry at the national level. The State Forest Enterprise (SFE) was established to manage forest exploitation and plantation. Since early 1989, 413 SFEs were established for managing 6.3 million ha of forest land in the country (Nguyen, 2005). This marked the rapid decline of natural forests.

Between 1980 and 1990, Vietnam lost an average of 100,000 ha of forest/year (FSIV, 2009). Between 1976 and 1990, Vietnam's forests decreased on average 185,000 ha annually (Sunderlin and Huynh, 2005). By 1990, only 9.175 million ha covering about 28% of the country's total land area remained (Pham et al., 2012; de Jong et al., 2006, Drollette, 2013a,b). The remaining forests were largely made up of degraded forest or plantations, and only 1%–2% of its primary forests remained (Pham et al., 2012; de Jong et al., 2006). By 1986, almost half of the SFEs had run out of forest to exploit (Nguyen, 2005). Logging peaked in 1992, when 1.2 million m^3 of timber were logged in Vietnam's forests. Since then the volumes of timber extracted have been drastically reduced, largely because only 107 SFEs have any forest remaining to log (Castrén, 1999).

To address the acute forest resource depletion problem, the government imposed a series of logging and export bans, the first in 1990, banning the export of raw logs. The control was further strengthened in 1992 by banning the export of raw cut and sawn wood, and reducing logging quotas by 88%. In the following year, a total ban on the export of forest products was imposed and almost all the forests in the north of the country were declared closed for logging (Sikor, 1998) The export ban was temporarily eased periodically in 1993 and reinstated in 1995, replaced with an export quota of 80,000 m^3 in early 1998 but, again revived in March 1998 (Lang, 2001). In 1999, the Ministry of Forestry set the logging quotas at 300,000 m^3 per year—a halving of the 1997 official forest harvest (Waggener, 2001; Barney 2005; World Bank, 2010; Castrén, 1999). The government also passed the Law on

Forest Protection and Development in 1991 (amended 2004) involving the local people and different sectors in forest protection and development (Nguyen, 2008).

Despite these regulating and controlling measures, illegal logging remains rampant. It is estimated that about 1 million m^3 of wood is logged illegally every year (To and Sikor, 2008). Between 2002 and 2009, some 62,000 ha of forest were lost each year (Pham et al., 2012). One of the most prominent examples is the illegal logging of 53,000 m^3 of trees worth more than $US 1.5 million from the protected forests in Tanh Linh district in the southern province of Binh Thuan between 1993 and 1995 with the abetment of government officials (To and Sikor, 2008). Forest law violations have become a common phenomenon in the forest sector in Vietnam. In 2000, for example, over 3700 forest law violations were detected (Lang, 2001). According to the Ministry of Agriculture and Rural Development, roughly 35,000 forest violations were recorded in 2006 where illegal conversion of forests to other land uses accounts for the largest amount of forest loss, followed by overexploitation and illegal destruction (IBRD and World Bank, 2011). However, the World Bank reported that forest law violations ranged from 30,000 to 50,000 cases/year. Moreover, these represent only a fraction of forest violations actually committed as many cases go undetected or the number is underreported (Castrén, 1999; IBRD and World Bank, 2011).

The main causes of deforestation and degradation vary from region to region. For example, in the Northeastern region, forest loss is mainly due to agricultural expansion while in the Mekong Delta, forests have been converted to shrimp farms and aquaculture production (Pham et al., 2012). Until today, deforestation remains a major problem in the country. In the Central Highlands, 5.4 million ha of natural forests in Tay Nguyen have been seriously affected because of unsustainable human activities. As conceded by the Deputy Minister of Agriculture and Rural Development, Ha Cong Tuan, in 2014, some 870 ha of forest were illegally logged—an increase of 7.7% compared with the previous year, and 3157 ha were destroyed by fire—an increase of 173.1% over 2013 (Vietnam News, 2013, 2015). Local deforestation also commonly occurs in the Central Coastal Region and the Southeast Region. However, forest degradation spreads throughout the remaining natural forests across the country. It is reported that over two-thirds of natural forests were considered as low quality forests and the rich and closed-canopy forests constituted only 4.6% of the total forest area in Vietnam (Pham, 2013). The GFW estimated that from 2001 to 2014, the country lost 1.5 million ha of forest.

Due to lax attitudes toward law enforcement of the local authorities and limited forest protection, nearly 10,500 ha of natural forest in the Tây Nguyên (Central Highlands) province of Đắk Lắk have been illegally destroyed since 2015 (Viet Nam News, 2018). Also despite signed commitment of nonforest exploitation of the Nà Pen Forest in the northern province of Điện Biên by the local communities in early 2018, large tracts of centuries-old trees in the region have been illegally decimated for house construction and farm conversion. Ineffective forest management, lax law enforcement and corruption contributed to worsen the forest misappropriation (Viet Nam News, 2019). Indeed, corruption is one the major reasons for weak protections and slack enforcement of forestry law (Nash, 2019).

Given that 30,000−50,000 forest violations are reported every year in Vietnam (Humphrey, 2018), it can safely be deduced that the scale of deforestation caused by illegal and unsustainable forest exploitation is substantial. This is also compounded by the fact that Vietnam's policies on forest governance do not adequately meet principles of good governance. Many regulations are only in the form of guidelines which are unclear, ambiguous and uncertain (Hoang et al., 2017). It is worth noting that Vietnam's rate of primary forest loss has doubled since the 1990s (Butler, 2019).

3.7 Global forest conservation: A bleak picture

The subject of deforestation, especially in the tropics, has entered the international political forefront since Stockholm, thanks to the unrelenting efforts of the United Nations' global call for greater political commitment to forest conservation. As discussed above, there has been significant progress made in many countries, especially in the tropics, in developing law, forest policies, or national forest management action plans in accord with the United Nations' environmental agreement and guides to promote sound forest management and sustainable forest resource use.

There has also been a marked rise in cooperative international efforts to sustain the momentum of global forest conservation across the globe particularly in the tropical rainforest regions. Despite this, as reflected clearly from the above assessment, on the whole, global natural forests are diminishing at an unprecedented scale due to uncontrolled human activities, and the global rate of forest destruction is escalating every year.

The above findings are further confirmed by the latest Five-Year Assessment Report on the Progress on the New York Declaration on Forests released in 2019 which reveals that global efforts to halt further deforestation still sound hollow (NYDF Assessment Partners, 2019). The state of the world's forests is now worse than it was when a broad coalition of governments, companies and endorsed the New York Declaration on Forests (NYDF) in September 2014 with the aim to end deforestation by 2030. Worse yet, the Report further claimed that there is little evidence to indicate that the goal of halving the rate of loss of natural forests globally by 2020 is possible.

Forestlands across the world including Madagascar and the Democratic Republic of the Congo (DRC) in Africa; Brazil, Bolivia, Colombia, and Peru in Latin America; and Indonesia, Malaysia, Cambodia, and Papua New Guinea in Southeast Asia, continue to be converted to other commercial land uses such as cattle ranching, soy and oil palm plantation development, and timber harvesting. Also in the DRC, despite significant financial investments into REDD + Readiness, the average loss of primary forest in the country was more than doubled in the last 5 years. On average, an area of forest the size of the United Kingdom was lost every year between 2014 and 2018. To note in passing, REDD + is divided into three phases. Phase 1 is the REDD + Readiness which includes developing national strategies, policies, and other capacity building activities. Phase 2 (REDD + Implementation) covers the implementation of these strategies or policies while Phase 3 (REDD + Payment) consists of results-based payments for emission reductions.

This poses far-reaching implications for the environmental integrity of our planet especially in relation to biodiversity loss. This will be systematically discussed in the following sections.

3.8 Biodiversity conservation: United Nations' ecological conservation initiatives in retrospective

It is by now clear from the above evidence that global forest conservation since Stockholm has, on the whole, been failing. The anthropogenic pressure on the planet forest systems has far reaching repercussions on animal and plant species which are declining at a magnitude faster than their regeneration capacity, and the world biodiversity loss is no longer within the ecological threshold level. This could start to threaten much of the Earth's biocapacity to support long-term human survival.

As discussed in the preceding chapter, global awareness of the rapidity of ecological degradation has grown substantially since the Stockholm Conference held in 1972. That eventually led to the publication of the World Conservation Strategy in 1980 by the International Union for Conservation of Nature (IUCN) and Natural Resources in collaboration with the United Nations Environment Programme (UNEP) and the World Wildlife Fund for Nature (WWF) in 1980 to strengthen the nature conservation movement among the member states. The preparation of the document involved governments, nongovernmental organizations and experts throughout the world and its main aim was to strengthen nature conservation efforts among the member countries of the United Nations (see Appendix 2.1). The United Nations' call for nature conservation was kept in momentum with the release of the World Charter in 1982 followed by the release of the Caring for the Earth by IUCN, UNEP and WWF in 1991 to secure the ethical commitment of the international community to manage and use natural resources within their safe planetary boundaries.

The United Nations environmental conservation initiatives reached another climax through the adoption of the legally binding CBD at the Rio Conference in 1992. With most countries (196 parties) having adopted the CBD, biodiversity conservation seemed to have become a great concern of the global community. Many countries have taken steps to enact environmental laws and regulations and action plans domestically to manage natural resources sustainably and to protect and conserve natural habitats and related biodiversity. As noted in Chapter 2, The United Nations' Journey to Global Environmental Sustainability Since Stockholm: An Assessment, a total of 190 of 196 (97%) Parties have developed NBSAPs in line with Article 6 of the CBD on General Measures for Conservation and Sustainable Use. In addition, since COP-10, a total of 168 countries have submitted new national biodiversity strategies and action plans (NBSAPS) with 143 Parties submitted revised versions, aimed at implementing the Strategic Plan for Biodiversity 2011−20 and addressing all threats to biodiversity and ecosystem services (CBD Secretariat, 2019).

One of the strategic means underlying CBD ecological protection initiatives is the promotion of in situ conservation through the creation of Protected Area (PA) to enhance the protection of ecosystems, natural habitats and the maintenance of viable populations of species in natural surroundings (Article 8[d] of CBD). PA is defined by IUCN as "a clearly defined geographical space, recognized, dedicated and managed, through legal or other effective means, to achieve the long-term conservation of nature with associated ecosystem services and cultural values" (IUCN World

Commission on Protected Areas, 2016: 4). Also the Aichi Biodiversity Targets 3, 10 and 15 explicitly indicate that establishing PAs is a potentially important means to halt biodiversity loss. Apart from the PA strategy, Aichi Target 11 also places great emphasis on the implementation of conservation measures in other areas which are equally important in contributing to enhancing the ecological health of the ecosystems.

In response to this legally binding requirement, the areas designated as PAs increased substantially globally since the Stockholm Conference as many countries took steps domestically to protect and conserve natural habitats and related biodiversity. Indeed, nearly every country has adopted PA legislation, and designated sites for protection, and new protection areas are being created to reinforce conservation efforts (IUCN, 2010).

The extent of national PAs created has increased gradually from a mere 1.1 km^2 (about 111 ha) in 1838 to about 128,860 km^2 (about 12.86 million ha) in 1972. Since then it has increased substantially to 1,506,714 km^2 (about 15 million ha) in 2017. The number of protected sites has also increased from merely 2 in 1838 to 11,337 in 1972 and, further increased to 103, 670 in 2017 (European Environmental Agency, 2018; Figs. 3.10 and 3.11).

It is evidently clear by now that in response to the United Nations' call for the reduction of biodiversity loss, global creation of PA networks in line with CBD has increased substantially since Stockholm. This contributes significantly to global biodiversity conservation efforts. The most pertinent question raised is: to what extent and how significantly has the establishment and management of PAs been able to contribute to the reduction of biodiversity across the globe? The following section will examine this critical question systematically.

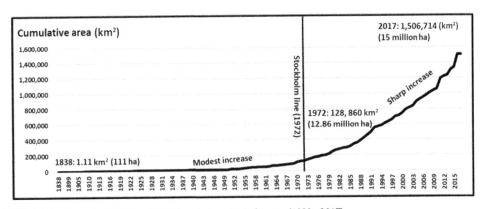

FIGURE 3.10 Growth of nationally designated protected areas (1838–2017).
Note: Since the Stockholm Conference held in 1972, the area of designated protected sites increased sharply. *Data from, European Environmental Agency, 2018. Increase in the number and size of nationally designated protected areas, 1838–2017. European Union. Available from: <https://www.eea.europa.eu/data-and-maps/daviz/growth-of-the-nationally-designated-3#tab-chart_4>.*

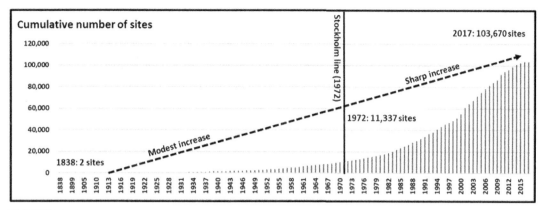

FIGURE 3.11 Growth of nationally designated protected sites (1838–2017).
Note: Since Stockholm (1972), the number of sites increased sharply. *Data from, European Environmental Agency, 2018. Increase in the number and size of nationally designated protected areas, 1838–2017. European Union. Available from: <https://www.eea.europa.eu/data-and-maps/daviz/growth-of-the-nationally-designated-3#tab-chart_4>.*

3.9 Biodiversity conservation and environmental degradation: The global reality in general

The degree of national priority accorded to biodiversity conservation varies widely across regions. Latin America accords a high priority to biodiversity conservation while Asia and the Pacific more recently accepted its legitimacy as an issue of both national and international concern. Both Europe and North America accord a high priority to biodiversity conservation, although the United States has not signed the CBD. Many African countries, which have been concerned about the economic potential of revenue from wildlife tourism, also recognize the need for biodiversity conservation. Apart from CBD, Agenda 21, which strongly emphasizes sustainable use of resources and biological conservation, also serves as an important guideline to promote a more environmentally sustainable mode of development of the member states including countries from Latin America, Africa, and Southeast Asia.

Despite this, various international biodiversity assessment reports, both old and recent, provide incontrovertible evidence that massive biodiversity loss remains one of the most serious environmental sustainability problems that threatens the very existence of our Planet. Extensive environmental degradation followed by widespread habitat fragmentation or destruction has resulted in massive biodiversity loss or species extinction. For example, according to the Millennium Ecosystem Assessment, roughly 10%–30% of the mammal, bird and amphibian species are threatened with extinction due to human activities (MEA, 2005). The Global Biodiversity Outlook 4 (GBO 4) also reveals that despite individual success stories, the average risk of extinction for birds, mammals and amphibians is still increasing. Also genetic diversity of domesticated livestock is eroding. Worse yet, more than one-fifth of the breeds are at risk of extinction while the wild relatives of domesticated crop species are increasingly threatened by habitat fragmentation and climate change (CBD Secretariat, 2014).

The global ecological quagmire is further confirmed by the findings of the IUCN which reveals that 21% of all known mammals, 30% of all known amphibians, 12% of all known birds, and 28% of reptiles, 37% of freshwater fishes, 70% of plants, 35% of invertebrates are at risk of extinction (Hilton-Taylor et al., 2009). It is also worth noting that for the past 40 years between 1970 and 2010, animal populations have declined by 52% due to habitat loss and degradation, driven by unsustainable human economic activities and consumption (WWF, 2014). Another new assessment by the WWF also indicates that species population has persistently declined further by 58% between 1970 and 2013 due to pressures from unsustainable agriculture, fisheries, mining and other human activities that contribute to habitat loss and degradation, overexploitation, climate change and pollution (WWF, 2016b).

Another updated assessment conducted by IUCN on more than 77,300 species against the IUCN Red List Categories and Criteria, indicates that despite increased efforts at conservation, our biological system continues to be subject to the threat of ecological extirpation. The species assessed include most of the known species of amphibians; birds; mammals; angelfish; butterfly fish; crocodilians; freshwater crabs and crayfish; groupers; gymnosperms (including cycads and conifers); lobsters; mangroves; marine turtles; parrotfish; reef-building corals; seagrasses; seasnakes; sharks and rays; tunas and billfishes; and wrasses. The result reveals threats to the continued survival of 63% of Cycads, 41% of Amphibians, 33% of Reef-building Corals, 34% of Conifers, 25% of Mammals and 13% of Birds. The Red List now includes 85,604 species, of which 24,307 are threatened with extinction (IUCN, 2015b).

The latest assessment, the 2019 Global Assessment Report on Biodiversity and Ecosystem Services, shows that the state of the global biodiversity is in great peril. The ecological resilience and environmental health of the global ecosystems are deteriorating more rapidly than ever. The Report warns that up to 1 million animal and plant species are now threatened with extinction in the coming decades—which is unprecedented in human history. It further reveals that:

- Three-quarters of the land-based environment and about 66% of the marine environment have been significantly transformed by human activities. In the European Union, for example, only 9% of marine habitat types show a "favorable conservation status."
- Native species in most major land-based habitats has fallen by at least 20%.
- More than 40% of frogs and amphibian species, almost 33% of reef-forming corals and more than a third of all marine mammals are threatened. In the European Union, for example, only 7% of marine species show a "favorable conservation status."
- At least 680 vertebrate species have been driven to extinction.
- More than 9% of all domesticated breeds of mammals used for food and agriculture had become extinct by 2016.
- The numbers of invasive alien species per country have risen by about 70% since 1970 (Díaz et al., 2019).

The Report further uncovers that the precarious state of the current ecological system is basically caused by changing land use patterns in many countries across the globe (Díaz et al., 2019). The findings of the 2019 Global Assessment Report are further reinforced by the Sixth Global Environment Outlook (GEO-6) assessment published in 2019 (UNEP, 2019b). The overall picture of the state of the global environment may be briefly noted below:

1. Africa: its biodiversity continues to face significant threats from illegal wildlife trade, monocropping, deforestation, the prevalence of alien invasive species and, air and water pollution.
2. Asia and the Pacific: ecosystems integrity and biodiversity are threatened throughout the region due to extensive agriculture development especially monoculture expansion such as oil palm and rubber plantation development, aquaculture, and illegal wildlife trade.
3. Europe: high resource footprint in the region arises from the overuse of natural resources which may exert a negative impact on the ecological resilience of the ecosystems. Biodiversity diversity loss and ecosystem degradation are increasing rapidly despite increased conservation and restoration efforts. The main driver for these ecological pressures is mainly due to land-use change which is instrumentally skewed toward utility maximization.
4. Latin America and the Caribbean: species continue to be lost across the region and ecosystems are increasingly subject to human assaults out of a multitude of driving forces such as changing land use practice, deforestation, and water pollution. Deforestation-linked biodiversity loss in the Caribbean may not be as severe as the other regions due to its increase in forest cover by 43% over the 1990 baseline.
5. North America: although much progress has been made with many individuals species, biodiversity on the whole, is at risk due to increasing pressures from land use change, invasive species and pollution. It is also worth noting that biodiversity conservation in the United States under the current "anti-nature" Donald Trump's administration has weakened substantially. More specifically, Donald Trump has systematically rolled back or revoked protections of biologically rich public land including the planet's largest remaining intact temperate rainforest, the Alaska Tongass National Forest, for economic exploitation. He has also significantly weakened the nation's bedrock conservation law, the Endangered Species Act, making it easier to remove a species from the endangered list and weaken its protection status in order to pave the way for new mining, oil and gas drilling, and development in areas which protected species strive and flourish. Donald Trump has also weakened the National Environmental Policy Act in favor of the oil and gas industry in order to speed up approval for various environmentally destructive projects such as pipelines and power plants (Friedman, 2019a, D'Angelo, 2020a, Newburger, 2020, see also, Chapter 4, Section 4.54).

As for the biological status of the ocean marine ecosystem, serious concerns have been raised about its ecological resilience or Holling stability. It may well be noted that over the past 100 years, 90% of all large fishes including tuna, marlin, swordfish, shark, cod, and halibut have disappeared from the world's oceans due to unsustainable commercial fishing practices (Walton, 2003). The world's fishery resources are also in an ecologically unsustainable state of development as a large percentage of the wild fish stocks with commercial value are either fully harvested or overexploited and many species have been severely depleted with some close to ecological collapse or extinction. It was reported that more than 75% of world fish stocks for which assessment information is available are already fully exploited or overexploited (FAO, 2007; Allsopp et al., 2007).

Recent findings by the Food and Agriculture Organization reveals that "the share of fish stocks within biologically sustainable levels decreased from 90% in 1974 to 68.6% in 2013" (FAO, 2016: 38). In other words, roughly 31.4% of fish stocks are being over-fished at a biologically unsustainable level. As a case in point, the overexploitation of the Bluefin tuna in the North Pacific Ocean is putting the species on the verge of extinction due to the drastic drop in its population by 97% (Center for Biological Diversity, 2016). The latest assessment indicates that in 2015, 33% of marine fish stocks were being harvested at unsustainable levels (Diaz et al., 2019). Also, most of the 10 most productive species including anchoveta, Alaska pollock, blue whiting, Atlantic herring, Chub mackerel, Gulf menhaden, and among others, which accounted for about 27% of the world's marine capture fisheries production in 2013, are either fully exploited or overexploited with some having no potential for increases in production (FAO, 2016).

Another factor which has contributed to the decline of the marine ecosystem is the expansion of dead zones in the world's ocean and coastal waters. Dead zones or Hypoxic zones are areas where oxygen levels have dropped too low to support most marine life. Dead zones are likened to oceanic deserts, devoid of the usual aquatic biodiversity (Howard, 2019). Dead zones in the coastal oceans are largely caused by the accumulation of nutrient-enriched (nitrogen and phosphorous) water from agricultural runoff discharged directly into the ocean. This results in eutrophication, that is, over-enrichment of water by nutrients followed by excessive blooms of algae that deplete underwater oxygen levels, leading to hypoxia—a phenomenon known as eutrophication-induced hypoxia. Simply put, hypoxia is a condition in which dissolved oxygen levels fall below ≤ 2 mL of O_2/L (less than or equal to 2 milliliter of oxygen per liter) (Diaz and Rosenberg, 2008).

It is reported that the open ocean lost about 77 billion tonnes of its oxygen over the past 50 years. Open-ocean oxygen minimum zones (OMZs) have expanded substantially by millions of square kilometer over the same period, roughly about the size of the European Union. The oxygen minimum zone, also known as the shadow zone, is the zone in which the oxygen concentration has dropped to its lowest level. For the past five decades, the volume of water completely devoid of oxygen (anoxic) has more than quadrupled, primarily caused by unsustainable human activities (Breitburg et al., 2018; Global Oxygen Network, 2018).

Dead zones have also developed in the continental seas, such as the Arabian Sea, Black Sea, Gulf of Mexico, Baltic, and East China Sea (Diaz and Rosenberg, 2008). For example, in the 1970s and 1980s, an estimated 60 million tons of bottom-living (or benthic) life perished from hypoxia in the Black Sea beach resorts in Romania and Ukraine (Mee, 2006). In another case, it was observed that piles of mussels washed onto the beach of Narragansett Bay located in Rhode Island in the United States after a dead zone event (Smithsonian, 2014). In the Gulf of Mexico, dead fish floated onto the coast due to the formation of a dead zone in the gulf each year. The Gulf of Mexico dead zone is the second largest human-caused dead zone in the world, averaging almost 6000 square miles in size (NOAA, 2015; Howard, 2019). The largest dead zone in the world is found in the Arabian Sea, covering almost the entire 63,700-square mile Gulf of Oman (Howard, 2019).

The number of dead zones on Earth has doubled every decade since the 1960s. Worldwide, there are now about 550 dead zones covering 250,000 km^2 (96,525 square miles) in both freshwater bodies including rivers, ponds and lake, and the salt water bodies, the oceans (Zimdahl, 2015; NOAA, 2015; Hand, 2016). Since the 1990s, phosphorus inputs to Lake Erie have increased substantially as a result of agricultural development, spurring an outburst of toxic algal blooms not seen since the 1970s (Greshko, 2017).

It is increasingly clear from all the global environmental assessment reports that our Earth's biodiversity has been and is being continuously subjected to an unending series of ecological despoliation. In other words, the entry into force of CBD and other international treaties and documents including the Agenda 21 to raise global awareness on global biological ecological degradation, to establish consensus on ecological conservation measures and to promote and support global environmental initiatives and programmes to arrest global ecological degradation, have met with relatively little success in halting the biological decline of our planet.

To better understand the impact of the United Nations environmental protection initiatives in promoting biodiversity conservation, a systematic examination will be presented in the following section of the biodiversity conservation efforts and the magnitude of biological degradation based on country-specific case studies in the tropical rainforest region and China—some of the most biologically diverse regions in the world. The following country-specific assessments also serve as an important basis for the ethical argument of environmental sustainability in the later part of the chapter.

3.10 Biodiversity conservation in the tropical rainforest region (specific case study): The status quo

3.10.1 The Brazilian Amazon: The status of biological diversity

Although covering less than 2% of the planetary system, the tropical rainforests are home to roughly 50% of all biological life on Earth. The tropical rainforests also contain far higher diversity on a per-area basis compared with subtropical, temperate, and boreal ecosystems. For example, compared with the temperate forests which are often dominated by a half dozen tree species or fewer that make up 90% of the trees, the tropical rainforest may have hundreds of tree species in a single hectare, or a single bush in the Amazon which has also more species of ants than the entire British Isles (Butler, 2006b). The highest tree diversity recorded to date is 1200 species in a 52-ha plot in Lambir Hills National Park in the state of Sarawak in Malaysian Borneo (Lee et al., 2002). Half of all the carbon stored in the world's forests is also found in tropical areas (GRID-Arendal, 2015).

Brazil, for example, has been identified by Conservation International in 1988 as one of the world's 17 mega-diverse countries among over 200 countries (UNEP-WCMC, 2014). It may be relevant to note that other global mega-diverse countries on the global-17 list are Indonesia, Malaysia, the Philippines, Australia, China, Colombia, Democratic Republic of the Congo (DRC), Ecuador, India, Madagascar, Mexico, Papua New Guinea, Peru, South Africa, the United States of America, and Venezuela. This group of countries, which together constitutes less than 10% of the global landmass, contains more than 70% of the earth's biological diversity.

"Mega-diverse" is a term used to describe the most biodiversity-rich countries of the world, with a particular focus on endemic biodiversity. To qualify as a megadiverse country, a country must have (1) at least 5000 of the world's plants as endemics and (2) marine ecosystems within its borders (UNEP-WCMC, 2014). Overall, the latest estimate ranked Brazil as the world's most diverse territory of the world's top 10 mega-diverse countries. It leads the world in plant and amphibian species counts. It ranks second in mammals and amphibians, third in birds, reptiles, and fish (Butler, 2016b). The world ranking for the rest of the top 10 countries are (2) Colombia, (3) Indonesia, (4) China, (5) Mexico, (6) Peru, (7) Australia, (8) India, (9) Ecuador, and (10) Venezuela (Butler, 2016b, see, also, UNEP-WCMC, 2014).

3.10.2 Biodiversity conservation in Brazil

Since becoming a signatory to CBD in 1992, Brazil has exerted much effort to promote environmental conservation and sustainable resource use. In 1994 when it ratified the CBD, it instituted the National Programme of Biological Diversity. The Programme underwent some adjustments in 2003 when the National Commission for Biodiversity, was created by the Presidential Decree 4703 to reinforce its commitments to protect its mega-diverse biodiversity, especially endangered species based on habitat conservation, and to promote sustainable use of the Brazilian biological components in compliance with Article 6 of CBD (Ministry of Environment, Brazil, 2004, 2016).

In 2002 the government established its National Biodiversity Policy-PNB through Decree No. 4339 along with its National Biodiversity Action Plan-PAN (BIO). In an effort to strengthen biodiversity conservation efforts, Brazil increased its PA by 5% between 2003 and 2009. This increased the total PA to 27.8% of the national terrestrial and marine areas in 2009 (CBD Secretariat, 2011). Since 2006, numerous public policies and new projects and programmes have been developed with the aim of promoting the three goals of the CBD, namely, biodiversity conservation, sustainable use of biological components and benefit sharing, as well as addressing various specific issues such as the conservation of species and ecosystems, the sustainable use of biodiversity, and mainstreaming biodiversity themes into different sectors (Ministry of Environment, Brazil, 2010).

Brazil has also put in place the Federal Constitution of the Republic of 1988 which guides the development of biodiversity conservation and its sustainable use in line with the CBD. The Constitution explicitly stipulates that the State is held responsible for the preservation of the diversity and integrity of the genetic patrimony of the country, to define PAs and to protect the flora and fauna. Other legal instruments which serve to enhance biodiversity conservation and sustainable resource use are Law No. 6938 (1981) which established the National Environment Policy and set up the National Environment System and the National Environment Council as well as laid down a number of regulations for environmental management. Also Law No. 4771 (1965) empowered the State to create National Parks, Biological Reserves, and National Forests while Law No. 7347 (1985) which defined civil public action for liability for damage caused to the environment helps control unsustainable environmental practices. The Fauna Protection Law enacted in 1967 aims to protect wildlife by prohibiting the use, persecution, destruction, hunting, or capture of wild animals (Ministry of Environment, Brazil, 1998).

In an attempt to protect its biodiversity, the Brazilian government has launched the biological corridor project within the ambit of the Amazonia Agenda 21, known as the Ecological

Corridors Project in the Amazon and the Atlantic Forest. Accordingly the following seven priority corridors have been planned under the project (Ministry of Environment, Brazil, 1998).

1. the Central-Amazon Corridor which includes the Mamirauá Reserve for Sustainable Development, the Anavilhanas Ecological Station, the Tefé National Forest, the Jaú National Park, the Adolfo Ducke Forest Reserve, 9 other PAs, and 14 indigenous areas;
2. the North-Amazon Corridor which includes the Pico da Neblina National Park, the Roraima National Forest, the Serra do Araçá State Park, and an additional 17 PAs and 20 indigenous areas;
3. the West-Amazon Corridor which includes the Serra do Divisor National Park, the Chico Mendes Extractivist Reserve, the Rio PretoJacundá Extractivist Reserve, and an additional 30 other PAs and 30 indigenous areas;
4. the South-Amazon Corridor which includes PAs in three states (Amazonas, Pará, and Maranhão), including the Tapajós National Forest, the Amazon National Park, the Gurupi Biological Reserve, and an additional three more PAs and 20 indigenous areas;
5. Corridor of the South-Amazon Ecotone (Amazon-Cerrado) which includes Araguaia National Park on the island of Bananal (state of Tocantins) and 17 indigenous areas in the states of Amazonas, Mato Grosso, and Tocantins;
6. Central Corridor of the Atlantic forest with areas of extremely high diversity and endemism in the states of Espírito Santo, Minas Gerais and southern Bahia. It includes the Sooretama Biological Reserve, the Linhares Forest Reserve, the Una Biological Reserve, the Monte Pascoal National Park, and the Serra do Caparaó National Park; and
7. the Southern Atlantic Forest Corridor (Serra do Mar Corridor). This includes 27 PAs covering the Serra do Mar Area for State Environmental Protection (São Paulo), the Serra da Mantiqueira Environmental Protection Area (Minas Gerais), the Serra da Bocaina National Park and the Itatiaia National Park (Rio de Janeiro), and the Guaraqueçaba Environmental Protection Area (Paraná).

Together, these represent 25% of the Brazilian rainforests, estimated to cover the geographical distribution of about 75% of the animal and plant species in the two biomes (Ministry of Environment, Brazil, 1998). These conservation measures are also in line with Agenda 21-Brazil which aims to promote sustainable development and environmental conservation. As reflected in the Brazilian second (2004) and third (2005) national biodiversity reports, biodiversity conservation initiatives and programmes have continuously been complemented and reinforced. The fourth national report (2010) introduced a more analytical format, presenting an assessment of the status and trends of biodiversity and ecosystems, as well as of the effectiveness of the national biodiversity strategy and degree of achievement of national and global biodiversity targets, among other related aspects. Continuing its efforts to fulfill the national commitments under CBD, Brazil updated in 2013 its National Biodiversity Targets (Ministry of Environment, Brazil, 2015).

3.10.3 The state of biodiversity in Brazil

Paradoxically despite the above comprehensive and encompassing environmental protection measures and programmes, deforestation as discussed in Sections 3.6 and 3.7

constitutes one the major drivers of biodiversity depletion in the Brazilian Amazon. For example, using a large data set of plants, birds and dung beetles (1538; 460 and 156 species, respectively) sampled in 36 catchments in the Brazilian state of Pará, a group of researchers found that a 20% loss of primary forest, the maximum level of deforestation legally allowed under Brazil's Forest Code, resulted in a 39%−54% loss of conservation value (Barlow et al., 2016; see also, Ochoa-Quintero et al., 2015; Franca et al., 2016). In another finding, after surveying 2000 species of plants, birds, beetles, ants and bees across more than 300 diverse sites in the Brazilian Amazon, researchers further confirmed that deforestation has doubtlessly caused a strong loss of biodiversity (Franca et al., 2016). While the Brazilian law requires individual landowners in the Amazon to retain 80% forest cover, this is rarely achieved or enforced (University of Cambridge, 2015; Ochoa-Quintero et al., 2015).

Indeed, extensive deforestation in the past for commercial agricultural expansions as discussed in Section 3.6 has exerted immense negative impact on the biodiversity. In this regard, it may further be added that field rescarch conducted in 31 landscapes in the northern half of the Brazilian Amazon using field observation and interviews with the local farmers by a group of researchers indicates that agricultural expansion in the recent decades has led to the loss of 41% of the original forest (about 2 million ha). The result indicates that one to two species could be affected by loss of forest cover in landscapes above the threshold (Ochoa-Quintero et al., 2015).

The "extinction debt," that is, the number of species heading toward extinction as a result of past deforestation followed by habitat fragmentation in the Brazilian forests was estimated based on a model developed by a group of experts using 50 by 50 km^2 of land (Wearn et al., 2012). The model draws on historical deforestation rates from 1970 to 2008 and animal populations including individual species data for vertebrates that depend on the tropical forest for food. The extinction debt modeling concluded that under the "business as usual" scenario, where around 160 km^2(62 square miles) of forest are cleared each year, at least 15 mammal, 30 bird, and 10 amphibian species were expected to die out locally by 2050, from around half of the Amazon. Under the Governance scenario which put in place some form of environmental regulations and sustainable resource management practices, the model projected that the local ecosystems could lose an average of 12 species and condemn 19 more to extinction.

Extensive habitat fragmentation caused by land conversion for agricultural expansion and cattle ranching and rapid urbanization has halved the population of Brazilian barefaced tamarins in 18 years. The continued survival of the white-cheeked spider monkeys, which feed on fruits high in the forest canopy, is also at risk. The endangered giant otter, found in the slow-moving rivers and swamps of the Amazon is also threatened because of water pollution from agricultural runoff and mining operations in the area (Sample, 2012).

3.10.4 The Brazilian Atlantic forest ecosystem dilemma—a human-induced ecological tragedy

Demonstrably the Atlantic forest ecosystem or the Mata Atlântica's ecosystems which is one of the most diverse and unique in the world is increasingly subject to ecological destruction due to severe biological alteration as a result of unrestrained and unsustainable human practices (Mendes, 2019). While more than 20% of the Amazon forest has gone, 93% of the Atlantic Forest has disappeared (Galindo-Leal and de Gusmão Câmara, 2003; Tabarelli et al., 2005). The remaining forest cover exists mostly in isolated remnants scattered throughout a landscape dominated by agricultural uses. In the states of the Central Corridor (Bahia and Espírito Santo) and Serra do Mar Corridor (Rio de Janeiro, part of Minas Gerais and São Paulo), for example, the remaining forest ranges from 2.8% in Minas Gerais to 21.6% in Rio de Janeiro (CEPF, 2005).

However, deforested areas in the Atlantic Forest have decreased recently. In the 2017–18 period an area of the biome totaling 11,399 ha was cleared which is 1163 ha less than the 12,562 ha recorded a year earlier. Despite this, nonetheless, the Mata Atlântica's ecosystems are still subject to intense pressure from agribusiness, timber harvesting, and the real estate market (see, e.g., Mendes, 2019). It may well be that despite many legal instruments in place to protect the Atlantic forest biodiversity, many species have been exterminated locally while a vast majority of the animals and plants are threatened with extinction due to logging, poaching of flora and fauna, illegal settlement, and commercial agricultural development such as coffee and soy plantation expansion and cattle ranching (Galindo-Leal and de Gusmão Câmara, 2003; WWF, 2016c).

The Mata Atlântica's natural environment is the country's most endangered biome (Mendes, 2019)—a human-induced ecological tragedy. It may be noted in passing that a biome is different from an ecosystem. An ecosystem includes all living things in a given area, interacting with each other, and also with their nonliving environments. A biome is a specific geographic area notable for the species living there. A biome can be made up of many ecosystems (FAO, 2013: 139). Returning to the ecological tragedy of Mata Atlântica, it may be reflected that of Brazil's 69 severely endangered mammals, 38 are from the Atlantic Rainforest and of 160 endangered birds and 20 endangered reptiles, 18 and 13 species of these, respectively, are from the region (Iracambi, 2016). The jaguars, lowland tapirs, woolly spider-monkeys, and giant anteaters, among the most ancient and threatened tropical mammal species on the planet, are now almost absent in Brazilian northeastern forests.

The white-lipped peccary, a species closely related to pigs, has been completely wiped out (Angelo, 2012). It is said that the woolly spider-monkey, symbol of the forest, is the most severely threatened of all (Iracambi, 2016). Flora species that are restricted to either rocky outcrops or scrub vegetation on sandy coastal plains also have the highest extinction risk among vegetation types (Leão et al., 2014). The extremely biodiversity rich Atlantic forests of Brazil which have been severely degraded is considered one of the most threatened global hotspots calling for urgent protection (Conservation International, 2010).

3.11 Africa

3.11.1 The Democratic Republic of Congo: The status of biological diversity

The Congo Basin in Africa spreads across Cameroon, Gabon, Republic of Congo, Central Africa Republic, DRC, and Equatorial Guinea. It is among some of the most biological diverse regions in the world. Notably the DRC, the largest country in the Congo Basin, has the highest number of species for almost all groups of organisms except for plants, for which it ranks second to South Africa. In addition, the DRC is classified as the world's fifth most diverse country in terms of plant and animal species richness (Debroux et al., 2007). Overall, it ranks 14th in the global mega-diverse list (Butler, 2016b). Its ecological treasures comprise 10,000 species of plants, 409 species of mammals, 1117 species of birds, and 400 species of fish many of which are highly endemic in the region (Debroux et al., 2007; Dipby Wells Environmental, 2015a,b). Although the repertoire of biological species in the DRC is not as high as in the Amazon or Southeast Asia, it is globally biologically unique because of its high endemism. Some of the endemic species include the okapi, Grauer's gorilla, bonobo, and the Congo peacock (UNESCO, 2010).

DRC also holds one of the world's largest reservoirs of carbon, second only to the Amazon Basin rainforest. It is estimated that there are roughly 23 GtC carbon (billion tonnes of carbon) stored in the humid tropical forests of the DRC with a mean carbon density of about 139 MgC/ha (tonne of carbon per hectare) (Saatchi et al., 2017). DRC is also the one of the most important countries in the world for tropical peat carbon stocks (Dargie et al., 2017). It may be noted that Congo Basin houses the world's largest tropical peatland, the Cuvette Centrale peatland complex, which covers 145,500 square kilometers (14.55 million ha) and extends across the Democratic Republic of Congo (DRC) and the Republic of Congo (RoC). With an area larger than England, the peatland complex is estimated to hold about 30 billion tonnes of carbon—equivalent to 3 years' worth of the world's total fossil fuel emissions (Dargie et al., 2017; Brown, 2017). Also, the DRC and RoC are considered as the second and third most important countries in the world for tropical peat carbon stocks (Dargie et al., 2017).

3.11.2 Democratic Republic of Congo biodiversity conservation policies

Since its ratification of the CBD in 1994, the DRC has been working on its implementation by holding consultations at the national and provincial levels. It has developed the National Biodiversity Strategies and Action Plans to promote conservation of its biodiversity and sustainable use of its components. In compliance with the terms set out in CBD, the DRC prepared its National Report to the CBD to provide updates on its progress in biodiversity conservation (Mpoyi et al., 2013).

The DRC has also ratified the Convention on Illegal Trade of Endangered Species of Flora and Fauna (CITES) in 1976 to ensure that international trade in specimens of wild animals and plants does not threaten their survival. Following the ratification of CITES, it issued an order on its implementation and entrusted the Institut Congolais pour la Conservation e la nature to serve as the focal CITES management center. It has also prepared and submitted a biannual report on the legislative, regulatory and administrative progress on CITES management (Mpoyi et al., 2013).

As a show of commitment to promoting global environmental sustainability, it has undertaken to ratify the World Natural and Cultural Heritage Convention to ensure the conservation and preservation of cultural properties including nature reserves. Some of the natural properties inscribed under its world heritage list include Garamba National Park (1980), Kahuzi-Biega National Park (1980), Okapi Wildlife Reserve (1996), Salonga National Park (1984), and Virunga National Park (1979). Also the DRC's constitution stipulates that all properly concluded treaties and international agreements take precedence over national laws (Seyler et al., 2010). In addition, together with the RoC and Indonesia, the DRC has also signed the Brazzaville Declaration at the third meeting of the Global Peatland Initiative, involving IUCN (International Union for Conservation of Nature), UN Environment and partners to conserve the Cuvette Centrale peatland complex for climate mitigation and other benefits (IUCN, 2018)

3.11.3 The state of biodiversity in Democratic Republic of Congo

Despite this, however, casual observation on the ground indicates that the biodiversity conservation efforts seem to have been overshadowed by humans' heedless actions and exploitative behavior. To start with, extensive habitat fragmentation largely caused by widespread deforestation and illegal logging, and overhunting, illegal bushmeat, and exotic wildlife trade are increasingly threatening the continued existence of many of the flora and fauna species at an alarming rate (Seyler et al., 2010; UNESCO, 2010). Even though around 18 million ha or 8% of the national territory have been designated as parks and protected or conservation areas, most of these exist on paper only (Seyler et al., 2010). For example, many forest concessions are in the biodiversity rich intact forests such as areas inside the Maringa-Lopori-Wamba landscape, which is a critical habitat for some of the last viable population of bonobos (Greenpeace, 2007). Although there are 72 threatened and endangered species, especially of mammals, that are completely protected by law in DRC, their protection in reality is very weak (Seyler et al., 2010).

Unsustainable activities in the country's World Heritage sites such as poaching of rhino, elephant, and buffalo by local hunters and highly armed and well-organized horsemen from Sudan in the Garamba National Park, poaching of elephant and commercial hunting for the bushmeat trade in Okapi Wildlife Reserve and Salonga National Park and poaching of large mammals, particularly hippo, in the central and northern parts of the Virunga National Park have all coalesced, threatening the continued survival of the diverse flora and fauna that used to flourish in the region (UNESCO, 2010).

The northern white rhino, which depended on the protection of its habitat in the Garamba National Park for survival has apparently become extinct in the wild due to widespread poaching and illegal trade in rhino horns (see, e.g., O'Neill, 2013). The northern white rhino species was once thought to spread from Chad to the Democratic Republic of the Congo. However, their population has declined from more than 2000 in 1960 to just 15 in 1984 (Zachos, 2018). Worse yet, with the world's last male northern white rhino' death in March 2018 from old age and poor health, the world is left with only two females to save the subspecies from extinction (Berlinger, 2018).

Elephant populations have also declined substantially due to the resurgence of illegal ivory trade in Asia (Seyler et al., 2010). For example, solid, precise data on the status of elephants based on aerial census revealed a low number of their population in northeastern DRC (Steyn, 2016). Indeed, high elephant density areas already became history in 2002 (Maisels et al., 2013). Overall, the total DRC elephant population (forest and forest-savanna hybrids) is likely to be under 20,000, down from a population estimated at over 100,000 elephants 50 years ago, and still dropping due to poaching spurred by the lucrative illegal ivory trade (Seyler et al., 2010).

The Congolese Giraffe, a rare subspecies of the Kordofan giraffe that only lives in the DRC could soon become extinct as there are only 38 animals left (Watkinson, 2016). The Grauer's gorilla (*Gorilla beringei graueri*), a subspecies of eastern gorilla, the world's largest ape, and confined to eastern DRC is also heading toward extinction. The species has been listed as Critically Endangered on the IUCN Red List of Threatened Species (WCS, 2016). It is also worth noting that the Cuvette Centrale peatland complex is currently under irreversible environmental threat. It may well be that in contradiction to their pledges under the Brazzaville Declaration, the governments of both the Republic of Congo and Democratic Republic of Congo have signed various oil exploration deals with oil companies in the Cuvette to extract its oil deposits. The Ngoki oil project in the Republic of Congo which is estimated to contain over 6,000 square kilometers of carbon rich peatland in Cuvette is a case in point (Global Witness, 2020).

3.11.4 The African status of biological diversity in general

Apart from the DRC, the biodiversity conservation efforts in the rest of the African region in general are not promising either. To begin with, Africa, which has 53 countries including the DRC, and one "non-self-governing territory" (Western Sahara), covers over 20% of the Earth's total land mass. It is the second largest continent in the world after Asia. Ecologically the continent is home to eight major biomes, that is, large-scale biotic communities characterized by distinctive assemblages of flora and fauna. These are the Mediterranean, Semi Desert, Dry Savannah, Moist Savannah, Tropical Rain Forest, Desert, Temperate Grassland, and Montane. To note in passing, the African classification of biomes may differ from others which may use broad classifications and count as few as five biomes. These are forest, grassland, freshwater, marine, desert, and tundra (UNEP, 2008).

Together, the biomes in Africa are home to some one quarter of the world's 4700 mammal species. Huge populations of mammals occur in the eastern and southern savannahs, including at least 79 species of antelope. It also has more than 2000 species of birds—one-fifth of the world's total—and at least 2000 species of fish, alongside 950 amphibian species. New species of amphibians and reptiles are still being discovered. The African mainland also contains between 40,000 and 60,000 plant species and about 100,000 known species of insects, spiders and other arachnids (UNEP, 2008).

3.11.5 The African biodiversity protection initiatives in brief

To conserve the regional biodiversities, the African governments have demonstrated their commitment through the ratification of various international conventions and treaties as discussed in Chapter 2, The United Nations' Journey to Global Environmental Sustainability Since Stockholm: An Assessment, and to set aside large portions of their territories as PAs (IUCN, 2005). Also most of the African governments have drawn up their national biodiversity strategies and action plans (NBSAPs) to promote CBD objectives (see, e.g., UNEP, 2010a). Particularly of the 54 African Parties to the CBD, 44 have submitted at least one NBSAP to the Convention of Biological Diversity since 1993 while three Parties are developing their first NBSAP (UNEP, 2016).

Through these conservation efforts, Africa has created over 1200 national parks, wildlife reserves, and other PAs covering more than 2 million km^2 or 9% of Africa's total land area (IUCN, 2005; Gelletly, 2014). Hunting in these PAs is prohibited by law. PAs have also increased substantially to enhance biodiversity conservation. These areas cover 16% of East and Southern Africa and 10% of West and Central Africa (UNEP, 2013; see also UNEP, 2010a, 2016). As discussed in the following sections, despite these environmental protection initiatives, however, biodiversity continues to decline in the African countries.

3.11.6 The state of biodiversity in Africa in general

3.11.6.1 Elephant: The poaching crisis

How horrific the decline has been will be brought home by looking at specific statistics of animals whose populations have dwindled to being close to extinction. The African elephant is one of these. Historically the African region as a whole may have been home to over 20 million elephants before European colonization, and later to 1 million in the 1970s (Hamilton, 1987; Milner-Gulland and Beddington, 1993). Between 1970 and 1990, thousands of elephants were killed for their ivory, leaving the African elephant population at an estimated 300,000−600,000 (Said et al., 1995). Between 1995 and 2006, the population dropped further to between 286,000 and 472,000 (IUCN, 2013a). In addressing its rapid population decline, in 2010, the African elephant range states drew up the African Elephant Action Plan to sustainably manage the elephant population level. The governments also declared poaching and illegal trade in elephant products as the top threat to elephants across the continent. The elephant subpopulation of Central Africa including some savannah populations in Chad and northern Cameroon were also classified as endangered under IUCN in 2008 (Maisels et al., 2013).

Despite this, the decline of the elephant population continued to remain a serious threat with poaching continuing unabated. Within 3 years, between 2010 and 2012, poachers across the African continent killed 100,000 elephants (Wittemyer et al., 2014), and in 2013, over 20,000 elephants were illegally killed (CITES, 2014). Spurred by the insatiable demand for ivory products in Asia especially in China, global ivory trade continued to increase in 2015, resulting in increasing levels of illegal killing of elephants. The situation was particularly serious in Central and West Africa, and was seen as posing an immediate threat to their continued survival (Milliken et al., 2016; CITES, 2016). It has been estimated that in central Africa, the elephant population has declined by 64% in a decade (Wittemyer et al., 2014; see also, Maisels et al., 2013). In 2011 less than 2% of the Central African forest contained elephants at high density.

In Central Africa Republic in Central Africa, rampant ivory poaching in the 1970s and 1980s had resulted in the drastic decline of elephant populations, dropping from around 35,000 in the 1970s to 4000 by the mid-1980s across the country. In 2005 conservationists found 929 elephants thriving in the region, but 5 years later there were just 68. An aerial survey conducted in 2017 indicated that elephants may have been wiped out in northern Central African Republic (Popescu, 2018). Even for Gabon, in the year 2011, high density populations were found in only 14% of the forest, representing a decline of over 18% between 2002 and 2011. Despite the effective ban of worldwide trade in ivory in January 1990, market-driven illegal trade for ivory was over three times larger than in 1998 and had more than doubled since 2007 (Lawson and Vine, 2014).

In a recent study based on Great Elephant Census (GEC) aerial surveys conducted in 18 countries in 2014−15 covering an area of 218,238 km^2 (about 24% of the total ecosystem area), it was revealed that poaching has overwhelmingly driven a huge decline in Africa's savannah elephants with almost 30% wiped out between 2007 and 2014. The 18 countries targeted include Angola, Botswana, Cameroon, Chad, DR Congo, Ethiopia, Kenya, Malawi, Mali, Mozambique, South Africa, Tanzania, Uganda, West Africa, Zambia, and Zimbabwe.

Around 144,000 animals were killed over this 7-year period. The total population of the 18 countries surveyed was disclosed at 352,271 elephants and the rate of annual population decrease was estimated at 8% primarily due to ivory poaching. The biggest drop in numbers was recorded in Angola, Mozambique, and Tanzania and low numbers were found in northern Cameroon and south-west Zambia. Tanzania and Mozambique had a combined loss of 73,000 elephants due to poaching in just 5 years.

It is noteworthy that in northern Cameroon, the survey team counted no more than 148 elephants, and many carcasses were found in the area surveyed. Indeed, the all-carcass ratio for the entire surveyed area was the highest in Cameroon at 83% which reflects that the regional population is facing the immediate threat of local extinction. In addition, the all-carcass ratio for Mozambique (32%), Angola (30%), and Tanzania (26%), reflecting declining population trends over the 4 years prior to the GEC survey (Chase et al., 2016; Steyn, 2016). Briefly noted, a carcass ratio above 8% generally reflects a population declining trend. On the whole, Africa's elephants are in crisis, with around 20,000 killed in 2014 alone for their tusks, at a rate faster than they are being born (Aldred, 2016; see also, Steyn, 2016).

Even the elephants that are protected in the conservation areas are threatened. For example, according to WWF, the elephant populations in Tanzania's Selous national park, a world heritage site and Tanzania's largest PA as well as one of the largest wilderness areas left in Africa, are heading toward an irreversible ecological collapse owing to uncontrolled mining activities and rampant ivory poaching. About four decades ago, the national park had nearly 110,000 elephants. But rampant poaching has reduced their numbers by 90% to about 15,000 today. Poaching was particularly acute between 2010 and 2013 with an average of six elephants being killed every day by poaching syndicates (WWF, 2016d). However, recent findings indicate that there has been a steady decline in poaching levels since its peak in 2011. An analysis in 2016 concludes that overall poaching trends have now dropped to pre-2008 levels. Despite this, elephant populations in Africa continue to fall as a result of continued illegal killing, land transformation and rapid human expansion (CITES Secretariat, 2017).

3.11.6.2 *The plight of the African rhino*

Similarly the rhino population in South Africa—home to 83% of Africa's rhino and 73% of all wild rhino worldwide—is heading toward the irreversible course of ecological extirpation due to escalating poaching and illegal trade in rhino horns. It was reported that by the beginning of 2011, there were 20,165 white rhinos and 4880 black rhinos in Africa. However, their numbers rapidly declined due to poaching. Between 2006 and September 2012, at least 1997 rhinos were poached mostly for their horns and over 4000 rhino horns have been illegally exported from Africa since 2009 (IUCN, 2013b). Also the poaching rates of the southern white rhino in South Africa have increased from 13 illegal kills in 2007 to 1215 in 2014 (Di Minin et al., 2015; Save the Rhino, 2019). During the rampant poaching of elephants and rhinos in the 1980s, the anthropogenic force of ecological exploitation was so destructive that wildlife could survive in Africa only in parks patrolled by armed guards, and often within fenced areas (Conniff, 2009).

In 2012 at least 745 rhinos were killed throughout Africa and of these, 668 rhinos were killed in South Africa alone (IUCN, 2013b). The number of rhinos illegally killed for their

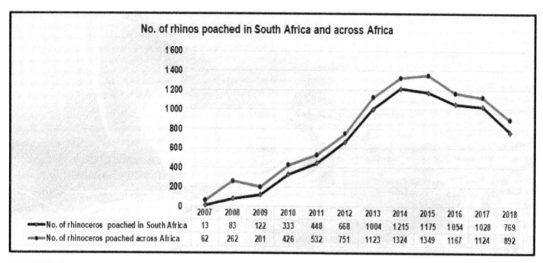

FIGURE 3.12   Number of rhinos poached in South Africa and across Africa. *Data from, Save the Rhino, 2019. Poaching Statistics. Save the Rhino International, London. <https://www.savetherhino.org/rhino_info/poaching_statistics> (accessed 10.12.19.).*

horns in South Africa reached 1004 in 2013—an average of three rhino were being killed each day, and almost double the number in 2012 (Chow, 2014). Rhino poaching in South Africa increased by more than 9000% between 2007 and 2014—from 13 to a record 1215 in 2014 (WWF, 2016d; Save the Rhino, 2019). Most of the illegal poaching activity occurred in Kruger National Park, a 19,485 km^2 protected habitat on South Africa's north-eastern border with Mozambique. The Park consistently suffered heavy poaching loses (Save the Rhino, 2019). The poaching rate for rhino in Africa continued to rise in 2015, with 1342 rhinos killed for their horns (Emslie et al., 2016). South Africa continued to lead as having the highest number of rhinos killed at 1175 heads compared with 1349 through Africa (Save the Rhino, 2019; Fig. 3.12). It is estimated that there were just 20,378 white rhino and 5250 black rhinos left in Africa in 2015 (Emslie et al., 2016). A recent study indicates that the number of rhinos killed for their horns in South Africa declined slightly in 2016, 2017 and 2018 (Fig. 3.12). Despite this, more than 1000 were killed (Save the Rhino, 2019). In 2019 South Africa further lost another 594 rhinos to poachers (AFP, 2020).

3.11.6.3 *The African wildlife crisis in general*

What is more, even common species of least concern have not been spared from the threat of extinction. The giraffe, for example, which is widespread across southern and eastern Africa has been recategorized from Least Concern to Vulnerable under the IUCN categories of species due to the dramatic decline of their population level by about 40% from approximately 151,702 to 163,452 in 1985 to 97,562 in 2015. The major causes were habitat loss, civil unrest and illegal hunting. The species is now facing a high risk of extinction in the wild in the medium-term future (IUCN, 2016a).

Similarly the Plains Zebra (*Equus quagga*) also known as the common zebra, which is the most abundant of the three species of zebra widely found on the grasslands of eastern and southern Africa has been recategorized from Least Concern to Near Threatened under the IUCN categories of species. The population has been reduced by 24% in the past 14 years from around 660,000 to a current estimate of just over 500,000 animals due to unrestrained hunting (IUCN, 2016b). In addition, the latest IUCN assessment reveals that the Eastern Gorilla has been listed as Critically Endangered due to drastic population decline of more than 70% in 20 years. It is estimated that its population has now dropped to fewer than 5000 individuals. The population of Grauer's Gorilla, a subspecies of Eastern Gorilla has also reduced by 77% since 1994, declining from 16,900 individuals to just 3800 in 2015. Although hunting of these endangered species is prohibited by law, illegal hunting remains the greatest threat to their continued survival (IUCN, 2016b). Three species of antelope found in Africa—Bay Duiker (*Cephalophus dorsalis*), White-bellied Duiker (*Cephalophus leucogaster*), and Yellow-backed Duiker (*Cephalophus silvicultor*)—have been recategorized from Least Concern to Near Threatened under the IUCN categories of species.

Other species such as lions, leopards, wild dog, buffalo, hippo, crocodile, and over 400 bird species are also threatened by large scale mining. Nearly 75% of the national park is covered by oil and gas concessions, and 54 mining concessions have been awarded. In 2012 the boundary of the national park was modified to enable the construction of a large-scale uranium mine in the southern area of the reserve (Vidal, 2016). It is projected that if these unsustainable activities are left unchecked, elephants could vanish from Selous by early 2022 (WWF, 2016d).

In addition, the vultures, the carnivorous birds which help stem spread of disease by eating carcasses that would otherwise rot are also at the brink of ecological extirpation. The Bird Life International found that six of Africa's 11 vulture species were at risk of extinction due to deliberate targeting by poachers, as the birds, which circle the sites where they feed, can alert authorities to the carcasses of illegally killed animals especially the African elephant and rhino for their ivory. Since the late 1980s, 98% of West Africa's vultures outside PAs have vanished, while half the population of the Gyps vulture species in Kenya's Maasai Mara park has gone. In South Africa, the number of Cape vultures has declined by 60%–70% over the past 20–30 years (Stoddard, 2015).

Similarly based on a scientific study, the lions, considered as "vulnerable" by the IUCN are declining rapidly to a critically endangered level as many of the species "are either now gone or expected to disappear within the next few decades" (Bauer et al., 2015). The African coastal and marine ecosystems are also not spared from human unsustainable practices as evidenced by decline in fish stock and increased rate of biological erosion caused by pollution and acidification from land-based sources. Coastal habitats such as mangroves, sea-grass beds, salt marshes, and shellfish reefs continue to decline, threatening biocapacity of many highly valuable ecosystem services that underpin human long-term existence (UNEP, 2010a).

The continued existence of the mandrills, the world's largest monkeys, which roam in the Congo Rainforest basin covering Cameroon, the Central African Republic, the Congo, the Democratic Republic of the Congo Equatorial Guinea, and Gabon, is at risk due to illegal hunting, habitat destruction caused by logging and mining activities. The wild vertebrate species, such as chimpanzees, Jentink's duiker, mandrills, and the pygmy hippopotamus, are threatened or in danger of extinction.

On a specific note, in Kenya, many wildlife species protected in the national parks and conservation areas are heading to the brink of extinction. For example, wildlife populations declined 63% in Tsavo East and West between 1977 and 1997 and 78% in Meru between 1977 and 2000. The overall decline of wildlife population in five national parks, namely, Tsavo East and West National Park, Amboseli, Nakuru, Nairobi, and Meru between 1977 and 1997 was 41% (Western et al., 2009). Poaching, including poaching across the border from Somalia, range loss and habitat fragmentation due to agricultural expansion are some of the major causes of wildlife decline in Kenya. Also extensive tracts of wetland ecosystems in Kenya have been converted to farmland resulting in irreversible ecological destruction. Unsustainable human activities have taken a heavy toll on the once dense Mau Forest, threatening the very existence of the ecologically and economically important Masai Mara Game Reserve, and the Sondu Miriu and Mara rivers (UNEP, 2010a). The unsustainable human activity is well exhibited in a cruel and heartless extermination of the white giraffes in Kenya. In a more specific note, two of only three white giraffes living in Ishaqbini Hirola Community Conservancy in Garissa County have been killed and butchered by armed poachers (Daly, 2020). This is a clear indication that the survival of wildlife species in Kenya is increasingly threatened by human destructive ecological activity.

Similarly in Madagascar, the lemurs, one of the unique species in the country which evolved roughly 150 million years ago, are facing the risk of extinction. Indeed, 15 species of lemurs have been driven to extinction (UNEP, 2010a). A new assessment conducted by IUCN revealed that the Madagascar Lemurs are becoming the most threatened mammal species in the world (Conservation International, 2012; Schwitzer et al., 2014). Many lemur species are on the brink of extinction and many are becoming critically endangered primarily due to habitat loss caused by the destruction of rainforests, accelerated illegal logging, illegal hunting and mining activities. C. Schwitzer together with 18 other authors estimated that 90 of the 101 known lemur species in the country are threatened with extinction, including 22 that are critically endangered (Schwitzer et al., 2014).

According to another new study funded by Conservation International's Primate Action Fund and the University of Victoria, the number of ring-tailed lemurs is thought to have plummeted to less than 2500 in Madagascar, the only place where they exist. The decline is primarily a result of habitat destruction, bushmeat hunting, illegal capture for the pet trade and open-pit sapphire mining activities (Scott, 2016). For the past 3 years, it has been reported that residents of Madagascar had stolen 28,000 lemurs from the wild, which is against Madagascar law prohibiting private ownership of the species as pets (Platt, 2015). Only three sites in the island country are known to harbor more than 200 species of the ring-tailed lemurs while 12 sites now have populations of 30 or less. However, there are 15 sites where ring-tailed lemurs have either become locally extinct or have a high probability of disappearing soon (Scott, 2016).

Overall, as revealed in the second edition of the "State of Biodiversity in Africa" published in 2016, biodiversity in Africa including the DRC continues to decline with ongoing losses of species and habitats driven by human unsustainable environmental practices such as deforestation, forest degradation and habitat destruction, illegal hunting, and wildlife trade (UNEP-WCMC, 2016). The Report further revealed that even though many African countries have already achieved their 17% terrestrial PA targets, the threat to species abundance has increased. In 2014, 6419 species of animals and 3148 species of plants in the African continent were recorded as threatened with extinction on the IUCN Red List.

In addition, 21% of all freshwater species are recorded as threatened while the bird population shows a decline over the past 25 years with many facing the risk of extinction. Also two African species of pangolins, the while-bellied (*Phataginus tricuspis*) and the giant ground pangolin (*Smutsia gigantea*), which were previously listed as "vulnerable" on the IUCN Red List have been moved to the "endangered" category due to drastic population decline caused by habitat loss and large-scale poaching (Mongabay, 2019). In general, the vertebrate population in Africa has declined by roughly 39% since 1970 with Western and Central Africa exhibiting the most rapid rate of ecological erosion. Also over the past two decades, mangroves, moist and seasonal dry forests and wetlands have all declined significantly.

The following is an assessment of the impact of CBD specifically on the biological conservation initiatives of the transboundary conservation region in Africa—one of the most biologically diverse territories on earth. The purpose is to reveal lucidly to what extent the legally binding CBD has been able to save some of the world's most important biological resources based on PAs or site-based conservation mechanisms.

3.11.6.4 Transboundary PAs: The W-Arly-Pendjari Parks Complex

The W-Arly-Pendjari (WAP) Parks Complex straddles the countries of Benin, Burkina Faso, and Niger. The Complex covers an area of 31,000 km² (3.1 million ha) or 50,000 km² (5 million ha) if riparian areas are included. This transboundary Complex which is one of the largest contiguous PAs in Africa, comprises a matrix of terrestrial, semiaquatic, and aquatic ecosystems (AU-IBAR, 2012). Respectively, Benin, Burkina Faso and Niger contribute 43, 36 and 21% of the land area to the formation of the WAP Parks Complex (UNDP, 2004). The biologically diverse ecological Complex is home to more than half of West Africa's elephant population. Its natural environment is also the only natural refuge remaining for most of the vulnerable and threatened animal species in the transboundary region (AU-IBAR, 2012). The Niger's W National Park lie within the Complex acquired World Heritage status in 1996. However, the Complex has become a new site on the World Heritage List in 2017 (UNESCO, 2017; Fig. 3.13).

The transboundary PAs in the region include Nyungwe forest (Rwanda)/Kibira National Park (Burundi); Great Limpopo Transfrontier Park (Mozambique, South Africa, and Zimbabwe); and the W-Arly-Pendjari (WAP) Parks Complex in Benin, Burkina Faso and the Niger, are becoming a focus for poaching, illegal cattle grazing and other unsustainable human activities. These increasingly threaten the ecological sustainability of the complex biological systems (UNEP, 2008, 2010a). In the WAP Parks Complex, for example, the "cotton belt" in northern Benin has markedly altered the natural vegetation over the last 20 years and protected lands of the Complex have become almost completely surrounded by agricultural land that is ecologically destructive (UNEP, 2008).

In particular, the part of the Complex lying within Niger which was listed as a World Heritage site in 1996 as noted above has been increasingly exposed to human threats such as agricultural expansion, poaching, uncontrolled bushfires, siltation and pollution of surface waters, and unsustainable harvesting of Non-Timber Forest Products, uncontrolled timber harvesting and overexploitation of fish resources (UNDP, 2004). Indeed, recent findings reveal that human pressures such as livestock grazing and gathering of forest products are

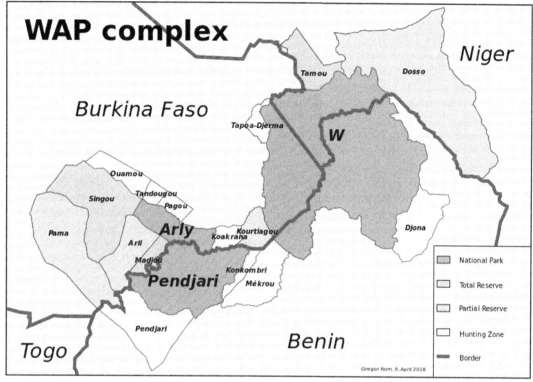

FIGURE 3.13 WAP Parks Complex (IUCN Protected Area). *IUCN*, International Union for Conservation of Nature; *WAP*, W-Arly-Pendjari. *Reproduced with permission from, Gregor Rom.* <*https://en.wikipedia.org/wiki/W-Arly-Pendjari_Complex#/media/File:WAP-Komplex_englisch.svg*>.

increasingly threatening the environmental integrity of the Complex (Harris et al., 2019). All these occurred despite the fact that the management of WAP Complex is supported by a number of national, regional and international laws such as United Nations Convention on Biological Diversity (UNCBD), United Nations Convention to Combat Desertification (UNCCD) and the Convention on International Trade in Endangered Species of Wild Fauna and Flora (CITES), among others, and has in place a well-established park management machinery comprising the Ministry Committee of Orientation, the Technical Committee of Control, and Parks and Wild Life Reserve Officers (Amahowé et al., 2013).

3.12 Southeast Asia

3.12.1 The United Nations environmental protection initiatives and its implications on the ASEAN-5's environmental conservation efforts

Despite covering only 3% of the Earth's surface, the Southeast Asia region is exceptionally rich in biodiversity. The region contains 20% of the planet's vertebrate and plant species (ASEAN Secretariat, 2009; Braimoh et al., 2011). In addition to its incredibly rich biological resources, the region also has an extraordinarily high rate of species discovery. For example, 2216 new species were discovered between 1997 and 2014 alone in the Greater Mekong Region which comprises Cambodia, Laos, Myanmar, Thailand, and Vietnam (WWF, 2015b). Also since 1997, more than 2500 species including the mountain horseshoe bat and the Vietnamese crocodile lizard have been discovered (WWF, 2017b). The region also has a remarkable concentration of species with high levels of endemism. More specifically it has the highest proportion of endemic bird and mammal species (9% and 11%) and the second highest proportion of endemic vascular plant species (25%) compared with the tropical regions of Meso-America, South America, and sub-Saharan Africa (ASEAN Secretariat, 2017; see also, ASEAN Secretariat, 2009).

Against this backdrop, the following is an assessment of the extent to which the United Nations' environmental protection initiatives and international agreements, in particular the CBD, have been able to help the region protect its natural environment and biodiversity. The analysis will confine its assessment to five selected countries in the region, namely, Indonesia, Malaysia, the Philippines, Thailand, and Vietnam, referred to here as the ASEAN-5.

To begin with, the ASEAN-5 region contains three of the world's 17 mega-diverse countries, namely, Indonesia, Malaysia, and the Philippines as identified by Conservation International in 1988 (UNEP-WCMC, 2014). Indonesia is notable; while occupying only 1.3% of the earth's surface, its forests have some 10% of the world's plants, 12% of the world's mammals, 16% of the world's reptile-amphibians, and 17% of the world's bird species (Ministry of National Development Planning, Indonesia, 1993; Margono et al., 2014). The latest estimates indicate that Indonesia is ranked in the world's top three out of the global-10 megadiverse countries after Brazil and Colombia (Butler, 2016b). Malaysia (ranking 15th on the global megadiverse list), Vietnam (16th), Thailand (18th), and the Philippines (24th) are also home to high concentrations of endemic species including various endangered species such as the orangutan, Javan rhino, Sumatran rhino, and Schomburgk's deer, among others (Vié at al., 2009; ASEAN Secretariat, 2009; Butler, 2013c, 2016b; Choy, 2015a).

Being signatories to the CBD, the ASEAN-5 countries have taken the initiative to protect their biological resources diversity to prevent significant reduction or loss of biological diversity and to enhance biodiversity conservation and sustainable use of its components as stipulated under the 1992 UNCBD. Each country has put in place comprehensive legal instruments to promote these legally binding environmental obligations (Table 3.6). In compliance with Article 26 of the UNCBD, each country has also submitted its national report to the Convention to provide information on the progress for the implementation of the Strategic Plan for Biodiversity toward achieving the Aichi Biodiversity Targets (CBD Secretariat, 2018). Each country has also ratified the CITES in an attempt to enhance biodiversity conservation (Table 3.7).

Recognizing the importance of environmental cooperation for sustainable development, the Southeast Asian leaders have, since 1977, cooperated closely in promoting ecologically sustainable development. Leaders in the region have also adopted various documents as listed in Table 3.8.

One of the most important documents which is the ASEAN Agreement on the Conservation of Nature and Natural Resources, the only legally binding agreement in the ASEAN. This document which formally came into force in 1995 provides some of the most holistic guidelines for the design of sustainable environmental planning and management

TABLE 3.6 National environmental laws in Southeast Asia (selected countries).

Indonesia	Malaysia	The Philippines	Thailand	Vietnam
Environmental Management Act (1997, revised 1982, 1990)	National Policy on Biological Diversity (1998)	Executive Order 578 (2006): a decree which seeks to "protect, conserve, and sustainably use biological diversity to ensure and secure the wellbeing of the present and future generations of Filipinos"	The Forest Act (1941, amended in 1948, 1982, and 1989)	Forest Protection and Development Law (1991, amended 2004)
Forestry Law (1999)	National Forestry Policy (1978, revised 1993)	Presidential Decree 1151 or Philippine Environmental Policy (1977): a decree which seeks to promote a harmonious relationship with nature, sustainable resource use and the principle of intergenerational equity	Wildlife Reservations and Protection Act (1960, amended in 1992)	Land Use Law (1993, amended 1998, 2003)
Conservation Law (1990)	Environmental Protection Enactment, Sabah (2002, amended 2004)	Presidential Decree 1152 or Philippine Environment Code (1977) (mandates the Department of Environment and Natural Resources to establish a system of national resource exploitation and wildlife conservation)	National Park Act (1961)	Environmental Protection law (1993, amended 2005)
Environmental Protection Law (2009)	Sabah Forestry Policy (2005)	Presidential Decree 705 or the Revised Forestry Code of the Philippines (1975) (primary legal instrument guiding the use and management of forest resources and biodiversity conservation)	National Forest Reserve Act (1964)	Law on Fisheries (2003)
Cultivation Law (1992)	Natural Resources and Environment Ordinance, Sarawak (1993, amended 1997, 2001)	Republic Act 7586 or the National Integrated Protected Area System Act, 1992	Reforestation Act (1992)	Ordinance on Plant Varieties (2004)

(Continued)

TABLE 3.6   (Continued)

Indonesia	Malaysia	The Philippines	Thailand	Vietnam
Plantation Law (2004)	Wildlife Protection Ordinance, Sarawak (1998)	Sustainable Forest Management Act (2005, 2009)	Tambol Council and Tambol Administration Organization Act (1994) (aims to strengthen the role of local government in sustainable natural resource use and management)	Biodiversity Law (2008)
Ministry of Agriculture Regulation (2007)	National Parks and Nature Reserves Ordinance, Sarawak (1998)	Republic Act 9147 or the Wildlife Resources Conservation and Protection Act (2001)	Thailand Constitution (2007) (seeks to promote systematic management and use of natural resources for the benefit of the public)	Decree 32 on the Management of Endangered, Precious, and Rare Species of Wild Plants and Animals (2006)

Based on, Choy, Y.K., 2015a. Sustainable resource management and ecological conservation of mega-biodiversity: the Southeast Asian Big-3 reality. Int. J. Environ. Sci. Dev. 6 (11), 876–882. Choy, Y.K., 2016a. Economic growth, sustainable development and ecological conservation in the Asian developing countries: the way forward. In: Indraneil Das, I., Tuen, A.A. (Eds.), Naturalists, Explorers and Field Scientists in South-East Asia and Australasia. Topics in Biodiversity and Conservation Series, vol. 15. Springer, Cham, Heidelberg, New York, Dordrecht, London, pp. 239–283.

TABLE 3.7   ASEAN-5 participation in international treaties and conventions.

	Signatory to the CBD by ratification	CITES by ratification
Indonesia	November 21, 1994	March 28, 1979
Malaysia	September 22, 1994	January 18, 1978
The Philippines	January 6, 1994	November 16, 1981
Thailand	January 29, 2004	April 21, 1983
Vietnam	February 14, 1995	April 20, 1994

CBD, Convention on Biological Diversity; CITES, Contracting party to Convention on International Trade in Endangered Species of Wild Fauna and Flora. Based on, CITES Secretariat, 2013. List of Contracting Parties. Convention on International Trade in Endangered Species of Wild Fauna and Flora (CITES). The Convention on International Trade in Endangered Species of Wild Fauna and Flora (CITES), Geneva, Switzerland. Available from: <https://www.cites.org/eng/disc/parties/chronolo.php>.

framework. Article 1 of the Agreement explicitly states that each member country should undertake to adopt measures in accordance with its national laws to protect the ecological life support system, to preserve genetic diversity, and to ensure the sustainable use of natural resources. It further states that member countries should undertake to integrate ecological conservation and sustainable management of natural resources in development planning process (Choy, 2015a,b).

Despite the abovementioned far-reaching environmental control initiatives, casual observation on the ground indicates that increasing environmental protection efforts do

TABLE 3.8 ASEAN environmental agreement.

ASEAN environmental agreement
Manila Declaration on the ASEAN Environment (1981)
ASEAN Declaration on Heritage Parks and Reserves(1984)
Bangkok Declaration on the ASEAN Environment (1984)
ASEAN Agreement on the Conservation of Nature and Natural Resources (1985)
Jakarta Resolution on Sustainable Development (1987)
Manila Declaration of 1987
Kuala Lumpur Accord on the Environment and Development (1990)
Singapore Resolution on Environment and Development (1992)
Bandar Seri Begawan Resolution on Environment and Development (1994)
ASEAN Vision 2020 (1997)
Jakarta Declaration on Environment and Development (1997)
Kota Kinabalu Resolution on the Environment (2000)
ASEAN Concord II (Bali Concord II) (2003)
Yangon Resolution on Sustainable Development (2003)
Joint Declaration on the Attainment of the Millennium Development Goals in ASEAN (2009)
Statement by the ASEAN Environmental Ministers for the eleventh meeting of the conference of the parties to the Convention on Biological Biodiversity 2012

Based on, Choy, Y.K., 2015a. Sustainable resource management and ecological conservation of mega-biodiversity: the Southeast Asian Big-3 reality. Int J Environ Sci Dev 6 (11), 876−882. Choy, Y.K., 2015b. From Stockholm to Rio + 20: the ASEAN environmental paradox, environmental sustainability and environmental ethics. Int J Environ Sustain 12, 1−25.

not seem to have been able to protect the regional biological environment from ecological impoverishment. The following discussion shows how the extensive deforestation discussed above (Sections 3.6 and 3.7), resulting in widespread habitat fragmentation, is increasingly threatening the continued survival of many unique, rare and critically endangered species in the ASEAN-5 region. This will be systematically examined below.

3.12.2 Indonesia: Biodiversity conservation efforts

In Indonesia, the evolution of a strategic concept for biodiversity started with the publication of a book entitled *Biodiversity for the Survival of the Nation* in 1989. This was followed by the Indonesian Country Study on Biological Diversity, completed in 1991. The study was conducted in response to the UNEP's need for documentation on biodiversity to be discussed at the UNCED in Rio de Janeiro in 1992. As an indication of its serious commitment in implementing the CBD, following the signing of CBD in 1992, it established the National Strategy for the Management of Biodiversity and Biodiversity Action Plan (BAP) in 1993 to guide sustainable use of the country's biodiversity for the benefit of the present as well as future generations (Ministry of National Development Planning, Indonesia,

1993). The Action Plan sets out several strategies for biodiversity conservation, namely, in situ conservation in terrestrial parks and PAs; in situ conservation outside the PA network (production forests, wetlands, agricultural lands, coastal, and marine conservation), and ex situ conservation. Indonesia ratified the CBD in 1994.

In 2003 a second document, titled the "Indonesian Biodiversity Strategy and Action Plan" (IBSAP) 2003–20, was developed which aimed: (1) to encourage changes in the attitude and behavior of Indonesian individuals and society, as well as in existing institutions and legal instruments, so as to increase concern about conservation and utilization of biodiversity, for the welfare of the community, in harmony with national laws and international conventions; (2) to apply scientific and technological inputs, and local wisdom when dealing with conservation issues; (3) to implement balanced conservation and sustainable use of biodiversity; (4) to strengthen institutions and law enforcement related to biodiversity conservation; and (5) to resolve conflicts over natural resources (Nalang, 2003).

Apart from the various environmental laws, shown in Table 3.6, to enhance environmental conservation, Indonesia has also put in place Law No. 5/1990 on Conservation of Biological Resources and Ecosystems to regulate ecosystems and species conservation especially in PAs. Also since 2013, IBSAP has been updated in accordance with the 10th CBD Conference of the Parties (COP) in Nagoya (Ministry of Environment and Forestry of Indonesia, 2014). Furthermore to strengthen biodiversity conservation efforts, the Indonesian government has established 566 national parks covering 36,069,368 ha consisting of 490 terrestrial PAs (22,540,170 ha) and 76 marine PAs (13,529,197 ha) as of 2010 (Partono, 2011).

3.12.3 Indonesia: The state of biological diversity

Despite this, the massive scale of environmental transformation as a result of deforestation largely caused by commercial oil palm plantation development (see, Section 3.6) is increasingly threatening the continued survival of many of the country's endangered species. In addition to the argument as expounded in Section 3.6, it was reported that from 2000 to 2009, dry primary forest shrank from 42.25 to 32.18 million ha and secondary swamp forest diminished while plantation areas were increasing (Ministry of Environment and Forestry of Indonesia, 2014). Extensive deforestation, habitat fragmentation and destruction in the past few decades have led not only to the extinction of the Balinese and Javan tigers in Indonesia but is also increasingly threatening the continued survival of a wide range of rare and critically endangered animal species such as the Sumatran tiger, the Sumatran elephant, the Javan Rhino, and the orangutan (WWF, 2014; Brown and Jacobson, 2005; Uryu et al., 2008; Choy, 2015a).

For example, in Riau, Sumatra, the critically endangered Sumatran elephant which thrived well in the once thick lowland forest in Tesso Nilo National Park, has declined in numbers by up to 84% from an average of 1342 in 1984 to approximately 210 in 2007, this being largely attributed to poaching, and aggravated by extensive habitat fragmentation caused by commercial agricultural expansion and timber harvesting (Uryu et al., 2008; Nicholas Jong, 2018). There are now less than 2000 Sumatra elephants in the wild in Indonesia (JG, 2018). The above unsustainable practices also led to the local extinction of the elephant population in Rokan Hilir, Kerumutan, Koto Panjang, Bukit Rimbang Baling, Tanjung Pauh, and Bukit Suligi by 2007 (Uryu et al., 2008).

Similarly habitat fragmentation caused by deforestation and uncontrolled expansion of oil palm plantations coupled with massive poaching resulted in an alarming rate of decline of the Sumatran tiger population in Riau by 70% from 640 in 1982, to 192 in 2007 (Uryu et al., 2008, Afrizal, 2018). It may be noted that the habitat centers of the tiger shrank from 29 pockets in 2010 to 23 in 2016. There are four Sumatran tiger habitats in Jambi, they are Kerinci Seblat, Bukit Tigapuluh, Berbak-Sembilang, and Harapan Forest. The Environment and Forestry Ministry suggests that the Sumatran tiger population currently stands at no more than 600 (Afrizal, 2018).

Also the rapid depletion of lowland forests, the native habitat for the orangutan, due to oil palm plantation expansion coupled with reckless hunting and poaching has led to a drastic decline of the iconic ape population. In Kalimantan, one of the prime regions for oil palm plantation development, for example, it is estimated that between 1950 and 3100 orangutans are killed annually (Meijaard et al., 2011). It is estimated that more than half of the orangutan population, about 148,500, has been lost from the forests in the 16 years between 1999 and 2015 due to habitat destruction and massive killing and hunting. The largest loss was found in Western Schwaner (42,700 individuals) followed by Eastern Schwaner (20,100 individuals) and Karangan (8200 individuals) (Voigt et al., 2018).

Currently Indonesia is considered as one of the top 10 countries in the world with the greatest number of threatened species, and is a global hotspot of great conservation concern (Hickey et al., 2004; Yeager, 2008; ACB, 2010). It is also relevant to point out that the Balinese, Caspian and Javan tigers have been declared extinct in Indonesia (see, e.g., Jackson and Nowell, 2008; Hanel, 2009). There are fewer than 400 Sumatran tigers left in Indonesia's wilderness and they are at serious risk of going extinct. Most of these animals are killed deliberately for commercial gain. It is reported that poaching for trade was responsible for almost 80% of Sumatran tiger deaths, that is, at least 40 tigers were killed per year (WWF, 2019).

Another case in point is the helmeted hornbill known as rangkong gading locally, the largest hornbill species found in Asia. Not counting the tail feathers, its body length is between 110 and 120 cm long and weighs about 3 kg. It is restricted mainly to the semievergreen and evergreen lowland forests up to 1500 m in Indonesian Borneo (Kalimantan) and Sumatra, Malaysian Borneo (Sarawak and Sabah), Brunei Darussalam, Myanmar and the far south of Thailand (EIA, 2016; IUCN, 2016c; BirdLife International, 2019). The hornbill is cited as a CITES-listed Appendix I species, meaning that the species is given the highest protection level in that all international commercial trade in the species is illegal and prohibited. Not only is it listed as a Near Threatened species by the International Union for the Conservation of Nature (IUCN) but it is also fully protected under various local laws.

Despite this, since 2011, hunting pressure caused by rampant poaching and unprecedented illegal trade in this species for its beak ivory, known as casques, coupled with habitat loss due to logging and oil palm plantation development, are driving the species to the brink of extinction (see, e.g., Krishnasamy et al., 2016; Laman, 2018, Arbi, 2019). In 2013 it was estimated that 500 helmeted hornbills were poached every month from three areas within Indonesia (EIA, 2016). However, other sources revealed that in that year, roughly 6000 helmeted hornbills were killed in West Kalimantan alone (Laman, 2018; Arbi, 2019). Between 2011 and 2016, about 1400 attempts to smuggle the hornbill's casques were foiled by the Indonesian authorities (Arbi, 2019).

Due to its ecologically vulnerable position, the species has been up-listed to the Critically Endangered category in 2015 (IUCN, 2016c). Despite this, the present trends of poaching and illegal trade are expected to remain unabated or even to escalate due to the increasing demand for Helmeted Hornbill products in recent years (Beastall et al., 2016). If the current level of illegal hunting pressure is not contained effectively, the continued existence of the hornbill will be jeopardized.

In acknowledging the critical ecological status of this endangered species, the government of Indonesia, in cooperation with the US Agency for International Development (USAID) through a project named Build Indonesia to Take Care of Nature for Sustainability, or Bangun Indonesia untuk Jaga Alam demi Keberlanjutan, is strengthening efforts to protect the helmeted hornbill. Also the Indonesian Ministry of Environment and Forestry has recently launched and ratified the new 10-year national conservation strategic action plan (2018–28) for the Helmeted Hornbill to intensify its biological conservation efforts for this critically endangered species (Jain et al., 2018). To what extent these are effective is still too early to assess.

3.12.4 Malaysia: Biodiversity conservation efforts

In Malaysia, the National Policy on Biological Diversity (NPBD) was developed after the Earth Summit in June 1992 and launched in 1998. In early 1993, the country established a National Steering Committee on Biological Diversity to coordinate implementation of the CBD. It set up the NPBD in 1998 to guide the implementation of the national biodiversity strategies and action plans to fulfill its obligation under CBD. As an indication of its serious commitment to implement CBD upon its ratification in 1994, the government started incorporating biodiversity concerns into its 5-year national development plans.

The Seventh Malaysia Plan (1996–2000), for example, explicitly stated that economic growth should not be achieved at the expense of remaining environmental and natural resources while the Eighth Malaysia Plan (2001–05) stated that the Biodiversity Action Plan (BAP) would be implemented in all the 13 states of Malaysia (Raman, 2002). The government further adopted the National Physical Plan 2 (NPP2) in 2010 to guide sustainable development at all levels of planning. To reinforce biodiversity conservation, the Wildlife Conservation Act was enacted in 2010 to replace the Protection of Wildlife Act 1972, imposing stronger punitive measures for unsustainable wildlife offences particularly in relation to illegal wildlife trade.

At the same time, a host of conservation programmes and action plans were developed to further strengthen ecological conservation efforts. These included the National Tiger Conservation Action Plan, The National Elephant Conservation Action Plan for Peninsular Malaysia, the Orangutan Action Plan 2012–16, and the Elephant Action Plan for Sabah 2012–16, among others. Malaysia is also committed to maintaining at least 50% of its land area under forest and tree cover in perpetuity as pledged under the 1992 Rio Earth Summit through the creation of protected forests and the application of SFM practices (Ministry of Natural Resources and Environment, Malaysia, 2014).

The NPBD which was updated in 2014/15 aims to promote the conservation of biological diversity and sustainable use of its components in compliance with Article 6 of CBD based on in situ and ex situ conservation. The policy gives directives and serves as a guide to all government agencies throughout the 13 states in Malaysia on the conservation and management of biological diversity. It outlines 15 strategies and 85 action plans. The strategies include: (1) enhancing sustainable utilization of the components of biological diversity; (2) strengthening the institutional framework for biological diversity management; (3) integrating biological diversity consideration into sectoral planning strategy; (4) reviewing legislation to reflect biological diversity needs; and (5) minimizing the impact of human activities on biological diversity, among others. To strengthen institutional capacity in biological conservation, the Biodiversity Council was established in 2001.

The institutional framework was enhanced further through the creation of the Ministry of Natural Resources and Environment in 2004 (Ministry of Natural Resources and Environment, Malaysia, 2006). Accordingly a total of 3.43 million ha of natural forest were designated as protection forest. In addition, 2.12 million ha of forest were demarcated as national and state parks, wildlife sanctuaries, turtle sanctuaries and wildlife reserves (Ministry of Science, Technology and the Environment, Malaysia, 1998). The conservation of terrestrial biological resources in production lands were strengthened and tightened considerably (Ministry of Natural Resources and Environment, Malaysia, 2005a). The country also formulated the National Physical Plan (NPP) which includes conservation of natural resources and the environment as a major element in the national physical development. The plan states that Environment Sensitive Areas shall be integrated in the planning and management of land use and natural resources to ensure sustainable development (Ministry of Natural Resources and Environment, Malaysia, 2005b).

It is reported in its Fourth National Report to the CBD that in 2007, of the 19.6 million ha of forested areas in Malaysia, 14.3 million ha (43.4% of total land area) would be gazetted as permanent reserve forest, and 1.9 million ha (5.9% of total land area) gazetted as national parks, wildlife, and bird sanctuaries (Ministry of Natural Resources and Environment, Malaysia, 2009). Various legislative ordinances as shown in Table 3.6 also serve to enhance biodiversity conservation efforts. In exhibiting its continued global commitment to achieving the objectives of CBD, it adopted its new NPBD (2016–25) to further enhance biodiversity conservation (Ministry of Natural Resources and Environment, Malaysia, 2016).

3.12.5 Malaysia: The state of biological diversity

Despite the above all-encompassing environmental protection measures, however, massive habitat loss in Malaysia as a result of widespread deforestation, uncontrolled logging activities (legal and illegal), mega-dam construction, and commercial oil palm plantation expansions as discussed above (see Section 3.6.4.1), are posing a significant threat to the continued existence of a wide range of endangered animal species such as the tiger, the pygmy elephant, the orangutan, the Sumatran rhino, and the clouded leopard (Greenpeace, 2004). The Indian gray mongoose and the Banteng have also become extinct. The leatherback turtle, the world's largest turtle, has also become extinct in the wild (Gates, 2013).

Also the rapid disappearance of natural habitat in the country has led to the extinction of the Javan or Lesser One-Horned rhino in Peninsular Malaysia, and the extinction of the Sumatran rhino in the state of Sarawak (Choy, 2015a). Poaching of the critically endangered Sumatran rhino in the state of Sabah has been a key threat to the continued survival of the species. In 2015 it was declared extinct in the wild in Malaysia due to continued illegal hunting and extensive habitat destruction (Zorthian, 2015). The last male Sumatra rhino which was captured and transferred to the Tabin Wildlife Reserve in the state of Sabah has died in May 2019 due to sickness, leaving behind only one female Sumatran rhino in the whole of Malaysia (Bittel, 2019).

The irreversible destruction of the Bikam Permanent Forest Reserve in the state of Perak which was degazetted for commercial use such as timber harvesting and oil palm plantation development also resulted in the extinction of keruing paya (*Dipterocarpus coriaceus*) on the

Peninsular Malaysia. Keruing paya is a large hardwood tree listed as Critically Endangered on the IUCN Red List. It also increasingly threatens the continued survival of a range of protected species including leopard, Malayan tapir, siamangs, and the great Argus pheasant thriving in the region (Choy, 2015a). Also, rampant and uncontrolled deforestation in the state of Kelantan over the last 20 years as discussed in Section 3.6.4.1 above has resulted in extensive habitat destruction which threatens the continued existence of a wide range of endangered species such as tigers, macaques and rare birds.

The increasing trend of environmental degradation as a result of changing land use practices in the country also posing a real threat to the continued survival of a wide range of key forest-dependent and endangered mega-fauna such as the Asian Elephant, the Malayan Sun Bear, and the Gaur (Ministry of Natural Resources and Environment, Malaysia, 2014). For example, The Malayan tiger is now on the brink of extinction as uncontrolled logging activities and the conversion of rainforest into oil palm plantations are destroying its natural habitat, rendering the animal more vulnerable to poachers. It is estimated that there are fewer than 250 tigers left in Malaysia. In the past, there were thousands of tigers roaming throughout the dense interior forests in Peninsular Malaysia, but they are now mainly confined to three PAs: Belum-Temengor in the north, Taman Negara in the center, and Endau-Rompin to the south (Mayberry, 2018). Overall, about 14% of the mammals in Malaysia have been identified as endangered under the IUCN Red List while 47 of the 218 species of amphibians are threatened with extinction (Hilton-Taylor et al., 2009).

Globally Malaysia is classified as a critical area for biodiversity conservation. Indeed, recent scientific research revealed that Malaysia top the list as a global hotspot where its species are most affected by human-induced threat arising from habitat alteration or destruction as a result of commercial land use practices such as oil palm plantation development. More specifically it has the highest average human impact score of 125 impacted species per grid cell (Allan et al., 2019).

3.12.6 The Philippines: Biodiversity conservation efforts

In the Philippines, in recognition of the irreversible ecological destructive impact of wanton habitat destruction caused by overexploitation of forest resources as discussed above (see Section 3.6.4.2), the government has strengthened biodiversity conservation efforts. It issued the 1990 Philippines Strategy for Sustainable Development. It acceded to the CITES in 1981 and became an active member in 1983. It also ratified the CBD in 1993 and enforced it in the same year. This was followed by the development of the 1993–98 Medium Term Philippines Development Plan. This paved the way for the establishment of the Philippine Strategy for Biological Diversity Conservation (PSBDC) in 1994 which consolidated the legal and institutional foundations for a concrete plan of action to implement Article 6 of the CBD (DENR-UNEP, 1997; DENR, 1998). The Strategy also aims to integrate biodiversity conservation and management in government and nongovernment sectors. It also seeks to strengthen human resource capability in biodiversity conservation and management through the establishment of two programmes, namely, Institutional Capacity Programme and Human Resources Development Programme (DENR-UNEP, 1997).

As a sign of its commitment to the Earth Summit, it established the Philippines Council for Sustainable Development in 1992 to promote the principles of sustainable development and to commit to integrate these principles into the country's national policies, plans and programmes that would involve all sectors of society. The Council was also mandated to develop the Philippines Agenda 21 which was adopted in 1996. The Council underwent some structural improvement in the same year through the creation of various Committees, one of which was the Committee on the Conservation and Management of Resources for Development to promote environmental protection and sustainable use of biological resources. The Philippines Agenda 21 represents a major sign of political commitment to the Earth Summit's agenda of sustainable development (DENR, 1998).

Two major programmes, similar to those in Indonesia and Malaysia, were created, under the PSBDC, to enhance biodiversity conservation efforts, namely, In situ Conservation Programme and Ex situ Conservation Programme. Under the In situ Conservation Programme, the PAs, and Wildlife Bureau was created to consolidate all government conservation efforts in the establishment of a network of PAs to conserve and protect representative ecosystems and habitats. To reinforce PA conservation measures, the NIPAS Law was enacted to provide the legal machinery for the promotion of biodiversity conservation of biologically important sites and habitats.

Under the NIPAS Law, 203 areas covering approximately 3.8 million ha or 12.8% of Philippines' land area were identified as PAs. These include 67 national parks/marine reserves, 8 game refuge and bird sanctuaries, 16 wilderness areas, 85 watershed forest reservations, and 27 mangrove swamp forest reserves (DENR, 1998). Various ex situ conservation programmes to conserve species out of their original location or habitats have also been developed to complement in situ conservation efforts.

The Philippines' government continued to update its action plans and programmes to reinforce its global commitment to enhancing the goals of CBD. It recently adopted its third action plan, the 2015−28 Philippines Biodiversity Strategy and Action Plan to further contribute to promoting environmentally sustainable development (DENR, 2016). More specifically the 2015−28 action plan reaffirmed the government's commitment to implement its commitment to implement the CBD Strategic Plan for Biodiversity 2011−20, including the 20 points Aichi Biodiversity Targets. In compliance with Article 26 of the UNCBD, it also periodically submitted its national report to CBD.

3.12.7 The Philippines: The state of biological diversity

However, the above environmental efforts do not seem to have halted the ecological destruction of the country's diverse ecosystems. To wit, the alarming rate of deforestation in the Philippines (Section 3.6.4.2) which has left the country with only 6 million ha of the forest from 17 million ha in 1935 has had destructive ecological impact (Vesilind, 2002; PRB, 2006; Choy, 2015a; Palma, 2016). Currently, the Philippines has the lowest forest cover in ASEAN (World Bank, 2013; Mongabay, 2011). This has negatively affected the continued survival of a vast range of endemic species including the Philippines eagle, the Philippines tarsier, tamaraw, kagwang, and the Philippines spotted deer (Maala, 2001; Duckworth et al., 2012; Conservation International, 2011).

The Philippines eagle which inhabits in primary dipterocarp forest has become a critically endangered species because of habitat loss caused by extensive deforestation (IUCN, 2016d). Many species such as the Cebu warty pig and Panay flying fox which could not survive outside of their natural habitat have become extinct (Alave, 2011). Also the critically endangered species such as turtles and sharks have been poached and sold illegally for their meat and fins in China and Taiwan where demand is high (Alave, 2011). Furthermore the

Philippine pangolin which is endemic to the Philippines has been uplisted from "endangered" to "critically endangered" on the IUCN Red List largely due to its rapid population decline caused by habitat destruction and large-scale poaching (Mongabay, 2019).

Overall, habitat loss caused by overexploitation, land grabs by large-scale mines, dam construction, reclamation, and other extractive and destructive projects constitutes some of the biggest causes of wholesale biodiversity loss in the Philippines (see, e.g., Mayuga, 2019). It is said that only 4% of the country's forest habitat is suitable for a large number of endemic species, including 6000 plant species and 1196 known species of amphibians, birds, mammals, and reptiles (Isaacson, 2011; Alave, 2011).

Massive loss of natural habitat has resulted in the Philippines having more endemic species that are severely threatened than any other country in the world (Oliver, 2006; ACB, 2010; Duckworth et al., 2012). It is reported that the rate of species extinction is 1000 times the natural rate (Isaacson, 2011; CEPF, 2014). Due to its extremely low natural habitat coverage, the Philippines is signified as "the hottest of the hot spots," (Alave, 2011). Unsustainable human practices such as destructive commercial and unregulated fishing activities over the past 50 years have also harmed its rich and healthy coral reefs extending over 27,000 km^2, leaving only less than 5% of the reefs in excellent condition and with just 1% in a pristine state (Alave, 2011). It is noteworthy that the Philippines is one of the few nations in the world that is, in its entirety, both a hotspot and a mega-diversity region as well as one of the top priority sites for global conservation (ACB, 2010; Conservation International, 2011; McGinley, 2013).

In addition, as the Department of Environment and Natural Resources (DENR) acknowledged in its report "Communities in Nature: State of Protected Areas Management in the Philippines," despite the enactment of various laws such as the NIPAS Act (1992) in compliance with CBD to ensure ecological protection through the creation of a system of PAs for biodiversity conservation, the country's biodiversity has in fact become increasingly threatened (DENR, 2012). It may be of interest to note that enactment of the NIPAS Act (1992) established a total of 240 PAs covering 5.4 million ha of land and sea to ensure biodiversity conservation.

3.12.8 Thailand: Biodiversity conservation efforts

Since the signing of CBD in 1992 followed by its ratification in 2003 and enforcement in 2004, Thailand has effectively used its provisions as guiding principles for biodiversity conservation and management of its rich ecosystems comprising 4591 species of terrestrial vertebrates (mammals, birds, reptiles, and amphibians) and 11,625 vascular plant species (OEPP, 2000; Ministry of Resources and Environment, Thailand, 2009). In the same year, Thailand listed and offered special protection for 15 animals considered to be rare and near-extinct under the Wildlife Protection and Preservation Act (1992), namely, the white-eyed river martin, Javan rhino, Sumatran rhino, kouprey, wild water buffalo, brown-antlered deer, Schomburgk's deer, the scrow, goral, black-breasted pitta, east sarus crane, marbled cat, Malayan tapir, Fea's barking deer, and the dugong (UNODC, 1992).

In 1993 the National Environmental Board established the National Committee on CBD to develop policies, measures and plans for sustainable conservation and utilization of biodiversity. Other environmental laws such as the National Park Act, National Reserved Forest Act, Plants Act, and National Environmental Quality Promotion and Preservation Act have also been enacted to guide environmental conservation efforts in Thailand. The government has also put in place various financial mechanisms to support protection, conservation, rehabilitation and sustainable biodiversity utilization activities (ONEP, 2019).

To reinforce its conservation efforts, Thailand adopted the First National Biodiversity Strategies and Action Plans (NBSAPs) in 1997 as "National Policy, Strategies and Action Plan on Conservation and Sustainable Use of Biodiversity 1998–2002" (Ministry of Resources and Environment, Thailand, 2009). The Policy has also been endorsed as the national administrative framework to ensure that its conservation efforts are prioritized to achieve Article 6 of CBD. In 2002 the Ministry of Natural Resources and Environment (MONRE) was established to strengthen institutional capacity in biological conservation. These conservation efforts are also in line with the Constitution of the Kingdom of Thailand. Section 57 (2) of the Constitution, for example, requires the country to "preserve, protect, rehabilitate, manage and use or make use of natural resources, the environment and biodiversity in a sustainable and balanced manner" (ONEP, 2019: 170).

Thailand increased its protected forest by an additional 18,097 km^2 (about 1.81 million ha) in 2000 compared with the area in 1995 and designated 37 additional PAs during 1997–2000. These comprised 20 national parks, 2 forest parks, 9 wildlife sanctuaries and 6 nonhunting areas. The total PA in the same period was roughly 91,231 km^2 (9.12 million ha) or 17.8% of the country's total area (OEPP, 2002). Upon the lapse of the First NBSAPs, Thailand continued its environmental protection efforts through the Second National Biodiversity Strategies and Action Plans (2003–07) and issued its Master Plan for Integrated Biodiversity Management (2015–21) to guide national strategies, plans or programmes for the conservation, and sustainable use of biological diversity (ONEP, 2015). Thailand has been working comprehensively to support the implementation of its 2015–21 biodiversity conservation plan through PA management systems (ONEP, 2019).

3.12.9 Thailand: The state of biological diversity

Despite the above environmental protection measures, and in spite of its compliance with Article 6 of CBD, however, habitat loss caused by widespread deforestation as discussed in Section 3.6.4.3 has adversely affected the continued existence of roughly 45 species of birds, 23 species of reptiles, and 72 species of fish (ESCAP, 2011). For example, from 2004 to 2012, the number of threatened mammal species increased from 38 to 57 species (Baillie et al., 2004; IUCN, 2017). Some of the species protected under the Wildlife Protection and Preservation Act have also become extinct. These include Schomburgk's deer, giant ibis, large grass-warbler, silver shark, Siamese flat-barbelled catfish, Siamese tiger fish, Eld's brown-antlered deer, kouprey, white-shoulder ibis, milky stork, and the Javan as well as the Sumatran rhinos (OEPP, 2000; Nabhitabhata and Chan-ard, 2005). Also conversion of forests to rubber and oil palm plantations has resulted in a 60% decline in bird species richness, with insectivorous and frugivorous birds declining more rapidly than omnivorous birds (Aratrakorn et al., 2006). For example, the Gurney's Pitta has become extinct recently due to habitat destruction caused by extensive forest land conversion for oil palm plantation expansions (see, e.g., RSPB, 2009; Anstee, 2019).

Overall, according to its Fifth National Report to CBD, the biodiversity status has worsened due to deforestation and habitat fragmentation or destruction. To compound the problem, the trafficking of wild animals and wild plants continues to threaten the ecological health of the country's biodiversity. Illegal logging activities in various PAs such as the Phu Wiang National

Park, Phu Phan National Park, Phu Sithan Wildlife Sanctuaries, Thap Lan National Park, Ta Phraya National Park, and among others continue to exert immense pressure on the forest ecosystems. Furthermore since 1961, more than 50% of the rivers, canals, and swamps have been destroyed, posing a severe threat to its aquatic biodiversity. Unrestrained human economic activities also resulted in drastic decline of wetlands. In the Northeastern region, for example, 40%–60% of the wetlands were lost compared with the 2009–12 data while in the North region, roughly 62% of the wetlands were destroyed in 2013 compared with the 2009–01 data, thus leading to the ecological impoverishment of the wetland ecosystems (ONEP, 2014). The marine and coastal biodiversity is also increasingly been threatened by illegal fishery such as off-season fishery and intrusion to commercial fishery in PAs (ONEP, 2019).

In its latest Sixth National Report on the Implementation of the CBD, Thailand revealed that wild plant species continued to be under threat from tourism and illegal trafficking. Out of approximately 11,000 plant species, 8.76% were classified as being under threat while two species, Sky Blue Vanda (*Vanda coerulescens Griff.*) and Amherstia (*Amherstianobilis Wall.*) have become extinct in the wild. In addition, of the 4731 vertebrate species found in Thailand, 12.03% were categorized as being threatened largely due to commercial land use practices (ONEP, 2019). Indeed, recent findings by scientists uncovered that Thailand is one of the world's major hotspots where species are most impacted by human-induced threats caused by habitat alteration or destruction (Allan et al., 2019).

3.12.10 Vietnam: Biodiversity conservation efforts

In 1962 Vietnam established its very first "prohibited forest," Cúc Pương Protected Forest (now Cúc Phương National Park), to protect it from exploitation (Suntikul et al., 2010). In 1985 it issued its National Conservation Strategy to promote sustainable resource use and biodiversity conservation (IUCN, 1985). In 1986 the government increased its PAs through the establishment of 73 Special-use Forests (SUFs) throughout the country, with a combined area of 769,512 ha (UNDP-GEF, 2009). In response to CBD, the government issued its Biodiversity Action Plan (BAP) in 1995 to reinforce its political commitment to biodiversity conservation (Socialist Republic of Vietnam, 1995).

Over the 10 years after the Earth Summit, Vietnam placed great emphasis on biodiversity conservation. It approved the BAP in 1995, and according to its Second National Report to the Convention on Biological Resources, the total forest area for conservation had been increased from 956,585 ha (1993) to 2,297,571 ha (2000), accounting for 6.7% natural land areas (National Environmental Agency, Vietnam, 2001). Since the ratification of CBD in 1995, the Vietnamese Government has made substantial investment in both human and financial resources to meet its commitment to achieving the CBD aim of biodiversity conservation and sustainable use of natural resources.

Its commitment to biodiversity conservation is well reflected in Vietnam's Agenda 21, National Strategy for Environmental Protection to 2010 and Vision toward 2020; Vietnam BAP 1995, the Draft Vietnam BAP to 2010 and Vision toward 2015; Management Strategy for a Protected Area System in Vietnam to 2010 (Ministry of Natural Resources and Environment, Vietnam, 2007). By 2006, the government had established 128 Special-use Forests (SUFs) covering 2.5 million ha or about 7.6% of the territory including 30 national parks, 48 nature reserves, 11 special/habitat conservation areas, and 39 landscape areas, all of which contain important forest ecosystems, endangered, rare, and endemic flora and fauna species (Ministry of Natural Resources and Environment, Vietnam, 2008).

3.12.11 Vietnam: The state of biological diversity

Despite the increasing number of PAs, however, the number of threatened species has increased in the country. According to the Vietnam Red Book 2007, the total number of endangered wildlife species had increased by 161 species to 882 compared with the previous Red Book published in 1992−96. Also nine animals and two Lady's slipper orchid (Paphiopedilum) species were believed to have gone extinct in the wild while many other valuable and rare species have been seriously decreasing (Ministry of Natural Resources and Environment, Vietnam, 2008).

The major factor contributing to biodiversity depletion in Vietnam is habitat fragmentation and destruction caused by unrestrained or illegal logging activities, coffee plantation expansion, and dam infrastructure development (Drollette, 2013a; Carew-Reid et al., 2010). Between 1990 and 2005, for example, the country lost 78% of its rich and closed-canopy forest (Mongabay, 2009). This massive closed-canopy forest destruction caused roughly 700 and 300 plant and animal species, respectively, including the banting, the Javan rhino, the tiger, the Asian elephant, and the saola, to be pushed to the brink of extinction, while nine rare animal and plant species including the Sumatran rhino, the sika deer, Eld's deer, and the kouprey became extinct in the wild (WWF, 1998; CNRES, 2000; Ly and Zein, 2006; Ministry of Natural Resources and Environment, Vietnam, 2008; Carew-Reid et al., 2010; Tordoff et al., 2012; Vietnam Net, 2013; Drollette, 2013b).

Poaching, habitat loss and degradation have wiped out the critically endangered Javan Rhino in Vietnam. The species was declared extinct in 2011, following the death of the last remaining Javan Rhino population in mainland Asia (Ker, 2011). Currently the Javan rhino only exist in a single population range between 58 and 61 individuals in Ujung Kulon National Park in West Java, Indonesia, making them extremely vulnerable to extinction (Emslie et al., 2016). The species is now restricted to only four isolated sites in Indonesia, with only between two and five animals each. The population of Sumatran rhino was estimated at 73 individuals in 2015 (Emslie et al., 2016).

Although the area of Vietnam's forest cover has increased, much of this increase has been due to commercial planting of production forest rather than natural forest. As a result, the quality of wildlife habitat is decreasing. Also land conversion and infrastructure development such as dam construction has significantly reduced the area of natural habitats. This threatens the continued survival of many of the rare and endangered species. Besides, both inland water and marine ecosystems are being degraded due to unrestrained human economic activities (Ministry of Natural Resources and Environment, Vietnam, 2014). Currently Vietnam is identified as one of the world's most endangered terrestrial eco-regions as well as one of the globally significant conservation sites (World Bank, 2005; Sterling et al., 2006; Hilton-Taylor et al., 2009).

In summary, despite putting in place well-established environmental policies and legislative frameworks to protect the natural environment in the ASEAN-5 region, its biodiversity is still increasingly under serious threat. This is fundamentally due to rampant deforestation followed by extensive habitat fragmentation and destruction, uncontrolled oil palm plantation development, unsustainable natural resource use and illegal wildlife trade and poaching, among others. As reflected in Table 3.9, Indonesia and Malaysia, two of the countries in Southeast Asia which have some of the most well-developed environmental protection frameworks exhibit the highest biodiversity loss in terms of the increasingly number of threatened species.

TABLE 3.9   Number of threatened species in Southeast Asia (ASEAN-5 region) as of 2017.

	Mammals	Birds	Reptiles	Amphibians	Fishes	Molluscs	Other invertebrates	Plants	Fungi and protists	Total
Indonesia	191	153	34	32	163	6	284	437	0	1300
Malaysia	72	55	31	48	87	37	227	717	1	1275
The Philippines	39	93	39	48	93	3	235	243	0	793
Thailand	59	54	28	7	107	15	196	153	0	619
Vietnam	56	46	49	43	82	30	122	204	0	632

Based on, IUCN, 2017. The IUCN Red List of Threatened Species. Version 2017-3. International Union for Conservation of Nature and Natural Resources (IUCN), Gland, Switzerland.

3.12.12 ASEAN transboundary regional environmental protection initiatives: The Heart of Borneo

As shown earlier, the Southeast Asia region is undoubtedly well noted for its unique and diverse flora and fauna. In particular, the island of Borneo, shared by Indonesian Kalimantan, the Malaysian states of Sabah and Sarawak (Malaysian Borneo), and Brunei Darussalam, the third largest island in the world after Greenland and New Guinea, covering an area of 74 million ha, has long been recognized as one of the most biological diverse habitats on Earth. The island which accounts for just 1% of the world's landmass houses roughly 6% of the global biodiversity. Its rainforests are home to roughly 15,000 species of flora of which 5000 species or 34% are endemic. Indeed, its flora is richer than that of the entire continent of Africa which is 40 times larger than the island (MacKinnon et al., 1996). There are also 100 amphibian species, over 420 bird species (37 endemic), around 210 mammal species (44 endemic), and 394 fish species (19 endemic) found in the island of Borneo (WWF, 2005).

Recognizing the importance of safeguarding the ecological health of the island's biological resources while optimizing their economic use, the governments of Indonesia, Malaysia and Brunei (hereafter the HoB member countries), with the cooperation and advocacy efforts of the WWF, formally signed the HoB Declaration (hereafter the Declaration) in 2007 and established a regional environmental protection framework known as Heart of Boneo Initiative (HoBI). The formal adoption of HoBI led to the mapping of a 23-million-ha of trinational and transboundary PA (almost a third of Borneo Island) in central Borneo known as the Heart of Borneo (HoB) for conservation and sustainable management purposes. Indonesia, Malaysia, and Brunei contribute roughly 16.8 million, 6 million, and 570,000 ha, respectively, to its formation. Each HoB member has also enacted various environmental laws and regulations to enhance the objectives of HoBI (Table 3.10).

The HoB members have also jointly developed a trilateral Strategic Plan of Action to put policies into action. Accordingly five sustainability programmes have been identified under the trilateral action plan, namely, Transboundary Management, Protected Areas Management, Sustainable Natural Resource Management, Ecotourism Development, and Capacity Building to reinforce environmental conservation efforts (Table 3.11).

TABLE 3.10 HoB member countries' environmental laws and instruments.

| Indonesia | Malaysia | | Brunei |
	Sarawak	Sabah	
Environmental Management Act (1997, revised 1982; 1990)	Statement of Forest Policy (1954)	Forest Enactment (1968)	Forest Act (1934, amended 2007)
Forestry Law (1999)	Natural Resources and Environment (Amendment) Ordinance (1993)	Forest Rules (1969)	The Wildlife Protection Act (1978, revised 1984)
Conservation Law (1990)	Forest Ordinance (1958, amended 1997)	Wildlife Conservation Enactment (1997)	National Forest Policy (1989)
Environmental Protection Law (2009)	Sarawak Forestry Corporation Ordinance (1995)	Cultural Heritage (Conservation) Enactment (1997)	The Wild Fauna and Flora Order (2007)
Cultivation Law (1992)	The National Park and Nature Reserves Ordinance (1998)	Park Enactment (1984)	The Fisheries Order (2009)
Plantation Law (2004)	Wildlife Protection Ordinance (1998)	Biodiversity Enactment (2000)	The Brunei Darussalam Long Term development Plan—Wawasan 2035
Ministry of Agriculture Regulation (2007)	Sarawak Biodiversity Centre Ordinance (1998)	Environment Protection Enactment (2002)	Marine Protected Area Network (2012)

HoB, Heart of Borneo.
Note: This compilation by the author from various sources is by no means exhaustive (see, Choy, 2018).

TABLE 3.11 Trilateral strategic plan of action programmes.

Programme	Aim
Transboundary management:	To address issues of management off natural resources and socioeconomic welfare of the local communities on the border areas
Protected area management	To enhance and promote effective management of protected areas especially those located on the common border
Sustainable resource management	To enhance sustainable management of natural resources outside the protected areas
Ecotourism development	To recognize and protect natural assets of special natural or cultural values within the HoB area
Capacity building	To ensure effective implementation of HoB initiative at all levels, both public and private sectors as well as the local community level

HoB, Heart of Borneo.
Based on, UNDP, 2012a. Project Document: Biodiversity Conservation in Multiple-Use Forest Landscapes in Sabah, Malaysia. United Nations Development Programme (UNDP), New York.

The above encompassing environmental protection initiatives serve as an important framework for the enhancement of biodiversity conservation of the HoB. In Indonesia, for example, its HoB area is legally recognized as an area of National Strategic Importance through the enactment of the Government Regulation No. 26 of 2008. Furthermore Presidential Decree No. 2 of 2012 was also created to prevent activities that may disturb the HoB while Presidential Decree No.3 of 2012 recognized the HoB as the "lungs of the world." The latter environmental initiative aims to achieve the national target of reducing greenhouse gases emissions by 26% by 2020 (Ariain, 2013).

In Sarawak, the land areas of national park, nature reserve and wildlife sanctuary have increased from 301,000, 945, and 192,000 to 417,000, 1900, and 206,000 ha, respectively (Sarawak Forestry Corporation, 2019). Also the state government is planning to expand the area of Totally Protected Areas (TPAs) in the HoB region to enhance forest connectivity between the landscapes and biodiversity corridor for wildlife (Chia, 2015). Similarly in Sabah, the state government has increased its TPA from 800,000 ha to 1.55 million ha or about 21% of the total state land area. The state has further pledged to increase its TPA to 30% or 2.1 million ha of its total land area over the next 10 years (Chong, 2015).

To reinforce conservation efforts, the state government has also institutionalized the application of a multiple-use forest landscape planning and management model to establish forest connectivity among the protection forest islands. A case in point is the proposed 261,264 ha of landscaping project which forms an important connecting landmass to three renowned PAs in Sabah, namely, the Maliau Basin Conservation Area (58,840 ha) to the West, the Danum Valley Conservation Areas (43,800 ha) to the East, and the Imbak Canyon Conservation Areas (16,750 ha) to the North (UNDP, 2012a). The state government has also established 19 field outposts within HoB landscapes for monitoring wildlife and tackling forest encroachment (Chong, 2015).

Meanwhile, in the small nation state of Brunei, the government has committed 58% (about 576,400 ha) of its total landmass as PAs for sustainable management and conservation purpose (WWF, 2010). Logging activities are banned in these areas. The management of biodiversity within the delineated HoB landscape is guided by the National Biological Resources (biodiversity) Policy and Strategic Plan of Action 2012. The plan outlines the strategic objectives and actions to conserve the biological richness in the HoB region (Ministry of Development, Brunei, 2015).

3.12.13 The Heart of Borneo: The state of environment

Yet, decades of environmental efforts seem to have failed to stem the tide of environmental decline as evidenced in the gradual depletion of natural forests, and the upsetting of the HoB ecological systems caused by illegal logging activities and spurred by cross border timber trade between Indonesia Kalimantan and Malaysian Borneo, uncontrolled oil palm plantation expansions involving clear cutting of natural forests, and dam infrastructure development within or in the vicinity of HoB involving extensive habitat transformation (Choy, 2015b, 2018; Obidzinski et al., 2007; WWF, 2007a).

Despite the establishment of PA status, an estimated 1 million m^3 of timber are being smuggled out of the HoB area each year, leaving behind destroyed forests and threatened biodiversity (ADP, 2013). Since 2007, the HoB has lost 10% (2 million ha) of its forest cover despite having in place various all-encompassing environmental conservation policies and measures as discussed above (Wulffraat et al., 2012). The IndoMet Coal Project (ICP) in Indonesia mentioned earlier, located within the HoB in the particularly heavily forested Upper Barito Basin, will result in extensive habitat fragmentation, thereby endangering the continued survival of a large number of orangutans and many unknown scientific biological species thriving in the dense forest (CMD, 2014; Choy, 2018). Rapid commercial oil palm plantation expansion in Kalimantan in which the HoB is situated is also increasingly threatening the ecological health of the HoB. It may well be noted that oil palm plantation expansion is the primary driver of forest loss in Borneo especially in Indonesian Kalimantan and the state of Sarawak in Malaysian Borneo. By 2010, it is estimated that commercial oil palm plantation will cover 9% of Borneo Island (Linder and Palkovltz, 2016).

In Sarawak, uncontrolled commercial logging activities are decimating the HoB protected forests at the rate of 9 km^2 a month, causing damage to the once intact rainforest canopy. These logging activities also threaten the continued existence of the proposed Sungai Moh Wildlife Sanctuary in the HoB (Global Witness, 2015). Furthermore satellite imagery reveals that logging operations in river buffer zones, severe erosion and landslides from road construction, and severe canopy destruction caused by unrestrained timber extraction were observed in or near the HoB region. Large scale logging activities have also been observed in the totally protected forests, the Usun Apau National Park (Global Witness, 2015).

Forest concessions have also been allocated to logging companies to extract timber from the Danum Linau National Park and two extension areas within the HoB region. Thousands of ha of logging concessions lie within the HoB, including those areas adjacent to the Indonesian Bentuang Karimun National Park to the south, and bordering the Lanjak Entimau Wildlife Sanctuary (Sarawak) to the west, have also been given out to various logging companies for timber harvesting. These forests are home to a wide range of critically endangered species totally protected under Sarawak Laws and include orangutans, clouded leopard, the Bornean gibbon, Hose's langur, western tarsier, slow loris, and giant squirrel (Council on Ethics for the Government Pension Fund Global, 2013).

In Sabah, about 80% of its rainforests have been heavily impacted by logging activities or oil palm plantation development. The oil palm plantation development, which cuts across the rainforest landscape in Sabah, is taking a heavy toll on its natural environment and its biodiversity. For instance, despite the fact that conserving the mega-diverse pockets of rainforest in Danum Valley, Maliau Basin, and Imbak Canyon conservation areas is instrumental in connecting the biodiversity "crown jewels" of the HoB, they are not spared from human economic intervention (Butler, 2012a, 2014). What aggravates the environmental condition is that about 82,000 ha of forest reserve in this region has been approved for oil palm plantation development (Butler, 2012b; Rakyat Post, 2014). These commercial activities have generally contributed to the dramatic decline in orangutan populations in Sabah which decreased by 50% from 22,000 to 11,000 in 50 years due to extensive habitat destruction (Hance, 2009; Butler, 2010).

In contrast, Brunei Darussalam, a tiny oil-rich nation with a total area of 5265 km^2 and a small population of about 412,000 as of 2012, relies on petroleum and natural gas exports and associated energy resource services to support its economy, and so has been able to avoid extensive forest destruction even though its contribution to the HoB is negligible.

3.13 China's environmental protection and biodiversity conservation

3.13.1 China's environmental protection initiatives

China is ranked third after Brazil and Colombia in terms of overall richness of amphibian, bird, mammal, reptile, and vascular plant species (López-Pujol et al., 2011). In addition, the country is ranked eighth out of the 17 mega-diverse countries in the world. Geographically the Three Gorges region located along the Yangtze River is recognized as one of most important biologically rich regions in the world. It is home to over 6400 plant species (roughly 19% of the total number of species found in China), with 57% of them categorized as endangered; 3400 insect species (8.5% of China's total); more than 360 fish species, and more than 500 terrestrial vertebrate species (22% of China's total) (Ministry of Agriculture, Beijing, 1995; He and Xie, 1995; Xie, 2003; Wu et al., 2003; Tian et al., 2007; Zhang and Lou, 2011). The Three Gorges region also contains roughly 36% of all freshwater fish species in China (Xie, 2003). Twenty-seven percent of all the Chinese endangered freshwater fish is also found in the Yangtze River, spanned by the Three Gorges dam, with as many as 177 species classified as endemic (Yue and Chen, 1998; Xie, 2003; Fu et al., 2003, Zhang and Lou, 2011).

Some of the noteworthy rare species listed under the IUCN Red List as endangered are

1. The Chinese sturgeon, the largest form of the legendary prehistoric fish species which has been swimming in the Yangtze waters for the past 140 million years. It would swim 3000 km upstream during spawning season (Greenpeace, 2010; Deng, 2018). The Chinese sturgeon is considered a national treasure, and is now listed as Class I State protected animal under the Chinese law (Hu et al., 2009).
2. The Chinese paddlefish or the "King of Yangtze," the world's longest freshwater fish, is also placed under China's Species Red List, and is protected under the Chinese state law. According to scientists, this ancient fish has lived since the Lower Jurassic period around 200 million years ago. It was listed as a critically endangered species under the IUCN Red List in 2009 (Cheung, 2020).
3. The Yangtze River dolphin or Baiji, the world's rarest freshwater dolphin which has been thriving in the Yangtze River for the past 20 million years (Wu et al., 2003). The Baiji is also classified in the First Category of National Key Protected Wildlife Species in China.
4. Other state protected or IUCN Red List species found in the Three Gorges region include the Yangtze finless porpoise, Yangtze soft-shell turtle, the Chinese tiger, the Chinese alligator, the Chinese giant salamander, the Siberian crane, and the giant panda.

In acknowledging the importance of protecting its rich biological resources and in affirming the concept of sustainable development as emphasized under the Brundtland Report (1987), China elevated its State Environmental Protection Commission (SEPC) established in 1984 to the State Environmental Protection Administration in 1998, which finally became the Ministry of Environmental Protection (MEP) in 2008 (Choy, 2016a, 2018). At the same time, it enacted and revised various environmental laws, some of which are indicated below, to strengthen its environmental protection efforts:

1. Environmental Protection Law (1979, amended 1989),
2. Wildlife Protection Law (1986, amended in 2000 and 2004),
3. Fisheries Law (1986, amended in 2000 and 2004),

4. Water Pollution Prevention Law (1984, revised in 1996),
5. Protection of Terrestrial Wildlife Law (1992), and
6. Regulation of Aquatic Wild Animals (1993).

In the wake of the Earth Summit 1992, China ratified the United Nations CBD. This was followed by the formulation and adoption of the National BAP in 1994 which was aimed at building a greener China and at halting the loss of biodiversity by 2020 in line with the CBD. In the same year, it also adopted its local Agenda 21, also known as the White Paper on China's Population, Environment, and Development in the 21st century. The Agenda aimed to promote the implementation of Agenda 21 (Bradbury and Kirkby, 1996). In other words, it sought to reinforce its commitment to environmental preservation and sustainable resource use while pursuing economic growth and social development. Indeed, China was one of the first few countries to propose and implement sustainable development strategies, and to publish its first, second, and third national sustainable development reports in 1997, 2002, and 2012, respectively (NDRC, 2012).

In reinforcing its commitment to sustainable development, China has also committed to implement various environmental control and resource use laws and policies. Under the umbrella of the Scientific Outlook on Development, China laid out one of the most important principles on sustainable development: the creation of a harmonious society based on the integration of humans and nature (green development). This green initiative was implemented through the 11th Five-Year Plan (2006−10) and the 12th Five-Year Plan (2011−15) (Hu, 2014).

3.13.2 The paradox of China's environmental sustainability: Water pollution

The above far-reaching environmental protection framework and legal instruments have, however, generally failed to halt further environmental decline. To start with, despite having in place the Environmental Protection Law and the Water Pollution Prevention Law to regulate and control environmental quality, about half of the 20,000 petrochemical plants located by the bank of the Yangtze River release large amounts of industrial wastes including toxic wastes, heavy metals (cadmium, mercury, lead, and arsenic), chemical effluents and agricultural runoff and organic matter, into the river. The amount of discharge increased at an alarming rate from 15 billion tonnes in the 1980s to 33.9 billion tonnes in 2010, causing unprecedented destructive impact on the Yangtze aquatic ecosystem (Wong et al., 2007; Ting, 2011).

For example, an analysis of hazardous chemicals in Yangtze River fish conducted by Greenpeace revealed widespread presence of certain hazardous chemicals such as alkylphenols and perfluorinated compounds and heavy metals such as mercury, cadmium, and lead within wild fish from the upper, middle and lower sections of the Yangtze River (Greenpeace, 2010). Furthermore the latest report from China Water Risk uncovered that 6 of the 11 provinces on the Yangtze River economic belt are facing acute water stress due to contamination by industrial effluent and unrestrained pollutants or unsustainable waste management practices (Hu et al., 2019).

Indeed, the dumping of untreated waste water and animal wastes into rivers by industries has been widespread (Turner and Ellis, 2007). About one-third of industrial waste water and more than 90% of household sewage in China are released into rivers and lakes without treatment (Refkin and Cray, 2013). In 2010 industrial waste water discharge volume was estimated at 237 billion tonnes, while domestic sewage discharge volume was about 380 billion tonnes. The total sewage discharge was about 659 billion tonnes (Wu et al., 2014). Also it is estimated that 5850 tonnes of organic pollutants are released into Chinese waters everyday compared with 2750 tonnes in the United States, 1700 tonnes in Japan, 1150 tonnes in Germany, 1600 tonnes in India, and 300 tonnes in South Africa (Refkin and Cray, 2013). Water pollution is so acute that up to 70% of China's rivers and lakes are seriously polluted (Morton, 2006; WWF, 2012). It may well be noted that China has some of most polluted rivers in the world including Yangtze, Xi, Dong, Zhujiang (Pearl), Hanjiang, and Yellow river (see, e.g., Dhiraj, 2017).

Water pollution has been gaining considerable attention in China since 2013, when it reached its record high levels (Piesse, 2019). In the latest national budget for 2019, the Chinese government increased its spending by a strong 45.3%, reaching RMB30 billion, to eliminate and prevent water pollution (Chien, 2019). In 2018 the government unveiled several key policies and action plans for addressing water pollution problems, with the Yangtze River remaining the key target for new ecological protection. The Bohai Sea has also come under its environmental protection target (Table 3.12).

TABLE 3.12 China's latest water pollution policy, regulation, and law.

Policy/plan/regulation/law	Key features
Action Plan for the War on the Protection and rehabilitation of Yangtze River	Industrial and agricultural pollution control
Action Plan for the War on the Integrated Bohai Sea Rehabilitation	Gradually establish quota systems for pollutant discharge
Action Plans for the Supervision and Inspection of Ecological Environment Monitoring (2018–20)	To improve the accountability system for ecological environment monitoring data quality
Implementation Plan for the War on Urban Black and Smelly Water Body Control	Black and smelly water cleaning rate should be >90% for key cities by the end of 2018 and for other cities by the end of 2020
Resolution on Strengthening the Protection of the Ecological Environment in All Aspects and Promoting Winning the War on Pollution in Accordance with the Law	Establish the strictest legal system for environmental protection
Opinions on Further Enhancing the Supervision and Enforcement in Ecological Environmental Protection	Clarify responsible personnel; promote random inspections and advanced tech to identify environmental violations
Opinion on Comprehensively Strengthening Ecological Environmental Protection and Resolutely Winning the War on Pollution	Achieve Beautiful China by 2035
Notice on Resolutely Curbing the Illegal Transfer and Dumping of Solid Wastes and Further Strengthening the Supervision of the Entire Process off Hazardous Wastes	Resolutely curb the illegal transfer and dumping of solid waste
Environmental Protection Law of the People's Republic of China (2014 Revision)	Example: Article 28 stipulates that the local people's governments at all levels shall, according to environmental protection objectives and pollution control tasks, adopt effective measures to improve environmental quality

Based on, Chien, T.L., 2019. Key Water Policies 2018-2019. China Water Risk, Hong Kong. <http://www.chinawaterrisk.org/resources/ analysis-reviews/key-water-policies-2018-2019/> (accessed 12.12.19.). Chinalawinfo Co., 2014. Available from: <http://greenaccess.law.osaka- u.ac.jp/wp-content/uploads/2019/03/Environmental-Protection-Law-of-the-Peoples-Republic-of-China-2014-Revision.pdf>.

Guided by the plans and guidelines listed in Table 3.12, China has been gradually cleaning up its water after years of indiscriminate dumping of industrial and household waste, overmining, and overuse of pesticides and fertilizers (Reuters, 2018). Over 95% of the foul water in 36 major Chinese cities has been transformed into clean water and the country's overall water quality has improved (Zheng, 2019). In particular, the overall conditions of the coastal waters previously affected by the massive and unrestrained discharge of waste water into the ocean have improved. However, improvements have been "unbalanced," with some regions faring far worse than others, such as the city of Yichang in Hubei. Also although China has made progress tackling water pollution, it is still facing difficulties in its efforts at imposing unified standards throughout the country (Reuters, 2019).

Excessive use of fertilizers accompanied by phosphorus and nitrogen run-offs has been posing a great pollution threat to Fuxian in southwest Yunnan province—one of China's biggest freshwater lakes. Also urban waste continues to be dumped directly into the Yangtze River (Stanway, 2019). It has been estimated that a total of 200.7 million m^3 of waste was dumped into China's coastal waters in 2018—a 27% rise over the previous year and the highest level in at least a decade in 2018. Also the government discovered an average of 24 kg of floating rubbish per 1000 m^2 of surface water in the same year, 88.7% of which was plastic (Reuters, 2019). This poses a great threat to the ecological health of the ocean.

It may be briefly remarked that to avoid the future specter of "black and smelly" water facing the nation, regulations, policy or action plans alone are inadequate. What is needed on top of these is the promotion of environmental awareness and the ethics of resource use and environmental conservation. This will be discussed in the second-half of the book.

3.13.3 The Three Gorges environmental dilemma

Massive infrastructure development followed by extensive habitat fragmentation or destruction is one of the most important contributing factors to the growing ecological impairment of the Three Gorges ecological treasure, which the construction of the Three Gorges dam across the Yangtze River makes most explicit. It has set the stage for extensive physical land transformation and irreversible habitat destruction of the regional landscapes, thus taking a heavy toll on China's animals and plants in the region. While nonresident birds, insects, or animals in the Three Gorges region could have escaped to the adjacent habitats, those resident species including endemic plants, nonmigratory animals, and birds, would have been irreversibly affected when the water behind the Three Gorges dam was impounded. These include 47 rare plants and 37 plants unique to the Three Gorges region. The dam construction also involved the ecological destruction of the "living fossil," the ancient Cathaya Argyrophylla trees and dawn redwood trees dating back to millions of years (López-Pujol et al., 2006).

Extensive habitat fragmentation in the region also affected the continued survival of the Siberian white crane, the South China Sika deer, the Chinese tiger, and the red panda. All these species are classified as endangered or critically endangered either by the IUCN, US Fish and Wildlife Service, or the WWF. They are also totally protected by the Chinese national laws such as the Environmental Protection Law or the Wildlife Protection Act as noted above. Another disturbing feature of the Three Gorges development is the growing problem of pollution. As noted in the preceding section, the Yangtze River has been persistently contaminated by colossal loads of domestic and industrial wastes and pollutants, making it as one of the most polluted and endangered rivers in the world (WWF, 2007b; Dhiraj, 2017).

Unmitigated river pollution and extensive habitat degradation coupled with illegal and unsustainable bycatch by fishermen using rolling hook long-line fishing, gill nets, electrocution and dynamite, or other banned destructive fishing methods, have resulted in the extinction of the world's most critically endangered and rare cetacean, the Yangtze River dolphin (Baiji or goddess of the Yangtze). The evolutionarily distinct Baiji had been thriving in the Yangtze River for the past 20–30 million years. In the past, the relic species was commonly hunted in the local fisheries for meat, oil, and leather. Its population dropped drastically from a healthy 6000 in the 1950s to only one individual in 2004. It was declared extinct in 2006; making it the first dolphin that mankind directly drove to extinction. This happened despite having in place various ecological protection programmes and legal instruments to protect its continued survival (Ding et al., 2006; Turvey et al., 2007). The critically endangered Chinese paddlefish, protected under the Chinese state law as noted above has been declared extinct in 2019 due to overfishing and habitat fragmentation. It may be added that the construction of the Three Gorges Dam has blocked its migration route and prevented it from breeding in the upper reaches of the river (Cheung, 2020).

In addition, the state protected Yangtze finless porpoises are also increasingly facing the threat of ecological extinction, its population having declined drastically from 2000 in 2006 to about 1000 in 2012 (Lovgren, 2007; Qiu, 2012; Hance 2012, 2013). Likewise, other state protected species which are facing the irreversible threat of destruction brought about by extensive habitat fragmentation or destruction in the Three Gorges region include the Chinese alligator, the Chinese giant Salamander, and the Siberian Crane (WWF, 2004).

Illegal wildlife trade in China has also contributed to endangering the continued survival of many of its rare and endangered species. China is a top consumer country of illegal wildlife products and one of the world's hotspots for the illegal trade in wildlife and wildlife parts (Felbab-Brown, 2011). Furthermore compared with the average global rate of biological loss of 10%, the rate of biodiversity loss in China is about 15%–20%. The China Red List indicates that 40% of mammals, 7% of birds, 28% of reptiles, 40% of amphibians, and 3% of fish are vulnerable to ecological destruction (McBeath et al., 2014).

3.14 Global biodiversity outlook: Some comments

The CBD and various international agreements, treaties, and global conferences as discussed in the preceding chapter provide an important impetus for the international community to implement biodiversity conservation and sustainable resource management in national laws or policies. For example, as noted in Chapter 2, The United Nations' Journey to Global Environmental Sustainability Since Stockholm: An Assessment (Section 2.9.6) 97% of the CBD Parties have developed NBSAPs or PAs to promote biodiversity conservation and protection. Yet, despite these environmental initiatives, the above discussions highlight the fact that the overall global or regional environment has in fact worsened and new environmental threats and challenges continue to emerge. What is clear is that the targets set by the international community "to achieve by 2010 a significant reduction of the current rate of biodiversity loss at the global, regional and national level" continue to sound hollow. A total of 110 national reports were submitted at the CBD in 2010—the International Year of Biodiversity—but not even a single country, neither from the advanced North nor the developing South, was able to claim success in achieving its targets. Also no subtarget has been completely achieved and most biodiversity indicators are negative, with direct pressures on the natural environment remaining constant, or increasing (Christophersen, 2010).

Species that have been assessed as "Critically Endangered" under the IUCN Red List are verging on extinction. Worse yet, species that were of least concern such as the giraffes in Africa have ascended to the category of "Critically Endangered." Throughout the tropics, the major factors directly impacting on the ecological sustainability of biodiversity are deforestation, habitat fragmentation and destruction, over exploitation, illegal hunting and wildlife trade, and pollution—all are human-induced. Extensive PAs created worldwide especially in the developing South are increasingly exposed to anthropogenic destruction.

The international agreements agreed upon thus far and all the COP meetings held to date have not been able to halt the world's biodiversity loss. More particularly member states, especially in the Asian developing region, have made little headway to deliver their commitment to reduce pressure on the global commons. On the upside, a small measure of success has been achieved in some cases of species protection, including the downlisting to lower extinction risk of: 33 bird species, including Lear's macaw, since 1988; 25 mammals since 1996, including the European bison, and five amphibians, including the Mallorcan midwife toad, since 1980. Also the extinction of at least 16 bird species, including the black stilt in New Zealand, has been successfully prevented by conservation actions during 1994–2004 (Butchart et al., 2010). Similarly, the once "endangered" Giant Panda in China has been downgraded to "vulnerable" in 2016 as a result of effective conservation measures undertaken by the Chinese authority.

Despite these encouraging achievements, global biodiversity continues to be trapped under the Holling unsustainable mode of development for more than four decades due to environmentally destructive human activities such as overexploitation of the planet's resources including fish stocks; excess deposition of reactive nitrogen to the natural environment; habitat or forest fragmentation; alteration of riverine systems through dam construction, human-induced climate change, among others (Butchart et al., 2010). In digression, nitrogen, which is an essential building block for life is unusable for the vast majority of living organisms in its molecular form (N2). It must be transformed into reactive nitrogen before it becomes usable. However, when present in excess, reactive nitrogen causes a range of negative effects such as creating unnatural growth rates of plants, nutrient imbalances, and decreasing or altering biodiversity. The creation of reactive nitrogen occurs both through natural processes and through human intervention. The latter is associated with the use of nitrogen fertilizers in farming or fossil fuel combustion which results in significant release of reactive nitrogen (UNEP-WHRC, 2007).

To round up, the case studies discussed above, which are by no means exhaustive, reflect clearly how ongoing massive-scale biodiversity depletion highlights the failure of the CBD in promoting ecological sustainability. By and large, our global biological system is still far from Holling sustainable. Also despite having CITES in place, illegal wildlife trading in the Asian Pacific region including China, Africa, and Southeast Asia has led to regional extinction of tigers, rhino, and other Asian species as well as severe depletion of marine wildlife. The disappearance of Baiji in the wild is a stark indication of how unrestrained pursuit of economic growth and socioeconomic progress is causing irreparable change to the country's natural environment (Choy, 2016a, 2018). It also symbolizes the disruption of harmony of human beings with nature. This is excruciatingly clear particularly since the Baiji had long been recognized as the rarest and most critically threatened mammal species on earth, and despite China having expressed serious commitment to its ecological conservation by legally categorizing it as the First Category of National Key Protected Wildlife Species.

Human activities are increasingly threatening the Holling sustainability of the global biological system to the detriment of the present as well as future generations. The Holling ecological quagmire is an indication of the lack of respect for the rule of law or international conventions by the stakeholders including the general populace in various countries. It is also a result of the lack of national capacity or genuine political will to implement the CBD objectives as ratified by practically all countries except the United States and the Holy See. More importantly it is a sign of the growing disconnect between humans and the natural world.

3.15 The United Nations Framework Convention on Climate Change, carbon dioxide emissions, and atmospheric concentration: A global assessment

Anthropogenic climate change is often regarded as the single greatest environmental threat to the long-term existence of human beings. Despite the United Nations' relentless efforts over the past few decades at addressing this potentially devastating problem, overwhelming scientific evidence shows that global warming has become unequivocally worse due to increasing atmospheric concentration of GHGs especially CO_2. Greenhouse gases include CO_2, methane (CH_4), nitrous oxide (N_2O), fluorinated gases and water vapor, and CO_2 is the most prominent source of GHG.

The GHG remaining in the atmosphere long enough to become well-mixed will form a blanket surrounding the earth, causing global warming which results in extreme environmental conditions. Recent increases in the frequency and intensity of extreme events such as deadly heat waves, severe droughts, intense precipitation, and devastating floods are signs that our climate system is increasingly heading toward a more hostile state of possibly no return.

Considering the above, it may be noted at the outset that there is a subtle distinction between global warming and climate change. Global warming refers to the rise in global average surface temperature due to greenhouse effect mainly caused by the increase in atmospheric concentrations of GHGs such as CO_2 in the atmosphere. It is just an indication of the rise of the average global temperature. Climate change, on the other hand, is a broader and all-inclusive term that is used to describe more than just a change in surface temperature. In other words, it also covers the "side effects" of warming like extreme drought conditions, melting glaciers and rising sea levels, among others. In short, global warming is just one aspect of climate change. However, both terms overlap and interact closely with one another to result in various adverse weather or climatic conditions as noted above.

Against this backdrop, the following sections will examine to what extent the UNFCCC has been able to contain GHG emission trends from a global and specific case study perspective. The analysis will confine its assessment to CO_2 emission — the largest source of greenhouse gas contributor to climate change.

It may be pointed out at the outset that CO_2 emission data from different sources may provide slightly different figures due to the differences in methodology used in assessment. However, the differences may be negligible. Insofar as the present analysis on global CO_2 emission is concerned, it will draw from its data source from Emissions Database for Global Atmospheric Research from the Joint Research Center (JRC) of the European Commission and PBL Netherlands Environmental Assessment Agency, indicated in short as JRC-PBL, it provides more comprehensive and up-to-date country-specific dataset (Crippa et al., 2019, see also, Olivier and Peters, 2018). Also emission levels are based on fossil fuel use (coal, gas, and oil) and industrial process emissions (cement production, carbonate use of limestone and dolomite, nonenergy use of fuels and other combustion, chemical and metal processes, solvents, agricultural liming and urea, waste, and fossil fuel fires). Emission from land-use change is excluded from the computation.

To begin with, based on global CO_2 emissions from 2010, around 21% of global GHG emissions is from industrial processes primarily involving the burning of fossil fuels on site for energy generation while 25% is from the burning of coal, natural gas, and oil for electricity and heat production. The latter is the largest single source of global greenhouse gas (GHG) emissions (IPCC 2014: 7−8; EPA 2019, see also, Brown et al., 2012). For example, CO_2 emissions from fossil fuel combustion and industrial processes made up the largest share (78%) of the total GHG emission increase from 1970 to 2010 (IPCC, 2014: 359).

In light of the above, it may be noted that since the convention of the Stockholm Conference in 1972, followed by the adoption of the Kyoto Protocol as well as the conventions of various climate change conferences (COP-1 to COP-23), global CO_2 emissions from the combustion of fossil fuels and cement production have increased by about 129.7% from 16.5 billion tonnes in 1972 to a record high of 37.9 billion tonnes in 2018. Compared with the Kyoto Protocol reference year at 22.7 billion tonnes, total emissions increased by 66.9% in 2018 (Fig. 3.14). Global CO_2 emissions from fossil fuel use and cement production increased

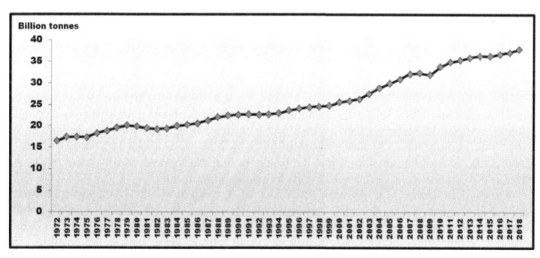

FIGURE 3.14 Global CO_2 emissions from fossil fuel use and cement production. CO_2, Carbon dioxide. *Data from, Crippa, et al., 2019. Fossil CO_2 and GHG Emissions of All World Countries. 2019 Report. European Commission, Joint Research Centre. Publications Office of the European Union, Luxembourg.*

TABLE 3.13 CO$_2$ emissions between 1972 and 2018 (selected countries).

		1972		2018			Cumulative from 1972 to 2018	
		Billion tonnes	% of global total	Billion tonnes	% of global total	Rate of increase/ decrease from 1972 to 2018 (%)	Billion tonnes	% of global total
Global total		16.5	100	37.9	100		1202.09	100
China		0.97	5.88	11.26	29.70	1,060.82	246.96	20.54
United States		4.81	29.15	5.28	13.93	9.77	233.78	19.45
India		0.24	1.45	2.62	6.93	991.67	46.08	3.83
Russia Federation		1.38	8.36	1.75	4.62	26.81	83.33	6.93
Japan		0.9	5.45	1.2	3.17	33.33	53.60	4.46
Big-5 total		8.3	50.29	22.11	58.35		663.75	55.22
EU-3	Germany	1.1	6.67	0.75	1.98	−30.09	44.78	3.73
	United Kingdom	0.65	3.94	0.37	0.98	−44.24	25.94	2.16
	Italy	0.35	2.12	0.34	0.90	2.29	19.36	1.61
EU-3 total		2.1	12.73	1.46	3.86		90.08	7.49
Australia		0.17	1.03	0.42	1.09	144.12	14.52	1.21
Canada		0.39	2.36	0.59	1.56	51.28	23.65	1.97
Mexico		0.13	0.79	0.50	1.52	284.62	16.21	1.35
South Korea		0.07	0.42	0.70	1.85	900.00	17.21	1.43
Lesser-4 total		0.76	4.60	2.21	6.02	184.8	71.59	5.96
Global-12 total		11.16	67.62	25.78	68.23		825.42	68.67

Note: Emission levels are based on fossil fuel use (coal, gas, and oil) and industrial process emissions (cement production, carbonate use of limestone and dolomite, nonenergy use of fuels and other combustion, chemical and metal processes, solvents, agricultural liming and urea, waste, and fossil fuel fires). *Data from, Crippa, et al., 2019. Fossil CO$_2$ and GHG Emissions of All World Countries. 2019 Report. European Commission, Joint Research Centre. Publications Office of the European Union, Luxembourg. See also Olivier, J.G.J., and Peters, J.A.H.W., 2018. Trends in Global CO$_2$ and Total Greenhouse gas Emissions 2018 Report. PBL Netherlands Environmental Assessment Agency, The Hague.*

almost persistently since 1972 after a slight drop from 36.33 billion tonnes in 2014 to 36.31 billion tonnes in 2015 and reached record high in human history to 37.89 billion tonnes in 2018. Total cumulative CO$_2$ emissions from 1972 to 2018 were about 1202 billion tonnes (Table 3.13). To avoid confusion, it may be briefly noted that there is no difference between tonne and metric tonne. The tonne is an official measurement unit.

As shown in Table 3.13, between 1972 and 2018, the Global-12 significantly emitted 825.42 billion tonnes CO$_2$ into the atmosphere. This accounts for 68.67% of the global cumulative emissions of 1202 billion tonnes. It is worth noting that the Big-5 led by China emitted a cumulative total of 663.75 billion tonnes in the same period, accounting for roughly 55.22% of the cumulative global total. China has the highest rate of increase from 0.97 billion tonnes in 1972 to 11.26 billion tonnes in 2018, an increase of roughly 1060.82%. This is followed by India which increased from 0.24 billion to 2.62 billion tonnes in the same period, an increase of about 991.66%. However, the emission standards for Germany and UK, two of the top emitters of all the EU's 28 member states, have dropped by 31.82% (from 1.1 to 0.75 billion tonnes) and 43.07% (from 0.65 to 0.37 billion tonnes), respectively, compared with their 1972 levels while the emission standards for the lesser-4, especially South Korea, have increased substantially. Indeed, South Korean has the highest increase of 900%, almost comparable to India (Table 3.13). To note in passing, the EU 28 comprises Austria, Belgium, Bulgaria, Croatia, Cyprus, Czech Republic, Denmark, Estonia, Finland, France, Germany, Greece, Hungary, Ireland, Italy, Latvia, Lithuania, Luxembourg, Malta, Netherlands, Poland, Portugal, Romania, Slovakia, Slovenia, Spain, Sweden and the United Kingdom.

The global CO_2 emissions between 1972 and 2018 as depicted in Table 3.13 are also reflected in Fig. 3.15 which shows the emission trends of various selected countries. More particularly it reveals that China overtook the United States as the world's largest CO_2 emitter in 2005, emitting 6.26 billion tonnes compared with the United States' 5.95 billion tonnes (Crippa et al., 2019). However, China's annual average emission standard between 1972 and 2018 is about 4.4 billion tonnes which is less than the United States at 5.14 billion tonnes. Also on a cumulative basis, China's total emission from 1972 to 2018 is about 213 billion tonnes while the United States reached 247 billion tonnes—the highest in the world. Japan's emission standards remain almost constant at an annual average of 1.12 billion tonnes. Russia's emission standards increased gradually since 1972 until reaching its peak in 1990 to 2.35 billion tonnes after which it dropped marginally to 2.31 and 2.13 billion tonnes in 1991 and 1992, respectively. Since then, its emission levels have remained almost constant at an average of about 1.7 billion tonnes between 1993 and 2018. The annual average emissions of the United Kingdom and Germany in the same period between 1972 and 2018 were 0.54 and 0.93 billion tonnes, respectively.

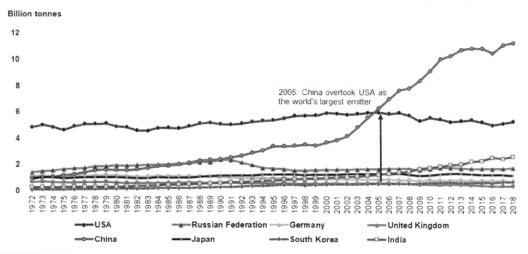

FIGURE 3.15   Historical CO_2 emissions (selected countries).
Note: Emission levels are based on fossil fuel use (coal, gas, and oil) and industrial process emissions (cement production, carbonate use of limestone and dolomite, nonenergy use of fuels and other combustion, chemical and metal processes, solvents, agricultural liming and urea, waste, and fossil fuel fires). CO_2, Carbon dioxide. *Data from, Crippa, et al., 2019. Fossil CO_2 and GHG Emissions of All World Countries. 2019 Report. European Commission, Joint Research Centre. Publications Office of the European Union, Luxembourg.*

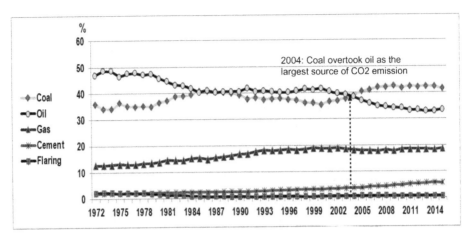

FIGURE 3.16   Sources of CO_2 emissions.

Note: The figures are drawn from Oak Ridge National Laboratory, US Department of Energy (Boden et al., 2016). Since its CO_2 emission data differ only slightly from Crippa et al. dataset of the JRC, the European Commission, its analysis is equally applicable to GRC-PBL CO_2 emission assessment as discussed above. CO_2, Carbon dioxide; *JRC*, Joint Research Center. *Data from, Boden, T.A., Marland, G., Andres, R.J., 2016. Global, Regional, and National Fossil-Fuel CO_2 Emissions. Carbon Dioxide Information Analysis Center, Oak Ridge National Laboratory, U.S. Department of Energy, Oak Ridge, TN. https://doi.org/10.3334/CDIAC/00001_V2016.*

However, on an interyearly basis, the largest increase in fossil fuel CO_2 emission between 2017 and 2018 is in India (7.2%) followed by Russia (3.55%), the South Korea (2.95%), the United States (2.87%) and China (1.52%). By comparison, Japan and Germany reduced their emission levels by 1.73% and 4.48%, respectively, while Italy reduced its emission by 3.1%.

The main source of CO_2 emission from 1972 until 2003 was from oil at an annual average of 42.84% (2453 million tonnes) of the global emission level. Since 2004, emission from coal burning has become the largest source accounting for 41.62% (3795 million tonnes) of the global emission standard. The annual average emission from gas accounts for 16.58% (1134.5 million tonnes) between 1972 and 2015 while cement and flaring account for 3.24 (234.82 million tonnes) and 1.02% (61.93 million tonnes), respectively, of the global emission standard in the same period (Boden et al., 2016, Fig. 3.16).

In 2004 coal overtook oil as the largest source of carbon emission, emitting 3.04 billion tonnes of CO_2. This is slightly more than oil emission level at 3.02 billion tonnes. The average carbon emission standard of coal between 2004 and 2015 is 3.79 billion tonnes while the average for oil is 3.14 billion tonnes. For the past 3 years between 2014 and 2016, the amount of CO_2 released into the atmosphere from burning fossil fuels, gas flaring and cement production was generally steady in that it did not increase or decrease significantly. This is partly due to less coal burning and more extensive use of renewable energy especially in the world's biggest CO_2 emitting countries such as China (Olivier et al., 2016, 2017, 2018, see also, IEA, 2019).

However, global energy-related CO_2 emissions rose by 1.6% in 2017, reaching 36.2 billion tonnes (Jackson et al., 2018). This is equivalent to the emissions of 170 million additional cars. The increasing trend is basically due to strong global economic growth of

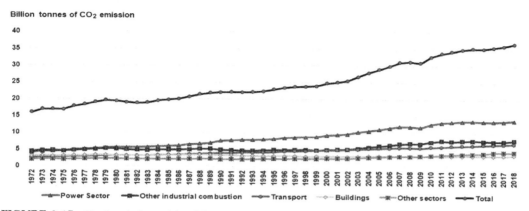

FIGURE 3.17 Total emissions of fossil fuels by sector from 1972 to 2018. *Data from, Crippa, et al., 2019. Fossil CO_2 and GHG Emissions of All World Countries. 2019 Report. European Commission, Joint Research Centre. Publications Office of the European Union, Luxembourg.*

3.7%, lower fossil-fuel prices which encouraged more coal burning and weaker energy efficiency improvement efforts. These contributed to increased global energy demand by 2.1% in 2017 (IEA, 2018). In 2017 coal continued to lead as the largest source of fossil-fuel-related CO_2. Overall, the direct drivers of CO_2 are combustion of coal, oil and natural gas, contributing to 89% of global emissions. Respectively, they contributed to 40%, 31% and 18% of the global total in 2017 (Olivier and Peters, 2018).

In 2018 driven by higher energy demand, CO_2 emissions from all fossil fuels continued to increase, with coal combustion alone surpassing 10 billion tonnes of CO_2 mostly in Asia (IEA, 2019). It may be noted that China, the United States, India, the EU 28, Russia and Japan accounted for 80% of the global fossil fuel consumption in the same year (Crippa et al., 2019). The share of CO_2 emissions mainly come from the power sector spurred by increasing energy demands of a growing population followed by other industrial combustion. Other sectors such as buildings and transport also contributed to the increasing levels of CO_2 emissions (Fig. 3.17).

As shown in Fig. 3.17, CO_2 emissions from the power sector increased substantially by about 246.5%—from 4 billion tonnes in 1972 to 13.86 billion tonnes in 2018. CO_2 emissions from industrial combustion increased gradually by 79.55% from roughly 4.4 billion tonnes in 1972 to 7.9 billion tonnes in 2018. However, CO_2 emissions from transport, including road transport, nonroad transport, domestic aviation, and inland waterways, increased significantly by 176% from 2.5 billion tonnes in 1972 to about 6.9 billion tonnes in 2018. However, CO_2 emissions from buildings remained more or less constant at an annual average of 3.2 billion tonnes in the same period. Other sectors, including industrial process emissions, indirect emission, agriculture, and waste, increased modestly from 1.98 billion tonnes in 1982 to 2.86 billion tonnes in 2005 after which they increased rapidly to 4.46 billion tonnes in 2018. Overall, the rate of increase between 1972 and 2018 for other sectors is 125.25%— from 1.98 billion tonnes to 4.46 billion tonnes.

Although CO_2 emissions leveled off from 2014 to 2016, there was a record increase in atmospheric CO_2 concentration. Indeed, the annual mean rate of growth of atmospheric

CO_2 concentration is not only increasing but is in fact accelerating. The annual mean growth of atmospheric CO_2 concentration in a given year is computed based on the difference in concentration between the end of December and the beginning of the January of that year. It represents the sum of all CO_2 added to, and removed from the atmosphere by human activities and by natural process (Dlugokencky and Tans, 2016). It may further be noted that data for atmospheric concentration are expressed in terms of part per million (ppm) which is derived by dividing the number of molecules of CO_2 by the number of all molecules in the air, including CO_2 itself after water vapor has been removed. Example, 0.000288 is expressed as 288 ppm (Dlugokencky and Tans, 2016).

During the preindustrial period ($\sim$ the year 1800), atmospheric CO_2 concentration stabilized in the range of 260–280 ppm for the preceding 10–12,000 years in the Holocene era. For example, atmospheric CO_2 concentration was 277 ppm in 1750, 279 ppm in 1775, 283 ppm in 1800, and 284 ppm in 1825 (Steffen et al., 2011). Fig. 3.18 shows that it increased from 327.45 ppm in 1972 to 366.7 ppm in 1998, surpassing the suggested threshold level of 350 ppm. Briefly noted, the Holocene era began when humans colonized new territories and the population swelled. It marked the onset of warmer and wetter climatic conditions at the end of the ice age, and was characterized by exceptionally stable climate and large sea level rise (see, e.g., Choy, 2016b).

As shown in Fig. 3.18, CO_2 concentration has been increasing gradually and persistently since 1972. For example, on an annual basis, CO_2 concentration increased from 1.22 ppm/year in the 1970s to 1.63 ppm/year in the 1980s and decreased to 1.53 ppm/year in the 1990s. However, it accelerated to 1.9 ppm/year in the 2000s and further increased to a significant level at 2.11 ppm/year during the period between 2000 and 2019. On the whole, global atmospheric CO_2 concentration has increased by an annual average of 1.75 ppm/year over the

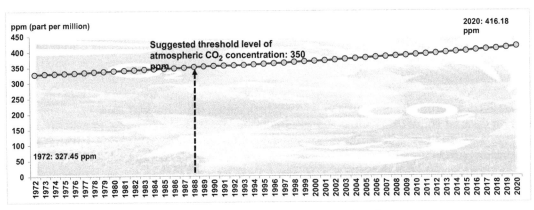

FIGURE 3.18   Global atmospheric CO_2 concentration from 1972 to 2020 (April).
Note: 2019 data is from Ian Tiseo, Statista while the preliminary data for 2020 (April) is from CO_2 Earth. CO_2, Carbon dioxide. *Data from, Tans, P., Keeling, R., 2018. NOAA ESRL Data. Scripps Institution of Oceanography. National Oceanic and Atmospheric Administration Earth System Research Laboratory Global Monitoring Division. CO_2 Earth. Available from: <https://www.co2.earth/>; Ian Tiseo, 2020. Global atmospheric concentration of carbon dioxide 2017–2019. Statista, <https://www.statista.com/statistics/1091999/atmospheric-concentration-of-co2-historic/>.*

past 48 years from 1972 to 2019, reaching 411.17 ppm in December 2019 April 2020 from 327.45 ppm in 1972. It further increased to 416.18 ppm in April 2020.

This means that the level of atmospheric CO_2 concentrationin April 2020 has increased by roughly 48.64% in April 2020 since the preindustrial period (1750–1850), during which the atmospheric concentration of CO_2 was about 280 ppm as noted above. It also breached the suggested threshold level of 350 ppm in 1998 for the first time when CO_2 concentration reached 351.57 ppm. The atmospheric concentration scaled further and hit the 400-ppm benchmark (400.83 ppm) in 2015 for the first time, primarily caused by human activities, especially the unrestrained combustion of fossil fuels.

It is evidently clear from the above that despite relentless efforts made by the United Nations and the world at large to stabilize global emissions within the Earth's threshold level, the overall growth rate has not changed much since 1972. Indeed, since the industrial revolution, human activities have been altering the carbon cycle in the atmosphere by adding more CO_2 and influencing the ability of the atmospheric sinks to remove additional CO_2. As it turns out, the current CO_2 abundance far exceeds the preindustrial period as discussed above.

The following section is an attempt to examine CO_2 emission trends based on specific case studies in China and the United States—the world's largest and second largest greenhouse gas emitters. This will allow us to see clearly to what extent the United Nations has been successful in halting the increasing global greenhouse gas emission.

3.16 Carbon emission country-specific case studies: China versus the United States

As has been discussed thus far, China and the United States are the world's two largest CO_2 emitters, contributing substantially to aggravating climate change. In recognizing the real and potential threat of climate change to the long-term existence of humanity, leaders from the world's two largest economies, President Xi Jinping from China and the former President of the United States, Barack Obama, vowed to commit to slow down and reduce the emissions of GHGs by formally ratifying the Paris Climate Agreement. Both leaders also pledged to keep emissions-cutting targets well below 2°C above preindustrial levels; and to pursue efforts to limit the increase to 1.5°C. In this section, we assess how the world's two largest economies and emitters have fared in their efforts to champion climate change mitigation.

The first-half of the main section examines the impacts of the Stockholm Conference on China's environmental initiative in fulfilling its Paris commitment. This is followed by a critical assessment of the extent to which China has been able to transform itself from a high carbon emitter to a low-carbon champion. This necessarily calls for the need to trace back its emission history since China's industrial revolution began during Mao Zedong's era. As will be made clear in the later section, this line of analysis will allow us to discern the different underpinning forces or real drivers behind their high carbon emission trends. The second-half of the main section focuses on the United States' efforts in delivering its Paris pledges from the Barack Obama era to the current administration led by President Donald Trump.

3.17 China's decarbonization initiatives: The road to Paris Accord

To begin with, as briefly noted in Chapter 2, The United Nations' Journey to Global Environmental Sustainability Since Stockholm: An Assessment, since the Stockholm Conference (1972), China began to develop environmental institutions to address the problem of environmental degradation. The first National Environmental Protection Conference was also held in Beijing in 1973 which led to the promulgation of the Guidelines on Environmental Protection and the Rules on Environmental Protection and Improvement. These documents are regarded as China's first environmental policies. In 1974 the Environmental Protection Committee of the State Council was established, culminating in the development of a national-level environmental administration (US Department of Commerce, 2002; Mol and Carter, 2007). It also set up the Office of Environmental Protection to promote environmental conservation and protection. This was followed by the enactment of the Environmental Protection Law in 1979 (amended in 1989 and 2014) (Mu et al., 2014). Environmental protection efforts were further intensified during Deng Xiaoping's administration. For example, the State Bureau of Environmental Protection was established in 1982 under the Ministry of Urban and Rural Construction and Environmental Protection, later restructured to become the State Environmental Protection Commission (SEPC) in 1984.

In 1987 the Law on Prevention and Control of Atmospheric Pollution was enacted. The law which deals with the supervision and prevention and control of atmospheric pollution especially by burning coal, was amended in 1995 and revised in 2000. In the following year in 1988, the SEPC was elevated to the status of State Environmental Protection Administration which finally became the Ministry of Environmental Protection (MEP) in 2008. Its main function is to formulate environmental protection guidelines, policies and laws to ensure sustainable development. Indeed, it has been asserted that environmental protection was among the "most heavily legislated sectors of public policy in the post-Mao period" (Ross and Silk, 1987: 3).

The publication of *Our Common Future* in 1987 also provided a remarkable conceptual framework and impetus in the post-Mao era and during Deng Xiaoping's era (between 1978 and 1992), for shaping policies based on the integration of environmental concerns into development policy in China. In the wake of the 1992 Earth Summit, the Chinese government established a Leading Group under the direction of the State Planning Commission and the State Science and Technology Commission, to formulate the local agenda—China's Agenda 21 as mentioned in Section 3.13.1.

With respect to carbon emission and climate change, Chapter 18 of the Agenda specifically deals with greenhouse gas emissions and climate change. It explicitly states that "China wishes to bring the emission of GHG under control, reduce growth rate of CO_2 emissions, study measures for reducing emissions of methane and nitrous oxide, maintain, strengthen greenhouse gas sinks and construct climate change monitoring, forecasting and service system" [China Agenda 21, Chapter 18 (31) quoted in Jiang, 2015: 171]. China's Agenda 21 also stipulates the "polluter pays principle" (PPP) to promote environmental protection. The PPP explicitly stipulates the polluters' responsibilities through normative and mandatory provisions (Qin, 2014).

A mixture of instruments including control and compliance procedures, emission charges and financial incentives to limit the residues from production processes have also been proposed under the Agenda (Bradbury and Kirkby, 1996). Meanwhile, as a sign of commitment to address the carbon emission and climate change problem, China also signed the Kyoto Protocol in 1998 and ratified it in 2002; although as a non-Annex I party, it is not legally bound by the treaty. China is also one of the first few countries to propose and implement sustainable development strategies, and to publish its first, second, and third national sustainable development reports in 1997, 2002, and 2012, respectively (see, e.g., Choy, 2016a).

In the year of the adoption of China's Agenda, the State Council issued a notice to all provinces, autonomous regions and municipalities encouraging them to implement the Agenda's guiding principles when formulating mid- and long-term socioeconomic policies, and particularly to integrate it into the Ninth Five-Year Plan (1996—2000) and the successive national 5-year plans relating to environmental protection (Tung, 2009). Under the Ninth Five-Year Plan and the Outline for Long-Term Target for the Year 2010 adopted in 1996, the concept of sustainable development was embraced as a major strategic guideline to implement China's modernization programme, while the 10th Five-Year Plan (2001—05) unveiled various specific SDGs for each development phase, including the implementation of some key ecological construction and environmental protection projects (see, e.g., Wang, 2000).

The successive 5-year plans with specific concerns about sustainability and climate change priorities are the 11th Five-Year Social and Economic Development Plan (2006—10), and the 11th Five-Year National Development Programme on Environmental Protection Law and Regulations. While the former is concerned with the promotion of energy conservation and emissions reduction, and addresses climate change as a long-term mission in achieving sustainable development, the latter sets out in detail the legislative programme for the amendment of current legislation as well as the enactment of new environmental laws (Tung, 2009; UNDP, 2012b).

In 2007 a political vision, "Ecological Civilization," which emphasized ecological quality and sustainability of China's economic development was introduced by the former president, Hu Jintou (Weng et al., 2015). In the same year, the Chinese National Development and Reform Commission (NDRC) unveiled the China National Climate Change Programme (CNCCP)—China's first and most comprehensive document for addressing climate change during that time. The CNCCP set out nonbinding but ambitious targets to reduce per-gross domestic product (GDP) energy consumption by 20% by 2010. In the same year, 2007, the NDRC issued the Medium and Long-term Development Plan for Renewable Energy, stipulating various goals for developing renewable energy and for increasing the share of renewable energy in total primary energy mix by 10% and 15% by 2010 and 2015, respectively (Tung, 2009; He, 2016). China's commitment to addressing climate change was also reflected in its switch from hardliner position to proactive stance in addressing global warming at the UN Bali Climate Change Conference held in 2007 (see, e.g., Liang, 2010).

One year later in 2008, the Chinese State Council Information Office issued its White Paper on China's Policies and Actions for Addressing Climate Change. This document reinforced the government's aim to combat climate change in line with UNFCCC and the Kyoto Protocol. It explicitly stated that "China is actively engaged in international efforts to address climate change and committed to the United Nations Framework Convention on Climate Change ... and the Kyoto Protocol" (State Council of the PRC, 2008: 1). In the same year, 2008, China amended the Energy Conservation Law of the People's Republic of China (1997) to promote energy conservation and renewable energy (Wang, 2008).

Other environmental laws enacted to promote sustainable development include the Cleaner Production Promotion Law (2002), which contains provisions encouraging firms to adopt cleaner production technologies to reduce pollution discharge (Catherine Yap et al., 2010), the Renewable Energy Law (2005) which came into force in 2006 and the Circular Economy Promotion Law enacted in 2008 and came into force in 2009. The latter two legal provisions aimed to address climate change through the promotion of the development and utilization of renewable energy, to protect the environment and to realize sustainable development. The Environmental Impact Assessment Law was also put in place in 2002 to assess and regulate the environmental impact of development and construction projects that would greatly affect environmental quality (see, e.g., Jiang, 2014). The introduction of an environmental impact assessment system provided the basis for the formulation and implementation of environmental protection counter-measures and the scientific management system (Qin, 2014).

It may further be noted that the rapid progress in the development of environmental law in China has been due to the government's belief that the most important and effective way to build an ecological civilization based on sustainable development is to strengthen environmental legislation. Consequently over the past few decades, apart from the above major laws, 50 relevant administrative laws and regulations on environmental protection, 660 local government regulations and 800 national standards related to environmental protection have been enacted (Wang et al., 2014).

In 2009 China committed itself at the UN Copenhagen Climate Change Conference to limiting its CO_2 emissions per unit of GDP by 40%−45% by 2020 compared with 2005 levels. It further agreed at the Cancun Climate Change Conference held in 2010 to abide by the international system to provide measurement, reporting and verification of mitigation actions (Jiang, 2014). In the same year, the 2009 version of China's Policies and Actions for Addressing Climate Change was issued. Again, China revealed its commitment to further integrating actions on climate change into its economic and social development plan (NDRM, 2009). Later versions of China's Policies and Actions for Addressing Climate Change issued from 2011 to 2015 also indicated China's political commitment to addressing climate change by promoting green, cyclical and low carbon, and by raising ecological awareness.

The Chinese government also acknowledges the importance of international conventions such as UNFCCC and Kyoto Protocol in addressing climate change. Article 46 of the Environmental Protection Law (1989), for example, explicitly states that "If an international treaty regarding environmental protection concluded or acceded to by the People's Republic of China contains provisions differing from those contained in the laws of the People's Republic of China, the provisions of the international treaty shall apply, unless the provisions are ones on which the People's Republic of China has announced reservations."

In 2010 the NDRC created a number of low carbon development zones covering five provinces and eight cities. The five provinces are: Guangdong, Liaoning, Hubei, Shaanxi, and Yunnan, and the eight cities are Tianjin, Chongqing, Shenzhen, Xiamen, Hangzhou,

Nanchang, Guiyang, and Baoding. These low carbon development zones are required to develop and implement low carbon action plans, focusing on industrial energy efficiency, and carbon mitigation (Wang, 2011). Furthermore in its 12th Five-Year Plan for National Economic and Social Development Programme (2011−15), China committed to intensifying efforts to mitigate climate change based on various measures including legislative improvement, provision of fiscal, and financial support to strategic emerging industries involved in energy conservation, clean energy, and environmental protection technologies (Jiang, 2014). More specifically it committed China to cut its carbon emissions by 17% in per unit of GDP as well as a 16% cut in energy consumption per unit of GDP. To achieve these carbon reduction targets, the State Council released a Work Plan for Controlling Greenhouse Gas Emissions during the 12th Five-Year Plan period to guide its carbon reduction processes (Sandalow, 2018).

It is succinctly clear from the above that since the Deng reform era, the Chinese government has put in place a comprehensive environmental protection framework with successive refinements to address CO_2 emission and climate change based on legal enactment and environmental policies. This raises the most pertinent question: to what extent the all-encompassing environmental regime established to combat climate change has been able to decarbonize its economy in tune with the Paris Accord? This will be systematically examined in the following sections.

3.18 China's pathways to decarbonization: The arduous journey

3.18.1 China's journey to becoming the carbon dioxide emission monster

To come to grips with an empirically observable carbon dynamics of China's transition from a carbon-emission regime to a carbon-constrained economy, it is necessary to trace back to Mao Zedong's-era march to socialism which went hand in glove with the country's rapid industrial take off accompanied by growing CO_2 emission levels. To start with, as seen in Table 3.12 above, the CO_2 emission standard of China is more than twice that of USA, eight times that of EU-3 (Germany, United Kingdom and Italy) and five times that of Lesser-4 (Australia, Canada, Mexico and South Korea). The increasing CO_2 emission trends in China leading to its current colossal emission standard may be traced back to Mao Zedong's era of economic modernization which was modeled after the planning system of the Soviet Union.

After taking over the political helm in 1949, Mao placed development and industrialization as the greatest priority to pull the nation out from its rural and agrarian past within three 5-year plans or slightly longer. The First Five-Year Plan took place from 1953 to 1957. However, a great deal of industrialization occurred during the Second Five-Year Plan (1958−1961), commonly known as the Great Leap Forward, as briefly noted in the preceding chapter. The promotion of heavy industrial development constituted the main agenda of Mao's Great Leap Forward economic modernization programme to catch up with the industrialized west. More specifically it aimed to catch up with Great Britain in 15 years and the United States in 20−30 years in the production of iron, steel and other main industrial outputs. In its attempt to achieve this aim, the Mao administration mobilized massive amounts of resources and manpower into heavy industries (Riskin, 1987).

The First Five-Year Plan set the springboard for the Great Leap Forward in early 1958. The Great Leap Forward (1958—62) took two forms: a mass steel campaign, and the creation of the people's communes to support the expansion of industry and construction. Between 1958 and 1959, nearly 2.6 million commune enterprises were established across China, each encompassing thousands of households in support of the promotion of rural industries (Mark et al., 2001). Within these communes, small backyard iron and steel furnaces were built to help to double China's steel production within 1 year. It was estimated that there were roughly 600,000 commune-run iron and steel operations across the country, registering an output of 2.4 million tons of iron and 530,000 tons of steel in 1958 and 1959 while other communes were involved in producing building materials, chemical fertilizers, farm machinery, and other goods and services. During the same period, it was estimated that 100 million pieces of farm machinery and repair factories were produced and established (Zhang, 1999).

Trees were recklessly felled and coal was extensively mined to provide energy to support the above industrial activities, especially iron and steel production. Extensive and uncontrolled construction of primitive small open-cast mines across the country had become commonplace in the Great Leap Forward economic scenario. By the end of 1958, there were some 110,000 pits in operation across China, engaging an incredible 20 million peasants. To ensure adequate supply of cheap industrial energy, the output of large mines had also expanded at a staggering pace as more than 400 new mines were established in 1958. It was estimated that 25.3 million tons of coal were produced during 1958 and 1959 (Smil, 2004).

The First Five-Year Plan fundamentally focused on the establishment of an industrial base. During this period, a great deal of construction such as railroad construction, plants, and factories were built and iron and steel industries were promoted. In the closing year of the First Five-Year Plan, China's steel output increased from 923,000 tons in the preliberation period in 1943 to 1,349,000 tons in 1953, and further increased to 5,340,000 tons in 1957, an increase of nearly almost four times and an annual average increase of 74% between 1953 and 1957. Meanwhile, pig iron production rose from 1,801,000 tons in 1943 to 1,878,000 tons in 1953 and further increased to 5,874,000 tons in 1957, an increase of almost four times from 1953 to 1957 and at an annual average increase of 53%. Between 1953 and 1957, coal output increased by 92.5% from 63,530,000 to 122,300,000 tons—an annual average increase of 23.12% (Niu, 1958). It is noteworthy that coal was China's only readily available and abundant source of energy, literally fueling its industrial transformation. As a result, between 1950 and 1952, 77 large- and medium-sized coal mines were restored or reconstructed to increase its coal output to support the First Five-Year Plan modernization programme (see, e.g., Smil, 2004).

Despite the industrial period, however, China's industrial output was still a century behind Britain's. Nonetheless, from the perspective of speed of development, China was far ahead of Britain: while the British industrial development increased less than 2% annually over the previous 100 years, the speed of industrial development in China in the short span of 5 years was 19% annually (Niu, 1958).

However, the Great Leap Forward turned out to be an economic and social disaster for the nation; it had not been based on sound economic planning but on misguided principles and untenable output targets set by Mao Zedong; inferior production technologies were used and communes were mismanaged. After its collapse in 1960, most factories in the communes were shut down. However, by the end of the 1960s and early 1970s, rural industrialization programmes in the communes had been revived, focusing on the development of five small industries, namely, iron and steel, chemical fertilizer, cement, energy (coal mining and hydropower development), and farm machinery (Zhang, 1999). During the cultural revolution (1966–76), in an attempt to increase agricultural output, farmers from the communes were mobilized to build terra fields in mountains, causing extensive ecological destruction to the ecosystem (Catherine Yap et al., 2010).

3.18.2 The Mao era of economic transformation and carbon dioxide emission

Although the evolution of the Mao era of economic transformation had been neither smooth nor trouble free, it set the structural direction for the modern economic transformation in China which focused on rapid heavy industrialization. This included the promotion of energy-intensive industries such as steel and iron manufacturing, the automobile industry, and coal mining. The Maoist growth ideology also had far-reaching implications for the Post-Mao development trends in China which placed great emphasis on GDP growth above every other factor, and the environmental factor was especially ignored.

It is necessary to emphasize that the promotion of iron and steel industry, the world's most energy-intensive industry and one of the largest industrial sources of CO_2 emission in China, is the pillar of the Chinese industrial modernization and economic development. This energy-intensive industry is also the largest consumer of coal in the industrial sector. Also coal plays a highly significant role in China's rapid economic growth (Li and Leung, 2012). It is thus logical to expect that the gradual and significant increase of steel production and economic growth would lead to a rising and significant increase in the use of coal and the level of CO_2 emissions. In the rest of the section, CO_2 emission trends are discussed in relation to iron and steel production trends in China. The sources of data used for analysis are as follows:

1. From 1950 to 1984: CO_2 emissions data compiled by the Earth Policy Institute (2015), which exclude emissions from cement production and gas flaring, while the data for iron and steel production are from a secondary source compiled by Rock and Toman (2015) from the Energy Research Institute, National Development Reform Commission of the People's Republic of China. During the early period, especially during Mao's reform era, data on the actual emissions from the industry may not be available. The main purpose of this part of the analysis is to demonstrate the relationship between steel production/economic growth and CO_2 emissions.
2. From 1985 onward, CO_2 emissions and iron and steel production data are again taken from a secondary source compiled by Rock and Toman (2015). Rock and Toman's data provide a clear indication of the contribution of the iron and steel industry to CO_2 emissions standards in China from 1985 to 2009.

To begin with, the First Five-Year Plan which had led to rapid growth in iron and steel output, coal production and other goods and services continued to generate growth in heavy industry even after the Great Leap Forward collapsed in 1960. Heavy industry had largely recovered from its collapse by 1965 and plants built during 1958–60 were producing at near full capacity for the first time since 1960 (Perkins, 1967). Overall, during the First Five-Year Plan period, rapid production levels, especially iron and steel output, began to steer the course of increasing CO_2 emission trend in China. As shown in Fig. 3.19, since Mao took the helm in 1949, CO_2 emission started to exhibit a rising trend from 1950 onward and peaked in 1960 at 781 million tonnes in the Mao era after which it declined modestly due to the collapse of the Great Leap Forward. However, since 1967, CO_2 emission has increased steadily (Fig. 3.19).

Although the cultural revolution, spanning a decade from 1966 to 1976 is considered a social/political movement rather than an economic reform movement, in a highly politicized country like China under Communism, it is difficult to stay clear of economic considerations. As Dwight H. Perkins pointed out succinctly, "It would be misleading, however, to leave the impression that the cultural revolution has foreclosed China's economic development for years ahead" (Perkins, 1967: 48). The launch of the Third Five-Year Plan (1966–70) which was overshadowed and disrupted by the unavoidable political and social chaos for 3 years from 1967 to 1969 (Perkins, 1991), may serve as a case in point. Overall, the rate of growth of output in the Third Five-Year Plan period was estimated at about 6% per annum with industrial production levels increasing substantially after the disruption period (Perkins, 1991). As shown in Table 3.14, for example, steel and coal output increased significantly from 1970 onward, indicating there was a close relationship between steel output and coal demand (Table 3.14).

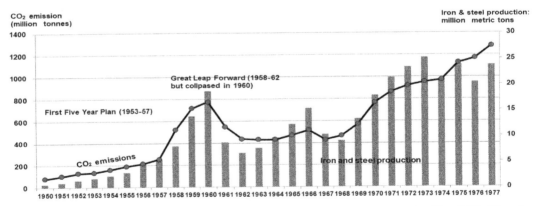

FIGURE 3.19 CO_2 emission during Mao's economic reform era from the First Five-Year Plan (1953–57) to the Great Leap Forward period (1958–62) and beyond.
Note: Emissions from cement production and gas flaring are excluded from the dataset. CO_2 emissions were significantly but not exclusively associated with iron and steel production. *CO_2*, Carbon dioxide. *Data from, CO_2 emission: Earth Policy Institute, 2015. Climate, Energy, and Transportation: Carbon Dioxide Emissions from Fossil Fuel Burning in Top Ten Countries, 1950–2012. Earth Policy Institute, Rutgers University. Available from: <http://www.earth-policy.org/data_center/C23>; Iron and steel production: Rock, M.T., Toman, M.A., 2015. China's Technological Catch-Up Strategy: Industrial Development, Energy Efficiency, and CO_2 Emissions. Oxford University Press, New York, Oxford.*

TABLE 3.14 Index of industrial output during the cultural revolution (1966 = 100).

Index of industrial output during the cultural revolution (1966 = 100)									
	1957	1962	1965	1966	1967	1968	1969	1970	1975
Electric power	23	56	82	100	94	87	114	140	237
Steel	35	44	80	100	67	59	87	116	156
Coal	52	87	92	100	82	87	106	140	191
Petroleum	10	40	78	100	95	110	149	211	530
Cement	34	30	81	100	73	63	91	128	230
Chemical fertilizer	6	19	72	100	68	46	73	101	218
Machine tools	51	41	74	100	74	85	156	253	319
Cloth	69	35	86	100	90	88	112	125	129
Bicycles	39	67	90	100	86	97	142	180	304

Adapted from Perkins, D.H., 1991 China's economic policy and performance. In: MacFarquhar, R., Fairbank, J.K. (Eds.), The Cambridge History of China. Cambridge University Press, Cambridge, pp. 473–539, 481.

3.18.3 Economic performance during Deng Xiaoping's reform period and post-Deng era

The cultural revolution paved the way for the Chinese institutional transition from a planned economy system to an open and market-oriented economy. After Mao's death in 1977, Deng Xiaoping, taking over as the new leader in 1978 began to reform the Chinese economy based on free-market principles which aimed to attract foreign direct investment and to promote economic growth (see, e.g., Chow, 2004). The sustained programme of economic reform and the open-door policy under Deng's leadership brought the country a period of explosive economic expansion (Naughton, 1993).

The commune system was dissolved and the name "township and village enterprise" (TVEs) was adopted, and farmers and herdsmen were given freedom to engage in industrial activities (Zhang, 1999; Catherine Yap et al., 2010). Since then, TVEs have become an important force of the national economy, and rural industrialization during this period led to rapid urbanization and transformation of thousands of villages into towns, and towns into cities (Zhang, 1999). State-owned enterprises could operate and compete on free market principles. Special economic zones were established along the coast for the promotion of more flexible market-based economic policies and to attract foreign direct investments, boosting exports, and importing high-technology products (Catherine Yap et al., 2010).

After Deng took over the political helm, China persistently achieved and sustained high economic growth with its GDP growth from 1978 to 1997, reaching an annual average of 10%/year compared with the prereform growth of 5.4% between 1961 and 1978. Remarkably its economic growth was three times the global average (Fig. 3.20). Within a decade, China was able to transform itself into a modern industrial state, a process that many developing countries took 100 years to achieve (Choy, 2016a). Iron and steel production also increased gradually and substantially since 1978 as shown in Fig. 3.21

GDP growth (annual %)

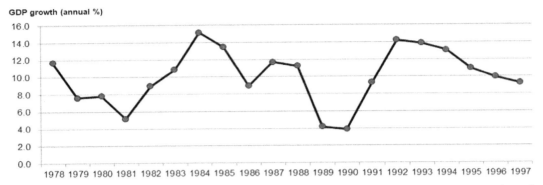

FIGURE 3.20 Economic growth during Deng's reform era. *Data from, World Bank, 2016a. Countries and Economies: China. World Bank, Washington, DC. Available from: <http://data.worldbank.org/country>. World Bank, 2018. Data. China. World Bank, Washington DC. Available from: <https://data.worldbank.org/country/china>*

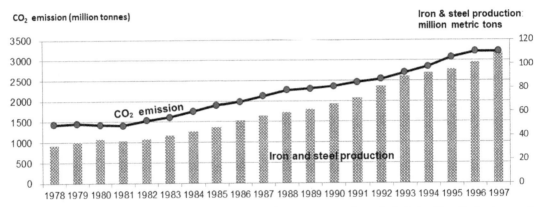

FIGURE 3.21 CO_2 emission trend during Deng's reform era spurred partly but significantly by iron and steel output.
Note: Emissions from cement production and gas flaring are excluded from the dataset. CO_2 emissions were significantly but not exclusively associated with iron and steel production. CO_2, Carbon dioxide. *Data from, CO_2 emission: Earth Policy Institute, 2015. Climate, Energy, and Transportation: Carbon Dioxide Emissions from Fossil Fuel Burning in Top Ten Countries, 1950−2012. Earth Policy Institute, Rutgers University. Available from: <http://www.earth-policy.org/data_center/C23>; Iron and steel production: Rock, M.T., Toman, M.A., 2015. China's Technological Catch-Up Strategy: Industrial Development, Energy Efficiency, and CO_2 Emissions. Oxford University Press, New York, Oxford.*

(see also, Figs. 3.22 and 3.23). Its production level had increased by slightly more than 240% from 31.8 million metric tons in 1978 to about 109 million metric tons in 1997. However, at the same time, the extraordinarily high increase in the production level contributed significantly to push up the CO_2 emission standard by 128.57% from 1.4 to 3.2 billion tonnes in the same period, and its impressive industrial performance and rapid economic growth was accompanied by severe environmental deterioration.

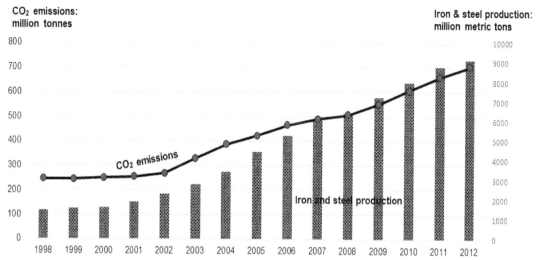

FIGURE 3.22 CO_2 emission trend from the iron and steel industry in the post-Deng reform era.
Note: Emissions from cement production and gas flaring are excluded from the dataset. CO_2 emissions were significantly but not exclusively associated with iron and steel production. CO_2, Carbon dioxide. *Data from, Rock, M.T., Toman, M.A., 2015. China's Technological Catch-Up Strategy: Industrial Development, Energy Efficiency, and CO_2 Emissions. Oxford University Press, New York, Oxford.*

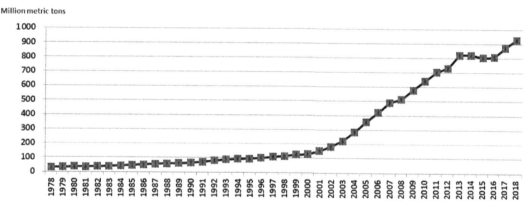

FIGURE 3.23 *China's steel production at a glance. Data from, World Steel Association, 2016. Steel Statistical Yearbook 2016. World Steel Association, Brussels. Available from: <https://www.worldsteel.org/en/dam/jcr:37ad1117-fefc-4df3-b84f-6295478ae460/Steel + Statistical + Yearbook + 2016.pdf>; World Steel Association, 2018. World Crude Steel Output Increases by 5.3% in 2017. World Steel Association, Brussels. <https://www.worldsteel.org/media-centre/press-releases/2018/World-crude-steel-output-increases-by-5.3--in-2017.html> (accessed 12.12.19.); World Steel Association, 2019. World Steel in Figures. World Steel Association, Brussels. Available from: <https://www.worldsteel.org/en/dam/jcr:96d7a585-e6b2-4d63-b943-4cd9ab621a91/World%2520Steel%2520in%2520Figures%25202019.pdf>.*

3.18.4 Industrialization and carbon dioxide emission in the Post-Deng era

After Deng's death in 1997, China continued to place its highest priority on promoting industrial modernization and economic growth. Average GDP growth in the post-Deng era from 1998 to 2012 reached 9.1% per annum. China had achieved an impressive near 9% GDP growth during the full-blown global recession in the year 2009. It was also the first of the major economies in the world which was able to withstand the effects of global recession. China also surpassed Japan as the world's second largest economy after the United States in 2010 (see, e.g., Choy, 2016a).

At the same time, China's steel output in the same period kept increasing at an unprecedented level, increasing from about 116 million metric tons in 1998 to 731 million metric tons in 2012. This helped to push up the CO_2 emission level from about 3 to 9 billion tonnes in the same period (Fig. 3.22).

Steel output continued to rise further at an annual average of 826 million metric tons between 2012 and 2018, reaching 928.3 million metric tons in 2018 (World Steel Association, 2016, 2018, 2019; Fig. 3.23). In the first 4 months of 2019, China's steel production increased further by 10.3% (315 metric tons) year-over-year, accounting for 52.5% of the global total (Richter, 2019). Its steel consumption in 2018 reached 835.45 metric tons accounting for 48.8% of the world's total (World Steel Association, 2019).

It is necessary to emphasize that the promotion of iron and steel industry, the world's most energy-intensive industry is the pillar of the Chinese industrial modernization and economic development. China has been the world's largest steel producer since 1996. It is also the world's largest steel consumer (Xu and Lin, 2016). The steel sector accounts for more than 50% of the country's energy consumption, mainly from coal—one of the largest sources of CO_2 emission in China (Xu et al., 2016). Coal or more specifically, coal-fired electricity has been playing a crucial role to sustain rapid economic growth in China (Li and Leung, 2012; Carbon Brief, 2019). China accounts for half of the global total coal consumption since 2011 (BP, 2011–2019; Fig. 3.24).

As shown in Fig. 3.24, compared with other countries such as the United States, Germany, India, and Japan, China's coal consumption increases gradually and substantially since the cultural revolution (1966–76). This coincide with the Third Five-Year Plan (1966–70) during Mao's era from 8.7% [122.4 million tonnes of oil equivalent (Mtoe)] of the global total in 1966 (1404.3 Mtoe) to 14.6% (234.2 Mtoe) of the global total in 1976 (1606.9 Mtoe). Since Deng took over from Mao as the new leader in 1978 after his death, coal consumption increased rapidly from 16.9% (282.8 Mtoe) of the global total in 1978 (1673 Mtoe) to 30.5% (696 Mtoe) of the global total in 1996 (2279.6 Mtoe). In the post-Deng era, coal consumption continued to increase to 50% (1904 Mtoe) of the global total in 2011 (3782.5 Mtoe).

Since then, coal consumption has remained at roughly 50% of the global total until 2018 (BP, 2018). More explicitly annual average coal consumption during the Mao' era between 1966 and 1976 is about 173 Mtoe while during Deng's administration (1977–96), it increased to roughly 454 Mtoe. In the post-Deng era between 1998 and 2018, its annual average increased substantially to 1452 Mtoe (BP, 2011–2019). For the past 10 years between 2009 and 2018, coal accounted for about 65.7% of China's energy mix on average (BP, 2011–2019). China overtook the United States in 1986 as the world's largest consumer of coal, burning 425.7 Mtoe of the "black rock" compared with the American consumption level at 413.2 Mtoe (BP, 2011–2019). China also surpassed the United States as the world's largest producer of coal in 1985.

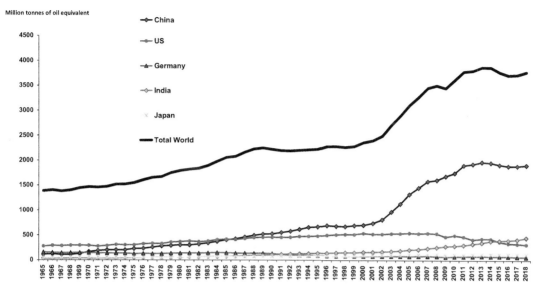

FIGURE 3.24 China's coal consumption trends: a global comparison.
Note: China's coal consumption contributes substantially to the global total. *Source: BP, 2011–2019. BP Statistical Review of World Energy (June 2011 to June 2019 issues). British Petroleum (BP), United Kingdom.*

In comparison, coal consumption in the United States accounted for about 17% of global average (2644 Mtoe) in the Deng and post-Deng' era (1977–2018) with annual average coal consumption at about 449 Mtoe. Generally coal consumption in Germany has declined from 128.9 Mtoe in 1977 to 66.4 Mtoe in 2018. Its annual average coal consumption in the same period was about 103 Mtoe, roughly 4% of the global average. As expected, due to increase in population and rapid industrialization process, coal consumption in India increased massively from 52.5 Mtoe in 1977 to 452.2 Mtoe in 2018 with its annual average coal consumption the same period at 183 Mtoe, contributing to about 6.9% of the global average. Coal consumption in Japan increased from 51.8 Mtoe in 1977 to 117.5 Mtoe in 2018. It has the lowest annual average among the above countries at 90.11 Mtoe which is about 3.4% of the global average.

The above discussion provides a clear explanation of the nexus of industrial production/economic growth, energy/coal consumption and CO_2 emission in China since Mao's reform era. It suggests a unidirectional trend from industrial output, economic growth, energy/coal consumption and CO_2 emission. In particular, as shown in Fig. 3.19 and Table 3.13, industrial output moves in the same direction with energy/coal consumption and CO_2 emission. This has far reaching implications for climate change and global warming as coal dominates the primary energy mix in China—the world's largest primary energy consumer (Fig. 3.25).

Another sector which contributed to the persistent increase in CO_2 emission in China is cement production—one of the world's most coal dependent industrial sectors. Here, it

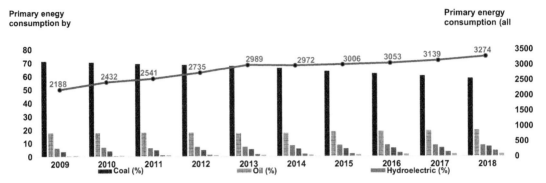

FIGURE 3.25   Energy mix in China. *Data from, BP, 2011–2019. BP Statistical Review of World Energy (June 2011 to June 2019 issues). British Petroleum (BP), United Kingdom.*

may well be noted that global cement production is the third largest source of anthropogenic emissions of CO_2 after fossil fuel combustion and deforestation/land-use change, contributing to 7%–8% of the of global carbon emissions (Andrew, 2018; Rodgers, 2018; Early, 2019). In China, the cement industry contributes roughly 7%–9% of its total annual carbon emission (Liu et al., 2015). It is also one of the primary drivers of climate change (Gregg et al., 2008).

Carbon emissions from the cement industry are fundamentally arise from process-related and direct fossil-related emissions. The former involves a calcination process, a chemical reaction which decomposes calcium carbonate (limestone) to calcium oxide (lime) and CO_2 at high temperatures to create clinker—the major component of cement. CO_2 emission from this chemical process cannot be avoided even with the deployment of advanced energy efficiency technologies, changing fuels or increasing factory efficiency (Early, 2019). The latter involves direct burning of coal in the production of clinker which emits CO_2. The emission standard in coal combustion is a function of coal's carbon content, heating value, oxidation value and combustion technologies (Liu et al., 2015).

It is estimated that about 1 kg of CO_2 is released for every kilogram of cement produced. The emission factor as reported by the NDRC for clinker production in China is 0.5383 tonne of CO_2 per tonne of clinker produced (Andrew, 2018) The overall emission factor of coal is 0.499 tonne CO_2 emission per tonne of coal consumption (Shan et al., 2019).

Against this backdrop, it may be noted that during the Mao reform era from the First Five-Year Plan to the cultural revolution through the 1950s, 1960s, and 1970s, small-scale county and commune cement plants multiplied, from around 200 by 1965 to over 4500 in 1980 (Global Cement, 2013). Annual output increased substantially since Mao's reform period from 0.45 million metric tons in 1949 to 0.8 million metric tons in the following year and further rose to 2 million metric tons in 1952 (USGS, 1952). During Deng's reform era, cement production increased by almost sevenfold from 65 million metric tons in 1978 to 493 million metric tons in 1997. It rose further to 2.48 billion metric tons in 2014, accounting for more than 60% of global production (Edwards, 2015; Fig. 3.26). However, since its production peaked in 2014, output has dropped slightly to 2.37 billion metric tons in 2018.

Million metric tons

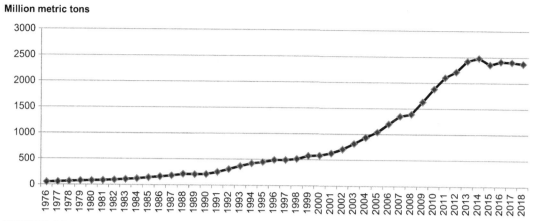

FIGURE 3.26 China cement production. *Data from, USGS, 1976–2019. Cement: Statistics and Information. U.S. Geological Survey (USGS), USA. Available from: < https://minerals.usgs.gov/minerals/pubs/commodity/cement/ >*

Million tonnes of CO₂ emission

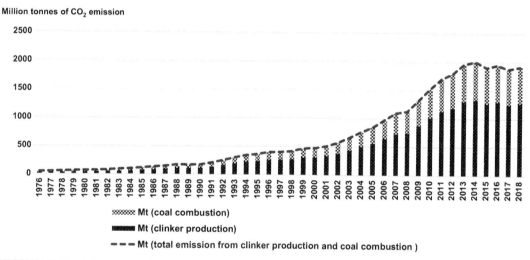

FIGURE 3.27   CO₂ emission from cement industry in China.

Note: The indirect emission from the consumption of electricity (electricity-related emission) from the crusher and grinder processes which do not emit CO₂ is excluded from the above computation. Although the crushing and grinding processes consume a substantial amount of electricity, it is practically impossible to ascertain the emission levels from the power plants which generate and supply the utility to the cement plants. Moreover, different sources of electricity generated from oil, coal, gas, hydropower, solar, or nuclear also have different emission standards. CO₂, Carbon dioxide.

The Chinese cement industry has been the largest in the world for the past two decades (Global Cement, 2013). It still remains the largest in the world today and contributes to aggravating the CO₂ emission problem in the country. Using the carbon emission factors as mentioned above, we are able to demonstrate the CO₂ emission levels of the cement industry in China. As shown in Fig. 3.27, the carbon emission from the cement industry increased substantially from 40 million tonnes in 1976 during the Deng or post-Deng era to reach its peak in 2014 at roughly 2000 million tonnes. This was spurred by rapid economic growth and urban development. Since then, the emission standard has dropped slightly to 1900 million tonnes in 2018. Also the decomposition of carbonates in the clinker production process constitutes the largest source of CO_2 emission from cement industry.

3.18.5 China: The paradox of climate change policies

It is increasingly clear that despite putting in place an all-encompassing environmental governance since Stockholm as discussed above, China has not been able to avoid massive anthropogenic disruption of the global carbon sink during the four decades of Deng's economic reform period. China's current emission levels are the outcome of its economic policy choices guided by the Great Leap Forward philosophy which placed high priority on economic growth over environmental conservation. The high emission standards also stemmed from policy, institutional and governmental failures, a common phenomenon not only in China but also in other developing as well as developed countries, including the United States. This will be discussed in the next section.

Mao's era of industrialization led to massive pollution just as England's Industrial Revolution did. In the post-Mao era and during Deng's open-market reform period, China, following the Stockholm Conference, had taken the initiative to put in place a comprehensive environmental regime to address its acute pollution problems. However, at the initial stage of the environmental regulation reform which was premised on the western models, China had little experience in the new field of environmental management and weak institutional capacity for environmental law implementation. For example, strategies for resolving enforcement problems were usually designed without considering other related issues. Also the system of regulations was extraordinarily complex because the guidelines for environmental protection had been established through many separate rules and regulations authored by different entities at different phases (Wang, 2010).

To compound the problem, China's international isolation in the 1970s and early 1980s had rendered the learning process a formidable task (Dunnivant and Anders, 2019). Also government bodies responsible for the implementation were hindered by weak institutional capacities and generally lacked the necessary experience in environmental management. In addition, the environmental protection departments or bureaus set up at the local levels in provinces, autonomous regions and municipalities lacked the authority, capacity and resources to monitor and enforce various pollution regulations. Besides, as at 2004, there were no environmental controlling monitoring stations and environmental supervision or inspection agencies at the local levels to address and monitor environmental pollution problems (Dunnivant and Anders, 2019).

For example, the specific approaches developed to mitigate GHG emissions and climate change such as establishing low-carbon development zones in various provinces and cities across China have not driven overall improvements in GHG emissions, reduction and energy-efficiency promotion (Shen, 2013). The reasons were that when the low-carbon zones were first created, there were no clear guidelines, no recognized definition of low carbon Industrial Zones, no quantifiable evaluation indexes, uniform regulatory standard or model guide policy design and implementation (ISC, 2012). Also some of the local governments were unaware of the low carbon concepts, let alone the will to execute it. In most zones, there was also a lack of government support and many companies had little low-carbon awareness and lacked motivation. Thus the green reality on the ground was far from ideal (Zhang, 2012).

China has traditionally developed "command and control" pollution regulations by first setting binding emission limit values and then attempting to achieve these targets by instituting prohibition and/or technology-oriented requirements (Raufer and Wang, 2003). While this form of regulatory environmental control may be effective in the advanced nations with strong central control, this may not be the case in a centrally planned economy and the state-dominated environmental governance in China (see, e.g., Bradbury and Kirkby, 1996). While in the western nations, environmental standards are enforced by law, China relies on incentive mechanisms embedded in the cadre management system to promote implementation of targets (Kostka, 2016). That is one of the major reasons for its ineffectiveness. Another factor is the subornation of polluting agents or manufacturers to increase production (Bradbury and Kirkby, 1996).

It may well be that local government officials benefit from higher levels of output in their region as they receive credit for economic development which help in their annual evaluations that form the basis for promotion and bonus (Chow, 2008; Kostka, 2016). Consequently it is often to the advantage of local governments to allow pollution to take place illegally. The proliferation of tens of thousands of spatially dispersed rural enterprises which are immune from urban-centered judicial regimes, ineffective emission monitoring systems, close ministerial connections and the political influence in the state industrial sector, and the willingness of polluting agents to incur fines over environmental protection, among others, are also some of the obstacles that hamper effective implementation of environmental controlling measures (Bradbury and Kirkby, 1996). Also the general environmental policy framework during the post-Mao era placed great priority on economic growth to environmental conservation, thus, compromising effective enforcement of environmental laws. It is thus clear why despite many of the laws and regulations enacted to arrest carbon emission, China has been unable to move from its position as the world's largest CO_2 emitter.

3.18.6 The reversal of China's climate change paradox? President Xi Jinping's green factors

It must be admitted that the high carbon emission levels in China are also a consequence of its enormous population of 1.43 billion and heavy reliance on coal to meet the energy needs of hundreds of millions of its citizens. China's high carbon emission standards are also caused by its massive investments in factories, infrastructure and housing which underpin the social welfare enhancement of its population. Yet, China's average per capita carbon emission of 3.44 between 1970 and 2018 is still lower than the global average of 4.42, and even far lower than the world's second largest carbon emitter, the United States at 19.89 in the same period. China's global ecological footprint of 3.62 global hectare (gha) is also far lower than that of the United States at 8.1 gha or of other advanced nations as shown in Fig. 3.28. From the resource consumption perspective, the ecological footprint measures the ecological assets that a given population requires to produce the

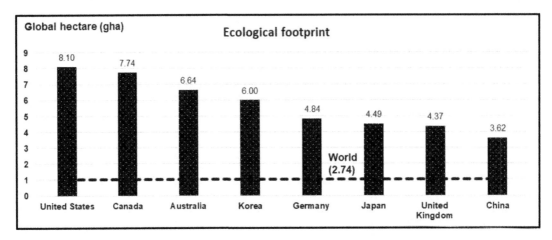

FIGURE 3.28 China's ecological footprint: a global comparison. *Data from, Global Footprint Network, 2019. Humanity's Ecological Footprint Contracted Between 2014–2016. Global Footprint Network. Available from: <https:// www.footprintnetwork.org/2019/04/24/humanitys-ecological-footprint-contracted-between-2014-and-2016/>.*

natural resources it consumes and to absorb its waste, especially carbon emission (Global Footprint Network, 2019).

Also compared with Deng's era of reform which placed greater emphasis on economic growth than on environmental conservation, the tide of environmental control sentiments has turned under the current administration led by President Xi Jinping. Under Xi's administration, China has made tremendous efforts to address climate change to fulfill its pledge to the Paris Agreement. In 2015 President Xi Jinping formally announced China's ratification of the Paris Climate Agreement and vowed to "unwaveringly pursue sustainable development" and addressing climate change in accord with the Paris deal with a novel aim to protect "the future of our people and the wellbeing of mankind" (Phillips et al., 2016).

More specifically in a US-China Joint Presidential Statement on Climate Change issued at the White House as part of President Xi's state visit to the United States under the then President Barack Obama, China reaffirmed its commitment to the Paris Accord and pledged (White House, The, 2015):

1. to reinforce the advancement of ecological civilization,
2. to promote green, low-carbon, climate resilient, and sustainable development through accelerating institutional innovation and enhancing policies and actions,
3. to peak its CO_2 emissions by around 2030 or earlier,
4. lower its CO_2 emissions per unit of GDP by 60%–65% from the 2005 level by 2030,
5. to promote low-carbon buildings and transportation, with the share of green buildings reaching 50% in newly built buildings in cities and towns by 2020,
6. to increase its forest stock by some 4.5 billion m^3 from 2005 levels by 2030 and, more importantly,
7. to launch a carbon reduction plan, emissions trading system (ETS) nationwide by 2017, covering key industry sectors such as iron and steel, power generation, chemicals, building materials, paper-making, and nonferrous metals.

Since the beginning of 2016, China has been implementing supply-side structural reforms, driving inefficient state firms to upgrade with new technologies while cutting capacity and leverage (Liao, 2017). In December 2017, China officially launched its National ETS as announced at the US–China Joint Presidential address in 2015 as mentioned above. The trading system, designed to enhance China's climate objectives in line with the Paris Accord, is poised to be the largest carbon market in the world when fully in operation. The initial national market is limited to the power-generation sector covering about 1700 companies which emit more than 26,000 tonnes of GHGs or consume more than 10,000 tonnes of coal equivalent (tce)/year (ICAP, 2019). The initial stage of the ETS would cover about 3500 metric tons of CO_2 equivalent ($MtCO_2$), which is almost twice as large as the 1839 $MtCO_2$ of carbon emissions covered by the European Union ETS (Baker, 2019). In 2018 China established a Ministry of Ecology and the Environment to replace the Ministry of Environmental Protection. The new ministry takes on the responsibility of addressing climate change formerly entrusted to the NDRC (ICAP, 2019).

In the same year, 2017, China also unveiled the "Blue Sky Defence Plans" for 2018–20. The plan places great emphasis on addressing pollution problems including greenhouse gas emissions, especially in the Beijing-Tianjin-Hebei region, and the surrounding areas, Yangtze River delta, and Fenhe-Weihe Plain, through a combination of economic, legal, technical, and administrative instruments and under the guidance of the "Xi Jinping's Thought on Socialism with Chinese Characteristics for a New Era" (Ministry of Ecology and Environment, 2017). In the same year, the state planner also announced that China's share of nonfossil energy sources would be increased to around 20% of total energy consumption by 2030 (Mason, 2017).

Also despite the United States' withdrawal from the Paris climate agreement in November 2019, China President Xi has shown no sign of wavering in his commitment to bring emissions to a peak by around 2030. Indeed, China has taken a variety of steps to fulfill the Paris Accord including a national carbon cap-and-trade programme, a green dispatch policy and a cap on coal consumption as part of its 13th Five-Year Plan for 2016–20 (Chen, 2017). Furthermore in its China's Policies and Actions for Addressing Climate Change (2018), the Chinese government asserted that it had always placed great importance on addressing climate change through green and low carbon transformation policies and sustainable development strategies (Ministry of Ecology and Environment, 2018). From 2018 onward, the important task of addressing climate change and emission reduction has been delegated to the Ministry of Ecology and Environment which is to be guided by "Xi Jinping's Thought on Socialism with Chinese Characteristics for a New Era," the spirit of the 19th National Congress of CPC, and "Xi Jinping's thought on ecological civilization" (Ministry of Ecology and Environment 2018: 2).

Some of the crucial measures undertaken by the government to achieve its greening process are (1) adjustment of industrial structure through promoting less-energy intensive industries such as the information technology (IT) industry; (2) optimization of energy structure through the promotion of clean heating and clean use of fossil fuel; (3) energy conservation through the promotion of energy saving and energy efficiency policies, and (4) controlling GHG emissions from nonenergy activities, among others (Ministry of Ecology and Environment, 2018). Indeed, China, under Xi's administration, has moved away from the fossil fuel-based and resource-intensive Great Leap Forward growth philosophy toward emphasizing the development of science and technology—two of Deng's "Four Modernizations" in his open-door policy. The other two Deng's "Modernizations" are agriculture and military.

Furthermore the country unveiled a state-led industrial policy, "Made in China 2025" in 2015, to guide China toward a global high-tech and service-based economy powered by the development of 10 advanced and high-tech industries. These include, for example, electric cars and new energy vehicles, advanced electrical equipment including computer and smartphones; new synthetic materials; advanced robotics, next-generation IT and telecommunications, agricultural technology, artificial intelligence, aerospace engineering, emerging biomedicine, high-end rail infrastructure, and high-tech maritime engineering. This innovative "new economy" constitutes a strategic course of China's transition from its "coal-economy" to "green development," thus exerting less pressure on the global carbon sink.

Also as reflected in Fig. 3.28, China's growing needs for energy are increasingly being met by natural gas and renewables such as solar energy while the use of coal has declined. China continued to be a global leader in clean energy projects and technologies in 2017. China is now the world's largest solar cell producer, accounting for 60% of the global market (Buckley et al., 2018). It also continues to dominate the world's renewable energy expansion, being responsible for a record of about 68 GW addition. This accounts for about 41% of total global renewable capacity additions of 165 GW. China's additional capacity is impressive compared with those of the United States and the EU which have only 24 GW (14.5%) and 21 GW (12.7%) to their renewable energy portfolio (Buckley et al., 2018). It is worth noting that as of June 2019, China's total renewable power capacity rose 9.5% to 750 GW (Reuters, 2019). Also in 2017 China had taken the initiative to reorganize its large state-owned power generators away from heavy reliance on coal (Buckley et al., 2018). Overall, the new less-energy intensive economic vision will not only help to promote sustainable development in China in line with SDGs, but also have momentous impact on drawing the Paris Accord closer to its carbon-limit realization.

Other climate actions taken by China which are no less important include the reduction of fossil fuel-CO_2 emission per GDP by 41% from 2005 levels in 2018 as against the global total at 19%. Its impressive record in decoupling carbon emission from economic growth is far better than that of the United States (29%), Germany (27%), Australia (26%); Canada (18%); Japan (14%), Korea (13%), or India (9%) (see Fig. 3.29). This in part due to the implementation of its national emission cap-and-trade scheme.

It is indisputable that since the Paris Accord, China has resolutely moved away from the Great Leap Forward's "growth-at-all-cost" philosophy and transcended Deng's "Modernization" philosophy in the pursuit of a green vision of development. Although it is still premature to gauge the real impacts of its decarbonization initiatives, the preceding observations reflect that it is on the right track toward fulfilling its pledges to the Paris Climate Agreement. In particular, if the ETS is systematically expanded and conscientiously implemented with strong political will, it could serve as a critical game changer in enhancing China's overall climate action efforts.

However, it must be stressed that given its massive population with an expanding ecological footprint in terms of increasing resource consumption, supply-side environmental policies alone are far from being a silver bullet. That said, the way to make real progress is molded upon the environmental attitudes and moral practices prevalent in the

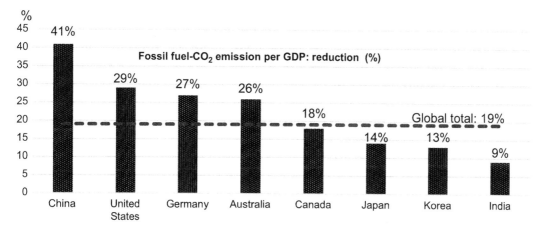

FIGURE 3.29 Per GDP CO_2 emission reduction in 2018 from 2015 levels. CO_2, Carbon dioxide; *GDP*, gross domestic product. *Data from, Crippa, et al., 2019. Fossil CO_2 and GHG Emissions of All World Countries. 2019 Report.* European Commission, Joint Research Centre. Publications Office of the European Union, Luxembourg.

socioeconomic systems. A new revolution of a green era requires a different kind of mindset on environmental belief and morality in association with everyday sustainable resource consumption practices which may be promoted through environmental and moral education—the demand-side solution, which will be examined in the subsequent chapters.

3.19 Carbon dioxide emissions and climate change: The American "carbon curse" and Donald Trump's anti- environmental attitudes and practices

3.19.1 The American carbon dioxide emission trends

To be sure, China is not the only major country to be blamed for its massive emission levels caused by its booming economy over the past few decades. Although China leads the world in its CO_2 emission standard as discussed above, on a cumulative basis, the United States is the world's largest emitter, emitting 317 billion tonnes of CO_2 between 1950 and 2018, while the cumulative CO_2 emissions in the same period for China is 223 billion tonnes. The cumulative emissions for Russia, Germany, Japan, India and the United Kingdom are 97, 63, 61, 49, and 39 billion tonnes, respectively (Fig. 3.30). In a related note, during the industrial revolution period, the United Kingdom emitted a cumulative amount of 8.68 billion tonnes of CO_2 between 1751 and 1949.

On a historical basis, the United States is also the world's largest emitter, emitting a massive 406 billion tonnes into the atmosphere since 1800 while the aggregate historical CO_2 emissions for China since 1899 is 225 billion tonnes. The aggregate historical CO_2 emissions for Germany, the United Kingdom, Japan, and India are 92, 78, 61, and 51 billion tonnes, respectively (Fig. 3.31). It is no exaggerated to lament that this single colossal source of commutative carbon concentration still lingering in the atmosphere connotes an American carbon curse rebounding in the form of the current worsening global warming scenarios.

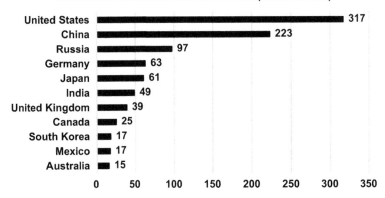

FIGURE 3.30   Cumulative CO_2 emission from fossil fuel burning and cement production, 1950–2018. CO_2, Carbon dioxide. *Data from, Boden, T.A., Marland, G., Andres, R.J., 2016. Global, Regional, and National Fossil-Fuel CO2 Emissions. Carbon Dioxide. Information Analysis Center, Oak Ridge National Laboratory, U.S. Department of Energy, Oak Ridge, TN. doi:10.3334/CDIAC/00001_V2016; Ritchie, H.R., Roser, M., 2019. CO₂ and Greenhouse Gas Emissions. Our World in Data. Available from: <https://ourworldindata.org/co2-and-other-greenhouse-gas-emissions>; Crippa, et al., 2019. Fossil CO₂ and GHG Emissions of All World Countries. 2019 Report. European Commission, Joint Research Centre. Publications Office of the European Union, Luxembourg.*

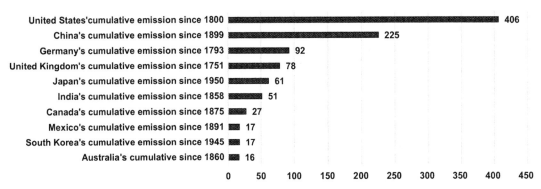

FIGURE 3.31   Cumulative CO_2 emission from fossil fuel burning and cement production: a historical perspective. *Note*: Data prior to 1950 are either unavailable or zero. For example, the data for China between 1751 and 1898 are zero as appeared in Ritchie and Roser 2019. Also the CO_2 emissions from the industrial process and cement production for China or India prior to 1950 are expected to be negligible as the Industrial Revolution is originated in the West. For consistency and for the latest update, all the data from 1970 to 2018 are based on Crippa et al., (2019). Data for Germany and Russia prior to 1970 are based on Ritchie and Roser 2019. Unless otherwise indicated, data for the rest of the countries prior to 1970 are based on Boden et al., (2016). Differences from the various sources of data are expected to be negligible as all enlisted their information extensively from more or less the same sources. Russia is excluded from the above figure due to data deficiency. CO_2, Carbon dioxide. *Data from, Boden, T.A., Marland, G., Andres, R.J., 2016. Global, Regional, and National Fossil-Fuel CO2 Emissions. Carbon Dioxide. Information Analysis Center, Oak Ridge National Laboratory, U.S. Department of Energy, Oak Ridge, TN. doi:10.3334/CDIAC/00001_V2016; Ritchie, H.R., Roser, M., 2019. CO₂ and Greenhouse Gas Emissions. Our World in Data. Available from: <https://ourworldindata.org/co2-and-other-greenhouse-gas-emissions>; Crippa, et al., 2019. Fossil CO₂ and GHG Emissions of All World Countries. 2019 Report. European Commission, Joint Research Centre. Publications Office of the European Union, Luxembourg.*

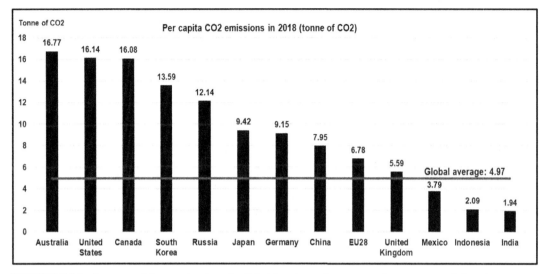

FIGURE 3.32 Per capita CO_2 emission from fossil fuel burning and cement production in 2018 (selected countries). CO_2, Carbon dioxide. *Data from, Crippa, et al., 2019. Fossil CO_2 and GHG Emissions of All World Countries. 2019 Report. European Commission, Joint Research Centre. Publications Office of the European Union, Luxembourg. World Bank, 2018.*

On a per capita basis, the United States is also one of the world's largest emitters. In 2018 its per capita emission was 16.14 tonnes of CO_2 as against the global average of 4.97 tonnes per capita, second only to Australia at 16.77 (Fig. 3.32). In comparison, the per capita emission for China was 7.95 tonnes of CO_2. It was even lower for Indonesia and India at 2.09 and 1.94 tonnes per capita, respectively. As reflected in Fig. 3.32, the per capita emission standards in 2018 were also very high for Canada, South Korea, and Russia compared with the global average.

However, on average, the United States leads as the most highly polluted nation compared with the rest of the countries under discussion, with an average of 19.89 tonnes/capita between 1970 and 2018 while the average per capita CO_2 emissions in the same period for the United Kingdom, Japan, China and India are 9.53, 9.16, 3.44, and 0.89 tonnes/capita, respectively (Fig. 3.33). Given that on average, the global per capita CO_2 emission is 4.42, the abovementioned advanced countries, especially the United States, have far exceeded the global average while the emission standards for India and China still remain comparatively low. The per capita CO_2 emission for Canada and Australia has remained persistently high at an average of 17.13 and 16.44 tonnes, respectively.

While the massive CO_2 emission in China is to a great extent caused by rapid industrialization, a huge population of 1.42 billion as of 2018 (roughly 18.71% of the world's total) and robust economic growth with a GDP worth 13,608 billion US dollars in 2018 (roughly 21.95% of the world economy) (Trading Economics, 2019a), its per capita fossil fuel (oil, natural gas, and coal) consumption was only 1.96 toe in the same year, slightly above the global average of 1.54 toe per person. China's annual average per capita fossil consumption between 1970 and 2018 is about 0.88 toe, which is below the global annual average of 1.38 toe per individual in the same period. It may be relevant to note that China's average per capita primary energy consumption (including fossil fuel consumption, nuclear

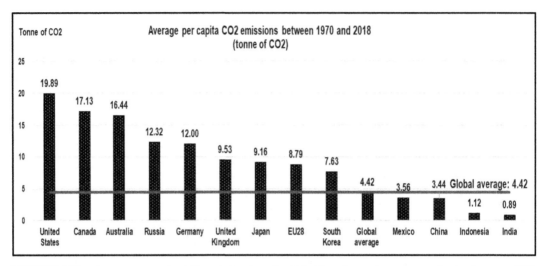

FIGURE 3.33   Average per capita CO_2 emission from fossil fuel burning and cement production between 1970 and 2018 (selected countries). CO_2, Carbon dioxide. *Data from, Crippa, et al., 2019. Fossil CO_2 and GHG Emissions of All World Countries. 2019 Report. European Commission, Joint Research Centre. Publications Office of the European Union, Luxembourg. Worldometers, 2019. Available from: <https://www.worldometers.info/population/>.*

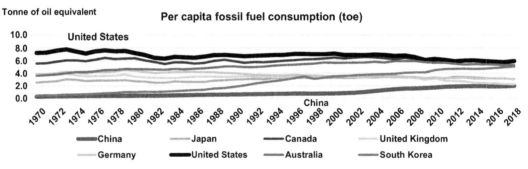

FIGURE 3.34   Per capita fossil fuel consumption: a global comparison. *Data from, BP, 2011–2019. BP Statistical Review of World Energy (June 2011 to June 2019 issues). British Petroleum (BP), United Kingdom. Worldometers, 2019. Available from: <https://www.worldometers.info/population/>.*

energy, hydroelectric, and renewables) between 1970 and 2018 is roughly 0.95 toe as against the per capita global average of 1.56 toe.

The United States, however, as the largest economy in the world, with a GDP worth $US 20,494 billion in 2018 (roughly 33% of the world economy) (Trading Economics, 2019b) and with a population of only 327 million, about a quarter of China's, recorded its per capita fossil fuel consumption in 2018 at 5.93 toe. This is more than three times that of China and almost four times compared with the global average of 1.54 toe (Fig. 3.34). Between 1970 and 2018, the United States' average per capita fossil fuel consumption was 6.76 toe. This is almost five times that of the global average of 1.38 toe in the same period. This is also the highest compared with many advanced economies such as Canada, Germany, or the United Kingdom as reflected in Figs. 3.34 and 3.35.

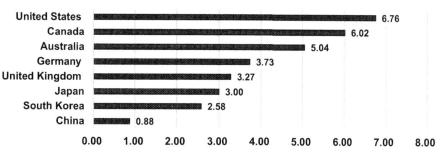

FIGURE 3.35 Average per capita fossil fuel consumption in figures. *Data from, BP, 2011–2019. BP Statistical Review of World Energy (June 2011 to June 2019 issues). British Petroleum (BP), United Kingdom. Worldometers, 2019. Available from: <https://www.worldometers.info/population/>.*

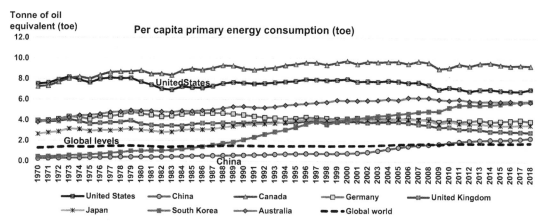

FIGURE 3.36 Per capita primary energy consumption (fossil fuels, nuclear energy, hydroelectric, and renewables): a global comparison. *Source: BP, 2011–2019. BP Statistical Review of World Energy (June 2011 to June 2019 issues). British Petroleum (BP), United Kingdom.*

In addition, although the United States accounts for only about 4.3% of the total world population in 2018, its per capita primary energy consumption in 2018 was 7.04 toe. This is about four times that of China at 1.82 toe. Per capita energy consumption in the United States is also more than that of Australia, Germany, United Kingdom, Japan and South Korea. But Canada, since 1974, has taken over the American position as the leading primary energy consumer. Between 1970 and 2018, the per capita average energy consumption for the United States was 7.54 toe while Canadian energy consumption level reached a whopping 9.04 toe. China's per capita energy consumption in the same period was about 0.85 toe, which is below the global average of 1.56 toe (Figs. 3.36 and 3.37).

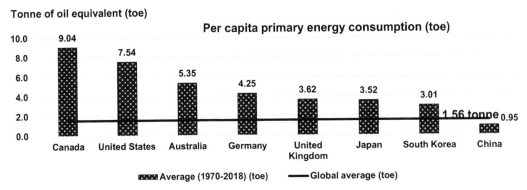

FIGURE 3.37   Average per capita primary energy consumption. *Data from, BP, 2011–2019. BP Statistical Review of World Energy (June 2011 to June 2019 issues). British Petroleum (BP), United Kingdom.*

Moreover, the United States, with a population 10 times more than that of Canada, accounts for 16.6% of the world's total primary energy consumption in 2018 while the Canadian share was only 2.5% (BP, 2011–2019). At the same time, China which has the largest population in the world accounts for 23.6% of the global share of primary consumption (BP, 2011–2019). However, given that its population is roughly four times that of the United States and about 40 times that of Canada, its absolute consumption level may be considered as low. Viewed from this perspective, it is more appropriate to use the per capita consumption and hence, per capita carbon emission as the measuring rod to reflect human carbon footprint which may be defined as the amount of CO_2 emissions induced by human activities especially in relation to fossil fuel use such as coal, natural gas, and oil for energy and transportation.

The high per capita energy consumption in the United States is directly associated with its position as the all-time largest cumulative carbon emitter on the planet as discussed above. Although China is now the world's largest carbon emitter, it has a much smaller share on a cumulative basis. Moreover, its high emission levels are fundamentally spurred by its huge population as noted above. Despite this, on average, its per capita carbon emission is vastly smaller at 3.44 toe compared with the United States at 19.89 toe as shown in Fig. 3.33 Moreover, global warming or climate change facing us today is inextricably linked to cumulative emissions—the amount of CO_2 accumulated in the atmosphere fundamentally caused by the combustion of fossil fuels. It is sobering to note that of the total amount of CO_2 emitted into the atmosphere, roughly 40% remains in the atmosphere for 100 years, 20% remains for 1000 years, while the final 10% will take 10,000 years to turn over, aggravating global warming and climate change (Union of Concerned Scientists, 2009).

It is instructive to note that most of the energy production in the United States involves the burning of fossil fuels. It is estimated that in 1 day, US per capita energy consumption includes 2.5 gallons of oil, 13.7 pounds of coal, and 234 ft.^{3} of natural gas. It is further projected that 80% of US energy consumption will derive from fossil fuels in 2040 (Center for Sustainable Systems, 2016a). Some the factors that contribute to the high per capita energy consumption in the United States are unsustainable consumption and lifestyle, and waste, which are also associated with its high GDP.

To demonstrate, an estimated 21% of edible food available is wasted at the consumer level. This is 50% more than in 1970 and accounts for roughly 15% of the municipal solid waste (MSW) stream. Expressed in monetary terms, this represents a loss of $US 455 per person/ year (Center for Sustainable Systems, 2016b). In 2013 the average American generated 4.4 pounds (lbs) of MSW per day with only 1.51 lbs recovered for recycling and composting. In comparison, MSW generation rates in Sweden, the United Kingdom, and Germany were 2.20, 2.98, and 3.71 lbs/person/day, respectively (Center for Sustainable Systems, 2016c).

The average daily calorie consumption of an American has increased from 2039 in 1970 to 2544 in 2010 (Center for Sustainable Systems, 2016c). Also food-related use accounts for nearly 16% of the national energy budget. About 10 units of energy primarily made up of fossil energy are consumed for every unit of food energy produced (Center for Sustainable Systems, 2016b). About 33% of the grain grown in the United States are used to feed livestock which otherwise may be used to feed more people (Center for Sustainable Systems, 2016b).

The unsustainable consumption patterns in the United States are also reflected by the fact that in 2000, per capita of all materials was 23.7 tonnes which was 52% more than the European average (Center for Sustainable Systems, 2016c). It is noteworthy that the United States is the single largest extractor of material resources with 6.5 gigatons (Gt) extracted in 2008—one-third of all materials extracted in OECD countries (OECD, 2015). This may be due in part to the changing residential living trends in the United States. Since 1950, the average residential living trends in the United States have been leaning toward owning bigger houses with fewer occupants. For example, the number of occupants per house has decreased by 25% and single occupant houses increased from 9% to 28%. Also living space per person has increased by 258% while house size has increased by 170% (Center for Sustainable Systems, 2016c).

It must be pointed out that despite carbon emissions in the United States falling slightly from 5.56 billion tonnes to roughly 5.13 billion tonnes in 2017, it is still the second biggest emitter of CO_2 in the world, consistently emitting more than 5 billion tonnes annually since 1988. Its emission level grew by about 2.9% in 2018 despite a steep drop in coal consumption by 4.9% from 332 to 317 Mtoe, the lowest level in 43 years since 1975 at 313 Mtoe. As a matter of fact, the drop in the consumption of coal was replaced by an increase in the consumption of natural gas which emits about half the CO_2 of coal. Natural gas consumption increased by 10.5% from roughly 636 to 703 Mtoe in the same period. The rise in carbon emission was fundamentally caused by an increase in energy demand 3.5% from roughly 2223 to 2301 Mtoe in the same period (BP, 2011–2019).

In closing, global warming and climate change is a cumulative problem caused by the increasing concentration of GHGs, especially CO_2, in the atmosphere. The United States is deemed to shoulder the greatest responsibility for this. It may well be that other high-emission countries such as the United Kingdom and Germany are also responsible for the current cumulative emission since the beginning of the Industrial Revolution, but the United States is the only nation in the world which has been racking up massive amounts of emissions since 1800 to a monstrous level of 406 billion tonnes as of 2018. This is the American carbon curse which is returning to haunt us. Thus the United States owes a moral responsibility not only to its citizens, but also all the world's citizens, to reverse its carbon curse as much as possible so that we may have a more sustainable world, for the benefit of present and future generations.

3.19.2 The breaking of the American carbon curse: The Barack Obama decarbonization legacy and the Donald Trump environmental syndrome

The American carbon curse inflicted on the world since 1800, if left unbroken, will continue to aggravate the greenhouse effects—one of the greatest threats to the long-term existence of humans and of other living things. Under Barack Obama's administration, climate change became a priority in the United States, despite strong opposition from Republicans in the Congress (Goldenberg, 2013). The Barack Obama administration failed to achieve much in its first term due to the formidable obstacles placed in its way by the Republican Members of Congress as well as the fossil fuel lobby and their political representatives, both from the Republicans and the Democrats. However, the President in his second term was able to push through many decarbonization strategies and policies despite the lack of a legislative majority or the support of Congress. He was able to achieve this basically through executive action and regulation (Lavelle, 2016).

Some of the President's most important decarbonization initiatives or climate change policies are (Lavelle, 2016; Tollefson, 2016):

1. The allocation of $US37 billion for clean energy research and development at the Department of Energy.
2. Introduction of higher fuel-efficiency requirements and the first greenhouse gas standards for passenger cars—the regulatory standards which the auto companies had opposed for 25 years.
3. Paris Accord Commitment: the Barack Obama Administration committed the United States to reduce carbon emissions by at least 26% below 2005 levels by 2025, and agreed to hold the average temperature rise to the 1.5°C−2°C above the preindustrial levels.
4. In recognizing that some of the fossil fuels are best kept in the ground, the Barack Obama administration rejected the construction of Keystone XL tar sands pipeline which would carry oil from Canada tar sands to US refineries—the first time a world leader rejected a fossil fuel project on climate grounds.
5. Entered into direct negotiation with President Xi Jinping of China to reach a climate deal on carbon emissions—an important global environmental initiative which underpins the success of the Paris Accord given that China and the United States are the only two largest carbon emitters in the world.
6. The introduction of the Clean Power Plan which was designed to cut carbon emissions from power plants—the largest source of CO_2 emission in the United States—by 30% from 2005 levels.
7. The introduction of the Appliance and Equipment Standards Program which aimed to save American household's energy through imposing minimum energy conservation standards on more than 60 categories of appliances and equipment. The program also serves as a strategic means to reduce carbon emissions by at least 3 billion metric tons by 2030—equivalent to the emissions of 524 million passenger cars in 1 year. According to the Office of Energy Efficiency and Renewable Energy, the programme has helped the American consumers to save $US 63 billion on their utility bills in 2015 alone.

The multipronged approach to addressing carbon emission and climate change problems initiated under the Barack Obama administration certainly provides important direction to the Americans at large to undo the dreadful carbon curse. Nonetheless, due to the lack of widespread popular support from the fossil-fuel vested interest groups and the ideological legislators in Congress, Barack Obama had not been able to go all out boldly to introduce and implement comprehensive decarbonization policy and climate change legislation to unlock the carbon dilemma which has been facing the United States since the 1800s.

Also the Barack Obama administration failed to create a sense of environmental urgency and moral imperative among Americans at large on the importance of decarbonization and sustainable development practices based on environmental science and ecological facts. As Al Gore pointed out rightly: "President Obama has never presented to the American people the magnitude of the climate crisis. He has simply not made the case for action. He has not defended the science against the ongoing, withering and dishonest attacks. Nor has he provided a presidential venue for the scientific community—including our own National Academy—to bring the reality of the science before the public" (Gore, 2011). As will be examined in the following chapters, without raising environmental awareness and promoting the ethics of sustainable resource use, any effort to enhance environmental sustainability is unlikely to be effective. This partly explains why, during the Barack Obama administration between 2009 and 2017, annual fossil fuel consumption standards or per capita fossil fuel consumption levels hardly improved due to the individualistic norms of unsustainable production and consumption practices prevailing in American society.

However, it is fair to say that despite all the political odds and congressional obstacles blocking his way in promoting environmental sustainability, President Barack Obama did successfully set the direction for the United States to strategically build a lasting and resilient green future. In particular, the Barack Obama Paris Accord legacy remains crucial to the achievement of a green future not only for Americans, but also for humanity at large. The upshot is that it provides the necessary impetus and the ethical impulse for the rest of the world, in particular the developing countries, to collectively and altruistically work toward a common goal of a low carbon global economic system for the benefit of humankind and of other living things.

3.19.3 The return of the American carbon curse: Donald Trump's environmental protection rollback

The Barack Obama green vision and efforts toward a low-carbon future for the United States are currently facing the biggest obstacles in the history of the United States—the statistically evident damage caused to the environment directly as a result of the environmental regulations that Donald Trump's administration has removed. To understand the recurrent nature of Donald Trump's environmental impediments, we may backtrack to the administration of President George Bush (43rd President of the United States from 2001 to 2009) as a backdrop for the contemporary ideologies of Donald Trump, the current president of the United States.

To begin with, maintaining and sustaining a mass capitalist production and large material consumption has always been a hallmark of American society. This is one of the main reasons why the United Sates, then under the George Bush administration, refused to ratify the Kyoto Protocol as this would impede its economic growth, and hence destabilize its traditional utilitarianistic production and consumption culture. Indeed, despite being the world's largest economy, the highest carbon emitter and the heaviest energy consumer, the United States refused to accept its responsibility for the world climate and expressed that it has "no interest" in its implementation. It further conceded that signing the Protocol "would cause serious harm to the economy" and hence it should be scrapped in favor of a new market-based accord that did not impose an onerous cost on the economy (Hovi et al., 2010: 130; Sanger, 2001). The major commitment of the Bush administration to the Kyoto Protocol was to push for more resources and new efforts into research to study global warming and its mitigating measures (Borger, 2001). The Kyoto Protocol was signed by the former Vice President Al Gore on behalf of the United States but it was never ratified by the Senate.

Also although the Barack Obama Administration ratified the Paris Agreement on climate change in 2016 and agreed to reduce US GHG emissions in 2015 by 26%−28% below 2005 levels, dissenting voices warned that it would increase local energy costs, slow economic growth and reduce per capita income growth (Groves at al., 2016; Loris, 2016). Furthermore in defiance of all scientific evidence, pointing to an urgent need for a powerful global movement to contain climate crisis, the present US Administration led by Donald Trump, in stark contrast to President's Xi's green vision as discussed in the preceding section, has signaled rejection of Barack Obama's GHG reduction policy. He has vowed not only to remove the United States from the Paris Climate Pact but also to end US funding for UN initiatives to control climate change, calling it "a hoax perpetrated by the Chinese" (Ohlheiser, 2016; Pritam, 2017).

Already, lawmakers under the Donald Trump administration are moving to eliminate Barack Obama's clean air initiatives, one of which was restricting the flaring of methane—a potent greenhouse gas—from natural gas wells on public lands. Also the US Environmental Protection Agency (EPA) website has changed its tone repeatedly to deemphasize climate change. In one instance, it belittled the importance of science-based rulemaking (Greshko et al., 2018). Worse yet, the Donald Trump administration, in reversing the Barack Obama stance on the restrictions on coal mining leases, has instead made the US fossil-fuels industry key in its pledge to revitalize the US economy (Greshko et al., 2018).

Furthermore in his anticlimate change stance, Donald Trump has appointed Scott Pruitt, a critic of the Clean Air Act which he deemed is in conflict with the economic interests of the fossil fuel industry, to head the US EPA (Pritam, 2017). The Act is Barack Obama's landmark regulation aimed at reducing CO_2 emissions in the electricity sector by 32% over the next 13 years (Lehmann et al., 2016). On November 4, 2019, the Donald Trump administration formally notified the United Nations of its intention to withdraw the United States from the Paris Agreement officially signed by the former president, Barack Obama. This anti-Paris Pact sentiment will not only bring back the "American carbon curse" into play but also, has far reaching political and practical implications in thwarting international efforts to curb global warming. However, at the time of writing, it is still too early to gauge how severe the impacts will be. Here, it is worth noting that that American negotiators have a history of repeatedly denouncing legally binding targets and accountability mechanisms in favor of voluntary action (Bennington et al., 2014). By leaving the Paris Accord, it will undermine the United States' reputation further as a trustworthy international partner and player.

By extension, Donald Trump has also played down the warning of the potentially cata-strophic impact of climate change including increasingly debilitating hurricanes and heat waves as revealed in the Fourth National Climate Assessment. He has simply and equani-mously assumed it away by stating that, "I don't believe it" (Nuccitelli, 2018). This tends to drive the American policy decisions on climate change mitigation and adaptation fur-ther into a cul-de-sac. To note in passing, the assessment report is congressionally man-dated report produced by the US Global Change Research Program which involves 13 federal agencies including the Department of Defense, the EPA, and NASA and more than 300 top climate scientists.

In summary, Donald Trump's anthropocentric view of nature and environmentally weak sentiments are well reflected in the media reports as summarized in Table 3.15. We may call this phenomenon the "Donald Trump anthropocentric proclivity *sui generis*"—a moral phenomenon which places immense importance on the economic interest and mate-rial progress of human beings (see Chapter 4: Greening for a Sustainable Future: The Ethical Connection, Section 4.2.1 for a detailed discussion of anthropocentrism).

The list of antiscience and environmentally malignant policies mentioned above is far from exhaustive. Nonetheless, they are of great importance to our understanding of the environmental assumption made by the Donald Trump administration about natural resource exploitation—the assumption that the American economy anthropocentrically dominates the natural environment. This means that the Donald Trump administration has an extraordinarily strong tendency to embrace nature as a means to an end. To Donald Trump, nature is a horn of plenty. It offers Americans a boundless and perpetual carbon sink gratis which allows them to spew limitless amounts of greenhouse gas into the atmosphere. It also provides them with almost limitless supply of oil, coal and gas which can be recklessly exploited to the fullest extent possible to establish American energy dominance and to "make America great again."

To put in his vivid phrasing: "We're here today to usher in a new American energy policy—one that unlocks million and millions of jobs and trillions of dollars in wealth … we have near-limitless supplies of energy in our country. Powered by new innovation and technology, we are now on the cusp of a true energy revolution … It's called Make America Great Again" (White House, The, 2017). Donald Trump's individualistic utilitarian American energy dominance doc-trine may be considered the most aggressive and violent polemic against the natural world.

Thus he declared the American battle against energy by rolling back all the environ-mental regulations initiated by his predecessor, President Barak Obama, which are meant to fight climate change. As President Donald Trump, who is indifferent to climate change, proudly claimed in front of his supporters, "Since my very first day in office, I have been moving at record pace to cancel these regulations and to eliminate the barriers to domestic energy production, like never before … I'm dramatically reducing restrictions on the development of natural gas. I cancelled the moratorium on a new coal leasing … We have finally ended the war on coal …" (White House, The, 2017).

By now, the stark contrast between the former president, Barack Obama and the current president, Donald Trump, is clear. While the former has ethically built a bridge of love and respect with nature through his environmentally benign policies and sustainable global com-mitments, the latter displays strong anthropocentric sentiments toward the natural environ-ment, treating it instrumentally as a means to an end with the view to make "America great

TABLE 3.15 A reflection of Donald Trump's anthropocentric view of nature and the "rolled back of environmental protection": a timeline of events.

Date of report	Headline	Source	Brief summary
April 30, 2020	Trump shows his cards on environmental protections— or a lack thereof	Revesz, R.L., The Hill	The last month has been an extraordinary one not just because of the COVID-19 crisis but because the Trump administration has now shown all its cards in its hypocritical crackdown on environmental protection. In three important proceedings, the Environmental Protection Agency has revealed unequivocally that it will employ any methodology at all to justify rolling back environmental regulations, even if the effect could be harmful and put lives at risk.
April 28, 2020	Five ways that Trump is undermining environmental protections under the cover of coronavirus	Tollefson, J., Nature	The US EPA turns 50 this year, but scientists and environmentalists see little reason to celebrate. In the past month alone, the agency has dialed down regulations on automobile emissions and fuel efficiency put in place under former president Barack Obama; it has weakened rules on mercury and other pollutants emitted by power plants; and it has shied away from strengthening standards to reduce fine-particle air pollution.
April 22, 2020	An Earth Day Reminder of How the Republicans Have Forsaken the Environment	Kolbert, E., The New Yorker	Today, as Earth Day turns fifty, it's hard to imagine more dolorous circumstances for the occasion. COVID-19 has forced online (or canceled) virtually all the celebrations and protests that had been planned for the anniversary. The Trump Administration has barely even taken the day off from gutting the nation's environmental regulations.
April 22, 2020	We can't let Trump roll back 50 years of environmental progress	Southerland, E., The Guardian	Under this (Trump) administration, the EPA has been transformed from an agency of environmental protection to an accommodating servant of special interests...The Trump EPA has repealed or weakened almost 100 environmental regulations...solely to maximize corporate profits... "
April 13, 2020	Trump Advances Massive Fracking Expansion on Colorado Federal Lands	Germanos, A., EcoWatch	The Trump administration on Friday released a new land use plan, "Approved Resource Management Plan" for southwestern Colorado that community and conservation advocacy groups warn is a "dangerous" pathway toward increased fossil fuel extraction that makes no "climate, ecological, or economic sense."
April 10, 2020	Trump administration is rushing to gut environmental protections	Garbow, A., CNN	"While our nation reels from the coronavirus pandemic, the Trump administration is accelerating a harmful agenda—rollbacks that dismantle critical health and environmental protections, and that will surely deepen the climate crisis."
April 2, 2020	Trump is aggressively pushing his antienvironment agenda amid a pandemic. It's inexcusable	Hayes, D.J., Washington Post	The coronavirus pandemic has virtually shut down the US economy. The lights remain on, however, at the EPA and the Interior Department, where the Trump administration is working to push through its radical reordering of our nation's environmental and conservation priorities—pandemic or not.

(*Continued*)

TABLE 3.15 (Continued)

Date of report	Headline	Source	Brief summary
April 1, 2020	Trump rollback of mileage standards guts climate change push	Associated Press	The Trump administration rolled back ambitious Obama-era vehicle mileage standards Tuesday, raising the ceiling on damaging fossil fuel emissions for years to come and gutting one of the United States' biggest efforts against climate change.
March 31, 2020	Three states pass antifossil-fuel protest bills in 3 weeks. It's not coincidental	Lewis, M., Electrek	While the United States is focused on the coronavirus crisis, states are quietly redefining fossil-fuel infrastructure as critical to prevent protests. More specifically at least three states (Kentucky, South Dakota and West Virginia) passed laws putting new criminal penalties on protests against fossil fuel infrastructure in just the past 2 weeks amid the chaos of the coronavirus pandemic
March 27, 2020	States Quietly Pass Laws Criminalizing Fossil Fuel Protests Amid Coronavirus Chaos	Kaufman, A.C., Huffpost	
March 26, 2020	Overnight Energy: EPA suspends enforcement of environmental laws amid coronavirus	Beitsch, R., The Hill (2020c)	The EPA issued a sweeping suspension of its enforcement of environmental laws Thursday, telling companies they would not need to meet environmental standards during the coronavirus outbreak
March 18, 2020	Not Even A Pandemic Can Stop Trump From Pushing Fossil Fuels	D'Angelo, C., Kaufman, A.C., Huffpost	As about 800,000 federal workers went without paychecks, the Interior Department worked to boost oil and gas development in the Alaskan Arctic, processed fossil fuel drilling applications and permits, and even brought back dozens of furloughed employees to ensure offshore drilling activities continued.
March 2, 2020	Trump Official Keeps Adding Climate Denial Into Scientific Reports, Says Report	Ross, J., The Daily Beast	An Interior Department official has led an effort to insert misleading language about climate change, including some already debunked claims, into official agency reports.
	The Trump Administration Is Just Flat-Out Lying About Climate Change	Lutz, E., Vanity Fair	
	Trump official inserted debunked climate change language into scientific documents: report	Klar, R., The Hill	
March 2, 2020	Trump Wants to Gut Crucial Environmental Regulation	Chen, M., The Nation	The rolling back the National Environmental Policy Act would prevent the public from learning about the consequences of massive construction projects.
February 25, 2020	Activists Slam Trump's Plan To Cut Public Out Of Environmental Review Process	D'Angelo, C., Huffpost (2020b)	The Trump administration has proposed sweeping overhaul of one of America's most important environmental laws to "streamline" a lengthy and burdensome federal permitting process so that the United States can swiftly build needed infrastructure. The change would allow "blatant conflicts of interest, leaving the door open for companies to engage in bald-faced self-dealing that benefits no one but themselves."

(Continued)

TABLE 3.15 (Continued)

Date of report	Headline	Source	Brief summary
February 20, 2020	Trump budget calls for slashing funds to climate science centers	Beitsch, R., The Hill (2020b)	Trump's fiscal 2021 budget would slash funding for the National and Regional Climate Adaptation Science Centers, eliminating all $38 million for research to help wildlife and humans "adapt to a changing climate."
February 19, 2020	EPA proposes additional rollback to Obama-era coal ash regulation	Frazin, R., The Hill	The EPA on Wednesday announced a new proposed rollback to an Obama-era regulation dealing with waste from coal-fired power plants known as coal as
February 18, 2020	With polar bear study open for comments, critics see effort to push drilling in ANWR	Beitsch, R., The Hill (2020a)	The Department of the Interior is pushing to open drilling in the ANWR, something the House has tried to block.
February 10, 2020	Trump budget slashes EPA funding, environmental programs	Beitsch, R., Frazin, R., The Hill	President Trump's proposed budget for fiscal 2021 calls for significant reductions to environmental programs at federal agencies, including a 26% cut to the EPA. Trump's budget would eliminate 50 EPA programs and impose massive cuts to research and development, while also nixing money for the Energy Star rating system.
January 9, 2020	White House unveils plan for major projects to bypass environmental review	Reuters, The Guardian, 2020	The Trump administration has unveiled significant changes to the National Environmental Policy Act that would make it easier for federal agencies to approve infrastructure projects like oil pipelines, road expansions, and bridges without considering climate change. This is one of the biggest deregulatory actions to narrow the use of environmental laws, especially assessments of how developments could exacerbate the climate crisis. He is "locking in permanent, irreversible damage to our environment through his irresponsible environmental policies, including his efforts to block progress on climate change" (Kann, 2020).
	Trump moves to overhaul the National Environmental Policy Act	Newburger, M., CNBC, 2020	
January 7, 2020	Report Detailing US Threats Ignores Climate Change	Frank, T.; E&E News, 2020	The Trump administration's 2019 National Preparedness Report, which describes the greatest threats and hazards such as natural disasters to the country, says nothing about climate change, drought or sea-level rise.
November 4, 2019	Trump Serves Notice to Quit Paris Climate Agreement	Friedman, L., The New York Times, 2019b	President Trump officially notified the United Nations that he would withdraw the United States from the Paris Agreement on climate change.
October 23, 2019	Trump Administration Sues to Block California's Climate Change Agreement With Quebec	Knickmeyer, E., Associated Press	The Trump administration sued to try to block California from engaging in international efforts against climate climate-damaging fossil fuel emissions.

(Continued)

TABLE 3.15 (Continued)

Date of report	Headline	Source	Brief summary
September 13, 2019	Trump opens protected Alaskan Arctic refuge to oil drillers	Holden, E., The Guardian (2019b)	The Trump administration is finalizing plans to allow oil and gas drilling in a portion of the protected Arctic National Wildlife Refuge—1.6 million-acre coastal plain which is home to threatened polar bears and other wild animals
August 29, 2019	Major Climate Change Rules the Trump Administration Is Reversing	Schwartz, J. The New York Times	The move to rescind environmental rules on methane emissions as noted below brings to 84 the total number of environmental rules that President Trump administration has worked to repeal.
August 29, 2019	EPA proposes rule easing regulation of methane emissions	Stracqualursi, V., Wallace, G., CNN	The Trump administration announced a proposal to rollback Obama-era climate rules on methane emission- a potent greenhouse gas which is 84 times more powerful than CO_2 at destabilizing climatic conditions within a 20-year period, and 34 times stronger over a 100-year period. The new rule absolves the oil and gas industry to install technologies that monitor and limit leaks from new wells, tanks, and pipelines.
August 29, 2019	Trump administration to seek rollback of methane pollution rule	Guillén, A., Lefebvre, B., Politico	The Trump administration seeks to roll back rules limiting methane pollution from oil and gas production initiated during the Obama-era. The move is the latest attempt of Present Trump to eliminate rules designed to fight climate change amidst rising temperatures, accelerated Arctic ice melting, and intensified forest fires around the globe.
July 26, 2019	War on science: Trump administration muzzles climate experts, critics say	Holden, E., The Guardian (2019a)	In its strong determination to fight maliciously against climate change, the Trump administration has foolhardily censoring facts and science on government websites and to exclude Obama-era climate change priorities from the government homepage. As a case in point, matters related to climate change have disappeared at an alarming pace since Trump took office in 2016. For example, the use of the terms "climate change," "clean energy," and "adaptation" plummeted by 26% between 2016 and 2018 across a wide range of government websites including the EPA and the USGS websites. The White House and its agencies also attempted to prevent their own experts from explaining how pollution from power plants and cars is increasing global temperatures, threatening both lives and economies.
May 28, 2019	The Trump administration escalates its fight against climate science	Enking, M., GRIST	The Trump administration is seeking to limit what is allowed to report in its National Climate Assessment with the view to realign science in a direction that is consistent with their anticlimate and fossil fuel politics. The administration has also ordered government scientists at the US Geological Survey to stop projecting the effects of the climate crisis past the year 2040.

(Continued)

TABLE 3.15 (Continued)

Date of report	Headline	Source	Brief summary
May 2, 2019	Offshore drilling safety rules rolled back	Greshko et al., National Geographic	This is an attempt of the Trump administration to promote climate-killing projects with the aim of boosting oil and gas production and consumption. It is also Donald Trump's plan to overturn Obama's veto against the long-stalled 1700-mile Keystone XL oil pipeline project which would transport some 800,000 barrels of crude oil from tar sands in western Canada to the US Gulf Coast. The fossil fuel industry will benefit immensely from Trump's fossil fuel executive orders.
April 10, 2019	Trump Signs Orders Making It Harder to Block Pipelines	Superville, D., Freking, V., Associated Press (US News)	
March 29, 2019	Trump moves again to clear path for Keystone XL Pipeline	Monga, V. The Wall Street Journal	
March 30, 2019	Trump issues new permit for stalled Keystone XL pipeline	Daly, M., Associated Press	
February 28, 2019	Andrew Wheeler confirmed as EPA administrator	Greshko et al., National Geographic	Wheeler, a former coal lobbyist, has dismantled Obama's regulative control policy on emissions from coal power plants and automobiles. He has also dismissed a panel of scientific experts that advised the EPA on air pollution regulation and weakened the criminal enforcement arm of the agency.
January 9, 2019	Trump Nominates Andrew Wheeler to Permanent EPA Job	Greshko et al., National Geographic	Wheeler proposed to roll back fuel-efficiency and pollution standards for vehicles as initiated by Obama. He also Wheeler also unveiled a proposed replacement for the Obama-era Clean Power Plan.
October 19, 2018	US Pushes to end children's climate change suit	Greshko et al., National Geographic	In 2015 during Obama's administration, 21 children field a law suit against the federal government over its failure to protect the Earth from climate change. The trial was scheduled to be heard in October 29, 2018. In July 2018, the Supreme Court rejected an attempt by the Trump's administration to derail the case. In October, the Trump's administration again attempted aggressively to end the case but in vain. However, the case has been delayed after a federal appeal court acceded to Trump administration's request to consider halting the case.
October 12, 2018	Trump's EPA scraps air pollution science review panels	Reilly, S., E&E News, Science	Andrew Wheeler, the acting chief of EPA dismissed the scientific review panel that assists in ascertaining air quality standards for particulate matter. He also scrapped plans to form a similar advisory panel to aid in the assessment of the ground-level ozone limits.
October 11, 2018	EPA to disband Air Pollution Review Panel	Greshko et al., National Geographic	The EPA revealed that it will dismantle a scientific review panel that advises the agency about safe levels of air pollution.

(Continued)

TABLE 3.15 (Continued)

Date of report	Headline	Source	Brief summary
August 21, 2018	EPA rolls back Obama-era coal pollution rules as Trump heads to West Virginia	Diamond, J. & Kaufman, E., CNN	In an attempt to reverse Obama's Clean Power Plan which regulates carbon emissions from the coal industry, the Trump administration unveiled its plan to replace it with Affordable Clean Energy rule which will allow states to set their own emissions standards for coal-fueled power plants. The move is expected to boost coal industry and increase carbon emissions nationwide.
August 7, 2018	Trump admin sees grim climate outcome in car rule	Colman, Z., Waldman, S., E&E News reporters, E&E News	The Trump administration artfully uses dire forecast of CO_2 concentration in 2100 to support rolling back of Obama-era fuel efficiency rule of 46.7 miles per gallon and to replace it with the SAFE Vehicles Rule which imposes a lower efficiency requirement of 37 miles per gallon. The Trump officials further claim that the estimated atmospheric CO_2 under the new efficiency rule would be 789.76 ppm, compared with 789.11 ppm under the Obama plan. That represents only a small increase of 0.65 ppm from a baseline of 789.11 ppm.
July 17, 2018	The Trump administration scrubs climate change info from websites. These two have survived	Rainey, J., NBC News	From the Environmental Protection Agency, to the Energy Department, to the State Department and beyond, references to climate change, greenhouse gases and clean energy keep disappearing to keep the sentiments or environmental awareness of the dangers of carbon-driven climate change away from the American public. The EPA also strived aggressively to disconnect scores of links to information designed to help local governments to address climate change problem with Scott Pruitt, the administrator of the Environmental Protection Agency who has now resigned, ordered some information to be removed from the EPA's homepage. Also the Student's Guide to Global Climate Change and the EDGI were taken out from the climate change homepage.
May 9, 2018	Trump White House quietly cancels NASA research verifying greenhouse gas cuts	Voosen, P., Science	The Trump administration has ended NASA's CMS, a $10-million-per-year project which aids in the monitoring of global carbon emissions as well as the verification of the national pledge to carbon emission cuts as agreed in the Paris climate accord. The CMS is deemed to be an obvious target for the Trump administration in view of its link to global climate treaties. The Trump administration repeatedly calling for the cancelations of other climate missions such as the OCO-3 which allows scientists to reconstruct atmospheric concentration from space.

(Continued)

TABLE 3.15 (Continued)

Date of report	Headline	Source	Brief summary
March 16, 2018	FEMA eliminates mentions of climate change from strategic planning document	Savransky, R. The Hill	The FEMA strategic planning document which deals with the effects of disasters like hurricanes and floods for the next 4 years, has dropped references to topics including climate change, rising sea levels and global warming.
February 15, 2018	Four things to know about the Trump budget's environmental cuts	Busby, J., The Washington Post	In its 2019 budget, the Trump administration has proposed deep cuts to Environmental Protection Agency's climate-change and clean energy programs designed to study climate-change mitigating measures, renewable energy technologies and energy efficiency initiatives.
December 15, 2017	Trump Drops Climate Threats from National Security Strategy	Chemnick, J., Scientific American	In breaking away from President Obama's stance, President Trump conceded that the real threat to national security is not climate change but regulations that impede economic and energy dominance of the United States. The President introduced his first National Security Strategy from which climate change has been deleted from its security list. As he stressed further "Economic vitality, growth and prosperity at home is absolutely necessary for American power and influence abroad."
October 9, 2017	EPA Announces Repeal of Major Obama-Era Carbon Emissions Rule	Friedman, L., Plumer, B., The New York Times	The chief of the Environmental Protection Agency, Scott Pruitt, had announced to repeal Obama Administration's Clean Power Plan unveiled in 2015 to cut power sector's emissions by 32% by 2030 (870 million tonnes of CO_2), relative to 2005 to fight climate change. The repeal is to facilitate the economic optimization of energy resources and to reduce unnecessary regulatory impediments to their exploitation and development.
June 1, 2017	Paris climate deal: Trump pulls US out of 2015 accord	McGrath, M., BBC, London (2017a)	In claiming the Paris Agreement is economically disadvantage to the American economy, coting US$3 trillion lost in GDP and 6.5 million job creation, Present Trump announced that he will pull out from the Climate Accord while President Xi Jinping from China stood by the Agreement.
March 28, 2017	Trump Signs Executive Order Unwinding Obama Climate Policies	Davenport, C., Rubin, A.J., The New York Times	President Trump signed an executive order to abandon Obama-era climate change policy in addressing global warming, making clear that the United States has no intention to meet the commitment made by his predecessor, President Obama.

(Continued)

TABLE 3.15 (Continued)

Date of report	Headline	Source	Brief summary
March 24, 2017	Trump administration approves Keystone XL pipeline	McGrath, M. BBC, London (2017b)	President Trump signed an executive order grating approval to revive the environmentally destructive Keystone XL pipeline project, blocked by his predecessor, calling a "great day for American jobs."
March 14, 2017	President Trump to reopen review of Obama-era fuel economy standards	Overly, S., Eilperin, J., The New York Times	In an attempt to reverse the fuel economy standards introduced the Obama administration, the Trump administration seeks to review the policy that required automakers to meet the fuel efficiency standards of 54.5 miles per gallon, on average, across their fleets of vehicles by 2025 to reduce greenhouse gas emissions.
March 6, 2017	Trump budget would gut EPA programs tackling climate change and pollution	Milman, O. The Guardian	The Trump administration's budget (2018) is seeking to slash the budget for research that deals with climate change, pollution clean-ups and energy efficiency by 31% on its existing budget (a $2.6bn cut). The environmental cuts would remove funding for the Clean Power Plan initiated by former President Obama and scrap all climate change research programs and partnerships.

Note: The National Geographic is the official magazine of the National Geographic Society—one of the world's largest nonprofit scientific and educational organizations and a well-trusted institution. The magazine, which has been published continually since 1888, provides a compressive list of Donald Trump "environmental protection roll back," many of which appeared in or overlapped with the list as contained in the table (see, Greshko et al., 2018, "Running list of how President Trump is changing environmental policy"). The magazine provides a remarkably fertile and legitimate source of information covering a host of encyclopedic disciplines not only in the United States but also, the world at large. It is thus necessary to stress that the above list compiled by the author from various sources is by no means politically biased and it is only intended to demonstrate the timeline of the chorological events of "Donald Trump's environmental protection rollback." The above information is also easily available to the readers through Google search. Compiled by author form various sources.
ANWR, Arctic National Wildlife Refuge; *CMS*, Carbon Monitoring System; *EPA*, Environmental Protection Agency; *EDGI*, Environmental Data & Governance Initiative; *FEMA*, Federal Emergency Management Agency's; *GDP*, gross domestic product; *NASA*, National Aeronautics and Space Administration; *OCO-3*, Orbiting Carbon Observatory 3; *USGS*, US Geological Survey.

again." Indeed, as reflected in Table 3.15, his war against nature has been remarkably consistent. Conceivably President Donald Trump, more than any other figure in the environmental history of the United States, places human beings momentously at the center of concerns for sustainable development. In more specific terms, motivated by the principles of anthropocentrism, the president displays a strong egoistic inclination to advocate for the commodification and exploitation of nature for the sake of its economic benefit per se—a development strategy befits the corporate world of economic maximization.

Donald Trump's anthropocentric proclivity *sui generis* is well reflected by his relentless attempt to smother or overturn Barack Obama-era environmental policies without due regard for far-reaching environmental implications is astounding. More to the point, Donald Trump's relentless efforts to unearth nature aggressively to an excessive extent in the teeth of all highly important scientific evidence of climate change is making the already formidable battle against global warming even more formidable. Fundamentally Donald Trump's environmental philosophy exhibits clear echoes of instrumentalism and materialism incisively developed in the magnum opus of Francis Bacon—a staunch individualistic utilitarian par excellence. To provide a better understanding of Donald Trump's utilitarianistic environmental orientation, we now reflect on Bacon's environmental philosophy to explain his impersonal code of environmental behavior.

3.19.4 Donald Trump's environmental philosophy: The contemporary Francis Bacon

Donald Trump's wholesale attachment to the anthropocentric view of nature as explained above does have a whiff of Francis Bacon's utilitarian philosophy of "Dominion of Man over the Universe" which emphasizes human supremacy over nature (Bacon, 1964). Accordingly Bacon calls upon the "true sons of knowledge" to penetrate further into "the outer courts of nature" and "find a way at length into her inner chamber," (Bacon, 1901: 146). Bacon urged men "to make peace between themselves, and turning with united forces against the Nature of Things, to storm and occupy her castles and strongholds, and extend the bound of human empire" (Francis Bacon, *De Augmentis Scientiarum*, p. 372, quoted in Peltonen, 1992: 291). In this way humans would subdue "nature with all her children to bind her to your service and make her your slave" (Bacon, 1964: 62). This fundamentally established Bacon's philosophy of "Dominion of Man over the Universe" and laid the foundation of "human utility and power" to "subdue and overcome the necessities and miseries of humanity" and improve the life of humankind (Bacon, 2011: 21, 27; see, also, Fideler, 2014).

In Bacon's mechanical view of earth, nature has value as an object of human use for the betterment of life, that is, it has only instrumental value, and as an object of technological control and manipulation. Nature does not deserve any special moral consideration or reverence because it is not imbued with intrinsic value. An object is said to be intrinsically valuable when it is valuable in itself or for its own sake. Thus Mankind is free to technologically shake it to the foundation and instrumentally appropriate it to the fullest extent insofar as it serves to maximize "human utility and power." Technically speaking, the

Baconian world is a human-centered world or an anthropocentric universe exhibiting a profound separation between man and Nature.

Apparently Francis Bacon's imperialistic view of nature constitutes a solid philosophical frame of thought that exerts a far reaching anthropocentric force in commanding the environmental norms, beliefs, values and attitudes of Donald Trump. It is worth reiterating that these collective traits of environmental behavior or the shared way of understanding man's position in the natural world is what is called the Donald Trump anthropocentric proclivity *sui generis* as mentioned above. This anthropocentric tendency creates an acute kind of hypermaterialistic anthropocentric sense of environmental consciousness in association with human—nature interactions which are cut off from the moral ties or bonds of environmentalism. In other words, there is an absence of moral norms or some form of environmental ethical system in helping to guide and shape environmental behaviors when interacting with nature. Thus to Donald Trump, the natural world is conceived on individualistic and utilitarianistic assumptions, and interacting with nature is guided by utilitarian motivation and instrumental orientation (for a conceptual discussion of environmental philosophy, see Chapter 4: Greening for a Sustainable Future: The Ethical Connection).

This hyper or extreme form of anthropocentrism is a very dangerous phenomenon in that when individuals no longer believe their moral obligations or duties toward the natural world, they tend to revert to the Baconian imperial ambitions over nature. In effect, in their interaction with the natural system, humans tend to skew toward instrumentally optimizing the economic use of nature without taking into consideration how their unrestrained use of the environment impacts on the resilience of ecosystems or the intergenerational justice of future generations.

Since the United States is the world's second largest carbon emitter, the world's largest carbon polluter on a historical or cumulative basis, and a potential game changer having momentous influence in drawing other nations, especially the developing nations, closer into the fold of the Paris Accord, addressing climate change problems effectively will not be possible without its serious political environmental commitment. Viewed from this perspective, it is legitimate to claim that the United States, by cutting off its moral ties with the Paris Climate Agreement, is, to put in Paul Krugman's words, committing "an indefensible crime against humanity" (Krugman, 2019).

3.20 The United Nations' road to global environmental sustainability: Some United Nation success stories

It is beyond dispute that the growing international endorsement of United Nations documents has not been concomitantly matched by adequate or real commitment to policy implementation. Today, we are witnessing a crisis of sustainability caused by man's insatiable quest for economic growth and materialism despite massive environmental protection efforts initiated by the United Nations to thwart further environmental decline for more than 45 years. However, as reflected in the following examples, the United Nations has also brought many countries into the course of sustainable development.

3.20.1 Mauritania—environmental mainstreaming, sustainable resource management and poverty reduction

Mauritania, a largely desert country, is situated on the Atlantic coast of Northwest Africa. It is the eleventh largest sovereign state in Africa and one of the world's poorest countries. In southwest Mauritania, considerable environmental problems such as natural resource degradation have been affecting nearly two-thirds of its poverty-stricken people. To address these problems, the country developed its Poverty Reduction Strategy Paper (PRSP) action plan with the view to improve the socioeconomic conditions of the poverty-stricken local communities.

Furthermore since 2000, the government further developed many sustainable development policies to achieve the targets of the Millennium Development Goals in poverty reduction and environmental protection. For example, in its PRSP (2006–10), the government integrated "the environmental dimension in public policies and to strengthen institutional and regulatory framework to enhance enforcement" (IMF 2011: 64; see also, IMF, 2013; UNDP-UN Environment, 2019). Ironically however, they were not implemented as intended (SDGF, 2017). As a result, the poor continued to be trapped in a vicious cycle of poverty and the natural environment was under serious threat from overexploitation.

To address these problems, the Mauritanian government, with the support of seven UN agencies as shown in Fig. 3.38, implemented a joint programme known as "Mainstreaming local environmental management in the planning process" between 2008 and 2012 (SDGF, 2017). The Programme aimed to help the national government to

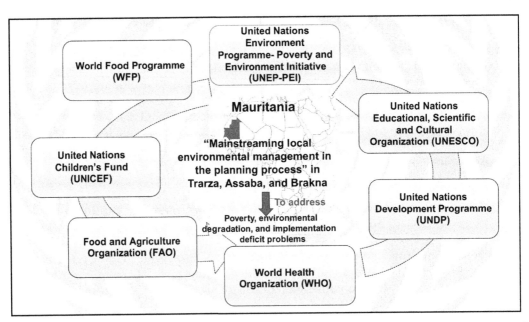

FIGURE 3.38 Composition of the "Mainstreaming local environmental management in the planning process" Programme.
Note: The joint programme is unique in that it involved seven international organizations to help Mauritania to solve its persistent poverty and environmental problems.

improve current and future efforts to integrate environmental considerations into central and decentralized planning processes. It also aimed to improve community's access to drinking water and sanitation, to reduce poverty and to enhance revenue of the poor through sustainable resource management. The local planning process was further enhanced through the following measures:

1. the establishment of a strategic environmental and social assessment framework with the view to strengthen national capacity to better understand the relationship between poverty and the environment,
2. the development of a national strategy to facilitate environmental policy integration, that is, the integration of environmental concerns into public policy (SDGF, 2017; see also, e.g., UNDP-UN Environment, 2019; Green Climate Fund, 2019), and
3. the integration of environmental considerations in the local planning process (SDGF, 2017, see also, e.g., IMF, 2018).

The Programme advocated a participatory approach toward promoting sustainable development. To note in passing, a participatory approach is a community-driven development approach which involves the local people collaboratively in development projects and programmes with an empowerment agenda. More specifically it is concerned about the commitment to create the necessary conditions which can lead to significant empowerment, capacity building, and improved wealth creation of those involved in the development projects or programmes (see, e.g., Rudqvist and Woodford-Berger, 1996).

The Programme was implemented in three selected regions, namely, Trarza, Assaba, and Brakna. These regions are three of the most poverty-stricken regions in Mauritania plagued by persistent and high poverty rates, natural disasters and environmental degradation. The Programme comprised a Management Committee of all UN agencies, the ministry, representatives of civil society and the private sectorcollectively called here the Programme management task force. It aimed to:

1. supervise, monitor, and evaluate programme activity implementations,
2. create synergies between all actors involved, and
3. propose corrective measures during the implementation stage.

Various tangible actions have been taken under the Programme to create the necessary conditions for the promotion of sustainable development. These included some of the following:

1. Creating favorable conditions for restoring natural ecosystems in the targeted areas and improving and diversifying the income of local communities. Six thousand households benefitted from the improvement of their revenue.
2. The local community, trained and supervised by about 30 NGOs, actively contributed to:
   a. the regeneration of 800 ha of gum trees,
   b. the development of 295 ha of silvopasture,
   c. the stabilization of 742 ha of dunes,
   d. the protection of 5800 ha of agricultural land, and
   e. the regeneration of 20 ha of mangroves in the area of Diawling National Park through the cultivation and transplanting of 40,000 plants by the villagers (Fig. 3.39).

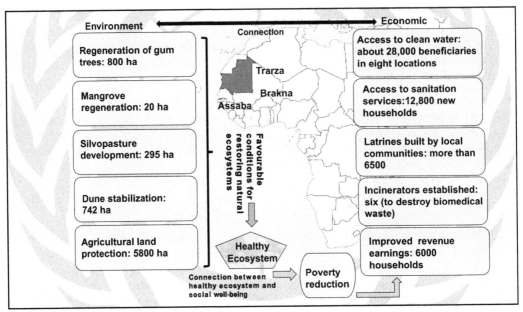

FIGURE 3.39   Programme mainstreaming success story. *Based on, SDGF, 2017. Mauritania: Mainstreaming Local Environmental Management in the Planning Process. Sustainable Development Goal Fund (SDGF), United Nations.*

The local people also benefitted immensely from the Programme in term of better access to clean water and sanitation, improvement and diversification of revenue earning from, for example, fishing, livestock rearing, food gathering, crafts, market gardening, and tourism (see, e.g., Fondation Ensemble, 2014, Fig. 3.39). Equally important, the Programme has also helped to raise environmental awareness of the policy makers and the local communities. Particularly it enables them to better understand the close connections between an ecologically healthy ecosystem, socioeconomic wellbeing and poverty reduction. This provides the necessary impetus to encourage greater national effort and local action to improve, manage, and protect the natural environment through reform and planning.

Overall, the Programme has been able to achieve its purpose in enhancing sustainable development in the targeted regions. Furthermore an important lesson that may be drawn from the Programme is that mainstreaming or integrating and institutionalizing environmental considerations into existing development policies constitutes an important step toward translating environmental sustainability into practice which ultimately leads to the improvement of the natural environment as well as the livelihoods of the people.

3.20.2 Project Predator and policing the global illegal wildlife trade—Tiger

As discussed in Section 3.12, tigers are listed under the IUCN Red List of Threatened Species as "Endangered" and protected under Appendix I of CITES. Near 70% of the world's remaining wild tigers are clustered within small, fragmented and often isolated landscapes in 13 range countries, namely, Bangladesh, Bhutan, Cambodia, China, India, Indonesia, Lao PDR, Malaysia, Myanmar, Nepal, Russia, Thailand, and Viet Nam (the 13

Range Countries or 13 TRCs) (Walston et al., 2010; see also, Verheij et al., 2010). India holds the largest wild tiger population estimated at about of 70% of the word's total (Kumar et al., 2019).

However, the endangered species which was once widely distributed in the above range countries is currently facing a high risk of extinction. As discussed in Section 3.12, deforestation, habitat fragmentation and destruction, and intense poaching pressure, have resulted in a dramatic decline of the tiger populations. Furthermore with the black market for illegal wildlife products worth $US20 billion a year, poaching and illegal wildlife trade continue to grow, pushing many species to the brink of extinction (INTERPOL, 2018a).

In a concerted drive to protect the wild tigers from the ubiquitous and mounting threat of extinction, INTERPOL launched Project Predator in 2011 at the 80th INTERPOL General Assembly in Hanoi, Viet Nam. It aimed to protect and save the world's last surviving wild tigers (INTERPOL, 2011). INTERPOL's Project Predator is supported by the World Bank's Global Tiger Initiative (GTI). GTI was launched in 2008 as a global alliance comprising the government bodies of the 13 range countries, international organizations such as INTERPOL and USAID, civil society, conservation groups, scientific communities, and the private sector. Its main aim is to save the wild tigers from extinction through cooperative efforts (World Bank, 2016b).

The INTERPOL Predator Project brings together authorities from the 13 TRCs, USAID, Defra, the World Bank, UK Department for Environment, the Smithsonian Institution, and INTERPOL (Fig. 3.40). The Secretariat of CITES provides support on enforcement-related

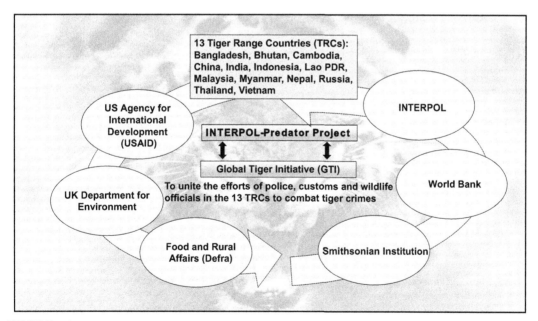

FIGURE 3.40   The structure of INTERPOL's Predator Project.
Note: The GTI includes the World Bank, the GEF, the Smithsonian Institution, Save the Tiger Fund, and International Tiger Coalition representing more than 40 nongovernment organizations (World Bank, 2016b). *GEF*, Global Environment Facility; *GTI*, Global Tiger Initiative.

issues in the GTI (Rosen, 2011). It is noteworthy that the project is not only confined to the protection of tigers. It also covered the protection of other Asian big cats such as leopards, snow leopards, clouded leopards, and Asiatic lions.

INTERPOL's Predator Project serves an important function in uniting and enhancing tiger protection efforts of the national police, customs, environmental agencies, prosecutors, wildlife authorities, and other specialized agencies such as anticorruption units in the 13 TRCs. More specifically INTERPOL provides member countries with various institutional, collaborative, or partnership support as shown in Table 3.16 to strengthen or streamline their enforcement capacities to combat tiger crimes. The Predator Project's

TABLE 3.16 INTERPOL investigative, analytical and collaborative partnership support.

	Aims
INTERPOL Environmental Security Programme (ENS) (with its activities later merged under Global Wildlife Enforcement)	To provide member countries with advanced, intelligence-led investigation techniques to assist them: (1) to implement national, regional and global environmental policy efficiently or (2) to enforce national and international laws and treaties effectively
INTERPOL Wildlife Crime Working Group (WCWG)	To provide an international platform to enhance cooperation among member countries in stemming wildlife smuggling activities and other related matters
National Environmental Security Taskforce (NEST)	To unite the national law enforcement agencies responsible for enforcing environmental laws
Regional Investigative and Analytical Case Meetings (RIACM)	To provide a platform for investigators from member countries to meet face-to-face to review case files and share intelligence and analysis to further their investigation
Investigation Support Team	To provide specialized law enforcement experts to support local law enforcement authorities in their investigations
Capacity Building and Training	To provide extensive training and capacity building activities for law enforcement agencies, with the aim to enhance the ability of member countries to undertake targeted law enforcement operations
Analytical Support	To provide both operational and strategic analytical support to wildlife crime-related projects, investigations, and operations
International Consortium on Combating Wildlife Crime (ICCWC)	To stimulate coordinated law enforcement response to environmental crime and to ensure transnational criminal networks are identified and disrupted
Wildlife Crime Working Group	To provide strategic advice on relevant issues and to harness global support. The working group serves as for law enforcement officials meet to discuss new strategies, share experiences and expertise as well as initiate and lead projects to detect and address organized wildlife crime

Based on, INTERPOL 2018a. Global Wildlife Enforcement. Strengthening Law Enforcement Cooperation Against Wildlife Crime. INTERPOL, Lyon, France. Available from: <file:///C:/Users/choy3/AppData/Local/Temp/WEB_Wildlife%20ProspectusMarch2019.pdf>.

initiative also encourages member countries to establish National Tiger Crime Task Force to address organized transnational environmental crime with coordinated, collaborative and international efforts (INTERPOL, 2011).

For the first time, the Predator Project's initiative has led to the mapping of eight criminal groups smuggling tiger parts from India to Nepal and the arrest of a number of wildlife criminals (INTERPOL, 2018a). In 2015 INTERPOL shared its information on the involvement of specific Bangladeshi pirate groups in tiger smuggling activities carried out in the Sunderbans region between India and Bangladesh with the authorities concerned through its two published reports. In the following year, the Government of Bangladesh implemented a strong enforcement policy against pirate groups. This led to the surrender of 12 pirate groups, leading to the overall decline in wildlife crime in the Sundarban region, including tiger poaching and trafficking. In 2017 the Government declared Sunderbans to be pirate free. It was further reported that tiger poaching and smuggling in the region has declined significantly (INTERPOL, 2018a,b).

To enhance enforcement capacity and efficiency, the INTERPOL's National Central Bureau in Kathmandu and the South Asia Wildlife Enforcement Network held a training session for the Nepalese authorities on the use of intelligence and information management in combating environmental crime. Following the training, the Nepalese Police, Department of National Park and Wildlife Conservation and Department of Forests of Nepal deployed enforcement officers to different parts of the country to stem out wildlife crime. This led to the seizure of two tiger skins, 53 kg of tiger bones and arrested four people alleged to be in connection to the crime (INTERPOL, 2013). Furthermore in 2018 the Indian police force in close collaboration with the Nepalese police authority through INTERPOL channels, arrested Lodu Dime, a tiger trafficker who was under INTERPOL Red Notice since 2013. Lodu Dime was the key suspect in the seizure of five tiger skins and 114 kg of tiger bones in Nepal in 2013 (INTERPOL, 2018b).

It is thus clear from the above that INTERPOL in collaboration with various international agencies has been instrumental in helping the TRCs as well as all member countries to address wildlife crime effectively through its various institutional, collaborative support and information sharing as listed in Table 3.15. Despite this, nonetheless, the above successes are by no means adequate to mitigate, let alone to solve the tiger crisis. It may well be that tigers are wide-ranging and conservation-dependent species which roam far and wide in large undisturbed tracts of habitat with ample prey in the wild to maintain long-term viable populations (see, e.g., Global Tiger Initiative Secretariat, 2011; Jhala et al., 2019).

It thus follows that without putting in place extensive environmental conservation programmes to protect the natural habitats, ecological connectivity, and foraging environment of the tigers, their long-term existence will be at stake. Also tigers are "walking gold" and worth a fortune on the lucrative black market (Guynup, 2014). Thus nothing short of developing a strong conservation ethic among policy makers, forestry management officers, biodiversity conservation enforcement staffs, and the local communities is adequate to protect the tigers and other Big cat species as well as a whole range of other animal species from being poached (see Chapter 4: Greening for a Sustainable Future: The Ethical Connection for the discussion of environmental ethics).

3.20.3 Sustainable forest management in Nepal: Community Forestry Development Programmes

Nepal has a total area of 147,148 km^2 with around 40% of the area covered with forest. Agriculture is the mainstay of the economy, providing employment, foods and shelter for almost two-thirds of the population. The local people depend heavily on the forests for their livelihoods. It may be said that in Nepal, forestry is not only about trees but also, about the people in that it serves as the life support system of its populace. In 2019 GDP per capita based on purchasing power parity (PPP) for Nepal was 3318 international dollars, making it as one of the poorest countries in the world (Knoema, 2020).

Nepal, building on UNEP-led Green Economy Initiatives, is committed to promote green economy to meet the development challenges facing the country (National Planning Commission, Nepal, 2011). Within the present context, Green Economy is defined as development that "results in improved human wellbeing and social equity, while significantly reducing environmental risks and ecological scarcities" (UNEP, 2010b: 5). The Government of Nepal embraced green economy as "an instrument for sustainable development, poverty reduction and inclusive and equitable economic growth" (National Planning Commission, Nepal, 2011: 2). More specifically, the Nepalese green economy initiative aimed to achieve sustainable resource use, sustainable environmental management, people-centered economic progress, and propoor development.

Despite various socioeconomic and technological constraints, Nepal has made considerable green progress in promoting sustainable forest management (SFM) and socioeconomic development through its Community Forestry Development Programme. Basically, the Programme started in 1987. It covered the management of 31.62% of the country's forest area (4.27 million ha). To delve deeper into the subject, in 1987 the forestry department convened a National Community Forestry workshop with the aim to reverse the unsustainable harvesting practices which have been facing the country since 1957. The workshop led to the formation of the Community Forest User Groups (CFUGs), the development of a Master Plan for the forestry sector and the enactment of the Forestry Act (Table 3.17).

The CFUG is a community-based, inclusive, member-driven and self-governed institution composed of diverse groups of people with diverse interest in the use and

TABLE 3.17 The evolution of sustainable community forests management system.

1987	National Community Forestry workshop	The development of the user group concept in forest management
1988	Master Plan for the Forestry Sector (MPFS)	To update existing forestry legislations (Forestry Bill, 1990; Forestry Development Rules, 1990; Leasehold Forestry Rules, 1990; Private and Religious Forestry Rules, 1990)
1993	Forest Act (1993)	Based on the forest policy of 1988 and building of the Master Plan (MPFS), the Forest Act (1993) formally included the concept of "user group" or "community forestry" in its legal structure, given rise to the evolution of Community Forest User Groups (CFUGs)

Shrestha, B., 1998. Changing Forest Policies and Institutional Innovations: Users Group Approach in Community Forestry of Nepal. International Workshop on Community-Based Natural Resource Management (CBNRM), May 10—14, Washington, DC.

FIGURE 3.41 Some grassroot members of CFUG at a general meeting. *CFUG*, Community Forest User Groups. *Reproduced with permission from, Anup, K.C., 2017. Community forestry management and its role in biodiversity conservation in Nepal. In: Lameed, G.S.A. (Ed.), Global Exposition of Wildlife Management. IntechOpen Limited, London, pp. 51–72.*

management of forest (Anup, 2017; Fig. 3.41). There are 22,266 CFs involving almost 2.9 million households in managing a total area of about 22.37 million ha of state forests based on community-based forest management agreement (Bhandari et al., 2019; see also, Nuberg et al., 2019, Figs. 3.42 and 3.43).

The government has refrained from setting any legal limits concerning the area and size of CFUGs. The community-based forest management agreement hinges on the willingness and ability of the community to manage the forest sustainably. In addition, as shown in Fig. 3.42, the CFUGs are given various institutional rights to form their structural organizations. Each CFUG draft its own constitution. However, its operational plans are required to be registered and approved by the District Forest Office. The CFUGs also define the social arrangements, responsibilities and rights of the group and to plan forest management based on various guidelines, practices, or norms (Anup, 2017).

To enhance sustainable forestry use and management, the CFUGs are given various direct or indirect forest utilization rights as shown in Fig. 3.43. Direct use refers to the actual use of the forestry products while indirect use is concerned about the use of funds derived from the sale of forest products for other investments. As depicted in Fig. 3.43, the CFUGs are given the authority and free will to use the forest resources sustainably to improve social wellbeing and to promote social equity. However, the government has imposed a condition on the groups to reinvest 25% of their revenue earned from fund investment for sustainable forest development (Upadhyay, 2013). Furthermore the CFUGs are required to pay taxes to the government for selling any forest products outside CFUG (Anup, 2017).

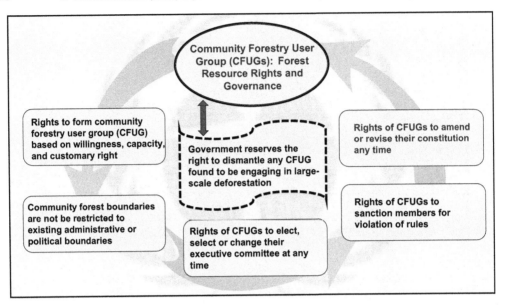

FIGURE 3.42 CFUGs forest resource rights and governance. *CFUG, Community Forest User Groups. Based on, Upadhyay, S., 2013. Community based forest and livelihood management in Nepal. In: Bollier, D., Helfrich, S. (Eds.), The Wealth of the Commons. A World Beyond Market & State. Levellers Press, Amherst, MA, pp. 265−270; Anup, K.C., 2017. Community forestry management and its role in biodiversity conservation in Nepal. In: Lameed, G.S.A. (Ed.), Global Exposition of Wildlife Management. IntechOpen Limited, London, pp. 51−72.*

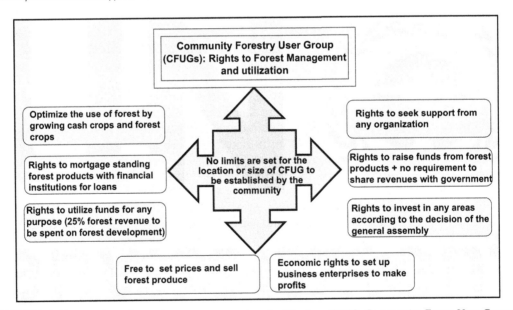

FIGURE 3.43 CFUGs rights to forest management and utilization. *CFUG, Community Forest User Groups. Upadhyay, 2013. Community based forest and livelihood management in Nepal. In: Bollier, D., Helfrich, S. (Eds.), The Wealth of the Commons. A World Beyond Market & State. Levellers Press, Amherst, MA, pp. 265−270; Anup, K.C., 2017. Community forestry management and its role in biodiversity conservation in Nepal. In: Lameed, G.S.A. (Ed.), Global Exposition of Wildlife Management. IntechOpen Limited, London, pp. 51−72.*

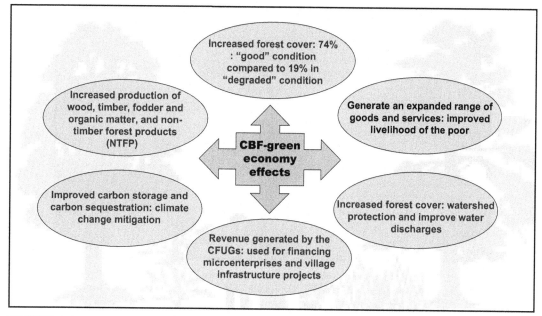

FIGURE 3.44 CBF-green economy effects. *CBF, Community-based forestry. Based on, Upadhyay, S., 2013. Community based forest and livelihood management in Nepal. In: Bollier, D., Helfrich, S. (Eds.), The Wealth of the Commons. A World Beyond Market & State. Levellers Press, Amherst, MA, pp. 265–270.*

The community-based forestry policy is not only environmentally beneficial but also, socially desirable and economically tenable and thus, contributing to green economic development in Nepal. Despite the lack of comprehensive national-level data, various studies show that the CBF model has the great potential to serve as an effective means to improve the livelihoods of the poor through poverty reduction and inclusive and equitable economic growth from optimizing the economic use of forest products—the CBF-green economy effects (Figs. 3.44 and 3.45).

On the whole, Nepal's green community forestry model plays an important role in enhancing the ecological resilience of the forest stands, rendering them more Holling sustainable. Various studies revealed that deforestation has generally declined and that forest areas in the country especially in the mountainous regions have improved due to community forestry intervention (DFRS, 2015). However, it has also been reported that the CFUGs are more concerned about optimizing the socioeconomic benefits from the "green gold" while paying less attention to the importance of biodiversity or forest conservation. This is partly caused by short-term socioeconomic interest and knowledge gap in the attributes of biodiversity, ecosystem functions, and services.

As it turned out, forestry monoculture of the economically valuable species is prioritized over the planting of natural forest. By and large, forests are conceived as "stocking of trees and sources of timber" rather than "the store house of biodiversity and sources of many ecosystem services" (Shrestha et al., 2010: 103). To compound the problem, in many parts of Nepal, there are still significant institutional, regulatory and policy barriers to the promotion of green economy (see, e.g., Nuberg et al., 2019). Thus to render the community-based forest

FIGURE 3.45   The CBF-green economy effects—increased forest cover and improved livelihoods of the local communities (a picturesque representation). *CBF*, Community-based forestry. Reproduced with the permission from *Anup, K.C., 2017. Community forestry management and its role in biodiversity conservation in Nepal. In: Lameed, G.S.A. (Ed.), Global Exposition of Wildlife Management. IntechOpen Limited, London, pp. 51–72.*

management system more effective, there is an exigency need to enhance an ethical sense of resource use and management not only among the community forest users but also, the policy makers (see, Chapter 5: The Nexus of Environmental Ethics and Environmental Sustainability: An Empirical Assessment and Chapter 6: The United Nations Environmental Education Initiatives: The Green Education Failure and the Way Forward).

3.21 The United Nations' success stories: Some remarks

For more than 45 years, the United Nations has embarked on a grand but daunting mission to bring humanity onto the racetrack of environmental sustainability. During those years it has had notable moments of success as reflected above. Numerous glorious moments of the United Nations environmental sustainability initiatives launched since Stockholm have been experienced elsewhere across the globe. The Euro standard (Euro 1), which entered into force in 1993 and amended regularly to improve emission standards is a case in point. Particularly, the revised Euro standard (Euro 6) for light passenger and commercial vehicles, which came into force in 2014 has contributed positively to reduce emissions of air pollutants and GHGs per unit of travel as well as measurable air pollution concentrations (UNEP, 2019b).

There can be little doubt that all these success stories spanning the globe, albeit piecemeal and sporadic in nature, contribute to promote environmental sustainability. However, they are far from adequate to fuel the determination of the global community at large to seek concrete solutions to the environmental crisis humanity is facing. Obstructionist attitudes driven by an anthropocentric view of nature and the lack of political will caused by the fervent belief in the pursuit of economic growth or material consumption per se often block the United Nations' journey to environmental sustainability. This brings us back to the fundamental issue of the ethical base of environmental sustainability. More directly what we need is a universal and ethically based solidarity involving everyone on our Planet to redress the impending environmental crisis caused by human abuse of nature.

In other words, we need a generalized attack on the global environmental ill based on human extension of moral considerability to nature. If the United Nations' environmental sustainability initiatives are to be operationally effective, as they should be, it is high time that they are raised on ethical grounds and moral consensus. If the sociopolitical context is not changed ethically, environmental sustainability will fall apart eventually amidst increased environmental efforts. This is the gist for discussion in Chapter 4, Greening for a Sustainable Future: The Ethical Connection, that follows.

3.22 Global environmental sustainability, biodiversity conservation and climate protection: The missing links

Despite the above success stories, it may generally be remarked that as discussed in Chapter 2, The United Nations' Journey to Global Environmental Sustainability Since Stockholm: An Assessment, to this end, the United Nations has, through numerous global conferences, including the CBD (COP-1 to COP-14) and the UNFCCC (COP-1 to COP-25), provided the international players, stakeholders, policy-makers, and the public at large, with strategic environmental insights and practical sustainable development policy solutions to address global biodiversity loss and climate change problems. In turn, this reinforced the UN's role in reaching and concluding global environmental agreements and treaties after undergoing numerous backdoor negotiations and a multitude of conferences and conventions as depicted in the previous chapter. The UN global environmental hindsight has been critical in raising concerns about the environmental implications of unsustainable development practices. At the same time, it has also been able to influence views and actions internationally on development discourse and environmental actions.

A growing consensus has emerged across the globe that the world is coursing on an environmentally unsustainable path of development and changes in development policies and the adoption of practical measures are indeed necessary to contain the heightening risk of a global environmental meltdown. Consequently revisions of legal and regulatory structures have been initiated, national environmental laws enacted and environmental policies and regulations promulgated across the globe to strengthen environmental protection efforts. By and large, as unveiled in the previous chapter, global environmental protection regimes have experienced remarkable growth and alignment since the 1972 Stockholm Declaration on the Human Environment.

However, as empirically demonstrated in this chapter, global environmental sustainability has deteriorated over the past few decades. This is reflected by rapid acceleration of worldwide biodiversity destruction and increasing global CO_2 atmospheric concentrations. The entire world—from Africa, Latin America to Southeast Asia continues to lose forest cover and biodiversity at an unprecedented rate, pushing many species to extinction or to the brink of extinction despite having in place a host of biodiversity protection laws and sustainable resource use and management instruments. This is basically due to nonimplementation and nonenforcement of, and noncompliance with environmental protection measures. As noted above, no country in the world including the developed countries has achieved the targets set under CBD to significantly reduce the rate of biodiversity loss by 2010—the International Year of Biodiversity. In short, the CBD has been less successful in providing a practical framework to bridge the gap between political commitment and policy implementation. Writing from the same vein, attempts to make progress on global CO_2 emissions for the past few decades to prevent the "dangerous anthropocentric interference" with the climate system have met with little success with the CO_2 atmospheric concentration reaching historically high levels.

Thus it must be admitted that although reaching international agreement is of critical importance to promote environmental sustainability, putting them into actual implementation is another question. Certainly moving to an ecologically sustainable world and a low-carbon economy is not achieved at the mere demonstration of awareness and concern through political commitment to environmental sustainability and ratification of environmental treaties or the enactment of environmental legislations. These environmental initiatives, no matter how sweeping and unambiguous, are only hortatory gestures unless they are implemented effectively.

It may well be that the missing factor in the United Nations environmental protection efforts is the recognition of how environmental and ethical beliefs would serve as an indispensable motivational force to stimulate environmentally sustainable behavior and actions, and hence effective implementation. Global environmental sustainability policies and practices are inherently ethical phenomena because their implementation in a real-world system is a result of human decision-making processes and this requires real environmental concern and changes in human behavior to accomplish. Certainly environmental concern is considered a prerequisite for global environmental sustainability but it alone is insufficient to achieve sustainable development because individuals who express concern may not indulge in environmentally responsible or protective behaviors (Maloney and Ward, 1973; Scott and Willits, 1994).

The root of the many environmental issues we are fighting go deep into our ethical and moral relationship with the non-human natural world—the ethical concern that is missing from the discourse in the United Nations approach to promoting global environmental sustainability. More specifically the United Nations' approach to international environmental negotiations does not consider the underlying and ultimate force that helps to advance efforts to transform the global system to a more sustainable world. To this end, environmental protection under the UN framework is fundamentally instrumental in nature and public discussion of environmental issues are in most cases confined to scientific and economic concerns.

Any realistic solution to the regional environmental problems will have to involve two interwoven factors, namely, environmental policies and legislation, and environmental ethics. It may thus be emphasized that the ethical aspects of environmentalism need to move more to central position in the sphere of global environmental sustainability. This calls for a fundamental systematic transformation of global society away from anthropocentrism toward a society grounded on the ethical commitment to the well-being of not only the present and future generations but also to the ecological integrity of our global commons. A theoretical discussion on this is in the next chapter followed by an empirical assessment of the practical effect environmental ethics is seen to have on environmental sustainability.

3.23 Concluding remarks

Over the past few decades since Stockholm, the United Nations has convened numerous environmental summits and conferences aimed to heighten global environmental awareness and to call on the international community to intensify actions to address global environmental decline especially in association with biodiversity loss and climate change. It has also provided a wide range of principles and programmes of action to prevent further irreversible damage to our environment. Practically all countries except the United States and the Holy See responded by acceding, ratifying or adopting a host of conventions, declarations and programmes dealing with environmental protection, biodiversity conservation and sustainable resource use practices.

Ironically however, empirical evidence as revealed above point to the direction that, from the first Convention on Biological Diversity (COP-1) to COP-14, the environment across the world, whether in developed or developing countries, has increasingly been put under serious pressure by the rapid growth of economic activities and intensified exploitation of natural resources. The situation is particularly worrying in many of the developing economies in the African, Latin American, and Southeast Asian region whose biodiversity planning and conservation often clash with the pursuit of economic growth and increased human consumption, leading to inadequate, ineffective, or biased policy enforcement. Also as revealed in all the case studies in these developing countries, nonimplementation and nonenforcement of, and noncompliance with environmental laws and regulations are considered the norm. It is also worth noting that the purge of environmental laws and regulations governing biodiversity conservation is also a common phenomenon in the United States under Donald Trump's administration.

On the issue of climate change, despite constant UNFCCC deliberations from the first Conference of the Parties (COP-1) held in 1995 to COP-24 held in 2018, CO_2 emissions have increased persistently from 1972 to 2018 with total cumulative CO_2 emissions of 1202 billion tonnes from burning fossil fuels and cement production. Moreover, COP-24 ended without firm and collective commitments to enhance climate action targets in line with the scientific advice of the IPCC. Similarly climate talks at COP-25 held in Madrid in 2019 collapsed without much sense of climate urgency especially among the developed nations including the United States and Australia. Worse yet, many developed and developing countries alike have all seen pushing back against climate change policies out of anthropocentric reasons.

Although China has surpassed the United Sates as the world's largest carbon emitter in 2005, on a per capita and cumulative basis; the United States is still the world's largest polluter—spewing 406 billion tonnes of CO_2 into the atmosphere since 1800. Despite this, the United States under the Donald Trump administration has officially informed the United Nations to go back on its Paris Accord commitment to carbon reduction. It is contended the United Sates' position on climate change is fundamentally shaped by its hyperanthropocentric worldview of nature, similar to its stance on its refusal to ratify the CBD. From Donald Trump's perspective and for anthropocentric reasons, the economic use of nature as against environmental conservation acquired a far greater importance in connection with his capitalist mission to make "America great again." Thus sustainable development, so to speak, has subtly slid from sustaining the ecological integrity of nature to sustaining the economic health of growth to the detriment of the natural system.

It is increasingly clear that the rising environmental awareness has been critically hindered by the lack of political will to accomplish change. This in turn is due to the hyperanthropocentric worldview of our society in relation to our natural environment. To be sure, without the United Nations, worldwide ecological and environmental degradation would have been even more rapid and far worse. However, the overwhelmingly large proportion of effort, time and resources incurred by the United Nations on a continuing basis in promoting global environmental sustainability has not been equitably met with more meaningful outcomes as reflected by the worrisome environmental status quo spanning the planet. For more than 45 years since Stockholm, its promise for a more sustainable world has largely evaporated.

Implicitly addressing environmental problems based on international treaties, global agreements, or national legislation alone cannot offer green hope for mankind. It is worth reemphasizing that nothing short of a human ideological transformation from the hyper-anthropocentric view of nature toward a society grounded on ethical commitment to the well-being of not only the present and future generations but also the ecological integrity of our global commons is required to turn the unsustainable tide around. In moving toward global environmental sustainability, the moral beliefs of key-policy makers, the ethical system of environmental governance, and an environmentally inclined society all play a part. This provides the vantage point for a reexamination of the human attitudes and values that influence individual behavior and government policy toward nature. This issue will be taken up in the following chapter.

4

Greening for a sustainable future: The ethical connection

4.1 Introduction

The preceding chapter has presented incontrovertible evidence that the current environmental conditions are as worrying as ever. For the past few 100 years, and especially in recent decades since Stockholm, humans, in their interaction with the natural environment, have significantly altered nearly all the Earth systems, especially the ecosystems, for our own material benefit. Today, the Earth systems, including biodiversity, are at risk of rapid deterioration, and the legally binding convention on biological diversity, national laws and environmental policies have failed to halt the continued loss of biodiversity. Furthermore, massive biodiversity loss followed by severe reduction in ecosystem services that underpin long-term human existence is increasingly becoming likely.

The implicit promises of the Kyoto Protocol in its battle against global warming have lost credibility. More specifically, two decades of United Nations Framework Convention on Climate failed talks have hampered progress in halting continued "dangerous anthropogenic interference" with the planetary climate system as reflected in the persistent increase in CO_2 emissions associated with increased frequency of life-threatening unusual and extreme weather conditions, and melting ice sheets. Waves of industrialization in the fast-developing Asian countries, especially China and the self-interested and noncommitting United States, Canada, and Australia continue to overburden the atmospheric, land, and ocean carbon sinks.

In a nutshell, our Planet is rapidly moving into a more biologically impoverished and much warmer state due to the failure of human beings to observe the biophysical limits to growth because of their unbridled quest for materialism and conspicuous consumption. It is also increasingly becoming clear that the unsustainable universal sequence of environmental change has been persistent for more than four decades since Stockholm.

Implicitly, in the earth's power structure, humanity has become the major geological force capable of manipulating, modifying, and transforming the Earth systems in the name of "sustainable development." More specifically, the prevailing scale of ecological degradation and environmental crisis highlights a growing disconnect between human systems and the natural world under the present anthropocentric economic system—a system which is overly concerned with the instrumental use of the natural environment for the pursuit of economic growth or material progress. Such natural resource use practices often entail the commodification and monetization of nature under one single monetary matrix, and this value monistic view of nature tends to promote social or political shifts in prioritizing resource exploitation over environmental sustainability. The slogan "Our Common Future" still sounds hollow in the contemporary world.

Although the present dire environmental conditions would not necessarily trigger an imminent collapse of human civilization, the environmental quagmire is alarming enough to demand serious action to halt further encroachment into the terrestrial limits of our Earth systems. Mapping forward to greening a sustainable future requires a new approach to conceptualizing sustainable development by placing greater emphasis on our ethical engagement with and stewardship of nature. Without a grasp of the philosophical foundations of the environmental problems confronting us today, it is difficult to perceive their root causes behind their crisis–ridden mode of deterioration.

With that in mind, this chapter begins with the examination of the longstanding ethical divide separating humans from nonhuman natural world in the contemporary global economic system dominated by the anthropocentric quest for economic growth and material progress. It is argued that nurturing a moral sense of environmental behavior and practices, and an ethical sense of ecological connection could be an important force to motivate actors at each level to undertake real actions to mitigate the perversity of the present environmental tragedy. In this regard, the study examines how human environmental attitudes and values help to influence individual behavior and government policy concerning resource management and conservation practices based on environmental ethics.

It is contended that matters of environmental policy often involve ethical questions. In other words, in our economic use of nature, human environmental value judgments influence individual ethical behavior and government policy on the natural system. This chapter affirms the importance of environmental ethics as an indispensable moral philosophy for our engagement with the environmental problems facing us today. The question here is how to define our place in nature, and what ethical principles should shape humans' environmental attitudes. This will be examined systematically in the following sections.

4.2 Environmental ethics and environmental sustainability: A theoretical assessment

Global sustainable development issues are philosophical in nature and must ultimately be resolved in environmental philosophy or more specifically, based on environmental ethics. Environmental ethics is concerned with the moral relationship between human beings and the nonhuman natural world. More specifically, environmental ethics manifests as an ultimate force in regulating value orientations of humans in their engagement with nature. Value orientation may be defined as "a generalized and organized conception, influencing behaviour, or nature, of man's place in it, of man's relation to man, and of the desirable and nondesirable as they may relate to man-environment and interhuman relations" (Kluckhohn, 1951: 411; Schwartz, 1992). Put differently, value orientation engenders the "moral dimension of the relationship between human beings and nonhuman, nature—animals and plants, local populations, natural resources and ecosystems, landscapes, as well as biosphere and the cosmos" (Gordon, 2017).

However, before entering into the core of its analysis, it is instructive to note that for the past few decades, the field of environmental ethics has developed into a wealth of highly insightful moral positions and distinct ethical principles to guide us in our interaction with nature. Yet, highly relevant and useful though these disciplines are, they are nonetheless entangled in epistemological argument. It may well be that each school of thought is entrapped in a welter of laborious arguments and counter-arguments over its avowed philosophical position, ranging from the utilitarian concern for human welfare to the moral concern for higher animal health or the ethical concern for the ecological integrity of the ecosystem or the biosphere at large.

The invitation is to inexorably achieve ethical-specific certainty and philosophical sophistication in its conceptual favor or philosophical glorification rather than directly establish a pragmatically clear and intelligibly comprehensible ethical frame of guidance. This tends to give rise to epistemological confusion and intellectual discomfort and preclude the grasping of all the relevant aspects of environmental ethics to practically guide human environmental behaviors and sustainable development policy making and practices. This argument obviously implies that rather than ceaselessly engaging in a battle of philosophical triumphs or conceptual certainty, we should opt for a pragmatic course of analysis to know in what direction our ethical analysis should proceed so that it could efficaciously yield practical results. Being pragmatic means opting for a flexible approach best suited to problem solving rather than engaging in rigorous philosophical debates to achieve conceptual credence or sophistication. This necessarily calls for the need to adjust the means available to achieve the end goal of solving real-life environmental problems.

Against this backdrop, the following section aims to systematically and critically examine the major school of environmental ethics with the aim of finding a clear direction toward instituting a pragmatic ethical framework comprising a set of unifying moral principles to guide human ethical engagement with the environmental problems facing us today. It is instructive to note that generally as revealed below, it is possible to distinguish three main categories of environmental ethics, namely anthropocentrism, biocentrism and ecocentrism.

4.2.1 Anthropocentrism: A conceptual analysis

In anthropocentrism, the main focus is primarily or exclusively on human beings with the natural systems subsumed to capitalist exploitation for the promotion of economic prosperity and social wellbeing. However, to provide a greater argumentative precision of anthropocentrism, this section seeks to conceptualize its moral principles into clearer and more distinct constructs. This will allow us to develop meaningful tests of them in promoting environmental sustainability.

To start with, anthropocentrism, which premised on Francis Bacon's philosophy of "Dominion of Man over the Universe" as discussed in the preceding chapter and further examined in relation to "Donald Trump's environmental protection rollback", is commonly understood as a theory of value which maintains that only humans have intrinsic moral value (Thompson, 2017). In other words, anthropocentrism "gives either exclusive or primary consideration to human interests above the good of other species" (Taylor, 1983: 240). John Baird Callicott affirms this position further by claiming that an anthropocentric ethic grants "moral standing exclusively to human beings and considers nonhuman natural entities and nature as a whole to be only a means for human ends" (Callicott, 2004: 757). Anthropocentrism holds that the planet, which is central to human welfare, exists for the sake of humanity and is also to be managed by human beings (Lamb, 1999). This is precisely Francis Bacon and Donald Trump's ethical positions on our natural system as discussed in the previous chapter.

Hence, only humans are considered as the morally considerable class of beings, and concern for ecosystems, species, or other nonhuman living organisms stems from the consideration of whether the socioeconomic wellbeing of humans will be adversely impacted by their depletion. The nonhuman entities have value only because humans value them, whether instrumentally or intrinsically (Martin, 1993). Thus humans are usually regarded as separate from and superior to the nonhuman world. They further consider themselves as having no direct moral duties to protect its ecological integrity. This self-centered attitude is mainly based on self-interests and economic rationale justifications. This may be regarded as a strong or hyper form of anthropocentrism as noted in the previous chapter (Section 3.19.3, see also, Thompson, 2017).

In contrast, a weak form of anthropocentrism is based loosely on human-centered value system. Individuals with weak anthropocentric inclination tend to exhibit a rational appreciation of the intrinsic value of nature which may be based on religious beliefs (religious value) or esthetic satisfaction (esthetic value), among others, and are concerned for their protection. The indirect anthropocentric extension of moral considerability to the nonhuman world based on these noneconomic values can contribute positively to the promotion of the ecological integrity of the environment. However, this intrinsically induced weak anthropocentrism may be unstable in that in case intrinsic value clashes with instrumental value, the latter takes precedence. In addition, weak-anthropocentric-induced moral extension to nature may not be as strong, sustainable, or effective compared to nonanthropocentric justifications which ethically extend direct moral duty to the nonhuman world to protect its ecological integrity, as in biocentrism and ecocentrism as discussed below.

4.2.2 Biocentrism

As distinguished from anthropocentrism, biocentrism rejects the notion of human superiority over other nonhuman species. It holds that all life in the natural world deserves equal moral consideration or has equal moral standing. Conceptually, biocentrism or biocentric ethics may be traced back to the philosophical underpinnings of Albert Schweitzer's ethics of reverence for life, Paul Taylor's ethic of egalitarianism, and Peter Singer's ethic of animal liberation. Basically, these ethicists have a different view of the logic of ethical deliberation. This will be systematically and critically examined in the following subsections with the view of providing a general pragmatic framework for environmental ethics to address the global environmental problems confronting us today.

4.2.2.1 Albert Schweitzer's ethics of reverence for life

Albert Schweitzer, conventionally considered as an early biocentric thinker and the precursor of modern or contemporary biocentric ethics, sought to advance his ethical principle based on the principle of reverence for life. It is contended that Schweitzer's ethics of reverence for life is deeply rooted in biocentrism as it locates inherent value in all living things (Martin, 1993). However, a systematic analysis of Schweitzer's biocentrism reveals that his ethic goes beyond the biocentric boundary of life-centeredness characteristics. I will make this point clear in the comments section below. In this section, I will confine my discussion of Schweitzer's ethics based on his conventional biocentric tradition—the ethic of reverence for life.

To start with, reverence for life, according to Schweitzer, means "to be in the grasp of the infinite, inexplicable, forward-urging Will in which all Being is grounded" (Schweitzer, 1949: 214). His approach is to re-establish a bond between ethics, starting with his recognition that "I am life which wills to live, and I exist in the midst of life which wills to live" (Schweitzer, 1998: 156). For Schweitzer, the recognition of this "will-to-live" is akin to the recognition of "the reverence towards all living things by human beings, who experience and, like all other living things, also wish to actualize their own will-to-live" (quoted in Palmer, 1994: 78).

It is noteworthy that Schweitzer's reverence for life is not a value in itself. Rather, it is an attitude arising from the awareness of existence. More specifically, it is the reflection of the will-to-live that establishes each being's value of life or the ethic of reverence for life but not the distinction within the value of life (Bessinger, 2000; Barsam, 2008). According to Schweitzer, all natural things have a "will-to-live" and all wills-to-live are of equal value, and humans are not in the position to judge the relative values of different species (Palmer, 1994). Indeed, Schweitzer dismisses the idea of putting a hierarchy of beings in nature in relation to the nonhuman world in reflecting varying degrees of intrinsic value with humans categorized as the most intrinsically valuable beings (Barsam, 2008). As he pointed out clearly, "The ethics of reverence for life makes no distinction between higher and lower, more precious and less precious lives...How can we know the importance other living organisms have in themselves and in terms of the universe?" (Schweitzer, 1965: 47). Schweitzer further affirms that "all life is sacred" (Schweitzer, 1948: 271) and "something possessing value in itself" (Schweitzer, 1923a: 94). However, Schweitzer qualified this statement by allowing a distinction to be made under the pressure of necessity (Schweitzer, 1948: 271).

In extending his ethics of reverence for life to the natural world, he claims that an ethical man "tears no leaf from a tree, plucks no flower, and takes care to crush no insect" (Schweitzer, 1949: 243). In applying the ethics of reverence for life to the animal world, he proclaims that, "Whenever an animal is in any way forced into the service of man, every one of us must be concerned with the sufferings which for that reason it has to undergo. None of us must allow to take place any suffering for which he himself is not responsible ... while animals have to endure intolerable treatment from heartless men, or are left to the cruel play of children, we all share the guilt" (Schweitzer, 1949: 253).

He further vows that, "We must fight against the spirit of unconscious cruelty with which we treat the animals. Animals suffer as much as we do. True humanity does not allow us to impose such sufferings on them. It is our duty to make the whole world recognize it. Until we extend our circle of compassion to all living things, humanity will not find peace" (quoted in Kemmerer, 2016: 157). According to Schweitzer, "Ethics grow out of the same root as world- and life-affirmation, for ethics too, are nothing but reverence for life. That is what gives me the fundamental principle of morality, namely, that good consists in maintaining, promoting, and enhancing life, and that destroying, injuring, and limiting life are evil" (Schweitzer, 1987: 79).

It is thus evident that Schweitzer's ethics of reverence for life is a fully positive philosophical concept in that it affirms that all life possesses intrinsic value simply because it exists. According to him, a truly ethical person "obeys the compulsion to help all life which he is able to assist, and shrinks from injuring anything that lives" (Schweitzer, 1949: 243).

However, it must be admitted that the ethics of reverence for life is not completely nonanthropocentric in that it does allow us to consider our self-inducement or egoistic value when the need arises. As Schweitzer claims, "Whenever I injure life of any sort, I must be quite clear whether it is necessary. Beyond the unavoidable, I must never go, not even with what seems insignificant" (Schweitzer, 1949: 252). At the Dale Lectures at Mansfield College, Oxford, Schweitzer reaffirms his ethical position: "The time is coming, however, when people will be astonished that mankind needed so long a time to regard thoughtless injury to life as incompatible with ethics" (Schweitzer, 1923b). Accordingly, "The fundamental commandment of ethics, then, is that we cause no suffering to any living creature, not even the lowest, unless it is to effect some necessary protection for ourselves, and that we be ready to undertake, whenever we can, positive action for the benefit of other creatures" (Schweitzer, 1987: 260).

The above claim signifies that it is permissible to inflict harm or kill animals out of necessity and with care for the sake of mankind. The condition he imposes is that whenever human beings need to injure life of any sort, they must be quite clear whether it is necessary. If it is necessary, they must then "take extra care to mitigate as much as possible the pain inflicted" (Schweitzer, 1949: 252; Schweitzer, 1929).

However, despite the above contradiction, there is no doubt that Schweitzer's ethics of reverence for life is a noble one for it pragmatically draws and encourages the human race to discard human-centered sentiments and begin to meet the needs of all including those of the nonhuman living entities, and to return civilization to its ethical and moral grounds.

4.2.2.2 Paul Taylor's ethics of egalitarianism

Taking a similar ethical stance as Albert Schweitzer, Paul Taylor considers humans as members of the Earth's community of life, and as integral elements in a system of ecological interdependence (Taylor, 1983). He further considers all nonhuman natural entities as having equal worth as "teleological centers of life, striving to preserve itself and to realize its own good in its own unique way" (Taylor, 1981: 210). This is the view of species egalitarianism which considers that all species have equal moral considerability (Schmidtz, 1998). This means that every nonhuman living organism has a "good" or inherent worth of its own and deserves to be respected. However, according to David Schmidtz, a species egalitarian may or may not believe that living things all have equal standing. For example, an endangered species may be accorded a higher moral standing compared to a nonendangered entity (Schmidtz, 1998).

Technically speaking, they are intrinsically valuable and deserve humans' extension of moral considerability. This ethics of respect for nature is a life-centered system of environmental ethics in which human beings have "prima facie moral obligations that are owed to wild plants and animals themselves as members of the Earth's biotic community. We are morally bound (other things being equal to protect or promote their good for their own sake" [but] "from the perspective of the biocentric outlook..." (Taylor, 1981: 140, 209; see also, Taylor, 1986; Rolston, 2003).

As Taylor claims further, "we could no longer simply take the human point of view and consider the effects of our actions exclusively from the perspective of our own good" (Taylor, 1981: 198). The kinds of moral obligation extended to the natural world under Taylor's standards of character of "fairness and benevolence" may be expressed in terms of respect and care or concern and responsibility for the protection of the biocapacity or biological function of the nonhuman living entities (Taylor, 1981: 203; Mathews, 2010). This is certainly in stark contrast to the human-centered anthropocentric perspective of nature as discussed earlier.

Taylor provides four basic duties under his life-centered system of environmental ethics. They are: (1) duty of nonmaleficence, that is, duty to do no harm to any entity in the natural environment that has a good of its own; (2) duty of noninterference, that is, duty to refrain from placing restrictions on freedoms of individual organisms and the observation of a "hands-off" policy with regard to whole ecosystems and biotic communities as well as individual organisms; (3) duty of fidelity applies to individual wild animals which are capable of being deceived or betrayed by moral agents. It prescribes that humans are required to refrain from breaking the trust an individual animal places in us; and (4) duty of restitutive justice, that is, humans who harm other living organisms have to make some kind of compensation to those organisms in recognition of the inherent worth destroyed (Taylor, 1983).

The above ethical framework allows human beings to receive favorable treatment over nonhuman living things despite proclaiming that we are all equal. Although Taylor insists that all life forms have inherent value equally in cases of conflict of interest between human and nonhuman living things, or out of necessity and self-defence, it is morally permissible under principle (4) to kill or harm a nonhuman species to protect human vital interests. As William C. French pointed out succinctly this necessity for self defence "oddly suggests that ethical principles ought to be consigned solely to some ideal sphere of pure theory, while our concrete decisions about human action ought to be made outside the sphere of moral view and governed strictly by concerns of power and raw necessity" (French, 1999: 128; Anderson, 1993). This criticism raises concern over the authenticity of Taylor's egalitarian position with Schmidtz claiming that his ethical stance is arbitrary in nature (Schmidtz, 1998).

Despite the above criticism, Taylor has nonetheless contributed immensely in his attempt to overcome the anthropocentric tradition of ethics which has dominated global society since Stockholm. Practically, his ethics of respect for nature offers a remarkable conceptual criterion for distinguishing between moral judgments and principles and anthropocentric ethics.

4.2.2.3 *Peter Singer's ethics of animal liberation*

Peter Singer, in picking up the protagonist impulse of Schweitzer's animal ethics reflected in the ethics of reverence for life, has developed a theoretical framework of the ethics of animal liberation (Singer, 1974, 1975). Singer's animal ethics is fundamentally based on Jeremy Bentham's moral concern for the plight of animals and the moral principle of equal consideration.

To emphasize, in his argument for the welfare of animals, Bentham states his ground for human extension of moral consideration by posing this criterion: "the question is not, Can they reason? Nor, can they talk? But, can they suffer?" (Bentham, 1823: 311). In other words, for Bentham, the capacity for suffering is the essential criterion for equal consideration of moral interests. Bentham also incorporates the essential component of moral equality based on the Greatest-Happiness Principle: "everybody to count for one, nobody for more than one," so called the Bentham's dictum (Mill, 1985: 36; see also, Mill, 1907: 93), to reinforce his fundamental commitment to human equality. As against the aristocratic view that some people's lives were intrinsically more valuable than others, Bentham's dictum presupposes that one man is worth just as much as another man, and as John Stuart Mill, in making reference to this dictum, puts it, "All persons are deemed to have a right to equality of treatment" (Mill, 1907: 94; see also, Mill, 1985).

However, it is unclear whether Bentham's concept of equality was also extended to animals (Preece and Chamberlain, 1993). It has been contended that "each to count for one" utilitarian maxim is not about the equality of animals with humans as argued by Singer but about the inexcusable exploitation. It seems that Bentham, in proclaiming this maxim, has failed to go further to extend it to the nonhuman, animal kingdom. However, it is claimed that what Bentham meant by the maxim is that "since animals have the potential for suffering, their interest must be considered" (Preece and Chamberlain, 1993: 269). Notwithstanding this animal moral concern of Bentham's, it may be interpreted that the principle of equality is also extended to animals.

Regardless, following Bentham's moral reasoning, Singer claims that "our concern for others ought not to depend on what they are like or what abilities they possess" (Singer, 1974: 107). Endorsing this moral extensionism logic, Singer expounds that the criterion for the extension of moral considerability is animal sentience, that is, the capacity to feel pleasure and pain. As he stated, "If a being suffers, there can be no moral justification for refusing to take that suffering into consideration" (Singer, 1974: 107). Singer further reinforced his ethical stance by applying Bentham's "each to count for one" utilitarian maxim, stating that "No matter what the nature of the being, the principle of equality requires that its suffering be counted equally with the like suffering...of any other being" (Singer, 1974: 107−108).

In this connection, it is relevant to note that Jeremy Bentham incorporated the essential basis of moral equality into his utilitarian system of ethics based on the condition that the interests of every being affected by an action are to be taken into consideration and be given the equal weight as the like interest of any other beings (Singer, 1974: 106). In essence, Singer's principle of equal consideration gives "equal weight in our moral deliberations to the like interests of all those affected by our actions" (Singer, 2011: 21). However, in exception, "If a being is not capable of suffering, or of experiencing enjoyment or happiness, there is nothing to be taken into account" (Singer, 1974: 108).

To strengthen his animal ethical stance, Singer argues against what he calls speciesism, claiming that using species to discount or ignore the interests of animals was no different from racial discrimination against certain groups of humans on the grounds of race or sex (Singer, 1974). Singer powerfully claims that "mere difference of species in itself cannot determine moral status" (Singer, 2009: 567). The term speciesism was coined by Richard Dudley Ryder in 1970 (Ryder, 2010). It describes the indiscriminate exploitation of nonhuman animal species by humans which led to the widespread cruelty, abuse, physical pain, distress, and suffering of countless sentient beings (Ryder, 1975; 2010). Here, Ryder claims that "If we believe it is wrong to inflict suffering upon innocent human animals then it is only logical, phylogenetically-speaking, to extend our concern about elementary rights to the nonhuman animals as well" (Ryder, 2010: 2). Thus following Ryder's line of logic,

Singer signals that human beings and animals are equal, and we should accord the same respect to the lives of nonhuman animals as we do to the lives of humans (Singer, 1974).

Singer then juxtaposes his principle of equal consideration into Benthamite utilitarian calculus which may be specifically expressed in terms of producing the "greatest possible quantity of happiness" (Bentham, 1823: 311). Here, Bentham defines ethics as "the art of directing men's actions to the production of the greatest possible quantity of happiness" (Bentham, 1823: 311). According to Singer, the ultimate moral goal is to achieve maximum utility or greatest interest-satisfaction possible. It thus follows that it is legitimate to override the rights to moral status or interests of certain individuals for the sake of bringing the greatest satisfaction or happiness. Singer's central argument on the interruption of moral extensionism discussed so far rests on the Benthamite utilitarian maxim of the "the greatest good for the greatest number" (Bentham, 1907). The maxim is solely concerned with achieving the maximum good even if this may cause injustice to others (Smart, 1973; Babor, 2007). Thus in Singer's view, while we must extend equal consideration to sentient beings, we must also consider that such obligations are founded on the aim of bringing about "the greatest good for the greatest number" (Cochrane, 2017).

However, Singer argues that whether humans are required to refrain from killing a given animal painlessly will depend on whether the animal has the desire to continue to exist into the future. For Singer, since human beings have the interest to continue to exist into the future, they have the right to life while mere animals will lack this right (Singer, 2011). He argues that generally human beings are morally superior compared to nonhuman animals, especially "mere animals", that is, animals with no future-directed interests to live in the way that "normal" humans with moral significant interests have in continuing to live. This is made clear in an interview conducted by George Yancy for "The Stone" (New York Times) in which he reiterates that "normal humans have an interest in continuing to live that is different from the interests that nonhuman animals have" (Yancy and Singer, 2015). The *raison d'être* for his point of view is that "normal" human beings have the ability to be self-aware over time and plan for their future but mere animals do not (Francione and Steiner, 2016). Singer argues that even though some animals may be self-aware in a certain sense, "they are still not self-aware to anything like the extent that humans normally are" (Singer, 2011: 122).

Based on the above reasoning, Singer claims that we might be able to justify killing these mere beings or nonself-aware beings with the replaceability argument. This agreement claims that it is acceptable to kill the mere animals or nonself-aware beings if equally happy animals are created to take their place. In other words, nonself-conscious beings are replaceable and that killing such an individual being can be justified if such an act is necessary to bring about the existence of another (Wilson, 2017). Thus Singer's position is that the lives of nonhuman animals are qualitatively inferior and less morally valuable compared to human beings. For animals, raising them humanly and give them a good life can offset their quick, or early death.

It is also assumed that if we kill one animal, we can replace it with another as long as that other will lead a life as pleasant as the one killed would have led, if it had been allowed to go on living. Thus it seems to follow that breeding happy animals that will be prematurely killed can be a good thing overall as it benefits the consumers in terms of greater consumption and contribution to promoting greater welfare to the world in terms of greater happiness, pleasure, or preference satisfaction (Delon, 2016). Conceivably, Singer's stance on animal liberation is that if humans give nonhuman animals a reasonably pleasant life and a relatively painless death, then exploiting them may be morally acceptable.

His claim on replaceability is made clear when he states that, "the untimely death of a human being is a tragedy because there are likely to be things that she hoped to accomplish but now will not be able to achieve. The premature death of a cow is not a tragedy in this sense, because whether cows live 1 year or 10, there is nothing that they hope to achieve. Even those great apes who can use sign language do not talk to us about their plans for the distant future. Scrub jays hide food for the next day, but as far as we know, they do not embark on long-term projects that will pay off in the years ahead" (Singer, 2011: 103–104).

This is obviously Bentham's utilitarian stance of "the greatest happiness for the greatest numbers." It may be emphasized that, in the utilitarian world, the exploitation and treatment of nonhuman animals as mere receptacles, that is, mere means for an end such as the production of greater goods or welfare is acceptable if the happiness their exploitation causes is greater than the harm it causes. Its focus is the total sum of positive wellbeing and less suffering.

In a nutshell, Singer's animal ethics of moral considerability to animals is based on equal consideration of interests of sentient beings with the criterion of the ability to feel pain or experience suffering more than on their intelligence, and at the same time by taking into consideration Bentham's utilization maxim. Indeed, Singer views that animal suffering is acceptable if there is no other way to satisfy human needs. Thus Singer's animal moral considerability is not absolute but conditional.

4.2.2.4 Tom Regan's animal rights

The discussion on animal ethics cannot be completed without bringing to the fore the ethical stance of animal rights as eloquently examined by Tom Regan who has developed with great precision the deontological animal rights position. However, before assessing Regan's animal ethics, it is relevant to trace its conceptual ethical root which is linked to Immanuel Kant's moral philosophy (1724–1804). To begin with, in his moral assessment of actions, Immanuel Kant contends that it is necessary to consider the notion of autonomy as "the basis of the dignity of human and of every rational nature" (Kant, 1909: 54). The term "autonomy" has a long and complex history which it is not possible to delve into detail here.

Simply put, a being, or a rational being for that matter, who has autonomy makes decisions independently based on the maxim of all its will (the faculty of practical reason), and free from external interference or impulses. To conceive oneself as autonomous is "to think of oneself as determining one's own actions, in the sense that one's motives are not dictated by interests, sanctions, or incentives alien to one's core self" (Martin, 2006: 116). This autonomous quality of the will in itself has intrinsic worth. It gives a being dignity which is the basis of moral obligations and responsibilities. Thus for Kant, a rational and autonomous being has "an absolute worth" and is "an end in itself" (Mackellar and Jones, 2012).

For Kant, nonhuman animals are not self-conscious and they lack will, and therefore are not autonomous. This means that they have no intrinsic value. They merely serve as the means to an end, that is, they are only instrumentally valuable. Kant reiterates that "The end is man" (Kant, 1963: 239). Animals do not have direct status as human beings do because unlike humans, animals are not connected to rational nature (Sebo, 2004). Furthermore, Kant claims that human beings have no direct duties toward animals because they are the end in themselves. Despite claiming that nonhuman animals are mere beings with no intrinsic value or moral status, Kant does not genuinely believe that human beings could dispose of them any way they wanted. He is still open to some sympathy for our nonhuman animals by stating that "Our duties towards animals are merely indirect duties towards humanity. Animal nature has analogies to human nature, by doing our duties to animals in respect of manifestations which correspond to manifestations of human nature, we indirectly do our duty towards humanity" (Kant, 1963: 239).

Kant's indirect moral stance toward the animal kingdom is made explicitly clear by stressing that, "If a man shoots his dog because the animal is no longer capable of service, he does not fail in his duty to the dog, for the dog cannot judge, but his act is inhuman and damages in himself that humanity which it is his duty to show towards mankind. If he is not to stifle his human feelings, he must practice kindness towards animals, for he who is cruel to animals becomes hard also in his dealings with men. We can judge the heart of a man by his treatment of animals" (Kant, 1963: 240). Thus despite his indirect duty argument, Kant, in the above statement, explicitly reflects his real moral obligations toward his nonhuman fellows.

By a twist of logic, Regan reconstructed Kantian "end in itself" moral theory into a new concept of moral personhood that he calls "subject-of-a-life." To start with, recognizing that the fundamental wrong in our system is the common acknowledgment of the view that animals are "our resources, here for us—to be eaten, or surgically manipulated, or exploited for sport or money" (Regan, 2017: 106). Regan argues for the extension of moral rights enjoyed by human beings to nonhuman animals (Regan, 2004). Conceptually, he extends Kant's principle of dignity beyond human community. He considers that the view of indirect duty as in the Kantian case "fails to command our rational assent. Whatever ethical theory we should accept rationally, therefore, it must at least recognize that we have some duties directly to animals, just as we have duties directly to each other" (Regan, 2017: 109). Nonetheless, in line with Kant's moral stance on inhuman treatment of animals as noted above, Regan used the term "cruelty-kindness" to unveil his moral principle to animals. He emphasizes that "...we have a direct duty to be kind to animals and a direct duty not to be cruel to them... Cruelty in all its guise is a bad thing, a tragic human failing" (Regan, 2017: 109). Pain, according to Regan is intrinsically evil (Regan, 1975).

Regan concedes that to attempt to limit the extension of moral considerability to humans is defective. His reasoning is that: "Animals, it is true, lack many of the abilities human possess. They can't read, do higher mathematics, build a bookcase or make baba ghanoush. Neither can many human beings, however, and yet we don't (and shouldn't) say that they (these humans) therefore have less inherent value, less of a right to be treated with respect, than do others." He further argues that: "We want and prefer things, believe and feel things, recall and expect things. And all these dimensions of our life, including our pleasure and pain, our enjoyment and suffering, our satisfaction and frustration, our continued existence or our untimely death—all make a difference to the quality of our life as lived, as experienced, by us as individuals. As the same is true of those animals that concern us... they too must be viewed as the experiencing subjects of a life, with inherent value of their own" (Regan, 2017: 111–112; see also, Regan, 2002: 144). To note in passing, Regan contends that inherent value "belongs equally to those who are experiencing subjects of life" (Regan, 2017: 112).

However, it should be noted that Regan confers inherent values only to higher animals, that is, normal mammals above a year of age while lower animals and plants are excluded (Regan, 2004; Warren, 1987; MacDonald, 2004). Furthermore, it may be remarked that dividing between higher and lower nonhuman organisms, and higher and lower or without inherent value and moral consideration is arbitrary and problematic. As Mary Anne Warren points out clearly it would be surely arbitrary and implausible to draw a sharp line between some birds such as crows, magpies and parrots, and the higher mammals as the former appear to be just as mentally sophisticated as most mammals (Warren, 1987: 347).

For Regan, nonhuman animals are not only alive in the biological sense but also have mental capacities. In other words, they are the "subjects-of-life" and are involved in more than merely being alive and more than conscious (Regan, 2004: 243). For example, like human beings, the nonhuman animals also have "beliefs and desires; perception, memory, and a sense of the future, including their own future; an emotional life together with feelings of pleasure and pain; preference- and welfare-interests; the ability to initiate action in pursuit of their desires and goals; a psychological identity over time; and an individual welfare in the sense that their experiential life fares well or ill for them, logically independently of their utility for others, and logically independently of their being the object of anyone else's interests" (Regan, 2004: 243).

Also, Regan claims that some nonhuman animals resemble human beings in morally relevant ways in that "they possess a variety of sensory, cognitive, conative, and volitional capacities." Like us, they also "see and hear, believe and desire, remember and anticipate" and share the "physical pleasure and pain...fear and contentment, anger and loneliness, frustration and satisfaction, cunning and imprudence" with us (Regan, 2004: xvi). These psychological states and dispositions which Regan calls the property of nonhuman "subjects-of-a-life," exist in the animal kingdom, and are also the common property of all human beings (human subjects-of-a-life).

The above mental capacities for which we think we deserve rights are also prevalent in many animals especially the normal mature mammals such chimpanzees and dolphins. Thus if human subjects-of-a-life deserve moral rights, it is only fair to accord the nonhuman subjects-of-a-life with a similar moral standing (Wilson, 2017). As Regan claims, both the human and nonhuman subjects-of-a-life have "a basic moral right to respectful treatment" (Regan, 2004: xvii).

Regan's fundamental message in his argument for this respectful treatment is reflected in his statement that: "There are some who resist the idea that animals have inherent value. 'Only humans have such value,' they profess. How might this narrow view be defended? Shall we say that only humans have the requisite intelligence, or autonomy, or reason? But there are many, many humans who fail to meet these standards and yet are reasonably viewed as having value above and beyond their usefulness to others. Shall we claim that only humans belong to the right species, the species *Homo sapiens*? But this is blatant speciesism" (Regan, 2017: 112). So according to Regan, we cannot deny animals inherent value merely because of their lack of reason, autonomy, or intellect (Regan, 2017: 112).

Regan's criterion for ethical extensionism appears to be quite similar to Taylor's "ability to feel pain" criterion in that both hinge on consciousness and mental capacities. However, unlike Kant who limits inherent value to humans, and unlike Singer's animal ethics with its utilitarian focus on animals' pleasures and pains, Regan argues his case for animal ethics by extending inherent worth to nonhuman animals which he claims as "a moral theory that more adequately illuminates and explains the foundations of our duties to one another" (Regan, 2003: 77).

However, like Singer's conditional animal ethical position, Regan's animal rights are by no means absolute. In circumstances where the rights of different individuals conflict, then, some individuals' rights may have to be violated. Regan applies the following two principles to deal with the situation:

 (i) the miniride principle which states that: "Special considerations aside, when we must choose between overriding the rights of many who are innocent or the rights of few who are innocent, and when each affected individual will be harmed in a prima facie comparable way, then we ought to choose to override the rights of the few in preference to overriding the rights of many" (Regan, 2004: 305).

(ii) the worse-off principle which states that: "Special considerations aside, when we must decide to override the rights of many or the rights of the few who are innocent, and when the harm faced by the few would make them worse-off than any of the many would be if any other option were chosen, then we ought to override the rights of the many" (Regan, 2004: 208). Here, it may be remarked that the Miniride Principle which is utilitarian in nature is difficult to reconcile with Regan's stance of equality in the extension of inherent value to all subjects-of-life. Sacrificing the interests of a few to maximize positive outcome is akin to fulfilling Bentham's Greatest-Happiness Principle.

Given the above considerations, it is permissible to override the rights of most nonhuman animals in circumstances where our own rights or the rights of other moral agents are at stake. Thus while recognizing that nonhuman animals as the subjects-of-life have inherent value and deserve to be treated with respect, this right may be overridden in special circumstances where they are required to serve as a means or to be harmed to provide humans with food or for the purpose of scientific research to save human lives.

4.2.2.5 Biocentrism: Some comments

All forms of biocentric ethics as discussed above provide us with an ethical theory in our treatment of animals. However, they are by no means without their shortcomings or contradictions. Although all the above principles of animal rights and welfare are morally motivated and ethically based, they have yet to completely move away from anthropocentric sentiments. Humans still dominate the biological system. Indeed, nonhuman animals do not qualify for Schweitzer's claim that "all life is sacred" nor do they deserve our respect as he confesses that killing animals is permissible out of necessity. Apart from avowing that "I must defend against the existence which injures it," he (Schweitzer) has failed to provide a clear-cut definition of "necessity." Thus human beings still take precedence over nonhuman animals. On a practical note, while Schweitzer recognizes the need to exploit nonhuman living things out of necessity, he has stopped short of establishing a compromise between his reverence for life animal ethics and human practical requirement. Thus the term "necessity" is open to subjective interpretations depending on the kinds of needs or wants humans pursue.

Nonhuman animals also do not share equal rights with humans, as argued by Regan. By using the miniride and worse-off principles, Regan allows humans to take control of the animal kingdom. Based on these two Reganian principles, nonhuman animals clearly do not enjoy the same degree of moral considerability as enjoyed by humans. Also, Singer's utilitarian stance on animal ethics starts from the moral premise that the wellbeing of sentient animals matters, but ends with the utilitarian conclusion that all that matters is welfare maximization à la Bentham. Thus based on his replaceability principle, it is permissible to kill animals if such an act would bring the greatest good for the greatest number. In addition, if we give animals a pleasant life, no wrong is committed if the animal is killed, and as a result of killing, it will be replaced by another animal living an equally pleasant life. On the other hand, Taylor uses the duty of restitutive justice to justify the change of moral stance toward animals despite his absolute objection to the anthropocentric view of human dominance, thus raising questions on the integrity of his egalitarian principles.

In addition, all these biocentric theories are too narrowly focused in that they emphasize the values and rights of individual nonhuman living things. They are a kind of individualism and their individualistic characteristic is unlikely to contribute to promoting a global environmentally sustainable development. Their moral concerns only extend to individual beings rather than to the natural environment itself. Certainly, the biological integrity and hence the welfare of nonhuman living things also hinge on protecting their habitats. The nonhuman living things are only a part of the natural environment which comprises the sum total of surrounding things and conditions including nonsentient environmental entities such as land, water, air, sunlight, and humans as well as nonhuman living species, among others. This may be distinguished from an ecosystem which refers to an ecological unit in a specific area consisting of biotic components, that is, living components including humans and biodiversity and abiotic components, that is, nonliving components such as the physical environment and conditions.

According to the animal ethicists, the above nonliving holistic entities have no moral status because, in Regan's words, they are not subjects-of-life. Take the case of Taylor, for example, who claims that our duties with respect to the ecological integrity of the natural environment stem from the motive that such acts help to achieve and sustain a healthy existence of species for their own sake. He further claims that the "balance of nature" is not itself a moral norm (Taylor, 1981: 198). Regan also argues that maintaining the ecological integrity of the natural environment is akin to a deliberate subordination of the rights or wellbeing of all individual subjects-of-life. He confesses that the "rights view cannot abide" ecological holism because it "categorically denies that inanimate objects can have rights" (Regan, 2004: 362). He further emphasizes that "Environmental fascism and the rights view are like oil and water: they don't mix" (Regan, 2004: 362).

However, as the sentient beings in the biocentric world are taken to mean the nonhuman organisms including the higher animals referred to by Regan, it does not exclude, for example, invasive species which may threaten the life of another nonhuman living organism. An ecological system in the real world is governed by producer—consumer or predator—prey relationships, and the life of one nonhuman living organism may premise on the death of another (Callicott, 2001). Hence, as John Baird Callicot pointed out succinctly one could hardly justify the argument that our killing of fellow subjects-of-being of the biotic community is, prima facie, ethically wrong. "It depends on who is killed, for what reasons, and under what circumstances, and how" (Callicott, 2001: 210). Here, it is beyond dispute that although the nonliving natural objects do not have independent moral standing in the biocentric world, its protection is equally important in upholding the healthy functioning of the whole system. This in turn helps to protect the rights, interests, or welfare of the subjects-of-life.

The biocentricism advocates may not dismiss the need for environmental protection totally. Regan, for example, acknowledges the importance of habitat protection as sound ecosystems provide a healthy thriving condition for animals despite their having no moral standing (Remele, 2013). This may incidentally contribute to promoting environmental sustainability. However, the position of environmental protection from this indirect perspective is a weak or unsustainable one as no robust or sustainable protection may be offered to protect the nonmorally considered environment. In other words, environmental protection hinges mostly, if not entirely, on upholding the rights or welfare of nonhuman living organisms.

In addition, the equality argument which confers equal rights or moral consideration to all nonhuman living beings is also problematic. By considering all nonhuman living things equal, no distinction is made between the endangered or critically endangered species as listed under the IUCN Red List such as the Sumatran tigers in Indonesia which are facing a serious risk of outright anthropogenic extinction (discussed in the preceding chapter) compared to, for example, rats or cockroaches which may cause nuisance to the society, or an invasive species which is ecologically destructive. Thus biocentrism, while providing useful moral principles and ethical concepts in animal protection, does not do justice to our planet if these concepts and principles are not also pragmatically applied to embrace environmental holism to solve real world environmental problems. Nevertheless, ethical motivation and moral principles, especially those propounded by Schweitzer, serve as an important point of departure in mapping forwards a more holistic framework of biological and environmental protection, and hence a Holling sustainable Earth system. The following section will examine this.

4.3 Ecocentrism and Aldo Leopold's land ethic

Notwithstanding the limitations and conceptual contradictions of biocentrism as discussed above, it offers a very useful normative framework of ethical codes to inspire not only animal protection, but also environmental protection. No theory is perfect and there is bound to be disagreement among different schools of thought. However, engaging in endless arguments and counter arguments over conceptual inconsistencies does not contribute to solving the looming environmental crisis confronting us today. To be sure, conceptual consistency is important for it provides convincing argument over certain environmental issues at hand and sheds light on the practical mitigating solutions to contain these issues. However, what is more important is the conceptual potential which can be applied to solve real world problems. With this view in mind, the rest of this section seeks to elevate biocentrism to a higher level of conceptual underpinnings by transforming its individualistic outlook to a holistic approach based on ecocentrism.

To start with, the various biocentric approaches to animal protection as discussed above raises a wide variety of value and moral questions that establish the ethical norms on how we ought to treat nonhuman living things. More particularly, the philosophical foundations and moral principles of these philosophers, in particular Schweitzer, provide far reaching moral guidance and practical ethical implications in our engagement with the natural environment at various levels—individual, collective, or societal. Nonetheless, their life-centered ethics may be extended to apply to not only living organisms but also, nonliving or abiotic entities such as rivers, rocks, atmosphere, and other abiotic entities which are equally important in enhancing the ecological heath of the natural environment, and hence the welfare of the individual nonhuman living organisms. This brings us to our search for an Earth-respecting belief system which subscribes to the ethical principle of ecocentrism.

As contrast to biocentrism, ecocentrism or Earth-centered ethics take a much wider view of the world. It holds that a spectrum of holistic natural entities such as ecosystems and species and the abiotic entities as mentioned above are the central objects for ethical concern. Put differently, ecocentrism is a worldview that recognizes intrinsic value in ecosystems and the biological and physical or abiotic entities that they comprise (Gray et al., 2018). Ecocentrism offers a robust ethical analysis of the negative impact that humans exert on the natural environment including its biodiversity.

In articulating the ecocentric moral norms that govern our interaction with nature, we may take the vantage point of Aldo Leopold, an internationally respected scientist and influential environmental ethicist who is universally hailed as the "father or founding genius of recent environmental ethics" (Callicott, 2001: 311). Leopold's concepts of environmental ethics may be found in his best-known volume of environmental ethics, *A Sand County Almanac*. The sequence of Leopold's ethics may be traced back to his perception of nature. To begin with, in recognizing that "all ethics so far evolved rests upon a single premise: that the individual is a member of a community of interdependent parts" (Leopold, 1949: 204), that is, the bulk of the ethics that has evolved is concerned with the loci of value of humans and the management of human interrelationships or the welfare of community (Callicott, 2001).

More specifically moving beyond the anthropocentric as well as the biocentric view, Leopold holistically stretches the extension of moral considerability not only to animals, but also to soils, waters and plants, collectively called "the land" (Leopold, 1949: 204). Leopold's definition of land is an all-embracing term which comprises the sum total of biotic, abiotic, cultural, and philosophical elements that may collectively be called the "environment." Biotic elements include plants and animals while abiotic elements cover the physical features and surrounding natural conditions. The cultural components refer to economic, social, and political factors in association with natural resource use or environmental exploitation while philosophical factors are concerned about our ethical interaction with nature.

Leopold's holistic concern for the environment stems from his foresight of the environmentally destructive impacts of anthropocentrism which gives rise to human abusive instrumental use of the natural environment. As he pointed out succinctly, "we abuse land because we regard it as a commodity belonging to us" (Leopold, 1949: xviii). However, as Leopold suggests, "When we see land as a community to which we belong, we may begin to use it with love and respect" (Leopold, 1949: vviii). This holistic view is certainly in stark contrast to the biocentric views discussed earlier, in particular, Regan's uncompromising biocentric stance. For Leopold, the ethical principle for the extension of moral considerability of land should be based on human prudent use of nature. Such prudent use may be expressed in terms of an "extension of the social conscience from people to land" (Leopold, 1949: 223). Social conscience is associated with the expression of "love, respect, admiration and a high regard for its value" (Leopold, 1949: 223). By value, Leopold means intrinsic value, that is, value in the "philosophical sense" (Leopold, 1949: 223). The extension of moral consideration to the natural system in this manner serves to connect humans with the natural system ethically. This gives rise to what Leopold calls a land ethic which he defines as "the existence of an ecological conscience, and this in turn reflects a conviction of individual responsibility for the health of land" (Leopold, 1949: 258). In short, "a land ethic changes the role of Homo sapiens from conqueror of the land-community to plain members and citizens of it. It implies respect for his fellow-members, and also respect for the community as such" (Leopold, 1949: 240).

It is clear from the above that human beings constitute merely a part of the larger community, that is, the natural environment. Additionally, acknowledging that humans coexist with nature in that the wellbeing of humans hinges on the wellbeing of the natural system and vice versa, Leopold urges us to "quit thinking about decent land-use as solely an economic problem" (Leopold, 1949: 262). On the issue of human economic exploitation of nature, Leopold professes that it is necessary to examine whether such instrumental use of nature is "ethically and esthetically right" and "economically expedient" (Leopold, 1949: 262). For Leopold, "A thing is right when it tends to preserve the integrity, stability, and beauty of the biotic community. It is wrong when it tends otherwise" (Leopold, 1949: 262).

In a related note, John Baird Callicott, who explores extensively the philosophical foundations of Leopold's land ethic claims that "in the last analysis, the 'integrity, beauty, and stability of the biotic community' is the measure of right and wrong actions affecting the environment" (Callicott, 1989: 58). Following Leopold, Callicott further concedes that environmental ethics "locate ultimate value in the biotic community and assigns differential moral values to the constitutive individuals relative to that standard" (Callicott, 1989: 37). In other words, Leopold ethics as emphasized by Callicott is holistic or collective and based on value pluralism. This form of environmental ethic is called "ethical holism" (Klonoski, 1991). It differs from biocentrism in that in ethical holism, right and wrong are a function of the "integrity, beauty, and stability of the biotic community" rather than of its individual constitutive members. It also almost completely rejects the anthropocentric view of nature (see below for further clarification). Furthermore, Leopold claims that no significant change in ethics has ever been achieved without internally altering our "intellectual emphasis, loyalties, affections, and conviction." (Leopold, 1949: 210).

In other words, in our engagement with nature, Leopold's right principle requires us to undertake "obligations over and above self-interest" to preserve its integrity and continuous existence (Leopold, 1949: 209). Obligation, in Leopold's tradition, refers to "a limitation on freedom of action in the struggle for existence" (Leopold, 1949: 202). This refers to prudent use of nature as discussed above. It further means that we should accord a moral or ethical standing to the natural environment, one of the most important philosophical attitudes toward sustainable environmentalism.

However, it must be noted that Leopold's notion of land ethic or "ethical holism" in respecting the "fellow members" of the land community does not necessarily call on us to completely withhold any form of ecological disturbances of nature. Rather, it requires us to shoulder certain moral obligations to protect the environment when harnessing its economic use (Choy, 2014). This is well reflected in his assertion that, "Conservation means harmony between men and land" (Leopold, 1999: 207). That is to say, it is a "positive exercise of skill and insight" to keep "the resource in working order, as well as preventing overuse" (Leopold, 1999: 164). Thus in extracting the economic value from land resources, it is necessary to observe the ecological health of the natural environment. In addition, Leopold's land ethic is eloquently crafted to accommodate conservation concerns and preservation motive in the same direction as Schweitzer, although the latter stops short of extending his moral boundaries to cover abiotic entities.

To avoid confusion, it may be noted that as distinct from environmental conservation which aims to protect the environment for present and future human consumption, preservation is concerned about protecting the natural system against humans, and for its own sake. It may further be pointed out that Leopold's land integrity ideas found a correspondence with the ecological sustainability concepts eloquently examined by Crawford Stanley Holling as discussed in Chapter 2, The United Nations' Journey to Global Environmental Sustainability Since Stockholm: An Assessment (Choy, 2014). Holling sustainability view is that to maintain and protect the biological systems' adaptive capacities, it is necessary to preserve the complexities and diversities of their genetic and biological attributes and landscape structures through prudent use of natural resources. Such a caring-for-nature attitude helps to ensure that the ecosystems are ecologically healthy enough to withstand external disturbances and thus avoid being altered in fundamental ways (Choy, 2014).

4.4 Albert Schweitzer's reverence for life ethic: The Leopold connection and its implication for a pragmatic unification of environmental ethics

Environmental ethics as discussed so far has made significant progress in terms of theoretical sophistication, positions, and scope in assessing the values and attitudes that govern our moral relationship with the nonhuman natural environment. It provides a wide range of positions and principles ranging from the narrow predisposition of anthropocentrism which is concerned about environmental exploitation for human benefits to the non-anthropocentric ethics (biocentrism and ecocentrism) which emphasizes environmental preservation to guide us in our treatment of nature. Despite the robustness of the literature, the discipline seems to have been entrapped in a welter of arguments and counter-arguments over their theoretical assertions and philosophical scope of moral consideration as reflected in the above discussion.

More specifically, each discipline has individually isolated separate environmental treatment of the natural entities in the light of its particular objective. For example, anthropocentrism is familiar only with its narrow concept of utility maximization and biocentrism is concerned about animal rights and welfare such as injury, pain, or death caused to nonhuman living organisms or subjects-of-life, while ecocentrism emphasizes the protection of the ecosystem. Also, while the scope and points of views of biocentrism and ecocentrism at times are at odds with each other, both disciplines are particularly hostile to the human-centered grounds of anthropocentrism. Also, both schools of anthropocentrism and biocentrism are indifferent to the holistic concerns of the biotic community.

It is this long-running disagreement that has impeded the ongoing development of a coherent and practical philosophical framework that is critically important to enable us to address the current widespread morally and environmentally destructive practices in our beleaguered and already extremely stressed world. For this reason, if environmental ethics are to have a real impact on environmental issues, each discipline must move away from its endless theoretical argument and steadfast adherence to its ethical consistency as these are inconsequential to solving real world problems. The way forward does not lie in an attempt to rebut the theoretical disputes raised at each point of ethical enquiry but rather in embarking on a collective effort to develop an ethical theory which is more practical and generally acceptable in a real world system dominated by humans.

In search for a pragmatic conceptual order in this direction, each discipline must incline to transcend its theoretical and methodological confinement to reflect the complex interaction of a constellation of the social, economic, environmental, and ethical disciplines embedded in a real world system. This is grown out of the belief that, "...we have to solve practical, not theoretical problems; and we must adjust the ends we pursue to the means available to accomplish them...", to put in Mark Sagoff's words (Sagoff, 1988: 14). This demands that we have to develop a new ethical foundation that belongs neither solely to anthropocentrism nor to biocentrism, nor to ecocentrism.

With this in mind, we now attempt to pragmatically develop a morally justifiable and more generally acceptable approach to environmental ethics resting on various ethical principles from a variety of positions. To begin with, in harnessing a more practical approach to environmental ethics, we have to consider not only the ecological integrity of the natural environment or nonhuman living organisms, but also the long-term socioeconomic prosperity of humanity. However, to reject in total the anthropocentric view of nature in the contemporary industrial humanistic system is impractical. To lay the philosophical foundation from this perspective, we may return to Albert Schweitzer as our starting point.

Of all the biocentric advocates discussed earlier, Schweitzer's moral philosophy, especially his ethics of the reverence for life, is the most sophisticated and far reaching moral acknowledgment of the ethical position of humans in the natural world despite the fact that he argues for all life on earth. As Rachel Carson concedes, "To me, Dr. Schweitzer is the one truly great individual our modern times have produced. If we are to find our way through the problems that beset us, it will surely be in large part through a wider understanding and application of his principles" (Carson, 1963: 223).

Indeed, the ethics of reverence for life may be considered the cornerstone of Schweitzer's warning about our unsustainable relationship with nature which houses all life on earth. As he warns us, "The great fault of all ethics hitherto has been that they believed themselves to have to deal only with the relations of man to man. In reality, however, the question is what is his attitude to the world and all life that comes within his reach...The ethic of reverence for life, therefore, comprehends within itself everything that can be described as love, devotion, and sympathy whether in suffering, joy, or effort" (Schweitzer 1998: 188). He further claims that "The ethic of Reverence for Life is the ethic of love widened into universality" (Schweitzer, 1998: 235). Leopold takes more or less a similar stance as Schweitzer's in claiming: "That Land is a community is the basic concept of ecology, but that land is to be loved and respected is an extension of ethics" (Leopold, 1949: xix).

What is clear from the above is that Schweitzer's ethical stance indeed extends far beyond the biocentric tradition to universally embrace the world at large, that is, our planet. In other words, Schweitzer has consciously or unconsciously extended his moral boundaries to cover the biotic community as Leopold has, and his ecological conscience may be expressed in terms of "everything that can be described as love, devotion, and sympathy" in relation to all forms of life on earth including the environment they thrive in. For Schweitzer, the ethics of reverence for life considers that "All destruction of and injury to life, under whatever circumstances they take place, it condemns as evil" (Schweitzer, 1949: 251). Here, Schweitzer is in fact indirectly expressing his moral philosophy concerning the ethical relationship of humans to the environment and its nonhuman organisms and seeks grounds to define our obligations toward environmental protection.

Surely, the destruction of or injury to life such as the life of animals can happen, for example, because of human instrumental use of nature or anthropogenic destruction of habitat. To Schweitzer, such an act is evil and to Leopold, it is an "anti-social conduct" and there is a need to bring back a more ethically social form of conduct to halt further habitat destruction to protect the animals, which also constitute one part of the land community. Here, it may be of interest to note that philosophically, Leopold defines an ethics as "a differentiation of social from anti-social conduct" which is precisely in line with Schweitzer's moral principles (Leopold, 1949: 202). Thus Schweitzer is not only concerned about the welfare of the nonhuman living entities but also the environments in which they thrive. This is akin to Leopold's land ethic in enlarging the moral boundaries of the community to include not only nonhuman living things but also the abiotic entities which make up the "land." Schweitzer's claim also inspires and motivates individuals to do the right thing by extending "love, devotion, and sympathy" to all forms of life including its environment, or to use land "with love and respect" and with "limitation on freedom," as Leopold advocates.

However, both Schweitzer and Leopold also admit the anthropocentric view of nature out of necessity while extending moral considerability to the environment. He further concedes that whenever humans need to harm animals out of necessity, we must "look out for opportunities of bringing some sort of help to animals, to make up for the great misery which men inflict on them, and thus to step for a moment out of the incomprehensible horror of existence" (Schweitzer, 1949: 253−254).

4.4.1 The anthropo−bioecocentric ethics

Similarly, Leopold's land ethic does not necessarily call upon us to completely withhold any form of ecological disturbance of nature as humans may have to exploit and use the land for socioeconomic sustenance or other practical reasons as noted above. As Leopold claims, "A land ethic of course cannot prevent the alteration, management, and use of these 'resources,' but it does affirm their rights to continued existence, and at least in spots, their continued existence in a natural state" (Leopold, 1949: 204). Both Schweitzer and Leopold provide an ethical foundation for reconciling the anthropocentric and utilitarian view of nature that is incompatible with the nonanthropocentric theory of intrinsic value by advocating the precautionary principles in our economic dealing with nonhuman living entities.

Apart from Schweitzer's ethics of reverence for life for animal rights and welfare, his deep ethical insight for the environment is also well expressed in his concern for the damage that human careless destruction of nature would cause (Braybrooke, 2015). In a letter he had written to a beekeeper whose bees had been destroyed by aerial spraying of pesticides, he states that "I am aware of some of the tragic repercussions of the chemical fight against insects taking place in France and elsewhere and I deplore them. Modern man no longer knows how to foresee and to forestall. He will end by destroying the earth from which he and other living creatures draw their food. Poor bees, poor birds, poor men" (Free, 1992: 6; Carson, 1962: Preface; Harris, 2016: 1132). It is clear from here that Schweitzer was not only concerned about the ecological integrity of the bee population but also the natural environment in which they thrived.

To foresee and forestall is the basis of the precautionary principle (Raffensperger and Tickner, 1999: 1). It is the central theme for environmental protection and biodiversity conservation rooted in the elemental concepts of "to predict, prevent or reduce" the causes of environmental impact or biodiversity loss (Kravchenko et al., 2013). The precautionary principles are contained in various international agreements and conventions including the 1992 United Nations Framework Convention on Climate Change, Principle 15 of the 1992 Rio Declaration, Stockholm Convention on Implementing International Action on Certain Persistent Organic Pollutants and Agenda 21, among others (Kravchenko et al., 2013). To foresee and to forestall, it is necessary to endorse and put in practice Schweitzer's claim that "It is good to maintain and encourage life; it is bad to destroy life or to obstruct it" (Schweitzer, 1949: 242). Surely, maintaining and encouraging life, or the life of nonhuman living beings for that matter, calls for the need to promote an ecologically resilient environment, based, again, on Leopold's "limitation on freedom" principle. The reason behind such ethical logic is that the biological health of animal population and ecological integrity of natural habitats are inextricably connected. Indeed, as discussed in the preceding chapter, habitat fragmentation and destruction is one of the most important factors contributing to biological impoverishment.

It is evidently clear from the above that Schweitzer's conception of "biocentrism" goes far beyond the extraordinarily narrow view of Regan's animal liberation or Singer's utilitarian animal ethics. Like Leopold, Schweitzer also tailored his "biocentric" outlook holistically to accommodate preservation concerns not permissible in Regan or Singer. It is also far more practical than the more inclusive biocentric position of Taylor who persistently insists that humans have no obligation "to promote or protect the good of nonhuman living things" (Taylor, 1981: 198; Callicott, 2001; Caldwell and Shrader-Frechette, 1993). Both Leopold and Schweitzer also share a revulsion against environmentally unsustainable human practices which are detrimental to the biotic community. From a practical point of view, they both acknowledge that withholding ethical sentiments from the biotic community is permissible "under pressure of necessity" (Schweitzer, 1998: 236). More specifically, they highlight the needs to fulfill legitimate human interests vis-à-vis illegitimate or unsustainable environmental exploitation.

The all-embracing moral impulses and ethical sentiments articulated by Schweitzer and Leopold in their holistic engagement with the biotic community are undoubtedly remarkable and the most far reaching. They constitute a set of universal ethical rules for governing human behavior in our relationship with the natural environment worldwide. Their environmental philosophies also address some of the theoretical rigidity of biocentrism in its inability to admit the anthropocentric and holistic view of nature. The Schweitzer-Leopold philosophical framework which may be coined here as the anthropo-bioecocentric ethics also addresses the practical challenges the United Nations faces in promoting global environmentally sustainable development. Anthropobioecocentrism or anthropo-bioecocentric ethics may be defined as the application of bio-ecocentric ethics in practice, that is, in the anthropocentric exploitation or economic use of nature. This necessarily engenders the application of the precautionary principles when optimize the economic use of the natural system as discussed above. Here, it needs to be pointed out that although Schweitzer's moral philosophy has been traditionally signalled as biocentric, the above discussion reveals clearly that he is a biocentric extraordinaire in that the ethical sequence of his biocentric position, unlike Regan or Singer, is philosophically flexible and holistically expandable. He is indeed more than just a biocentric advocate as has been conventionally argued.

Also, both Leopold and Schweitzer advanced a code of environmental decency for human relationship with nature: Leopold based on the "right principle" and Schweitzer premised on the "foresee and forestall" insights. Together, these moral and precautionary principles lay a concrete universal philosophical foundation for initiating and nurturing social impulses and sentiments in environmental awakening—one of the most important conditions contributing to the evolution of ethics which connect humans with nature based on love, respect, and care. The ethical entry point here involves a systematic and gradual shift in our conception of nature and the recognition of the nature of value pluralism. On this count, the philosophy of reverence for life will certainly "put us in touch with ourselves, our inner thoughts, our inner hearts" (Free, 1992) to change our anthropocentric view per se and to move out from the biocentric rigidity to embrace the Schweitzer−Leopold anthropo-bioecocentric ethics toward harnessing the "ecological integrity, stability, and beauty" of our Earth System. This will help us to define our moral and ethical obligations toward the environment and a sustainable future.

As demonstrated in the following section, without moving away from the hyper or extreme form of anthropocentrism as defined above toward embracing Leopold−Schweitzer's anthropo-bioecocentric ethics, humans will turn into a real threat to nature and life on Earth. More particularly, it illustrates how hyper-anthropocentric motivations vis-à-vis anthropo-ecocentrism contributes to extensive and ruthless environmental destruction, highlighting the urgency of our moral reorientation toward embracing Leopold−Schweitzer's environmental philosophy and ethical principles in avoiding a looming global environmental crisis.

4.5 Environmental ethics and de-ethics in a real-world system: Some empirical reflections

4.5.1 The United Nations road to global environmental sustainability: The environmental paradox revisited

This section aims to demonstrate the nexus of ethics and global environmental sustainability in a real world system based on the ethical philosophies discussed earlier and the empirical assessment of the status of our global environment in the preceding chapter. It also introduces new examples from the Canadian extractive industry to reinforce this line of analysis. This method of assessment provides a clear and convincing line of reasoning to explain why more than 45 years of the United Nations' far reaching international environmental protection initiatives and efforts since Stockholm as discussed in Chapter 2, The United Nations' Journey to Global Environmental Sustainability Since Stockholm: An Assessment, have been less successful in motivating the world leaders to heed Schweitzer's "foresee and forestall" insights or to do the Leopoldian "right things" to reverse the unprecedented trend of global environmental decline.

To begin with, the Brundtland Report, the basis of the 1992 Rio Earth Summit, claims that "Humanity has the ability to make development sustainable" (WCED, 1987: 8). However, as demonstrated clearly in the preceding chapter, environmental degradation, forest destruction, and biodiversity depletion have assumed a global dimension in the contemporary era despite the United Nations' momentous efforts at arresting the environmentally destructive practices discussed in Chapter 2, The United Nations' Journey to Global Environmental Sustainability Since Stockholm: An Assessment. It may well be that the economics of sustainable development is still to a large extent skewed toward the maximization of economic growth and the optimization of human socioeconomic welfare. The United Nations' call for sustainable development since the Stockholm Conference is increasingly open to moral inquiry and ethical interrogation.

As revealed in Chapter 2, The United Nations' Journey to Global Environmental Sustainability Since Stockholm: An Assessment, to this end, countless global conferences have been convened by the United Nations to promote environmental sustainability. Hundreds of international treaties and agreements have also been endorsed or ratified by countries around the world to reinforce their commitment to ensure conservation and sustainable management of the global environment. Furthermore, as discussed in Chapter 3, The United Nations' Journey to Global Environmental Sustainability Since Stockholm: The Paradox, a wide range of national environmental laws and policies have been promulgated by government bodies across the world to promote sustainable resource use and environmental conservation. Yet, despite these environmental initiatives, leaders across the world are no closer to genuinely implementing the policies or translating definite commitments into actions that can slow or reverse the rapid acceleration of worldwide environmental degradation.

The fact is that setting in motion worldwide concern for environmental sustainability and establishing environmental frameworks, guidelines or the like to address environmental problems is one thing, translating them into real commitment is another. To be sure, sustainable development does not happen in a vacuum. It happens only as a result of human commitments followed by environmental initiatives and political will, and the latter will emerge only when humans embrace the ethical principle of resource use. Otherwise, all the commitments made at the meetings are mere rhetoric or hortatory proclamations.

Clearly, these global, regional, and national paper accomplishments fail to convey coherently environmental values and ethics which humans ought to pursue to turn the tide of rapid environmental degradation. Still more important is the claim that the United Nations' call on the global leaders to protect our planet because it underpins human long-term socioeconomic existence is by itself an ethical compromise from the very beginning. This is because of its explicit focus on human needs premised on the Benthamite philosophy of utility maximization rather than the moral responsibility humans owe to nature as well as to future generations. The United Nations has thus failed to undertake concerted efforts to perform this ethical lift to motivate international leaders or the global community to embark on a gradual shift from the current dominant anthropocentric view of nature or the dominant human-centered utilitarian ethical framework toward a more environmentally benign conception of our Earth system.

It is worth reiterating that the anthropocentric view of nature is closely associated with Francis Bacon's philosophy of "Dominion of Man over the Universe" as discussed in the preceding chapter. The Baconian philosophy views nature instrumentally as the repository of resources for human exploitation and as a means to an end. It gives humans the mandate to conquer and subdue her and to "shake her to her foundations" for the benefits of humanity expressed in term of the Benthamite utilitarian maxim of "the greatest good for the greatest number."

The Baconian metaphoric view of Nature and the Benthamite philosophy of utility maximization may be traced back to the Industrial Revolution when in all over Europe, a flurry of nature excavating activities began, including tunnelling into the earth's crust for coal and iron ore for the promotion of heavy industries and for the economic and material betterment of life. Inevitably, the world at large has been pulled by the colonial forerunners into the Baconian and Benthamite perspectives of a never-ending process of economic development and material progress for humanity at large even at the expense of the natural environment as discussed in this chapter and Chapter 3, The United Nations' Journey to Global Environmental Sustainability Since Stockholm: The Paradox.

The development of this anthropocentric view of human nature in itself is a form of utilitarian and selfish stance, with all moral principles taking their character and meaning from the means and ends perception—the idiosyncratic pursuit of rapid economic growth, wealth maximization, sustained material progress, and high consumerism. Indeed, in the anthropocentric world, people make decisions based on self-interests, and environmental protection arises only in so far as it is necessary to protect people's interest. The colonial anthropocentric development ideology came to frame in continuity the development discourse for countries especially the developing countries from the earlier colonial era to the later globalization period and the contemporary human-centered economic system.

To emphasis, the contemporary economic system is dominated by an acquisitive and egoistic human race where every member of society seeks to pursue the highest possible wants and material progress at the expense of the environment. The ethic of human conquest and mastery over nature in the name of sustainable development has persisted and still constitutes as the basis of development policy formulation in many countries with its resultant global and often irreversible impacts on our earth system. Consequently, as revealed in the previous chapter, many of the natural systems, including forests or biodiversity in Latin America, Africa, Southeast Asia, and China, are protected only on paper or in name.

It is explicitly clear that the various environmental laws, policies, and regulations crafted in these countries have failed to intrinsically reflect the kinds of values especially nonmonetary values associated with the natural systems they are protecting. There is also a sense of disquiet in the rationales, moral principles, or ethical imperatives underlying environmental protection. Thus in the absence of this corpus of environmental value concepts and ethical principles of nature protection which constitute a strong guiding force that shapes human attitudes and behavior, all the canonically and rigorously formulated environmental protection tools are unlikely to have a significant practical impact in containing morally destructive environmental practices. The following discussion will demonstrate this more clearly.

4.5.2 Anthropocentrism: The Baiji tragedy and the African common Zebra ecological quagmire

An obvious case which is worth repeating is the ecological tragedy of Baiji—one of the most critically endangered, threatened, and rarest mammal species on Earth as revealed under the IUCN Red List, the Guinness Book of World Records, the US Endangered

Species Act, and the Zoological Society of London, United Kingdom. Despite according this critically threatened and unique species the strongest commitment for its protection and preservation by classifying it as the First Category of National Key Protected Wildlife Species in China, there was a significant disjuncture between policy and real action.

For one thing, the ethical dimension of its preservation and the unique intrinsic value of Baiji which is worth preserving is completely absent from its ecological protection policy and real practices. More specifically, as revealed in the previous chapter, its preservation status did not comport with human ethical or moral intuitions in that the extraordinarily rare and intrinsically legendary species had been driven to extinction by human destructive economic and environmental practices. Biocentrically, the extensive use of rolling hook long-line fishing, gill nets, electrocution, and dynamite or other banned destructive fishing methods to hunt the species indicates that humans in association with such unsustainable practices had no "love, devotion, and sympathy" for the Baiji, in Schweitzer's words. Those involved in its extinction failed to do the Leopoldian "right thing" to reverse its real threat of extinction because of their insatiable quest for material progress—the dominant attribute of human-centered utilitarianism.

The disappearance of Baiji in the wild is a stark indication of how unrestrained pursuit of economic growth and socioeconomic progress in the anthropocentric world is changing irreparably the country's ecosystem. It also symbolizes the loss of harmony of human beings with nature. This is excruciatingly clear particularly since the Baiji had long been recognized as the rarest and most critically threatened mammal species on earth, and despite China having expressed serious commitment to its ecological conservation (Choy, 2016).

It is also noteworthy that for similar reasons, the once very common, most abundant, and least endangered Zebra in Africa also could not escape the fate of being categorized as a near threatened species due to the unsustainable hunting practices of humans. There is very little sign of any ethical intuition of biocentrism for animal love and respect in the anthropocentric and utilitarian impulse of humans' vested interests. Thus it is difficult for them to conceive why sentient animals warrant human moral consideration, let alone have the ethical responsibility of avoiding pushing endangered species to the brink of extinction.

4.5.3 The Asian environmental philosophy in a human-centered world system

Other cases of moral nullification of nature for economic gains include: (1) the commercial poaching of rhinoceros, elephant, and buffalo in Africa; (2) extensive destruction of forests in Latin America and Southeast Asia for commercial timber resources or large-scale plantation development involving widespread habitat destruction and biodiversity degradation; (3) the environmentally destructive IndoMet Coal Project in Indonesia; and (4) the degazettement of the Bikam Permanent Forest Reserve in the state of Perak as well as the extensive and irreversible destruction of vital forest haven for rare biodiversity in Gua Musang in the state of Kelantan in Malaysia as discussed in the previous chapter.

These unsustainable practices reflect how those individualistic utilitarian economic agents with vested material interests regard our Earth system. Driven by an anthropocentric worldview of nature, these instrumentally inclined vested interest groups see themselves as separate from the natural system, and as dominating it. For them, the natural environment matters only because they are instrumentally valuable as a means to an end, that is, as a means to enhance wealth accumulation and material progress. It is this lack of ethical dimension of resource use that thwarts any human initiative to confer intrinsic value, and hence moral consideration on nature—a moral initiative which may gradually translate into a "positive exercise of skill and insight" when optimizing its instrumental use to preserve its "working order". The fervent quest for economic growth and material progress has also led to abusive exploitation of nature, especially of forest resources as in the case of Latin America, Africa, and Southeast Asia as examined in the previous chapter.

On the one hand, it appears that leaders in these countries have been unable to foresee the long-term harm and irreversible damage to nature and biodiversity resulting from their exploitation of the forested environment. On the other hand, the anthropocentric view of nature is so deeply ingrained that even awareness of the continued environmental and ecological destructive impacts of extensive habitat fragmentation on biodiversity as a result of their uncontrolled forest exploitation have not be able to evoke feelings of moral obligation toward nature, let alone to ascribe some responsibility to mitigate their environmentally damaging policies in the interests of future generations. Proenvironmental behavior will be activated only insofar as such sustainable practice is necessary to maintain and sustain economic growth or material progress. Till today, the gap between environmental concerns and real sustainable action is still alarmingly wide across the world dominated by the anthropocentric worldview of nature.

4.5.4 The American world of utility maximization: Donald Trump's anthropocentric decimation of nature

The barely conscious moral attitudes to the Earth system is also clearly reflected in the advanced countries, especially the most economically and politically powerful country in the world—the United States as noted in Section 3.19. While China, the world's second largest economy and currently the world's largest emitter of greenhouse gases has, out of humanistic and biospheric altruism, responded positively to the global carbon treaties to reduce its CO_2 emission, the United States, motivated by the mechanics of utility and the economic imperative for growth has right until the last Climate Convention, almost persistently refrained from endorsing global carbon treaties of any form (Chapter 2: The United Nations' Journey to Global Environmental Sustainability Since Stockholm: An Assessment). To put matter into perspective, it is worth reiterating that from an environmental philosophical point of view President Donald Trump generally perceives that only humans have intrinsic moral value while the nonhuman natural world has only instrumental value for the betterment of mankind expressed in term of large production and large consumption lifestyles (Section 3.19 especially Subsection 3.19.4). It is along this line of reasoning that Donald Trump is by and large aligned with the Baconian ethical philosophy of anthropocentrism. From this philosophical standpoint, humans are regarded as separated from the Earth system and nature is perceived only as an object for instrumental use. Thus for Donald Trump, nature merely serves as a means to an end, that is, to utilitarianly promote and sustain his carbon economy to "Make America Great Again."

As a staunched climate change denier Donald Trump has continuously rolled back Barack Obama's environmental regulations to put the country into the perspective of progress (Table 3.14, Chapter 3: The United Nations' Journey to Global Environmental Sustainability Since Stockholm: The Paradox). To wit, informed largely by a wordview of anthropocentrism and unguided by the moral philosophy of bioecocentrism, the Donald Trump administration embarked on a bold capitalist programme to claim greater economic decimation of nature by allowing oil and gas drilling in the protected, fragile, and ecologically sensitive Arctic National Wildlife Refuge—one of the last intact wilderness left on Earth which is home to more than 270 species of wild animals including the threatened polar bear, Porcupine Caribou Herd, musk oxen, Arctic fox, wolves and migratory birds (Holden, 2019). In a related note, the Donald Trump administration has also reauthorized the use of the chemical traps to kill coyotes, dogs, foxes, and other wild animals across the United States with M-44 devices. The devices trap wildlife with bait before releasing sodium cyanide into their mouths to kill them (Lewis, 2019).

Inexorably, the above discussion adds a great deal that is important to our understanding of the link between human perception of nature and environmental sustainability. More precisely, when nature is instrumentally commodified through privatization, marketization, or monetary valuation, it will be stripped of moral considerability and subjugated to human unrestrained economic exploitation to the detriment of its ecological integrity. From an environmental philosophical viewpoint, the discussion reified the importance of the following worldviews in promoting environmental sustainability: (1) our perceptions of and place in nature; (2) our mental states pertaining to the kind of world human civilization might flourish and within which humans exist; and (3) our moral views of ethical responsibility with respect to the natural environment as well as the future generations.

4.6 The Canadian "anthropocentric conquest" of nature: The power of anthropocentrism

It is relevant to first focus on the whole pattern of interaction in the human separation from and transcendence in the Earth system based on the Canadian anthropocentric conquest of nature à la Donald Trump/Francis Bacon's philosophy of "Dominion of Man over the Universe" which was launched by the former Conservative government led by Stephen Harper. Such extended line of analysis provides a definitive and all-encompassing treatment and overall picture of the whole discourse on the inextricable link between the human anthropocentric view of nature discussed thus far and destructive environmental practices. The Canadian supreme example also momentously reflects the reality of the troubling future of our individualistic age of modern capitalism. It also epistemologically adds concrete contents to the environmental philosophical dimension of human−nature relations as briefly raised in the preceding section.

Still more important is the claim that the Canadian case study brings into perspective how the Canadian political moral nullification of nature with the view of instrumentalizing it to serve human ends has led to the spectre of a looming environmental crisis in the country and across the borders. Incontrovertibly, it provides a deeper, precise and more concrete reflection, and a clear understanding of the complex interaction of several systems or factors involving humanity, economics, environment, biodiversity, pollution and health, among others, which adequately supports all the empirical studies examined so far. It also allows us to see clearly the anthropocentric fault lines along which modern civilization separates human beings from nonhuman living entities to define its optimal resource utilization trajectory.

As will be made clear in the subsequent analysis, this anthropocentrically dictated environmental discourse fundamentally alters the human relationship with nature, making it more instrumentally inclined to commodify and dominate nature for the sole purpose of serving human needs and economic interests. Quintessentially, such anthropocentric judgment of value works to shape man—nature relationships utilitarianly toward an exploitative mode of resource utilization, thus triggering an irreversible and hostile transformation of our Earth system.

By treating the environment as having no moral significance of its own, the former Canadian Conservative government led by Stephen Harper set its anthropocentric arms "to destroy the environment as fast as possible," in Noam Chomsky's words (Germanos, 2013; Nadeau, 2010; McDonald, 2013, Frampton and Redl, 2015). We now provide a full authoritative exploration of Chomsky's statement on the Canadian "anthropocentricconquest" of nature. This will not only reinforce the argument expounded thus far but will also allow us to see clearly how profoundly the human anthropocentric view on nature has affected the Earth system in terms of the degree and scope of environmental changes. When the Conservative government took the helm in 2006, it set in motion a grand Baconian mission to anthropocentrically unlock the natural environment and instrumentally turned it into the slave of mankind so that it could better serve human interests and material comfort. To lay the Baconian foundation for its athropocentric conquest of nature in the interests of industrialization the Harper government withdrew from the Kyoto Protocol on climate change in 2011, claiming that signing the Protocol was one of the previous government's biggest blunders as it had impeded job creation and economic growth (Guardian, 2011). As a matter of fact, Stephen Harper had been opposing the Kyoto Protocol since he became Prime Minister in 2006 (Walsh, 2011). This was basically attributed to the government's belief that the Protocol presented an obstacle to its utilitarian pursuit of advanced capitalist economy.

Harper's anthropocentric conquest of nature is well reflected in the pervasive and profound excavations deep into the earth's crust in the province of Alberta for tar sands which are deposited below the boreal forests and wetlands in an area about the size of Florida, to promote what is lamented as "the most destructive project on earth" (Hatch and Price, 2008; Bruno et al., 2010). Tar sands are a mixture of sand, clay, water and bitumen. Bitumen is the fossil fuel component of tar sands. It is a heavy and extremely viscous tar-like substance that must be treated through intensive processing to be converted into crude oil or gasoline. Tar sands have been dubbed the most environmentally destructive form of fossil fuel which threatens human health and long-term existence (Kalman, 2015; see Box 4.1). In the following sections, we shall systematically uncover an appreciable dose of environmental impacts of the tar sands development in Canada to better reflect the anthropocentric force of environmental destabilizations.

4.6.1 Tar sands mining: Irreversible environmental transformation, global warming, and biodiversity impoverishment

The tar sands industry, which involves in situ mining and surface mining, results in extensive and irreversible transformation. The Syncrude Aurora tar sands mine in the boreal forest north of Fort McMurray, for example, has completely and irreversibly

BOX 4.1

Tar sands—the most environmentally destructive form of fossil fuel.

Producing and processing a barrel of tar sands oil releases some 14%–20% more carbon dioxide than from processing a standard barrel of American oil (Schultz, 2013; Cushman, 2017). In situ mining releases between 99 and 176 Kg of CO_2 equivalent per barrel of sand oil while surface mining emits between 62 and 164 Kg of CO_2 equivalent per barrel of sand oil (Grant et al., 2009). In situ (drilling) method is used when tar sands resource or bitumen deposit is buried too deep beneath the surface for mining It is estimated that greenhouse gas emission standard for upgrading ranges between 52 and 70 Kg of CO_2 equivalent per barrel of sand oil (Droitsch et al., 2010). However, the figures may differ from different sources. Greenhouse gas emissions from tar sands have increased from 121% to 37.2 million tonnes from 1990 to 2008 (Droitsch et al., 2010). In 2011, the industry produced 1.8 million barrels per day of tar sands oil, emitting roughly 47.1 million tonnes of CO_2-equivalent into the atmosphere (Muradov, 2014: 9–10).

Tar sands also requires significant amounts of capital, natural gas, and water to extract and process. In particular, its highly water-intensive recovery process is not only excessive but also environmentally destructive. About four to six barrels of fresh water are used for every barrel of oil produced which is four times higher than that for producing a barrel of oil from conventional sources (Schneider, 2010; Bruno et al., 2010). For example, in 2011, the industry consumed 170 million cubic meters of water in its operations. This is equivalent to the annual usage of Barcelona (Friends of the Earth Europe, 2015).

Most of the water used in the production process is drawn from the Athabasca River—one of North America's longest free-flowing rivers (Greenfield, 2015). Only about 10% of the waste water is returned to the river while the rest which contains huge amounts of toxic waste called "tailings" comprising sand, silt, clay, contaminants, and hydrocarbon, which is too contaminated or toxic to be released into the water systems, is stored in tailing ponds or holding dams located in the vicinity of the Athabasca River. These human-made toxic waste sinks are increasing in volume by 200 million liters, or 80 Olympic-sized swimming pools, every day, and in 2010 covered an area of 170 Km^2, containing 840 billion liters of tailing waste after 40 years of mining (Dyer, 2010; Bruno et al., 2010; Schneider, 2010; ERCB, 2010).

The contaminants in the tailings include naphthenic acids, polycyclic aromatic hydrocarbons (PAHs), phenolic compounds, ammonia, mercury and other trace metals that are acutely toxic and hazardous to human beings, aquatic ecosystems and mammals (Bruno et al., 2010). For example, the water bodies of the Athabasca River which eventually flow into the Arctic Ocean through the Peace, Slave and MacKenzie Rivers, have been found to contain elevated levels of toxic PAHs including benzo(a)pyrene, a chemical which is linked to cancer, genetic damage, reproductive disorders including birth defects, and organ damage (NRDC, 2014).

BOX 4.1   (cont'd)

A study conducted with the local community in Fort Chipewyan located in the area where the Athabasca River empties into Lake Athabasca indicates that the cancer rate of the local people was 30% higher than what would typically be expected for that period of time. In particular, certain types of cancer such as biliary tract cancer, blood, and lymphatic cancers, lung cancers in women, and soft tissue cancers in the community occurred at rates higher than expected. Scientists have also reported that the Alberta waterways and landscape are contaminated with increasing levels of methylmercury which is a potent neurotoxin that causes developmental and behavioral problems including lower intelligence quotient in children, and cardiovascular effects in adults (NRDC, 2014). Also, the water-borne PAHs are toxic to embryonic fish at concentrations as low as one part per billion, and the release of toxic pollutants into the environment has also resulted in tumors and mutation of some fish and game animals (Hatch and Price, 2008).

The tailing ponds are also a grave source of toxic pollution. As revealed in the 1997 Decision report for the Application for Amendment for Approval No. 7632 for Proposed Steepbank Mine Development, 1.6 million liters of water leak from the Tar Island Pond into the Athabasca River every day (Grant et al., 2009). The current leakage rate of the Tar Island Pond is roughly 6 million liters a day (Price, 2008). In 2008, the Environmental Defense Canada revealed that as much as 10.97 million liters of toxic laden water leaks into the environment every day, causing extensive environmental degradation, especially ground water contamination (NRDC, 2014). It may be noted that tailing

lakes seep toxins such as naphthenic acids (toxic water-soluble carboxylic acids) into the ground water below causing far reaching pollutant migration and contamination through the ground water systems, which is extremely difficult and at times impossible to clean up. Naphthenic acids associated with oil sands do not break down easily in the natural environment and are acutely toxic to aquatic organisms and mammals (Grant et al., 2009).

The tailings ponds which are situated on an important migratory bird pathway pose a great risk to millions of geese, ducks, swans, loons, and dozens of other species which migrate northward in spring and southward in the fall (Steward, 2015). They are in fact death traps for the migratory birds which mistake the ponds for fresh water lakes especially in spring when the natural fresh water bodies are still frozen, and the tailings ponds, heavily loaded with chemicals, are not completely ice covered (Steward, 2015). In April 2008, for example, 1600 ducks were found dead after landing on the Aurora tailings lake of Syncrude (Dyer, 2010). In 2010, another 350 ducks died after landing on Syncrude's Mildred Lake tailing pond (National Post, 2012).

The tar sands industry also spews out substantial amounts of toxic carcinogenic pollutants including benzene and styrene into the environment from its upgrading process to convert the raw bitumen into synthetic crude oil using chemicals and through heating. These pollutants which are associated with leukemia and cancers of the lymph and the blood-forming systems were found at elevated levels near major upgrading facilities just north of Edmonton (NRDC, 2014). Other pollutants which are hazardous to the environment as well as to

---

BOX 4.1   *(cont'd)*

human beings and other nonhuman living organisms include nitrogen dioxide (a smog maker), sulfur dioxide (an acid rain promoter), volatile organic compounds (an ozone developer), and particulate matter (a lung and heart killer) (Nikiforuk, 2010).

---

altered cthe physical structure of the natural environment (Guardian, 2010). More specifically, the unrestrained extraction of the tar sands which are deposited below the boreal forests, has resulted in irreversible transformation and destruction of the forest resources including peatlands—one of the world's most important forest carbon storehouses. It may be noted that the boreal forests store almost twice as much carbon per acre as in a tropical rainforest. It contains 703 billion tonnes of carbon compared to the tropical rainforest at 375 billion tonnes (Carlson et al., 2009). The Canadian boreal forests store about 71.4 billion tonnes of carbon in forest ecosystems, 136.7 billion tonnes in peatland ecosystems, and 0.6 billion tonnes in numerous lakes located in the Boreal Forest region (Carlson et al., 2009). This means that the depletion of boreal forest and peatlands contribute directly to aggravating the global warming problem by destroying the globally important carbon storehouses.

The boreal forest depletion also led to habitat destruction and fragmentation wich threatens the biological resilience or continued existence of millions of migratory birds including long-distance migratory songbirds and waterfowl, and northern mammals such as woodland caribou, bears, wolves, moose, and the endangered species like the whooping crane, trumpeter swan, Peregrine Falcon, and Piping Clover, among others (Wells et al., 2008; Carlson et al., 2009). It is estimated that over the next 30−50 years, between 6.4 and 166 million migratory birds could be lost only to the destruction of staging and breeding areas caused by unrestrained surface and in situ mining, but also to mistaken landings on the toxic pollutant-filled tailing ponds as mentioned earlier (Wells et al., 2008). The Boreal woodland caribou, an iconic, and threatened species in Canada which is sensitive to human-related environmental changes, is increasingly facing the threat of extinction due to extensive destruction of its critical habitat. As a case in point, the species population in Beaver Lake Cree Nation's traditional territories was found to have declined by 70% since 1996 (Linnitt, 2014).

4.6.2 The politics of anthropocentrism: The Canadian libricide

To speed up its anthropocentric conquest of nature and to eliminate those impediments to its self-interests, the Harper government embarked on a libricide to curtail the free flow of scientific information and to destroy scientific records and books, especially those related to climate change—the worst-faring archives of the Harper's administration (Doctorow, 2014; McDonald, 2013; Hayden, 2014; Zhang, 2017). For example, it closed down more than a dozen federal science libraries run by Fisheries and Oceans Canada and Environment Canada. These libraries contained hundreds of thousands of documents on fisheries and aquatic science including important information on historical fish counts and water-quality analyses (Owens, 2014; Sowunmi, 2014). The libraries closed down include the century-old St. Andrews Biological Station in New Brunswick and the library at the Freshwater Institute in Winnipeg, Manitoba. The government's libricide indicates it strongly regards environmental science as a threat to its unrestrained resource exploitation at the expense of the environment (Doctorow, 2014).

4.6.3 The politics of anthropocentrism: Transboundary environmental degradation

The Harper administration's anthropocentric transgression of the Earth systems is not confined within its own borders. Being the world's leading mining nation, it has also extended its instrumental path of resource exploitation globally to Argentina, Chile, Colombia, El Salvador, Guatemala, Honduras, Mexico, Panama, and Peru. In exploiting weak environmental regulations, low tax burdens, and neoliberal economic policies in these developing regions and taking advantage of the 1994 North American Free Trade Agreement, the Canadian multinational companies, often with government political, legal and financial support, signed agreements with these countries to facilitate extensive excavation of the earth's crust for its natural resources. Currently, between 50% and 70% of all mining in Latin America is carried out by Canadian firms (COHA, 2014; Hill, 2014). There are 22 large-scale projects operated by 20 companies across Argentina, Chile, Colombia, El Salvador, Guatemala, Honduras, Mexico, Panama, and Peru (COHA, 2014). The global anthropocentric conquest of nature is causing serious environmental impact by destroying glaciers, contaminating water and rivers with heavy matter such as iron, aluminium, magnesium, arsenic and other toxic pollutants, and cutting down forests (Ismi, 2009; Hill, 2014; COHA, 2014).

For example, the Argentina–Chile border's Pascua Lama open-pit mine which is located within the UNESCO-protected San Guillermo Biosphere Reserve and overlaps a water reserve composed of glaciers is causing contamination and the shrinking of three glaciers, Toro I, Toro II, and Esperanza (COHA, 2014). Similarly, 54 ha of the old growth and gallery forest in the Petaquilla's Molejón Gold mine in Panama, situated in the Mesoamerican biological corridor that is designated as a protected area by local authorities has been clear cut for infrastructure development and processing plants (COHA, 2014). Other countries which are also the targets of the Canadian instrumental mission of environmental exploitation include Guyana, Ghana, the Philippines, Papua New Guinea, the Democratic Republic of Congo, Burma, India, Kyrgyzstan and Romania, causing extensive irreversible environmental transformation and environmental pollution (Ismi, 2009; Oxfam, 2014; Hill, 2014; Jamasmie, 2015).

River and water pollution is especially serious in Honduras, Argentina, and Papua New Guinea. For example, the Canadian Goldcorp mining operations in the Siria valley in Honduras are contaminating the local water sources outside of the San Martin mine. Likewise, the Canadian Barrick Gold's Veladero mine in Argentina spilled more than a million liters of toxic cyanide in September 2015, contaminating five rivers and affecting the mental health, safety, and lives of the people (Chewinsk, 2016). Likewise, in Papua New Guinea, the Porgera gold mine, one of the world's largest gold mines operated by Canada's Barrick Gold, has dumped tons of poisonous wastes directly into rivers and streams, a practice that is illegal in Canada (Ismi, 2009).

In northern Peru, the Lagunas Norte, operated by the Canadian Barrick Gold, has caused serious contamination problems in the headwaters of the Perejil, Chuyuhual, and Caballo Moro rivers (Hill, 2014). Gold mining remains a major cause of deforestation in Peru with Madre de Dios being the hardest hit region, with whole areas transformed into veritable deserts and wastelands. It may well be that deforestation in Madre de Dios caused by unregulated gold mining activities increased by 400% between 1999 and 2012— from 10,000 ha to more than 50,000 ha (Hill, 2016). The gold mining activities also pollute the rivers with heavy metal such as mercury which can cause malfunction of the nervous, digestive and immune systems, and of the lungs, kidneys, skin and eyes (Chow, 2016). Roughly 41% of the population (about 50,000 people) in Madre de Dios are exposed to mercury health risks such as acute mercury poisoning (Fraser, 2011; Chow, 2016). The Peruvian government was forced to declare a 60-day public-health emergency on 23 May 2016 to address the escalating mercury pollution problem caused by unregulated gold-mining along the Madre de Dios River (Fraser, 2016).

4.6.4 The Canadian anthropocentric conquest of nature: Rounding up

It is clear from the above that the Harper administration, driven by the anthropocentric quest for economic growth and material progress, conceived that it was entitled to instrumentally embark on an extensive scale of nature exploitation and destruction for its tangible benefits without constraint. More specifically, under the Harper human-centered value system, the natural environment rich in tar sand deposits has no intrinsic value apart from the instrumental value it provides to enhance economic progress and industrialism. This assumption came to frame the development and environmental discourse of the former Canadian government where economic growth, material progress, and industrialism are at the center of concern, even at the expense of the natural environment as illustrated by the irreversible destruction caused to the already stressed natural environment. More particularly, this environmental vandalism reflects that viewing the natural world as a horn of plenty and a limitless waste and carbon sink, environmental conservation and sustainable mining practices are getting nowhere because they are incompatible with the Harper government's utilitarian philosophy of profit or growth maximization.

Viewed from this perspective, it may further be remarked that the Harper government's relation to the natural world was far from biocentric, let alone ecocentric. It was also non-altruistic. In other words, there was almost an absence of environmental and personal ethical norms that were contemplated or activated in its administration when optimizing the instrumental use of nature to address the destructive environmental and human impacts. As it turned out, the local people affected by the tar sands industry are not even treated as ends in themselves deserving moral standing. Obviously, the Harper government's attribution of moral standing was so exclusive that it excluded not only all the nonhuman natural entities, but also human beings, especially the local communities living near the tar sands mining and processing regions.

There was also an absence of moral deliberation to uphold the principle of intergenerational equity in ensuring that future generations will not be worse off than the present generations in terms of inheritance and enjoyment of an ecologically resilient, intrinsically valuable, biologically diverse, and resource rich natural environment. As a result, moral commitment was not extended to protect and preserve the ecological integrity of the life-support functional value of the boreal forest habitats and the ecosystems which contribute to support the long-term existence of humans (weak anthropocentrism and indirect environmental protection) as well as nonhuman living organisms into the indefinite future. To the Harper government, all nonliving entities have no inherent worth and thus deserve no moral consideration.

The government's as well as corporate unsustainable use of nature stems from their hyper anthropocentric philosophy that nature is there for their use—as a means to an end. This institutional, national, and corporate self-interest characterized by the egregious quest of materialism and consumerism, negated the extension of moral considerability to the natural system. Thus, there is no obligation to act with care, foresight and restraint or forbearance in any human engagement with nature if other human-centered values, such as economic growth or material progress, are at stake. The obvious conclusion is that such widely endorsed precept of anthropocentrism has the disturbing effects in driving the country to an ecologically fragile state, rendering it more susceptible to the risk of being severely affected by natural disasters such as forest fires, flood and extreme weather conditions (see, e.g., Hayden, 2014; Sanger and Saul, 2015; Casselman, 2018).

It is increasingly clear from the above as well as the case studies as examined in the previous chapter that there are strict causal relationships between environmental attitudes, ethical behavior, and moral concern for environmental conservation on the one hand, and in the denouement, the ascription of environmental responsibility and individual will of environmental conservation and sustainable resource use, on the other. In other words, in the final analysis, environmental sustainability ultimately hinges on the dictates of environmental responsibility and will on the part of those with vested interests, or leaders to undertake the moral obligation to act with care, foresight and restraint or forbearance as emphasized by Albert Schweitzer and Aldo Leopold. For this to happen, however, there is a need to reexamine our attitudes and values in relation to our Earth systems by consciously enlarging our purview of ethics beyond anthropocentrism and embracing our natural world as worthy of moral consideration, biocentrically or ecocentrically.

4.7 Sustainable environmental governance and the ethics of sustainability: The Nexus

As discussed in Chapter 2, The United Nations' Journey to Global Environmental Sustainability Since Stockholm: An Assessment, since Stockholm, virtually all nations across the globe have enacted a plethora of environmental legislation and policies to manage our global environment in accordance with the United Nations' environmental sustainability initiatives. Yet, Chapter 3, The United Nations' Journey to Global Environmental Sustainability Since Stockholm: The Paradox, reveals the devastating truth that despite this, the overall global environment has in fact worsened and new environmental threats and challenges continue to emerge. This chapter further highlights the incontrovertible fact that this environmental paradox is basically driven by the anthropocentric tendencies of the global community to implacably prioritize economic growth over environmental protection. As reflected in Chapter 3, The United Nations' Journey to Global Environmental Sustainability Since Stockholm: The Paradox, there is also the lack of strong and genuine political will to put environmental policies into action—the touchstone of institutional and governance failures (Choy, 2005). To compound the problem, society in general being motivated by individualistic impulses for material progress, displays an egoistic predisposition for social wellbeing maximization over sustainable consumption practices.

These observations signal the need for the establishment of a sustainable environmental governance to circumvent the intractable institutional barriers to policy implementation as well as the obstruction to moral proenvironmental behavior and action. That said, environmental governance is an essential component of sustainable development. It provides the mechanism for the constitution of long-term commitment to the end goal of reversing global environmental decline. Environmental governance also allows coordination and cooperation among different institutions and actors in relation to policy design and implementation.

It is important to clarify at the outset a couple of the concepts used in the analysis. First, environmental governance comprises a concatenation of rules, regulations, management practices, and policies that regulates the processes of global environmental protection (Najam et al., 2006; Yoshida, 2012). It is in fact the sum of the many ways the state and nonstate actors manage their common environmental affairs (Commission on Global Governance, 1995: 4; Fig. 4.1). However, the formal institutions and organizations such as the Environmental Ministry and local government will play the key role in enforcing environmental compliance (Fig. 4.1).

Second, it is also instructive to distinguish between "governance" and "government." While "governance" focuses on acts and actions carried out by the administrators or law enforcers, "government" focuses on the mechanism of rule-making (Yoshida, 2012). Also, "governance" or environmental governance, for that matter, is not limited to the state or institution but also covers the nonstate actors such as the economic agents (e.g., manufacturers), civil society and non governmental organizations (NGOs), among others. As shown in Fig. 4.1, a member country, in response to the United Nations' environmental initiative, establishes an environmental ministry (institution) under the charge of the Ministry of Environment (state actor) which is involved in environmental law enactments and policy formulations, and ensures a proper sequence of policy making.

In the operational realm, the policy implementation processes proceed downwards to the local governments or law enforcers while civil servants may be involved in administrative work. Finally, the environmental ministry itself may undertake the task of compliance monitoring to oversee and ensure effective policy implementation. Proceeding further down to the micro-level, the incompatible production and consumption or environmental practices of the nonstate actors are subject to bureaucratic surveillance and control of the

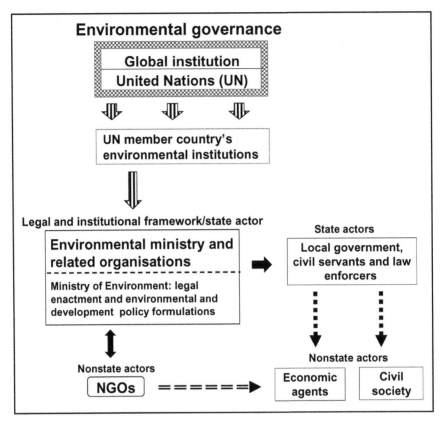

FIGURE 4.1   A simple structure of environmental governance.

local governments or law enforcers. In the chronological order of policy implementation, the NGOs may play their role in inspiring or motivating the state and nonstate actors (economic agents and civil society) to act in ways that reflect the biophysical and ecological basis of economic activity through information disclosures. However, the NGOs in the institutional web are not completely free to conduct their activities as they are often subject to institutional control and at times, even repression. This is especially common in developing countries.

However, putting in place the institutional blocks of environmental governance system does not, *ipso facto*, ensure a more sustainable mode of development. The problem is that the implementation of environmental policies lies in the hands of the heads of governments, especially the environmental ministries and the local government or law enforcers as described above. Furthermore, sustainable development plans may meet all the criteria of environmental sustainability on paper only to fail in practice due to differences in economic and environmental priorities. Thus the existence of an implementation gap between the ability of environmental policy formulation and the capacity for policy implementation resulting in policy failure is a commonplace (Robinson, 2010). It thus follows that to bring the ability policy formulation *pari passu* with the capacity of policy implementation, a more sustainable system of environmental governance needs to be designed, referred to here as the bioecocentric-inclined environmental governance which readily translates policy into action.

To begin with, the fundamental problem impeding effective policy implementation as noted above is basically attributed to the phenomenon of institutional environmental parochialism. Institutional environmental parochialism is typified most conspicuously by the anthropocentric proclivity of the state or institution to embrace the natural environment instrumentally as a means to an end. Thus in many cases, there is often a lack of political will to put environmental policies effectively into practice (Chapter 3: The United Nations' Journey to Global Environmental Sustainability Since Stockholm: The Paradox) mainly because effective policy implementation is anthropocentrically hampered by the state's narrowly construed short-term economic goals over the long-term interests of environmental sustainability. The institutional shift of environment–economic interest inexorably leads to the deterioration of the "environmental" governance into a structural form of economic growth—per se-oriented governance.

An economic growth—per se-oriented governance may be defined as a structural environmental controlling system unconstrained by the ethical norms of ecological sustainability (Choy, 2005). In this ethically deficient mode, the states or institutions tend to resent placing environmental consideration over economic interests, especially when facing a social dilemma in which economic interests conflict with environmental interests (see, Chapter 6: The United Nations Environmental Education Initiatives: The Green Education Failure and the Way Forward, Section 6.9.2). Thus there is no genuine political will to fulfil their commitment to environmental sustainability as pledged under various United Nations declarations. Furthermore, within the spectrum of this ethically flawed governance system, nonstate actors such as the economic agents are anthropocentrically driven to opt for an environmentally destructive mode of capital accumulation process. Likewise, the society is also individualistically skewed toward unsustainable modes of lifestyle and consumption habits—the case of profligate consumerism in American society as discussed above is an instructive illustration.

What is needed to address the above institutional challenges is to bring back the ethics of sustainability, used here interchangeably with the ethics of bioecocentrism, into the heart of the environmental governance system. The main reason for this argument is that genuine political will for policy implementation and law enforcement cannot grow firmly devoid of ethics. Failing to incorporate the ethics of sustainability into its institutional structure is likely to lead to ineffective policy implementation or unsustainable production and consumption practices. The gist is that the ethics of bioecocentrism as comprehensively discussed above, if properly infused, promoted, and embraced, can inexorably transform environmentally destructive behavior into more ecologically friendly conduct based on "love and respect" when optimizing the economic use of nature—an anthropo-bioecocentric resource use practice (Fig. 4.2).

To reinforce man's embrace of the ethics of sustainability, it is vital to promote environmental and moral education because it has momentous importance in the enhancement of an environmentally conscious and ecologically moral society. An environmentally conscious society can, to some extent, exert a certain degree of influence on the state directly or indirectly to refrain from overly unsustainable development practices. Furthermore, the impersonal orientation of environmentalism (ethics of sustainability) may again be cultivated through environmental education which is discussed extensively in Chapter 6, The United Nations Environmental Education Initiatives: The Green Education Failure and the Way

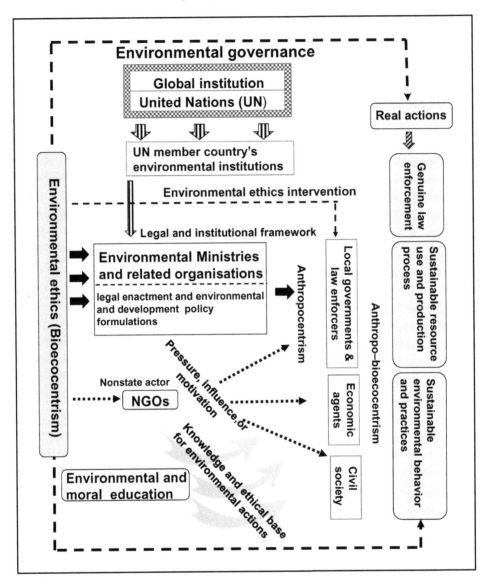

FIGURE 4.2 Environmental governance: a diagrammatic representation.
Note: Environmental ethics intervention refers to the infusion of the ethics of bioecocentrism into the policy implementation process as well as production and consumption processes which serve as a springboard to induce
materialization of real actions. This may be enhanced based on the promotion of environmental and moral education as discussed in Chapter 6, The United Nations Environmental Education Initiatives: The Green Education
Failure and the Way Forward. The application of the ethics of biocentrism in sustainable resource use and consumption processes is termed anthropo-bioecocentrism.

Forward (Sections 6.14–6.15.2; Fig. 4.2). The ethics of sustainability infused into the structure
of environmental governance give rise to new combinations that result in enduring changes
in the state and nonstate actors' moral perception of nature based on "love and respect" or
the "reverence of life." These are the most powerful loci of the countervailing force of

anthropocentrism *par excellence*, and the most distinctive and universally applicable ethical force of environmental sustainability orientation.

To sum up, deeply ingrained anthropocentrism is the greatest impediment to environmental sustainability as it tends to promote unrestrained resource exploitation, profligate use of resources, and unsustainable production and consumption practices. It thus follows that without moral intervention in attitudes and institutes based on a shared set of environmental values, ecological beliefs, and ethical norms collectively called here the ethics of bioecocentrism (Section 6.10.2), it is unlikely that the institutional gap between policy making and implementation can be closed effectively; rather it would further weaken the political will of policy implementation. Worse yet, the gap may continue widening with excessive environmental exploitation in the name of sustainable development. In the same vein, it is also unlikely that society will subscribe to a more sustainable mode of consumption pattern in the absence of a binding ethical force to constrain unsustainable human behavior and practices.

Universally, the ethical force of environmental sustainability is compellingly topical and relevant. Once established and ingrained into the hearts and minds of individuals, it will constantly remain a lodestar to guide and drive environmental sustainability (Sections 6.14–6.15.2). Indubitably, the ethical prism of this moral guiding thread is the same tomorrow as today and is the same in the North as in the South—it knows neither present and future in time, nor North and South in geological demarcation. Thus to lead the global community back to the race-track for environmental sustainability, we must unroll the limitless horizons of environmental ethics to activate and drive the human race (state and nonstate actors alike) under a sustainable environmental governance that is capable of leading the human race toward our common mission—The Future We Want.

4.8 Concluding remarks

The gravity of the unprecedented, long-term, and ever pervasive global environmental decline as heretofore discussed signals the loss of harmony of human beings with nature. That being said, the global environmental treaties and the like as promulgated by the United Nations are only a part of what is needed toward committing the international community to undertake significant remedial measures to address this looming environmental crisis. Moreover, political commitment to environmental sustainability is only a mere acknowledgment of the deep systematic environmental crisis plaguing our Earth systems; it does not amount to real and concrete action. To be sure, commitment to shared environmental responsibility given by the international community cannot be administered or enforced effectively through an ethical vacuum. It thus follows that there is an urgent need to elevate the critical role of environmental ethics as a vital element and a forward-looking approach to effective decision-making process, and as a motivational fulcrum for translating sustainability visions or environmental aspirations into real actions.

As argued, environmental sustainability hinges on the ethical underpinnings and moral obligation to act with care and restraint in dealing with the Earth systems. It thus follows that in healing the global environmental wound, the way forward is to usher in the onset of a harmonious human–environment relationship resting on the philosophical principles of Albert Schweitzer or Aldo Leopold and deontic conceptions of biocentrism or ecocentrism as the case deems fit. Indisputably, if global environmentally sustainable development is to take shape, it must embrace the ethical dimension of sustainability.

The inexorable conclusion is that environmental ethics or anthropo-bioecocentric ethics for that matter, as a practical discipline, constitutes the *conditio sine qua non* for the reconstruction of morally justifiable and sustainable environmental practices or actions. The following chapter will reveal an appreciable amount of highly important evidence on the inextricable link between environmental philosophy and environmental sustainability, gathered from extensive field research on real-life indigenous relationship with the natural environment conducted in the tropical rainforest region in Bornean Sarawak, Malaysia. The field research provides essential evidence of practical import to reify the practical and paramount importance of environmental ethics in promoting a sustainable world. How the conviction to environmental sustainability can be accomplished will be the subject of Chapter 6, The United Nations Environmental Education Initiatives: The Green Education Failure and the Way Forward.

The nexus of environmental ethics and environmental sustainability: An empirical assessment

5.1 Introduction

As examined in the preceding chapter, with regard to healing the human−nature divide, a wide variety of positions of ethical worldviews and philosophical bearings of considerable practical importance are available. More specifically, in addressing this ethical question, the way forward is for us to extend our moral concern to nature following the philosophical logic of Aldo Leopold's land ethic and Albert Schweitzer's ethics of reverence for life and to translate our anthropocentric motives into ecocentric actions (the anthropo-bioecocentric ethics). To what extent can the perceived separation between humans and nature be addressed by applying environmental philosophy? What are the implications of the causal link between environmental philosophy and human−nature relationship for environmental sustainability? These are definitely not merely academic questions. In particular, does the way we view the world have any impact on our moral value orientations in reality? Does moral value orientation lead to proenvironmental attitudes and behaviors in a real world system? Do proenvironmental attitudes and behaviors lead to real environmental actions in practice? What other kinds of value orientation have real impact on enhancing proenvironmental attitudes and behaviors? And also, more importantly, how can people be convinced? These questions will be systematically examined later.

5.2 Environmental philosophy—the epistemological disputes

It turns out that epistemic peers or individuals may often hold different concepts or assumptions in confronting these questions. From an academic point of view, as revealed in the preceding chapter, even the theoretical and conceptual discussions of environmental philosophy and ethics are deeply entrenched under the epistemology of disagreement such as Paul Taylor's ethics of egalitarianism as against Peter Singer's ethics of animal

253

liberation. These two schools of thought are often entrenched in a series of intramural argument and counter-argument over the conceptual differences and terminological distinctions between the narrowly crafted biocentric/animal ethics and the environmentally encompassing ecocentric ethics in considering the best way of providing a philosophical basis in our engagement with the Earth system. Thus epistemic peers often lock horns when the answers differ from their separate prior beliefs in core environmental or moral values. For one thing, there is no right or wrong answer over peer disagreements; the difference really depends on which perspective one's view is taken from.

The epistemological disputes which stem from the wide varieties of ethical positions and moral theories put into question their practical importance and real impact in influencing and shaping individuals' environmental perceptions and morality. It may well be that individuals, especially the nonintellectual or uninitiated, may find it difficult to arrive at an intelligible understanding as to which normative positions of the peer disagreement are right or wrong. Clearly the troubling status of the translucid peer disagreement raises problems in negating human efforts in addressing real world environmental problems. This does not do justice to environmental philosophy, as a practical discipline, in addressing our looming planetary crisis. As noted in the previous chapter, we are faced with an urgent challenge to address practical problems, not theoretical disagreement.

To break this longstanding armchair theoretical impasse and intellectual confusion, it is necessary to cut deeper into the core of the philosophical framework so as to allow individuals to sort out which of the philosophical disciplines is directed toward practical solutions in addressing environmental problems in a real world system. That being said, a conceptual or theoretical epitome alone will not suffice. It is necessary for us to connect philosophical enterprise with evidential support gathered from empirical research. This will allow the philosophical disciplines as elucidated in the previous chapter to become more intelligible, coherent, convincing, and justifiable.

Furthermore viewed from a human cognition perspective, individuals are driven by the desire for an accurate view of the world around us (Alexander et al., 2017). Thus when one's philosophical argument or an environmental worldview is subject to scrutiny or provocative suggestions from critics, one should be capable of defending it in an intelligible, conciliatory, and steadfast way. It may well be that critics may be convinced when the counter-argument is consistent, reasonable, and appealing. The theoric—empiric approach provides that key.

The main purpose of this chapter is to test the ethical analysis and philosophical thought as extensively expounded in the previous chapter by drawing from empirical evidence gathered from ethnographic fieldwork conducted among the forest-dwelling indigenous people in Malaysia. More particularly, the fieldwork aimed to empirically examine to what extent the indigenous ethical engagement with nature serves as a critical motivational force in shaping moral attitudes toward nature, and how these environmental moral sentiments help to stimulate human long-term commitment to foster mutually enhancing human—earth relations, and hence, sustainable resource use practices. It also examines how the indigenous codes of ethical and environmental conduct relate to the theoretical and conceptual studies as discussed in the previous chapter. This serves to provide scientific philosophical support for the theoretical concepts and their practical importance in sustainable land resource use and environmental conservation. More particularly, it reveals immense insights into how we human beings should redefine our relationships with nature in a highly mechanized world, and how changing environmental value judgments may help us to ethically realign our sense of duty and obligation to nature.

Another salient feature of the present study is that it is fundamentally grounded in first-hand accounts of the evidence of indigenous ethical and environmental behaviors gathered over several years of actual experience with the local communities rather than on the extractions of aboriginal documentary narrations or testimonials. Such a distinctive framework of empirical study constitutes a more reliable contemporary ethnographic disclosure of the remnants of indigenous ethical and environmental legacies of their remote ancestors of a few centuries ago still being retained and observed by the present generations.

5.3 Indigenous land culture in brief

It needs to be made clear at the outset that through their daily interactions with their natural landscapes, the indigenous people have developed moral beliefs about the natural environment, in particular their ancestral land and forests, referred to as the ancestral domains or Datuk Nenek Moyang Temuda in the local term. More specifically, Datuk Nenek Moyang Temuda refers to land on which the indigenous people have lived since time immemorial. Historically it contributes to defining the cultural identity, spiritual attachment, and moral obligations of the communities to their traditional land. This human—nature cultural and moral orientations are deeply embedded in the cultural fabric and socioeconomic systems of the indigenous people. Particularly notable are the Penans who normally do not practice sedentary farming or cultivate land as other tribal groups do. They are traditionally hunters and gatherers but are now mostly seminomadic or settled. Nonetheless, because of their hunter—gatherer tradition, they still depend heavily on the forest environment for their cultural and socioeconomic sustenance compared to other nonhunter—gatherer tribes. This makes them extraordinarily vulnerable economically, socially, and culturally to land and forest degradation caused by various development activities.

Traditionally the indigenous people are bound by their adat to use the natural resources sustainably based on a set of ethical principles of land resource use. The term adat, as commonly used by the local communities, refers to customs or a moral code which has evolved over the past few centuries. More specifically, it refers to the oral traditions, cultural beliefs, rights and responsibilities, and customary practices that were created, nurtured, and preserved by previous generations based on their daily interactions with nature. These normative unwritten rules have been inherited wholly or partially and further developed by successive generations over the past few centuries (Colchester, 1993; Choy, 2014, 2018a). Adat, which is obligatory, not only governs the social behaviors of the local communities but also prescribes certain ethical norms in the use of the natural environment in such a way as to avoid massive irreversible destruction of the natural surroundings. The adat also lays down the authoritative moral principle of intergeneration equity which imposes upon the local communities the duty to protect the ancestral land and forests for the benefit of future generations. To the local communities, their surrounding natural environment is spiritually alive and they are an integral part of it. Hence they owe a moral duty toward protecting and preserving its ecological integrity.

Conceptually the indigenous adat is inextricably linked to the concept of usufructuary land rights system. Usufruct refers to the reserved rights of future generations to utilize the natural environment, commonly known as community or communal land and forests (ancestral domains), for their own cultural benefit. Under the usufructuary right system, private ownership of land and forests of the usufructuaries (the heirs) are not recognized because these environmental assets are held in common by all the inhabitants at large. In other words, usufructuary rights are considered as tribal rights rather than individual rights, which belong to the past (the dead), the present (the living), and the future generations (Colchester, 1993; SAM, 1996). These rights, which are considered as native customary rights are inviolable and unsubstitutable. The indigenous communities are bound by their adat to preserve the continued inheritance of these rights. However, private ownerships are recognized on the noncommunal land and forests in which the local communities are allowed to optimize their instrumental values for their socioeconomic sustenance.

The local communities are also guided by their adat to utilize land resources wisely based on sustainable natural resource use system generally known as Pusaka. It may be noted that other tribal groups may use different terms for their land use patterns, but basically all of them follow the same trend. The Pusaka system consists of three types of sustainable resource utilization patterns, namely, (1) hunting and gathering in the forested region known as Pemakai Menoa; (2) farming in agricultural land called Temuda, which is located in the vicinity of the tribal longhouses; and (3) preservation of old-growth forest termed as Pulau Galau. This preservation initiative is to protect the catchment areas, medical plants, or fruit trees found in the areas (Gabungan, 1999; Choy, 2004, 2014). The Pusaka land use system enables the local communities to live within the carrying capacity of the ecosystems. In other words, it is Holling sustainable (Choy, 2004, 2014).

Taken together, Pemakai Menoa, Temuda, and Pulau Galau constitute the ancestral domains of the indigenous communities. One of the ways the indigenous people conserve the ecological health of their ancestral domains under the Pusaka system is by practising their traditional farming known as swidden agriculture or shifting cultivation. This agricultural practice is one of the defining components of the local communities' traditional culture. Accordingly a relatively small plot of land is temporarily cleared for hill paddy planting. This farmland is used over and over again after it is allowed to fallow and regenerate. This traditional subsistence agricultural practice only involves temporary clearings of a small plot of forest within the forest landscape. In this way, the local peoples have been able to optimize the economic use of the natural environment without at the same time causing extensive or irreversible destruction to their native forests (Fig. 5.5).

5.4 Targeted areas of study—some basic facts

This section examines the nexus of environmental ethics and environmental sustainability based on lessons drawn from extensive field research conducted between 2007 and 2011 with the indigenous people in the tropical rainforest in the state of Sarawak in Malaysia. Geographically Sarawak is located in the north—west of Borneo Island. It is the largest state in Malaysia (Fig. 5.1 and Table 5.1). Its forests, which are some of the

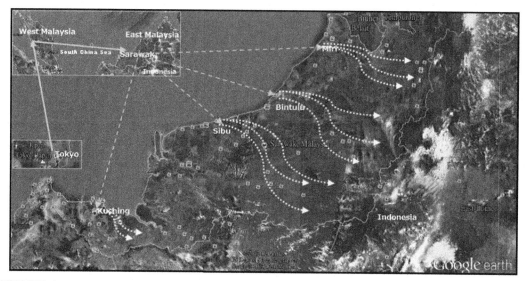

FIGURE 5.1   Geographical indication of targeted areas of study.

TABLE 5.1   Targeted areas of study in the state of Sarawak.

Year	Month	Name of longhouse/tribe	Location
2007	May	Mudung Ambun (Kenyah)	Bintulu
	May	Terbila Tubau (Kenyah)	Bintulu
2008	February	Ado Bilong (Penan)	Bintulu
	May	Long Bala (Kenyah)	Bintulu
	May	Long Apok (Penan)	Bintulu
	May	Rumah Anthony Lerang (Kenyah)	Bintulu
	August	Rumah Bagong (Iban)	Bintulu
	August	Rumah Jalong (Kenyah)	Bintulu
	August	Long Biak (Kenyah)	Bintulu
	August	Kampong Gumbang (Bidayuh)	Kuching
	August	Tanah Mawang (Iban)	Kuching
	August	Nanga Entawai (Iban)	Sibu (Song)
	August	Kulleh Village (Iban)	Sibu (Song)
	October	Rumah Amit (Iban)	Bintulu
	October	Rumah Mulie (Iban)	Bintulu
	October	Rumah Kiri (Iban)	Bintulu
	October	Uma Sambop (Kenyah)	Bintulu
	November	Rumah Akeh	Miri

(Continued)

TABLE 5.1   (Continued)

Year	Month	Name of longhouse/tribe	Location
2009	January	Long Lawen (Kenyah)	Bintulu
	January	Long Wat (Penan)	Bintulu
	January	Long Pelutan (Penan)	Bintulu
	January	Long Peran (Penan)	Bintulu
	January	Long Jek (Penan)	Bintulu
	July	Long Koyan (Kenyah)	Bintulu
	October	Rumah Sekapan Pitt (Kenyah)	Bintulu
	October	Long Dungun (Kenyah)	Bintulu
	October	Sekapang Panjang (Kenyah)	Bintulu
	October	Rumah Aging Long (Penan)	Bintulu
	November	Kampong Sg. Entulang (Iban)	Miri
	November	Kampong Sg. Buri (Iban)	Miri
	November	Long Laput (Kayan)	Miri
	November	Long Tutoh (Kenyah)	Miri
	November	Long Ikang (Kenyah)	Miri
	November	Long Banyok (Kenyah)	Miri
	December	Long Miri (Kenyah)	Miri
	December	Long Na'ah (Kayan)	Miri
	December	Long Pillah (Kayan)	Miri
	December	Long Kesih (Kayan)	Miri
2010	February	Arur Dalan (Kelapit)	Miri (Bario Highland)
	February	Bario Asal (Kelapit)	Miri (Bario Highland)
	February	Ulung Palang (Kelapit)	Miri (Bario Highland)
	August	Rumah Busang (Iban)	Miri
	November	Rumah Ranggong, Sungai Sah (Iban)	Miri (Niah district)
	November	Rumah Umpur (Iban)	Miri (Niah district)
	November	Rumah Ampan (Iban)	Miri (Niah district)
	November	Rumah Usek (Iban)	Miri (Niah district)
	November	Rumah Tinggang (Iban)	Miri (Niah district)
2011	February	Batu Bungan (Penan)	Mulu (near Miri)
	February	Long Iman (Penan)	Mulu (near Miri)
	February	Long Terawan (Berawan)	Mulu (near Miri)

Notes: Sarawak is the largest state in Malaysia, with an area of roughly 12.4 million ha mostly covered with forest.

oldest and most diverse in the world, have long been home to various indigenous communities such as the Kayan, Kenyah, Penan, and Iban among others. These forest-dwelling communities have developed, over hundreds of years, an intimate cultural relationship with their land and forests. In addition, the communities depend on these natural resources for their socioeconomic needs and sustenance in the form of hunting and fishing grounds, forest produce, and agricultural land for cultivation, shelter, spiritual well-being, and psychological comfort directly from the forest within which human—nature interactions occur.

The areas selected for empirical study were the indigenous settlements located mostly in the forest interiors in Bintulu, Sibu, Miri, Mulu, and Kuching in Sarawak. Special attention was paid to the research in the indigenous settlement in Long Lawen in Bintulu which was covered by almost undisturbed forests at the time of the field studies. Throughout the empirical studies, an expert driver who was familiar with deep forest track routes was hired, and a local guide who spoke various indigenous dialects was also engaged to aid in the field surveys.

5.5 Field trip physical environmental conditions in brief

Many of the field trips required long travel times of over 10 hours, particularly during the rainy seasons when dirt tracks became partly impassable. For instance, it took approximately 12 hours to travel from Bintulu to Long Lawen during a field trip in 2009 due to severely poor road conditions caused by heavy rain a day before our journey. Additionally accessing the rugged mountain terrains in the forest interiors without a proper and extensive network of transportation was almost impossible. Some of the field trips were particularly risky because of steep, treacherous, and winding mountain slopes, flooding, and degraded road conditions. For example, in another field trip from Bintulu to Long Lawen in 2010, the vehicle transporting the research team nearly overturned into the valley because of steep, muddy, and slippery dirt roads caused by heavy rain the day before.

In addition, field trips in the forest interiors necessarily engender a wide range of factors that contribute to unexpected situations. This required us to change in ways which met the demands of the unexpected events to proceed with the planned field trip. As a case in point, for example, at the time of almost reaching the remote site of Long Miri settlement after long hours of driving, we came across a small broken wooden bridge which was practically impossible for the land cruiser to cross. In view of this, we were forced to change our mode of transport by boat which doubled our total traveling time (Fig. 5.2).

Besides during the dry season, earth logging tracks were churned into clouds of dust by logging trucks, obscuring visibility, and making driving treacherous. In various settlements such as those in the upper Baram river bank in Miri, Sekapan Panjang longhouse, and Sekapan Pitt settlement, which can only be partially accessed by land-based transport, boats were hired as a means of access. At times, these boat trips could be risky due to high water levels and swift current conditions (Fig. 5.3).

FIGURE 5.2   Field trip—unexpected event (an illustration).
Note: The severely damaged wooden bridge with broken planks was uncrossable.

FIGURE 5.3   Field work environmental conditions at a glance.
Notes: Our way to Long Wat was blocked by a broken tree and the travel guides needed to remove it before we could proceed. Also during the wet season, swift currents can be dangerous when traveling along the river by small boat as shown in the figure. Additionally the treacherous road conditions necessitated the engagement of an experienced driver skilled in maneuvering the acute environmental situations.

FIGURE 5.4 Physical environmental conditions at site (some picturesque photographs).
Notes: According to my guide cum driver, the Long Apok settlement was a red area for dengue fever. Moving around the site can be a problem in hilly areas such as the Long Usek settlement.

In addition, conducting research deep in the forest interiors necessarily required enduring inhospitable environments, including the lack of basic needs such as clean drinking water and electricity, poor sanitary conditions, and the danger of viral infections such as dengue fever (to which the Penan settlements such as the Long Apok settlement are particularly prone) among others (Fig. 5.4).

5.6 Fieldwork and interviews

At the outset, it is relevant to point out that the indigenous people in Sarawak who have been interacting with their native land and forests for the past few hundred years are considered fully informed competent agents in defining what is good for them. In addition, they are assumed to have a settled preference for an ecologically healthy natural environment which generally overrides material benefits (Choy, 2004, 2014; Choy and Onuma, 2014). It is also relevant to note that the local communities, including the remote tribal groups in the deep forest interiors, are fully aware of the nonindigenous forms of resource use and management as practiced by the nonindigenous developers in their surrounding areas such as those in oil palm plantation development or dam construction. Thus the local communities are fully aware of the empirical relation between the environmentally destructive commercial land use practices and their environmentally sustainable traditional land use systems. Many of the indigenous peoples who travel to town to purchase their daily needs and other material goods are also familiar with the cash economy in the townships and can be said to have a sound appreciation of the purchasing power of money. Thus the problem of capacity limit of the local respondents to articulate their preference due to low educational level or poor numeracy skills as found by Venn and Quiggin in their Australian case studies does not arise (Venn and Quiggin, 2007).

Fieldwork and interviews with the indigenous people, including village headmen and old and young men and women, were carried out in the period between 2007 and 2011. The field research covered 46 longhouses located across the forested regions in Sarawak (Table 5.1). The tribal groups include 186 Kenyah, 146 Iban, 66 Penan, 57 Kayan, 30 Kelabit, 24 Bidayuh, and 6 Berawan (Table 5.1). A random sample of roughly 10–15 people from each longhouse were interviewed, depending on whether the prospective interviewees were in their longhouses or the surrounding vicinity or they were temporarily away at their farms (which could be located far from their settlements) and also on their willingness to be interviewed. Interviews were primarily but not exclusively conducted in the Malay language through random house visits and encounters on the farm and on the road. All the translations from the Malay language into English were done by me.

The main purpose of the field research was to examine the following: (1) the ways in which the natural environment has been construed in philosophical, cultural, and ethical terms; and (2) their relationship with environmental sustainability. This was achieved by documenting the indigenous environmental worldviews gleaned from data from face-to-face interviews with 496 people through random house visits, and encounters on the farm and on the road. Unless otherwise specified, all the names of the interviewees are withheld to protect their privacy as agreed.

The following are some of the issues of interest brought up in the interview process:

1. Moral concern for environmental protection for the benefit of future generations (the principle of intergenerational equity or justice).
2. Environmental preference: willingness to accept compensation for the loss of ancestral land and forests.
3. Sense of place/human relationship to nature. Sense of place is defined as the meanings of and the attachment to a place held by an individual or a community (Semken, 2005).
4. Place attachment—to compare local people's attachment to their forested environment and to the town area, that is, whether they prefer to live in a forested environment or town area and whether they are psychologically happier and satisfied or esthetically pleased to live in a forested environment.
5. Place identity—to examine the local people's emotional attachment to their natural surroundings.

Based on the findings on the aforementioned issues, we attempted to establish a cluster of moral or ethical inclinations and environmental values of the indigenous people underlying their environmental belief system. We further examined the implications for environmental sustainability through field observations.

The following are some excerpts from selected interviews. The first is an interview conducted with the seventh-generation longhouse chief from the Kenyah tribal group in Long Bala in May 2008 concerning his environmental worldview:

> We settled here at least 500 years ago. The land and forests here are very important to us. We will not sell them irrespective of the amount of money the government is willing to pay us. We feel happy living together with nature, and we owe a duty to our ancestors and future generations to protect them from degradation...

The rest of the respondents including Roslin (41), Mering (41), and Usen (48) concurred with their Chief, and all said that they felt very happy to be with the environment even though they were not materially rich. They also indicated a strong preference for living in their forested environment rather than in the town.

In a separate interview conducted with the Penan tribal group in Long Apok in May 2008, the longhouse chief, Junie Lating, said

> ...surely, we love and show due respect to our ancestral land and forests. We will not sell them irrespective of the amount of money an interested party is willing to pay. Whenever I look at the trees planted by our ancestors in our surrounding area, I rejoice and think of them. I feel very happy to live side by side with our natural surrounding....

Other respondents including Jaya Udau (30), Jackson Lavang (39), Joy Bunyi (30+), and Alo Jackson (20+) also revealed a strong belief that nature and humans are interconnected and the relationship has inherent values that cannot be compensated with money or other material benefits. They also revealed a strong preference to live in their ancestral land although they occasionally traveled to town to buy groceries or other goods.

Interviews conducted with the Penan tribal people in the Long Wat longhouse, including Saran (43), Baya (50+), Juman (50+), Labang (32), Jakun (22), Edin (22), Dywa (24), and Sati (26), also revealed an exceedingly strong preference for the land and forest. They refused to accept any tradeoff between their ancestral environment and money. They further indicated that culturally and traditionally, they were obliged to protect it for the benefit of future generations.

In another interview conducted in 2009 with the Kenyah community in Long Lawen concerning the local community's environmental perceptions, the community chief, Gara Jalong, replied:

> Lands and forests are our Datuk Nenek Moyang Temuda (ancestral domain). We have a total area of 21,700 hectares (ha) of forests, and out of these, 11,900 ha are marked as completely preserved areas. These are our communal forests and constitute our cultural identity. We owe a responsibility to our ancestors to protect them at all costs for the benefit of our future generations. Indeed, I have been jailed a few times when I tried to protect our forests from encroachment by the private developers. Every one of us feels disturbed over the destruction of environment in our surrounding area caused by oil palm plantation development. We also feel distressed when we see the rivers that flow through our territory being polluted by mud caused by land clearing by the oil palm companies... (Fig. 5.5)

At this juncture, Juk Nyok Along and Siting Selong interjected and claimed that they had also been jailed for attempting to protect their ancestral land from the logging companies.

The longhouse chief further added that "In order to use our land resources sustainably, each household, depending on the size of its family, uses about 10–20 acres of the secondary forests for shifting cultivation (swidden agriculture). The same piece of land is used over and over again after it is allowed to replenish. In this way, we are able to manage our lands and forests in a sustainable way...also, we go hunting, fishing and gathering forest produce near the secondary forests." (Figs. 5.5 and 5.6)

The above land use practice is structurally based on the Pusaka land use system.

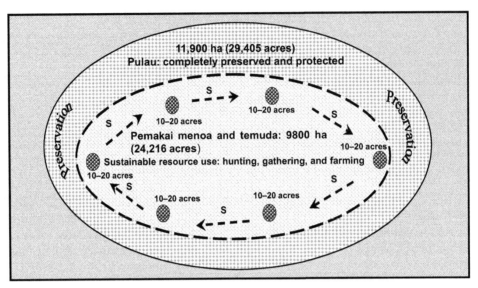

FIGURE 5.5 Indigenous land use patterns in Long Lawen. *S* indicates shifting cultivation; arrows indicate rotation; and, ⊛ the land under shifting cultivation. *From Choy, Y.K., 2004. Sustainable development and the social and cultural impacts of dam-induced development strategy—the Bakun experience. Pacific Affairs 77 (1), 50-68.*

FIGURE 5.6 Indigenous land use patterns in Long Lawen (photographic representation). Picture 1: land use pattern of the local community in Long Lawen; Picture 2: durian fruit trees by the river bank; Picture 3: the long-house Chief with a durian fruit picked from the river bank; and Pictures 4 and 5: environmental degradation caused by oil palm plantation development in the surrounding area of the local community in Long Lawen. *Note*: Photographs taken by the author in 2009.

During a boat trip with the longhouse chief from Long Lawen to Long Anau, he pointed at the durian fruit tree on the river bank planted by their ancestors and said "Look at the durian tree over there; it is a gift from our forefathers. It belongs to the community and everyone has the right to harvest fruit from it. We feel happy seeing the tree growing healthily until today and we think back of our ancestors when we see it..." (Fig. 5.6).

It is particularly revealing that during a farm encounter near Long Lawen, Unit Liah, in her mid-30s, was asked whether she would give up her land resources for, say, 1 million Malaysian Ringgit (roughly US $234,000 as of July 13, 2020 exchange rate), which would allow her to buy anything she wanted such as a big, modern house, and other luxurious items. Her reply was

> Of course not! Land resources are the most important things to us. They are not meant for sale. What is money without land? The dense-forested place here also gives our people pleasant and comfortable living conditions compared to Bintulu town...You just listen; the sound of the birds is so sweet to hear. How can you experience this in Bintulu? Although we are not rich, we are happy and satisfied living here...Our adat (custom) also requires us to protect the land for the benefit of our children... (2009).

The traditional wisdom in environmental conservation has helped the local community to preserve extensive tracts of communal forests in an ecologically healthy state. More specifically, the traditional land use practices and subsistence activities which exert relatively minimal pressure on the natural environment, and the mapping of traditional forests as communal forests for the benefit of future generations have been extraordinarily effective in preserving the ecological integrity as well as preventing extensive habitat destruction and fragmentation of the surrounding area. This contrasts with the commercial land use patterns such as oil palm plantation development of the economic agents which results in extensive habitat fragmentation and irreversible destruction to the forest ecosystem (Fig. 5.7).

Field studies were further conducted in the Long Lawen settlement in 2010 and 2011 regarding local socioeconomic situations and cultural orientation amidst a degree of integration into the mainstream economy and against the background of the socioeconomic impact of various rural economic development activities being carried out in the surrounding areas. It may be noted that various rural development projects such as the Bakun and Murum dam constructions and oil palm plantation development, carried out in their surroundings had brought a range of socioeconomic benefits such as better communication networks, provision of health and education facilities, and employment opportunities, among others.

The field studies indicated that the local communities were steadfast in upholding the moral codes of proenvironmental behaviors previously elucidated. As a case in point, the community chief revealed at an interview conducted in 2011 that despite the attractive offer from industrialists of improved road conditions and monetary benefits in exchange for part of their ancestral land for development, the local communities remained unmoved. Instead, they became even more determined to protect their land resources from external acquisition (Choy, 2014). This reflects that the ethical force of adat has consistently exerted an authoritative influence on their collective consciousness

FIGURE 5.7   Indigenous and commercial land use patterns: a comparison.
Note: Photographs for Long Lawen were taken in 2009 while pictures of oil palm plantation development were taken in 2010 during a flight from Bintulu to Bario Highland by the author.

in upholding the moral principle of intergenerational equity and the tradition of harmonious human–nature relationships.

Interviews were also conducted with the Kenyah community in Long Dungun in 2009 concerning their environmental worldview and morality. In summary, all those interviewed including Laing Laeng (longhouse chief), Antuk Iban (36), Lirong Lambong (40+), Ampit Lusat (50+), Majit Nuen (38), Iban Angit (60), Kojan Kavan (83), Lorong Ipui (48), and Ingan Usa (60), among others, concurred that traditionally, they nurtured an intimate relationship with their land and forests which constitutes their cultural identity. They also revealed that they own a traditional responsibility to uphold their adat to protect the land and forests for the benefit of future generations. Until today, their culture of respect for nature has not declined despite the changing socioeconomic environment and lifestyle trends in terms of easier access to the town area and better modern facilities such as the supply of electricity and the possession of electronic appliances. All of them were very concerned about the degradation of their natural surrounding areas especially river pollution and extensive accumulation of mud by the river bank caused by oil palm plantation development, logging activities, and dam construction (Bakun dam) in the vicinity of their settlement or the upstream region (Fig. 5.8).

It may be noted that, according to the longhouse chief, the local community originated from Long Busan in the upper Balui River near the Indonesian border more than 30 years ago for educational and health facilities available in the nearby area. In the former settlement, the children took one day to travel to school which was very inconvenient.

FIGURE 5.8 Long Dungun: river pollution and mud accumulation.
Note: Photographs taken by the author. The river was choked with yellow mud caused by oil palm plantation development and logging activities in the vicinity of the local settlement.

However, the local community still maintains their ancestral roots in Long Busan and some of them, including Antuk Iban and Lirong Lambong, would return occasionally to recapture their ancestral sense of identity from their original and traditional territories. At the time of the interview, both of them were considering returning to Long Busan due to their strong ancestral attachment.

It may further be disclosed that interviews conducted with the Sekapan community in Sekapan Panjang settlement and Kayan community in Rumah Aging Long settlement in 2009 also revealed the same environmental worldview and concern as the local community in Long Dungan settlement. Hagoh Gatah (43) from Sekapan Panjang, for example, claimed that

> Our community has been living here since James Brooke's period. That's the reason why we have culti-vated a close relationship with our natural surroundings. Surely, land and forests are the most important things to us because they are harta Datuk Nenek Moyang (ancestral property) and, we will not sell them for any amount a person is willing to pay. I am happy to live side by side with our forests and I feel very pleased when I see the plants and trees planted by my ancestors are growing healthily. We owe a responsi-bility to protect our ancestral land for our children and grandchildren. However, we feel very sad that the dam construction and logging activities upstream are destroying our environment... (2009).

To note in passing, Sir James Brooke was a British soldier and adventurer who founded the Kingdom of Sarawak. He ruled the Kingdom as the first White Rajah of Sarawak from 1841 until his death in 1868.

Ugu matu (55) and Lorraine Unyi (35), among others also shared the same sentiments as Hagoh Gatah. In Rumah Aging settlement, Lujah Sepak (73) has the following to say:

> Our community has been living here for the past 200 years. We belong to our land and forests and we are required under our adat to protect and preserve them for our future generations. We feel happy living here. For me, I will never sell any part of land for money. It is the most important asset to me. The problem our community is facing is river pollution caused by the Bakun dam construction... (2009)

Other respondents including Levoh (49), Bawe (35), and Lapong (53), among others, also revealed the same environmental sentiments about morality and their preference for forest living.

In a further interview conducted with the Iban tribal people in 2010 from Rumah Ampau longhouse, Kudang Enpaleng (57), Bagol Neyang (60), Malin Lankai (60), Teggong Sigas (64), Buki Gasinh (48), Angoh Duman (29), Angang Asan (32), and others indicated a strong relationship between traditional environmental attitudes and values. Similar to the rest of the interviewees as noted above, their value of the environment was based on social-altruistic attitudes and deontological behavior. Their relationship to land and forests involves a sense of place and belonging. According to them, to move away from their ancestral environment or to relocate to a different geographical location is akin to destroying their socioeconomic fabric and cultural identity.

Interviews were also conducted in 2010 with one of the smallest tribal groups in three of the settlements in Bario Highland, the Kelabit tribal minority group. All the interviewees including Laju Balang (46), Sina Riput Bala (59), Mujan Agan and Paran matu (both 82), among others from Barrio Asal settlement; Rining Turun (45), Sinar Madu (67), Fendi Gjafar (36), among others, from Ulung Palang settlement; and Sinah Do Ayu (60), Agustina Kaya (41), Rosly Meria (44), Christina Harrison (24), and Charles Migong (56), among others, from Arur Dalan settlement, concurred that they shared an intimate relationship with their land and the natural environment. Although they were living in an isolated highland which is basically and conveniently accessible only by air, they felt happy to be in their ancestral territories which they considered awe-inspiring natural landscapes. All of them shared a collective responsibility as required under their adat to preserve the ecological integrity of their land and forests to uphold the traditional duty of intergenerational justice.

Interviews conducted with the Penan community in 2011 in Batu Bungan longhouse also revealed the same environmental beliefs and value system. For example, as Mamat Beti (34) disclosed, "I will not sell off my land irrespective of the amount an outsider is willing to pay because by tradition, I have a responsibility to preserve it for the sake of my children and grandchildren. In addition, land and forests form a part of our cultural identity and we cannot live without them...".

It may be noted that all the interviewees above expressed a great sense of grief over the oil palm plantation development in their surroundings which destroyed the natural environment. They were also concerned about the severe river pollution caused by the deposits of significant amounts of mud and silt from land clearing by the oil palm companies as well as chemical pollution to their surrounding natural landscapes caused by extensive use of pesticides in oil palm plantations, depriving them of the supply of clean water from the river (Figs. 5.9 and 5.10).

FIGURE 5.9   Oil palm plantation development in the forest interiors: some evidence.
Note: Photographs taken by the author.

FIGURE 5.10   Environmental degradation caused by oil palm plantation development.
Note: Photographs taken by the author.

To furnish a large sampling and a qualitative index of the overall studies, field research was also carried out in other areas of the forest interiors in the settlements listed later. Similarly the communities were surveyed about their environmental morality, cultural beliefs, and way of life. The following revelations from the interview excerpts are worth noting:

1. Mudung Ambun (Kenyah tribe)—The chief of the longhouse: "How can we live without land? It is a part of our life. We are born here and live here. You just look at those people in Sungai Asap. Many of them feel a sense of emptiness inside them because they have been forced to separate from their ancestral lands and to give up their traditional way of life..." (May 2007). (To note briefly, giving up their traditional way of life includes giving up their traditional farming, namely, swidden agriculture which is an important part of the indigenous culture. Sungai Asap is a resettlement area for mostly the Kenyah people who were forced to relocate because of a dam construction project.)

2. Long Bala (Kenyah tribe)—The chief of the longhouse, Lenjau Lian, who claimed to be the seventh generation's chief, said that "I can't remember when we settled down here but it must be at least more than 700 years ago. We have been living together with our lands and forests throughout our life. Our forefathers are also buried here. Surely these traditional resources are irreplaceable. They cannot be bought or sold as we like because they belong to our community... we owe a duty under our adat to use them properly for the benefit of our children..." (May 2008). Other local settlers such as Steven Usen, Roslin Era, and John Mering, all in their early 40s, also shared the same views as their chief and further reaffirmed their traditional land custodial duties.

3. Long Apok (Penan tribe)—The chief of the longhouse, Junie Lating, in his early 60s, has the following to say: "Surely we love our lands and forests because we cannot be separated from them... This is our ancestral home and it gives us most of the things we need. We have to take very good care of them or else our children will blame us in the future..." (May 2008). Other local inhabitants such as Jaya Udau, Jackson Lavang, and Joy Bunyi also expressed similar sentiments and conceded that traditionally they cannot live without their lands and forests.

4. Rumah Jalong (Kenyah tribe)—Ejam, in her early 60s, revealed that "...every one of us must have a piece of land...without land we will not be happy in life... We belong to this place as we have been living here since our ancestors came here long time ago... Our forefathers also 'live' (buried) in our surrounding area..." (August 2008). Lerah (48) interjected that they could not live just with money alone because they would feel insecure in life and psychologically distressed without any land which they considered as a part of their social fabric.

5. Kampong Gumbang (Bidayuh tribe): among the persons interviewed, the local settlers such as John and Rasang; Anyan and Jipep revealed a strong affection for their ancestral land resources. For instance, Jipep (56) who had 20 acres of land claimed that "I will not sell off any part of my land no matter how much the developers are willing to give. It is a very important resource to me. Although I am poor, I am happy living here because this place had been created and passed to us by our ancestors..." (August 2008).

6. Nanga Entawai and Rumah Kulleh (Iban tribe)—The respondents such as Matu, Senin, and Alexius and, Andy and Jawi, from their respective villages affirmed their traditional duty to preserve the land resources for their children's benefit. Alexius, for instance, a 17-year old

with 5 acres of land has the following to say: "Land surely is very important to us...our traditional way of life depends on whether we are able to keep our lands in proper state and then pass them to our children for their use in the future..." (August 2008).

7. Rumah Sambob (Kenyah tribe): at the interview, the longhouse chief said that "...an indigenous individual who has lost his ancestral land is just like a ship without a captain" (October 2008).

8. Long Peran (Penan tribe): the chief of the longhouse affirmed that the local people are bound by their adat "to use the natural resources sustainably for the benefit of our future generations...and we feel happy roaming around in the forested areas and listening to the birds' sound...But we are now getting worried because the Murum dam project will force us to leave our traditional place and to settle in a different area which we have no connection with..." (January 2009). (It may be noted that the Murum dam which was under construction in the forest interior at the time of the field research would result in forced relocation of all the Penan communities in the surrounding areas. The dam was completed in 2016.)

9. Bario Asal, Arur Dalan, and Ulung Palang in Bario Highland, Miri (Kelapits tribe): all the respondents including Laju Balang, Elvis Lagit; Mujan Agan and, Renai Uding from Bario Asal, Sinah Ayu, Agustina Kaya; Steward Pian and, Charles Miong from Arur Dalan and, Fauziah Abdullah, Juaing Katu, Fendi Ghafar and Raja Tepun from Ulung Palang unanimously agreed that they could not be separated from their land resources which they considered an important part of their life. Fauziah Abdullah for instance, has the following to say "...lands and forests are a part of our culture and we must ensure that they are kept in a healthy state so that they can be passed down to our children for their use again..." (February 2010).

10. Batu Bungan and Long Iman (Penan tribe) and Long Terawan (Berawan tribe): these local communities exhibited the same line of ethical and environmental reasoning expressed by the rest of the tribal groups. Ukau Lupung (52) from Batu Bungan, for instance, disclosed that, "...without lands and forests, we will feel lost...we must be careful not to allow the developers to enter into our territory to develop the areas or else our children will suffer in the future..." (February 2011). Other interviewees from Long Iman (such as Raymond Lejau, Bulan Osong, and Pengiren Juga) and Long Terawan (such as Nicholas Ugum, Paulus Banda, and Michael Ugum) also share the same land perceptions of the local communities in Batu Bungan (February 2011).

On the whole, the field studies reveal clearly the moral and environmental positions of the local communities in fostering carefully managed and sustainable uses of their ancestral resources. This may simply be summarized as follows:

1. All unanimously concurred that they were bound by adat to use their land resources sustainably for the benefit of their descendants. Particularly the community members saw themselves as stewards for the communal land resources inherited from their ancestors.

2. All the local communities showed due respect and passion for their ancestral domains which had also been the home of their forefathers as well as the base of their socioeconomic and cultural sustenance.

3. All those who had traveled to town areas felt that they were happier living in the forests than in the town because the forested areas were their original home to which they belonged.
4. All the local people agreed that they represented a part of the natural environment in which they lived.
5. Although most of the local people were poor, practically everyone interviewed concurred that psychologically they were happy to be with nature which also served as their life-support system.
6. All local communities were very protective about their ancestral land and forests which constituted a part of their cultural identity. They would strive to uphold their traditional environmental stewardship responsibility which had evolved over the past few centuries. They would defend it against outsiders especially the private developers and logging companies encroaching into their territories for commercial exploitation.
7. All interviewed unanimously revealed that they would feel concern and distress if their natural surroundings were degraded or destroyed by the oil palm plantation developers or logging companies. In other words, they were extraordinarily ecologically sensitive.

Here, the forces that underscore the shared understandings of land ethic and environmental attitudes of the local communities may not readily be apparent, but clearly they require deeper exploration and explanation as this will unpack a wealth of novel principles to stimulate the practice of environmental ethics in ecological conservation. This is taken up in the following section.

5.7 Empirical findings and the indigenous land use philosophy: Some conceptual underpinnings

One may argue that just as individuals vary in degrees of conformity to their norms of social behavior; all members in a given community likewise, may not collectively adhere strictly to the cultural attributes as embedded in their social structure. For instance, while some members may exhibit a strong tendency to follow firmly their standard habits of practice, others may choose to march to the beat of a different drummer. Nonetheless, while accepting the premise of such behavioral norms among members in a given community, there are also exceptional cases. For instance, in the case where the virtue of social conformity to certain moral principles is historically and cohesively observed without fail due to their potential cultural importance, an abiding fear for being irresponsive to collectivistic observance persists among the community members (Hsu Francis, 1948; Markus and Kitayama, 1994; Fiske et al., 1998; Choy, 2014; 2018b).

Conceptually these cohesive cultural norms constitute the collective representation among members of the local community. Within the present context, collective representation refers to moral ideas, ethical beliefs, and environmental values collectively embraced by members of a given society and are not reducible to individual self-interested values, beliefs, or goals (Durkheim, 1915). The indigenous people are said to act morally when they work "towards goals superior to, or beyond, individual goals", that is, beyond self-interested goals (Durkheim, 1915: 69). This moral norm is an obligatory source of social and cultural solidarity which is deeply embedded in the indigenous adat. Consequently all the members of the community are obliged to collectively uphold the adat-based cultural norms of moral practice in their interaction with the natural environment.

Thus as empirically revealed above, despite the diversity among indigenous groups, tribal languages, and geographic locations, all the respondents consistently recognized the cultural importance of the forest landscapes that were once the dwellings of their remote ancestors. The local communities expressed deep reverence for the forest landscapes or the ancestral domains, imbuing them with an array of sentimental qualities such as cultural values, social morality, passion, spiritual significance, and psychological drives, among others. The belief in this environmental value pluralism gives rise to a unified system of moral beliefs and environmental practices of the local people in relation to their interaction with land and forests.

More specifically, following Durkheim's conceptual argument, this environmental moral system of which cannot be altered according to individual taste has been internalized and deeply embedded in the inner thoughts and hearts of the local people and it serves to shape their ethical behavior toward the natural environment in terms of right and wrong actions affecting the forest landscapes. These ethical principles serve as a guide in determining how much of their ancestrally defined forest landscape they intend to preserve intrinsically, how much culturally modified landscape they can instrumentally afford, and in what ways the land should be modified to foster mutually enhancing human—nature relations.

Demonstrably the Pusaka land use system is a good illustration. Accordingly the Pusaka land use philosophy comprises a conceptual matrix of weak anthropocentrism, biocentrism, and ecocentrism. More specifically, it consists of the instrumental or anthropocentric use of land and forests for socioeconomic sustenance and esthetic/recreational satisfaction, and the moral consideration for nonhuman living things such as birds (biocentrism) and land ethics (ecocentrism) in relation to land and forests. In short, this land use philosophy reflects both the anthropocentric and nonanthropocentric view of nature— the anthropo-bioecoentric environmental worldview. As demonstrated in the land use pattern in Long Lawen, for example, the local people instrumentally utilize a relatively small portion of their land of roughly 10—20 acres for farming/socioeconomic sustenance (instrumental or anthropocentric use of nature) while setting aside 11,900 ha out of a total of 21,700 ha of land and forests as totally protected and preserved area (intrinsic value orientation or bioecocentric view of nature).

This totally protected area created for environmental protection and preservation is not considered as a commodity which belongs to the local people to be dealt with or disposed of according to their preference but a commodity to which the local people belong. The local people are required by adat to protect its "integrity, stability, and beauty" based on "love, respect, admiration and a high regard for its value", for the benefit of future generations. Such environmental philosophy is also in consonance with Aldo Leopold's "limitation on freedom" principle in the economic use of nature as discussed in the previous chapter. It also resonates well with Albert Schweitzer's precautionary principle in taking anticipatory action to prevent irreversible harm to the natural environment.

Conceptually the "integrity, beauty, and stability of the biotic community" is the measure of right and wrong actions affecting the environment (Callicott, 1989: 58). Accordingly protecting and preserving the traditional land and forests for the benefit of future generations is the right and desirable thing for the local people to do because it conforms to their adat—the traditional fulcrum for morality of the local communities. The indigenous adat shows a significant affinity with Durkheim's argument of moral rules in that it is invested with a special authority by virtue of which it is obeyed simply because it commands (Durkheim, 1953: 35–36).

More particularly, the indigenous people are bound by their adat to obey the moral rules inherited from the past because the adat commands what it commands and such command is deemed good and desirable insofar as it concerned about the preservation of the cultural identity of the local communities. Put differently, the indigenous people submit themselves to such command not because certain actions are negatively or positively sanctioned by preestablished cultural rules but because of their submission to the obligatory and binding authority of their adat. In this respect, it may be noted that according to Durkheim, a sanction "is the consequence of an act that does not result from the content of that act, but from violation by that act of a preestablished rule" (Durkheim, 1953: 43).

Indeed, as revealed from the field study, the indigenous people are required to abide by their adat to refrain from performing certain acts which are detrimental to the natural environment and to the interest of the future generations because such acts are customarily forbidden. Furthermore to the local people, observing this centuries-old adat is an irreducible and distinct duty. This traditional moral belief must be obeyed in all situations and circumstances irrespective of any consequences for human welfare that might arise from doing the "right" action (Choy, 2014, 2018a). Thus individual freedom (as in the Long Lawen's case) can be sacrificed for doing the "right" action. This obligatory or duty-based moral principle is one of the fundamental characteristics underlying the indigenous bioecocentric environmental worldview. This may further be conceptually examined in the next section.

5.8 The indigenous duty-based moral principle: The Kantian categorical imperative

The obligatory or duty-based moral principle of the indigenous people as discussed earlier is not without its conceptual underpinnings. Conceptually it is well reflected by the deontological moral theory as propounded by Immanuel Kant (1724–1804), one of the most influential philosophers in modern philosophy. According to Kant, what makes an action "right" (as defined by Leopold.) in itself is its conformity with moral law rather than consequences and that it has moral worth in that it is done out of respect for the moral principle (Kant, 2002). In Kant's view, the moral worth of an action rests behind the motive rather than the outcome that is achieved by such an action (McCormick, 2018). Kant devises his supreme principle of morality or moral law known as categorical imperative from which our duties and obligations are derived.

Accordingly acts done in accordance with the supreme principle of morality apply to everyone, universally, absolutely, and unconditionally. In other words, such moral acts are premised on objective necessity and do not depend on one's own goal, desire, ulterior motive, or end: it is a goal in itself and must be obeyed in all circumstances. That is to say, the categorical imperative commands us what we, as rational beings, are morally obliged to do without exception. Accordingly Kantian categorical imperative constitutes the most important principle of deontological ethics according to which the rightness or wrongness of actions does not depend on their consequences. Rather it depends on whether it fulfils the moral law of conduct on which all rational beings should unconditionally act in accordance with, that is, "followed even against inclination" (Kant, 2002: 33).

To explain Kantian supreme principle of morality in the modern context and from the indigenous perspective, it means that the protection of ancestral land and forests at any cost including personal sacrifices for the sake of future generations is morally right if and only if the intention of carrying out this obligation by a member in the community is an obligatory act which everyone in the community would be willing to carry out for the same reason. This is because it is considered the right way for all members in the community to behave. It would be wrong for any member of the community to contravene this moral principle solely for self-interest.

Thus as reflected clearly in the Long Lawen case, the local people, when acting in accordance with their adat, were concerned only with the maxim of their act, not at how much misery is likely to be caused by the act of protecting their land and forests for the benefit of future generations. Furthermore the maxim is universal because everyone in the community would act without inconsistency to protect the interests of future generations. This is precisely the basic tenet underlying the indigenous adat (custom) which can be applied at all times to every moral agent in the community (Choy, 2014, 2018a).

The second formulation of Kant's categorical imperatives is the Humanity Formula which states that "Act so that you use humanity as much in your own person as in the person of every other, always at the same time as an end and never merely as means" (Kant, 2002: 46−47). The Kantian means-end principle places great emphasis on human dignity in that it contends that it is wrong for us, as rational beings, to treat others merely as means. According to Kant, "a human being is not a thing that can be used merely as a means but must in all his actions always be considered as an end in itself" (Kant, 2002: 47). In other words, a human being is intrinsically valuable as an end in itself. Kant further concedes that ethically we have a duty of beneficence toward other persons (Kant, 2002: 41−49), and each rational being must "aspire, as much as he can, to further the ends of others" (Kant, 2002: 155).

It thus follows that we must strive not to cause harm to others based on our best rational judgment. We must also respect their rights and concern about their needs, welfare, or happiness and must not turn them into objects of manipulation or an instrumental means for the attainment of personal goals or purposes, no matter how good those attainments are (Rachels, 1997; Lazier, 2010). Thus in effect, Kant is imposing a supreme limiting condition of every rational being's freedom of action in the use of means to every end (Kant, 2002: 55−56). This resonates well with Leopold's land ethics defined as "a limitation on freedom of action in the struggle for existence" as discussed in the previous chapter.

To put Kantian humanity ethics in the indigenous context, it may be that all members in the tribal community including those in the distant future have equal dignity and the proper treatment they are accorded is an end in itself. That is to say, they deserve equal treatment and moral respect as enjoyed by the present generation. Viewed from the Kantian perspective, to treat the unborn members of future generations merely as a means in accord with one's own will is to involve them in discretionary use to further an end to which they could not in principle consent, and is morally wrong.

Being yet to be born, they are also not free to say "no" to being robbed of the opportunity to make informed choices. No matter how beneficial the consequences of overexploiting or disposing of the ancestral land and forests, these actions are not morally permitted under the indigenous custom. In addition, such an act would violate the traditional obligation to care for the future generations and obviously affect the welfare and interests of the community (Choy, 2018a). The rule of the game is that treating others as ends requires that each person's rights or freedom be respected. Thus the local peoples are bound by their adat to preserve "the integrity, beauty, and stability" of their ancestral domains for the interests of future generations.

This traditional principle is the supreme principle of morality in association with the concept of intergenerational equity or justice. It constitutes an integral part of a coherent system of ethical thought (collective representation in Durkheim) of the local people where the value of humanity as eloquently revealed under the Kantian formula of humanity must always be respected. Based on this maxim which may seem to be an inviolable ancestral command, the local people are morally and culturally forbidden by their adat to indulge in any action that treats the future generations merely as a means to an end such as the disposal of land and forests in exchange for the attainment for material progress.

In other words, they are commanded by their adat to treat the future unborn members of the society as ends in themselves. It is noteworthy that the indigenous adat establishes an absolute and binding system of morality for all local forest-dwelling tribal communities in the whole of Sarawak in relation to the protection and preservation of their ancestral land and forests for the benefit of future generations. Such a moral principle must "universally" be adhered to all the time without exception and any member of the community who is guilty of its violation is considered immoral and irrational. It may further be added that the preservation of "the integrity, beauty, and stability" of the ancestral domains for the benefit of future generations is also spurred by the local people's bioecocentric behavior toward the natural environment.

5.9 Indigenous bioecocentric environmental worldview: The nexus of values, environmental attitudes, and moral actions

With reference to the questions raised in the Introduction section, it is by now supremely clear that the conception of morality as deeply embedded in the indigenous adat primarily albeit not exclusively stems from the nonanthropocentric concern for the protection and preservation of "the integrity, stability, and beauty of the biotic community." Such environmental worldview, called here the anthropo-bioecoentric environmental belief system is inextricably linked to the cultural sustainability, socioeconomic fabrics and the equity of future generations.

To the local communities, land and forests are objects of awe and wonder which provide them with a sense of belonging. This environmental inclination coincides with the awareness of man's ethical relationship with nature based on reverence, respect, admiration, and love. This environmental belief influences and guides the local communities ethically to order and interpret the natural environment nonanthropocentrically. In other words, it ensures that all members of the society share a common bioecocentric belief in protecting and preserving "the integrity, stability, and beauty" of their ancestral land and forests, as well as the surrounding natural landscapes. To the local tribal communities, it would be wrong if they did "otherwise"—the benchmark of the rightness of action as prescribed by Aldo Leopold discussed in the previous chapter. As revealed from field observations of the tribal communities' surrounding environment, such environmental inclination has helped them to protect and preserve the ecological integrity of the tribal communities' ancestral land and forests for the past few hundred years.

The environmental attitudes and behavior or the moral basis of environmental decision making of the indigenous people as elucidated above are themselves a reflection of a host of values. In other words, values serve as the criteria the indigenous people use to justify their moral actions or deontological behavior although they themselves may be unaware of the conceptual underpinnings of their actions. To emphasize, when the indigenous people think of ancestral land and forests as important to their life and hence need protection and preservation, they are indeed placing a series of values on them. For example, in fulfilling their adat-based supreme principle of morality, they are in fact placing a high value on their traditional duty of intergenerational equity. In other words, they are said to have a strong cultural-value orientation. However, value orientations are not mutually exclusive as each orientation may lead to another orientation. Thus an individual may hold several orientations which may collectively lead to the enhancement of proenvironmental behaviors (Stern et al., 1993; Stern and Dietz, 1994).

More specifically, the adat-based moral belief system of the indigenous people provides the fundamental basis in guiding the local communities' deontological environmental attitudes, behaviors, and actions. Particularly it serves to influence the local communities to interpret and order their ethical position and moral duty toward nature. As the empirical studies reveal succinctly, all the tribal communities (100%) are culturally obliged by their adat to morally protect and preserve the ecological integrity of the ancestral landscape environment for the benefit of future generations.

These proenvironmental attitudes and deontological beliefs are in turn linked to a matrix of human welfare enhancement values which further reinforce the local communities' determination to do the "right" thing from Leopold-Kant's perspective when optimizing the instrumental use of nature. These values as shown in Table 5.2, Table 5.3, and Fig. 5.11 may be briefly listed as follows:

1. Sense of place value: value derived from a set of meanings (e.g., cultural meanings) of and attachments to places (localized meanings) that are held by local people.
2. Kinship value: value placed on local people's emotional attachment to their natural surroundings and sentimental feeling (sentimental value) including reverence for the trees, mountains, and rivers (environmental value).
3. Spiritual value: the local landscapes are seen as places that the local communities connect emotionally and spiritually with the ancestors (sentimental value and spiritual value).
4. Bequest value: the local people are bound by custom to protect the communal land and forests (cultural value, communal value, and conservation value) so that future

TABLE 5.2 Values subscribed by the indigenous people.

Tribal group	Total number of respondent	Cultural value		Esthetic value		Psychological value		Sentimental value		Kinship value		Moral value		Ethical value		Bequest value	
		Yes	No	Yes	No	Yes	No	Yes	No	Yes	No	Yes	No	Yes	No	Yes	No
Kenyah	166	166	0	166	0	166	0	166	0	166	0	166	0	166	0	166	0
Iban	146	146	0	146	0	146	0	146	0	146	0	146	0	146	0	146	0
Penan	66	66	0	66	0	66	0	66	0	66	0	66	0	66	0	66	0
Kayan	57	57	0	57	0	57	0	57	0	57	0	57	0	57	0	57	0
Kelabit	30	30	0	30	0	30	0	30	0	30	0	30	0	30	0	30	0
Bidayuh	24	24	0	24	0	24	0	24	0	24	0	24	0	24	0	24	0
Berawan	6	6	0	6	0	6	0	6	0	6	0	6	0	6	0	6	0
Yes (%)		100		100		100		100		100		100		100		100	
No (%)			0		0		0		0		0		0		0		0

Cultural value Satisfactions derived from being able to live together in harmony and peace with nature (land and forests)

Esthetic value Satisfactions derived from the enjoyment of scenic beauty of nature including natural green, plants and flowers, flowing river and streams as well the happiness derived from the enjoyment of the songs of birds and the sound of river water, among others.

Psychological value Satisfactions derived from the feeling of tranquility and peacefulness from the forests or psychological health benefits derived from the forested environment

Sentimental value Satisfactions derived from the indigenous people's emotional ties with their land and forest. It is a symbolic value which contributes to social well-being measured in term of happiness and can only be sustained through sustainable resource use and forest conservation.

Kinship value Kinship value is the value of wholeness or totality. It refers to the indigenous people's kinship ties with their land and forests. More specifically, they see themselves as a part of nature in which they belong and the natural world is embraced as an integrated system of human—nature relations. Put simply, the indigenous people derived their satisfactions from being able to live side by side with their ancestral land and forests.

Moral value Satisfactions derived from fulfilling the moral responsibility of land and forest protection as required under the indigenous adat (custom).

Bequest value Satisfactions derived from being able to pass on ancestral land and forests to future generations.

Existence value Satisfactions derived from being able to protect the land and forests to enable them to exist in an ecologically healthy state.

Notes: The indigenous people were asked about their feelings toward nature expressed in terms of values such as moral value or esthetic value as defined. "Yes" means nature exerts a positive impact on their emotions. "No" means they are indifferent, that is, nature does not impact on their feelings or emotions. All indigenous people interviewed reflect a positive relationship between human and nature with 100% said Yes while 0% said No. The values as shown in Table 5.2 may further be summarized in Table 5.3.

TABLE 5.3   Summary of value and ethical orientations of the indigenous people.

Tribal group	Number of people interviewed for each tribal group	Average age group	Hard-to-define environmental values attached to land and forests				Remark
			Cultural value (%)	Esthetic value (%)	Psychological/ sentimental/ kinship value (%)	Moral, ethical, and bequest values (%)	
Kenyah	186	48	100	100	100	100	The hard-to-define values influence ethical behavior of the local communities and order a strong preference for environmental goods compared to material gains, i.e., intrinsically skewed or more specifically bioecocentrically inclined
Iban	146	45	100	100	100	100	
Penan	66	37	100	100	100	100	
Kayan	57	57	100	100	100	100	
Kelabit	30	50	100	100	100	100	
Bidayuh	24	51	100	100	100	100	
Berawan	6	33	100	100	100	100	
Summary	495	46	100	100	100	100	
	Total number of people interviewed and age groups (all tribal groups)	Total average: 100% (all tribal groups)					

generations can inherit an ecologically healthy natural environment (environmental value) as enjoyed by the present generation (instrumental value).

5. Existence value: the protection of land and forests in (4) also gives rise to existence value. Existence value may be defined as satisfaction gained from knowing that the ancestral domains will continue to exist without onsite use.
6. Esthetic value: pleasure derived from nature appreciation.

All of the above values collectively contribute to the life-satisfactions, happiness, mental health and hence, psychological well-being of the indigenous people. This may be collectively termed as psychological value. This specific value is associated with valuable mental health benefits or happiness derived from fulfilling the adat-based moral duty of intergenerational equity or justice (intergenerational relatedness) and individual emotional attachment to ancestral land and forests (human−nature interconnectedness/feeling of belonging). It also includes the local people's esthetic experience of the scenic beauty or the intrinsic worth of nature as well as the sensory experience of the sweet trills of bird songs. Thus as empirically revealed from field research, all the respondents are happy living in the forests, the place to which they belong and which is essential to their being and cultural identity. All these values collectively underpin the bioecocentric value orientation of the local communities—a predominant motivational force that leads to the activation of individual's perceived moral obligation to act proenvironmentally (Fig. 5.11).

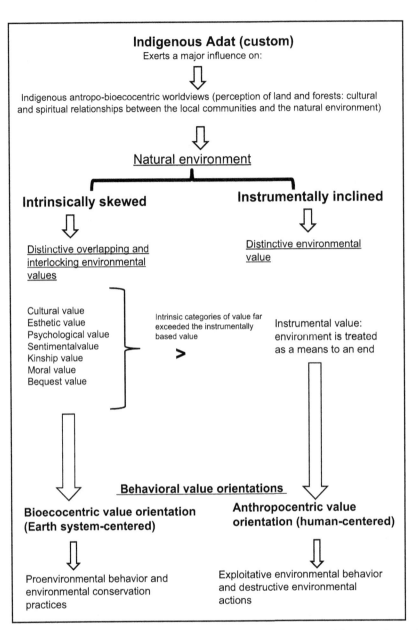

FIGURE 5.11 The indigenous adat, anthtropo-bioecocentric environmental worldview and sustainable environmental behavior—the connections.

Notes: The distinctive overlapping environmental values, that is, cultural value, moral value, and communal value, among others added together to motivate a strong bioecocentric value orientation, that is, nature is intrinsically embraced. At the same time, nature is also anthropocentrically or instrumentally viewed as a means to an end—an anthropocentric or instrumental value orientation which stimulates an exploitative mode of environmental behavior and destructive mode of environmental action. However, as shown in the figure, the bioecocentric value orientations of the local people are far greater than their instrumental value orientations. This gives rise to an Earth-centered belief system of the local communities which places great emphasis on environmental protection or conservation over the instrumental exploitation of nature.

Nonetheless, as argued elsewhere (Choy, 2014, 2018a), despite the ethical and intrinsic value orientations, the natural environment is not exclusively accorded with full moral status. Anthropocentrically but reasonably, it is ethically permissible under adat to instrumentally exploit a part of the natural environment for socioeconomic sustenance—an anthropocentric value orientation in the instrumental use of nature (Fig. 5.11). However, as reflected in Fig. 5.11, the degree of anthropocentric value orientation pales in comparison with the level of bioecocentric value orientation. (For the definition of value orientation, see, Chapter 4: Greening for a Sustainable Future: The Ethical Connection, Section 4.2).

Viewed in this way, the indigenous environmental worldviews and value orientations are logically intelligible and coherently practical in the sense that it requires that moral consideration be given to land resources yet permits the natural surroundings to be "sustainably" degraded for socioeconomic survival needs—anthropo-bioecocentric use of nature (Choy, 2014, 2018a,b). Thus the local communities have been able to strike a sustainable balance between intrinsic motivation to protect nature and anthropocentric inclination to exploit the instrumental value of land and forests (anthropo-bioecocentrism). This is precisely in accord with Aldo Leopold's land ethics and Albert Schweitzer's ethics of reverence for life—a more pragmatic ethical approach to promoting environmental sustainability.

5.10 The ethics of sustainability (bio-ecocentrism): The nonindigenous people versus the "urbanized" indigenous communities

5.10.1 The urban nonindigenous people's view of nature

The analysis will not be complete without a brief discussion based on a comparison between the environmental inclinations of nonindigenous people in the urban areas and the influence of urban culture and modernity on the "urbanized" indigenous people's environmental worldview. "Indigenous people" here refers only to the forest dwellers in Sarawak, in particular in Bintulu insofar as it is relevant within the present analysis. The environmental inclinations of urban dwellers may be reflected only by deduction (discussed later), as it is beyond the scope of the present analysis to provide an accurate evidence-based account, which would require extensive field work and interviews. Notwithstanding this limitation, this section will provide an indication of the environmental inclination of the urban dwellers whose economic activities are inextricably and significantly linked to the use of the natural environment, hence their clear-cut objective of utility maximization.

Generally the nonindigenous urban people in Bintulu can be said to have been molded permanently into a material basis of life. They work as small business owners, restaurant or small hotel owners, traders or general workers whose daily transactions and habitual behaviors do not involve close interaction with the forested environment as in the case of the rural indigenous people. In addition, their environmental dispositions are not imperatively determined and governed by indigenous adat (customs). It may be accurately deduced that, in general, they are unlikely to have a strong intrinsic morality of environmentalism as that of the indigenous rural communities. This is not to say, however, that the urban nonindigenous populace has no keen sense of environmentalism; but merely

FIGURE 5.12   Anthropocentric view of nature by the urban nonindigenous people.
Note: The above figure shows the decimation of rainforests by the "urban" economic agent. *Modified from Rhett A. Butler/Mongabay with permission from, Cannon, J.C., 2017. Leading US plywood firm linked to alleged destruction, rights violations in Malaysia. Mongabay. Available from: https://news.mongabay.com/2017/10/leading-us-plywood-firm-linked-to-alleged-destruction-rights-violations-in-malaysia/#: ~ :text = Leading%20US%20plywood%20firm%20linked%20to% 20alleged%20destruction%2C%20rights%20violations%20in%20Malaysia,-by%20John%2C C&text = An%20investigation %20has%20found%20that,logging%20and%20indigenous%20rights%20violations.*

that their appetitive faculties of environmental passion, love, and will are not as deep and enduring as the forest dwellers'.

To clarify the environmental predispositions of the urban dwellers, we may safely begin by examining the natural resource use patterns of the economic agents. Generally speaking, this group of "environmentally embedded" urban entrepreneurs view nature anthropocentrically as a source of material wealth, devoid of intrinsic value, meant for human exploitation (Fig. 5.11). For example, by the end of 2014, one of largest timber companies, Shin Ying, belonging to an urban business man, had anthropocentrically decimated the proposed Danum Linau National Park located within the Herat of Borneo transboundary conservation area at the rate of 9 km^2 per month (Global Witness, 2012, 2014; Cannon, 2017; Choy, 2018b). Its reluctance to recognize the biophysical limit of resource exploitation and the ethical imperative for environmental conservation resulted in extensive ecological damage to the once intact rainforest canopy as reflected in Fig. 5.12.

Similarly driven by the utilitarian impulse for profit maximization, the Cipta Sawit Plantation company and the Stone Head company controlled by urban dwellers encroached into the indigenous land and forests in Ulu Kelawit Tatau for rock quarry mining and oil palm plantation expansion, both of which activities are detrimental to the ecological heath of the surrounding environment, including the water source of the village (Borneo Project, 2018). Near Bintulu, it was reported that a highly "urbanized" palm oil plantation company, Golden Hope, bulldozed the land and forests belonging to the rural indigenous people for oil palm plantation development, allegedly by fraudulent means to optimize the economic benefits of nature without taking into consideration its destructive environmental impacts (Erenberg, 2015).

The earlier mentioned case studies illustrate how the urban view of nature is cemented by the economic motivation of profit maximization. Governed by the material authority of progress and driven by the egoistic motive for insatiable wealth accumulation, nature, is by and large, conceptualized as a physical source of wealth. The emotionally centered concepts of environmentalism and the moral orders of environmental protection, that is, the ethics of sustainability, do not play into the social dynamics of the urban dwellers who are overwhelmed by the ethics of anthropocentrism. It is this anthropocentrically driven urban culture and the egoistic self-absorption of utilitarianism that contribute to much of the environmental decline not only in Sarawak, but across the material universe in general.

5.10.2 Rural indigenous people and urban mainstreaming: The changing environmental face?

To provide a more complete picture of the nexus of urban culture and environmentalism, we may now turn to investigating the impact of urbanization on the indigenous conception of nature in a material environment setting. With increasing integration and mobility between urban and rural regions, thanks to the improvement of the road transport systems, will the rural—urban movement of the tribal people concomitantly drive them into embracing the anthropocentric view of nature? This can be a highly interesting subject of study. For one thing, it will leave the field clear for the treatment of the qualitative aspects of the indigenous environmental worldviews in the dynamic matrix of rural—urban divide.

Field work conducted in the forest interiors revealed that many local people, especially young people, were increasingly exposed to the outside world through the rural—urban movement in search of new employment opportunities and to improve their income. In an interview conducted with the Kenyah community in Long Bala, for instance, Lenjau Lian (7th Generation Longhouse Chief), said that

> ...many young people have moved to town areas for work. One of the main reasons for this is financial. The local people cannot market their agricultural products if they were to cultivate their land. Hence, they are left with no choice but move to the town to work to earn enough to finance the cost of education of their children...But they will come back to the village whenever they have holidays... (May 2008).

Field work conducted with the Kayan communities from Rumah Aging Long Taah in October 2009, also revealed the same situation as in Long Bala where many young people had left the village to look for work in the urban areas. The rural—urban migration was basically caused by population increase and the lack of employment opportunities in the rural interiors. In the new urban settings, these rural dwellers who work in the Bintulu town center have been "mainstreamed" into the modern urban economy dominated by nontribal people, especially the Chinese community. Many of those from native households work in the nonagricultural sector as construction workers, cashiers, waiters/waitresses, factory workers, or laborers. They work, interact and live side by side with other dominant nonindigenous "economic" men and have become increasingly exposed to the urban culture, market economy, and modernity.

Conditioned by the town lifestyle, they are no more dependent on their ancestral land and forests for their livelihoods. They are also forced to give up their centuries-old swidden agriculture which is closely associated with their cultural identity, social identification, and spiritual distinctiveness. Their unique traditional human—forest relationship came to an end at the urban stage of modernity. Many of them live in small rented concrete rooms—missing the fresh forest air, the sounds of free flowing and gushing streams, and of chattering night birds common in the heart of mother nature, surrounded by luxuriant pristine foliage, back in their ancestral land which they call home.

It may be postulated that continual exposure of these "urbanized" natives to the complex urban network of social environment and human relations set the stage for their utilitarian and linear transition from the ancestral past toward a modernized and materialistic future. More specifically, the progressive adoption of a continuum of modern ideas, materialistic norms, and new behaviors in the new urban economic setting may have led to rapid changes to their traditional ethical beliefs and environmental worldview.

To be sure, it is fallacious to claim that these "urbanized" indigenous people remain unchanged despite their rural—urban detachment from traditional culture. First, by mainstreaming into the urbanized world of modernity, the indigenous people have to orient their conduct and alter their livelihoods in general consistent with the urban landscape and the needs of an industrialized society. Essentially such a change could not take place without unsettling traditional beliefs or breaking old habits in a new model of capitalist society. Nonetheless, while such a logical argument is cogently valid and well-conceived, a different epistemological position is possible.

It may be reiterated that the adat of the indigenous people comprises shared cultural practices and an environmental belief system deeply entrenched and firmly organized in the hearts and minds of the indigenous people and has been collectively embraced by the tribal communities for more than a few centuries. For this very reason, it is highly unlikely that the local people will easily and submissively succumb to moral debility despite the advancement of urban life, for this will lead to rapid erosion of their traditional cultural values which are highly regarded (Sections 6.14—6.15.2 for a detailed discussion of morality).

That said, mainstreaming into the urban economy permits them to organize themselves socially in association with others, to modify their traditions and culture through: (1) a greater understanding of the economic norms and behavioral patterns of the heterogenous urban beings; and (2) a more critical observation or appreciation of the essence of the interrelationships between modern and traditional aspects of social, cultural, and economic life in a rural—urban divide. Based on the existing relational networks of adaptation, the indigenous people produce a new form of social life in a new urban environment which, unlike the forested environment, is managed by the capitalist impulses and forces of economic progress. Thus in contrast to the psychological influences of environmentalism predominating the cultural fabric of the local communities in the forested environment, the social dimensions of human affairs in the new urban setting must necessarily and pragmatically be resolved into a moral command of economic life. In this urban setting, there is no agricultural land to farm, no hunting ground to game and no forest to protect, nothing that permits the practice of their traditional ways of life. Thus it is inevitable that they instrumentally adapt their moral values for a closer alignment with economic reality in the urban landscape.

That said, it is highly significant that once back in their forested environment, the sentimental dimensions of environmentalism which have been deeply conserved in the inner-self of the indigenous people will be emotionally resurrected as circumstances and conditions fit into their traditional and cultural process of life. For these reasons, it must not be construed that urbanization would disastrously weaken or imperil the force of indigenous environmentalism which has evolved for more than a few centuries. What is more important in the claim is that, as noted in Section 5.3, adat is not an immutable entity: it is wholly or partially inherited and further adapted and developed through successive generations under given situations or circumstances.

To substantiate what has been said so far, some authentication is necessary. In the following excerpt from a telephone interview conducted on March 5, 2020, with Sem Kiong, the former Chairman of the now defunct NGO, the Indigenous People Development Center, he revealed that

> Many local people from the villages moved to the urban center mainly to look for jobs to supplement their household incomes especially to support their children's education. Although it is not too difficult to survive in the forests as they are able to get most of their daily sustenance from farming and hunting and gatherings from the forest, they still need cash to buy groceries and diesel to run their power generator and motor boat. It is difficult to sell their farm products in the forest as everyone is planting their own. Also, hunting has increasingly become difficult now compared to the past due to forest destruction caused by logging and oil palm plantation development. Despite this, although they are in the urban center, their hearts are still with their forest settlement. Most of them will return to their villages sooner or later because they still hold high regard for their traditional place of origin, that is, the forests. Surely, they have not and will not forget their *adat*...

In another interview conducted with an "urbanized" individual from a Kenyah tribal group, Usat Ibut from Uma Kulit village in Sungai Asap resettlement area in May, 2008, revealed that

> I have planned to move to Bintulu and have bought a double-storey terrace house. However, I only stay there for one or two days or not at all whenever I need to travel to town to buy my stuffs. I do not like to live in the town because I prefer and am happier to live side by side with the forest. Although the resettlement area is not the same as my original settlement in the upriver in Balui River which is covered with thick forest, it is still a green place surrounded with trees and forests in the surrounding area compared to Bintulu (see, Fig. 5.13). The green here gives me some sense of belonging and peace of mind compared to Bintulu town.

It may well be that although highly urbanized and educated, the respondent has a strong emotional attachment to land and forests. The natural surrounding is existentially central to his cultural identity and is imbued with a strong sense of place meaning. Because of all the traditional values passed down from generation to generation, he still holds a strong and durable moral bond with nature to treat it with "love, respect, and care" amidst living with a keen sense of urbanization in his social and moral environment (Fig. 5.13).

FIGURE 5.13   Rural and urban scenes: Sungai Asap and Bintulu town.
Note: Compared with the indigenous settlement in Sungai Asap, the town center is less environmentally lively and soothing.

In another interview with the same respondent via telephone on March 7, 2020, about the rural—urban divide of environmentalism, he revealed that

> Most of the people moved to the urban areas basically because of economic reasons. With population increase coupled with the lack of job opportunities, they have no choice but to migrate out of necessity. They are there for economic reasons. They need to support their families especially their children's education. However, this does not mean that they have lost their interest in the natural environment. They still love and respect the forests as before...

On the same topic, Steven (48) from Long Bala, a sole proprietor who spent most of his time in the town areas said

> Whether in town or in the village, our hearts are still with our ancestral land. In fact, we show due respect and love for the surrounding area including the trees here. Whenever the durian trees come into my sight, I will think of my *Datuk nenek-moyang* (ancestors). We still observe our *adat* until today. I will not sell my land and forests for any sum of money because they are very important to us. We owe a responsibility to our ancestors to protect them for the benefit of our children and grandchildren... (May 2008)

It is increasingly clear from the above that the emerging new urban culture characterized by the impersonality of market behaviors and utilitarianism has not necessarily subjugated the old culture imbued with the preestablished rules of morality and environmentalism. The two cultures could exist without conflict, adapting to serve as a remarkable *tour de force* to order tribal man's moral temperament and conduct in different settings and situations. They denote a logical edifice integrating the extant knowledge of Adam Smith's invisible hand into Leopold or Schweitzer's moral principles of environmentalism.

What is more, the fact that as societies become more complex due to changing social, economic, and physical environmental conditions, it will be difficult for morality or adat, for that matter, to operate as a purely automatic mechanism. It may well be that, according to Émile Durkheim, under changing circumstances, the rule of morality must be applied intelligently. Society is continuously evolving due to changing circumstances. In view of this, the immanent impulses of morality must by themselves be sufficiently flexible to adapt concomitantly in line with socioeconomic reality (Durkheim, 1972). It is thus increasingly clear that the encroachment of urban culture into the traditional socioeconomic milieu of the indigenous people does not necessarily drive their original cultural belief of environmentalism into a *cul-de-sac*. Essentially it will be the same as today as in many years to come.

5.11 Concluding remarks

This chapter has used empirical evidence of the indigenous land ethic from the Borneo tropical rainforests in the state of Sarawak in Malaysia to reinforce the theoretical analysis as expounded in Chapter 4, Greening for a Sustainable Future: The Ethical Connection. It is succinctly clear that the philosophical context of the indigenous land ethic is evidently a collective body of traditional and empirical experience summed up in the term *adat* handed down to the present generation by previous generations over a few hundred years. The *adat* is invested with a collective moral force and a special deontological authority based on the principle of intergenerational equity, the intrinsic value, Émile Durkheim's moral philosophy, the Kantian deontological ethics, Aldo Leopold's land ethics, and Albert Schweitzer's ethics of reverence for life.

These intrinsic beliefs, philosophical, and moral principles shape the local communities' preferences and priorities for collective representation over self-interest or utilitarian motive. They also express the ways the local communities think of themselves in relation to the natural surroundings. Accordingly the indigenous people rightfully conceive of themselves as an inalienable part of nature which represents their cultural identity and spirituality. These environmental worldviews regulate indigenous natural resource use patterns by guiding and restraining individuals' collective consciousness and environmental behaviors which are detrimental to the ecological integrity of the natural system and equitability of future generations. They have laid down the biospheric limits of resource exploitation (based on the traditional Pusaka land use systems), beyond which individuals could not deviate without bringing about unpleasant social repercussions, resistance, or sanctions. The indigenous environmental worldviews also collectively unite and combine the indigenous people's moral sentiments to affirm a sense of custodial responsibility to protect the "stability, integrity and beauty" (Holling sustainability) of land resources based on a culturally viable and mutually beneficial relationship with nature when instrumentally optimizing their economic use.

It is noteworthy that despite the above intrinsic value orientations and ethical dimensions, the natural environment is not exclusively accorded with full moral status. Anthropocentrically but intelligibly, it is ethically permissible under the indigenous adat to instrumentally exploit a part of the natural environment as a self-enhancement means for socioeconomic sustenance. Viewed from this perspective, the indigenous environmental worldviews are logically intelligible and coherently practical in that it requires moral consideration be given to the intrinsic protection of communal land resources, yet admitting the instrumentally valuing attitudes toward nature (noncommunal land and forests) to serve socioeconomic survival needs—an anthropo-bioecocentric environmental worldview. The quintessence of this land use vision and human—nature relationship is an excellent example of Holling's concept of sustainability as discussed in Chapter 4, Greening for a Sustainable Future: The Ethical Connection.

It is increasingly evident that the present analysis has successfully established a direct link between environmental ethics and environmental sustainability, empirically and theoretically, and has derived a morally justifiable deliberation of environmental ethics in a real world system. Empirically it is shown that there is an irrefutable ethical connection between the ways the environmental resources are conceived and the state in which their ecological surroundings should be held and the habits of practice, attitudes, responsibilities, and obligations the local communities, as land custodians, morally acknowledge, and deontologically enforce.

By welding together the empirical evidence with the theoretical constructs as expounded in Chapter 4, Greening for a Sustainable Future: The Ethical Connection, the empirical analysis lends immense force to the credentials, importance, and practicality of environmental ethics. More importantly, it allows us to see the practical importance of environmental ethics in promoting proenvironmental behaviors, and sustainable development practices and actions clearly, intelligibly, and convincingly. In short, beyond the shadow of a doubt, environmental philosophy, ethics, and morality from the founders of modern environmental philosophy such as Aldo Leopold, Albert Schweitzer, Émile Durkheim, and Immanuel Kant serve as a practical discipline to address the prevailing moral disorder of the looming planetary crisis.

It is certainly true that there is no single quick fix or policy choice that will completely redress any given environmental ill. Nonetheless, the present study, empirically researched and conceptually substantiated, has succeeded in making an irrefutable case that sustainable natural resource use and environmental conservation are not necessarily contradictory if we are able to recognize our moral intuitions and ethical obligations when harnessing the economic use of nature. In today's highly anthropocentric world, the need for such pluralistic, virtue-oriented natural resource use is more acutely felt than ever, particularly in the Asian resource-based developing economies, and even more so in the world's fastest growing economy, China and the world's largest anthropocentrically skewed economy: the United States under Donald Trump's administration as discussed in the previous chapter. In Chapter 6, The United Nations Environmental Education Initiatives: The Green Education Failure and the Way Forward, I shall examine how the culture of environmental ethics may be enhanced and established.

The United Nations environmental education initiatives: The green education failure and the way forward

6.1 Introduction

As has been examined in Chapter 3, The United Nations' Journey to Global Environmental Sustainability Since Stockholm: The Paradox, and Chapter 4, Greening for a Sustainable Future: The Ethical Connection, our Earth systems are struggling against unprecedented assaults caused by unsustainable human socioeconomic practices and materialistic consumerism. At the very core of these anthropogenic afflictions of nature is the increasing human disconnection from the natural system, which in turn has led to unsustainable human behaviors and activities with their resultant destructive environmental impacts. For example, as noted in Chapter 3, The United Nations' Journey to Global Environmental Sustainability Since Stockholm: The Paradox, in the advanced nations, especially in the United States, high production and high consumption lifestyles sustained by enormous use of fossil fuels and material inputs with high levels of waste production and large environmental impacts have hitherto dominated its socioeconomic system, putting its natural system in a precarious state of development. Moreover, with an excessively materialistic outlook, human beings tend to embrace a lifestyle that is hostile to the natural environment.

Furthermore, the fervent quest for rapid economic growth and material progress at the expense of the environment in developing countries, as well as in developed nations such as the United States and Canada, has become so pervasive and profound that it is increasingly and irreversibly pushing our Earth systems toward a Holling unsustainable path of development. A case in point worth reiterating is the Canadian "tar sands" conquest of nature as discussed in Chapter 4, Greening for a Sustainable Future: The Ethical Connection, which is not only irreversibly altering the Earth's structure but also destabilizing the healthy functions of our Earth systems to the detriment of long-term human existence.

Why has the global community at large hitherto steadfastly adhered to the above unsustainable path of development in the teeth of clear scientific evidence of its destructive environmental impacts as discussed in Chapter 3, The United Nations' Journey to

Global Environmental Sustainability Since Stockholm: The Paradox? As indicated in Chapter 4, Greening for a Sustainable Future: The Ethical Connection, this business-as-usual approach to environmentalism is fundamentally attributed to the human hyper-anthropocentric perception of nature characterized by a profound shift in man's relationships with the natural environment. This human−nature mismatch is spurred by the general belief among many of the globalized societies that human-induced global environmental change will not be severe or rapid enough to trigger major or irreversible disruptions to the global socioeconomic system, or to threaten human long-term existence as in the case of Covid-19 or coronavirus pandemic. This optimistic view of environmentalism is further underpinned by the general belief that science and technology will be able to circumscribe any major environmental disruptions.

For example, numerous scientific and technological solutions such as resource efficient or recovering technologies including solar technology or waste recycling techniques, energy efficiency technologies, pollution abatement techniques, and cleaner manufacturing processes, among others, have been deployed to reduce greenhouse gas (GHG) emissions or human impact on the Earth system. By the same token, the use of geographical information system (GIS) and remote sensing technologies in wildlife conservation and animal tracking has dramatically improved ecological monitoring and surveillance against poachers (Wall et al., 2014).

Despite these, as discussed in Chapter 3, The United Nations' Journey to Global Environmental Sustainability Since Stockholm: The Paradox, GHG emissions have increased persistently while wildlife, especially rare and endangered species, are increasingly facing the threat of extinction. It is also ironic to note that digital and tracking technologies have enabled poachers to employ a more sophisticated way of locating animals by hacking into the Global Positioning System (GPS) collars and killing the tracked down animals—what is known as "cyber poaching" (Cooke et al., 2017; Lewis, 2017). For clarification, GPS uses satellites that orbit Earth to send information on wildlife distribution to a ground-based GPS antenna and receiver. GIS is a software program that helps people to use the information to determine the location of the wild animals.

It may well be that scientific invention and technological innovations will not effectively help to propel us toward a sustainable path of development if people across the globe, especially those in the affluent west, fail to change substantially their unsustainable production and resource use patterns, lifestyle choices and material consumerism habits. In addition, as discussed in Chapter 4, Greening for a Sustainable Future: The Ethical Connection, environmental law or policies are unlikely to have lasting effects unless and until there is strong moral commitment by policy makers and society at large to protect the environment.

The task is fundamentally one of environmental and moral education which is concerned with the development of the mind's capabilities through the acquisition of ecological knowledge and moral principles to critically assess and evaluate the environmental "problematique" confronting us today and the moral causes which give rise to it. Without putting a well-defined environmental and moral education in place, the environmental issues and moral disciplines, and their relationships as discussed in the previous chapters, exist mostly in a vacuum in that the practical consequences they attest to will elicit no more than casual attention.

That said, environmental education (EE) going hand in hand with moral education allows us to apprehend our environmental crisis-ridden world and human moral attitudes and duties toward nature. This, in turn, leads us to discover by careful introspection the environmental obligations incumbent upon us when interacting with the natural system. Hence, this chapter attempts to pragmatically set forth a realistic environmental and moral education model—a model which interrelates sociological philosophy, ecosystem interconnectedness, environmental awareness, and human—nature relationship, among others, with the view of raising the urgency of environmental protection and inspiring human stewardship responsibility.

With this view in mind, the work will first critically examine the role played by the United Nations in promoting environmental literacy through environmental education since Stockholm, and to what extent it has been successful in achieving its aim. Drawing from Émile Durkheim's influential works on a number of topics including sociology, philosophy and education, and René Descartes' philosophical principles, it also examines how the United Nations' environmental education frameworks may be improved. It further demonstrates how environmental education should systemically be constructed with proper and clear direction, from simple to the complex based on empirical analysis on the interconnectedness between nature and humans. This line of analysis enables us to re-conceptualize our moral responsibilities toward the natural environment. Acknowledging that the United Nations has not been able to give adequate weight to the promotion of moral education which underpins human proenvironmental behavior, this work constitutes an attempt to fill this epistemological deficiency. This can lead us to substantiate moral claims about the ethical imperative to integrate environmental sustainability discourse with moral discipline.

6.2 Environmental education: The *raison d'être*

As reflected in the previous chapters, global environmental sustainability is not so much concerned about taking measures to solve environmental problems after they have surfaced as a result of human unsustainable activities; rather, it is more about changing our environmental attitude or behavior in order to avoid causing undue harm to our global commons, resulting in environmental problems. This changing environmental attitude or behavior is in turn influenced by the moral or ethical values we hold toward the natural system as reflected in the empirical assessment of indigenous environmental worldviews in Chapter 5, The Nexus of Environmental Ethics and Environmental Sustainability: An Empirical Assessment. The ability to acquire these ethical values or moral wisdom is premised on a pedagogically sound and well-designed environmental education.

Enhancing public environmental awareness has become acutely necessary in the contemporary scientific and digital world of distraction. A common sight all over the world today is that of people from all walks of life constantly staring or prodding at their iPhones/iPads while walking, commuting, dining, or drinking. Obsessed with the MacWorld fantasy, they are engrossed to the extent of becoming totally unaware about events happening around them. In this age of distraction, many people are losing their ability to sustain concentration or discussion on serious global issues in today's acutely faltering and fractured world such as the environmental crisis blighting the future of human long-term existence. Sadly, digital addiction is increasingly leading to human deficits in core mental, intellectual, and analytical skills in global environmental sustainability. Our present advanced technology may invent more sophisticated and cutting-edge machines, but these cannot replicate human thinking. To save the human mind from the deleterious digital waves of distraction, environmental education seems to be the most relevant to our purpose.

At all levels, the aim of environmental education ought to be to nurture learners' attitudes to the nonhuman natural environment, that is, from the anthropocentric or instrumental view of nature to one of concern, love, and respect—the bioecocentric system of environmental beliefs premised on Aldo Leopold and Albert Schweitzer's environmental philosophy as discussed in Chapter 4, Greening for a Sustainable Future: The Ethical Connection, and Chapter 5, The Nexus of Environmental Ethics and Environmental Sustainability: An Empirical Assessment. It is very unlikely that science and technology, however, advanced, or environmental laws and regulations, however, rigorously crafted and enforced, could induce such an ethical shift.

6.3 Environmental education and environmental literacy: The United Nations/UNESCO initiatives

6.3.1 Environmental literacy: Some basic concepts

The concept of environmental education (EE) was first formalized by the International Union for the Conservation of Nature and Natural Resources (IUCN) at a working meeting on 'Environmental Education in the School Curriculum' held in Nevada, USA in 1970. It defined EE as "the process of recognizing values and clarifying concepts in order to develop skills and attitudes necessary to understand and appreciate the interrelatedness among man, his culture, and his biophysical surroundings". EE also entails practice in decision making and self-formulation of a code of behavior about issues concerning environmental quality (IUCN, 1970: 26).

Following from the above and within the present context, EE may be defined as a learning process which enhances environmental literacy. Environmental literacy refers to the acquisition of environmental knowledge which enhances environmental awareness and concern about the functional properties of the natural world as well as its associated problems. This necessarily includes understanding the life support system functions of the natural system, the environmental tipping points, and their associated challenges as well as the capacity to appreciate the intrinsic value of nature. It also covers the behavioral change and commitment to environmental protection and the will to translate this commitment into real action. Environmental literacy also concerns the acquisition of problem-solving skills and motivation to work toward making informed decisions to respect, protect, and sustain the wellbeing of natural systems (NAAEE, 2010; OSSE, 2017). Thus environmental literacy covers three aspects, namely, knowledge, skills, and habits of mind which eventually lead to actions framed by awareness and attitudes (Krnel and Naglič, 2009).

An overlapping concept interrelated with environmental literacy is ecological literacy. This differs from the broader concepts of environmental literacy, which mainly focus on the environment as a series of issues to be resolved through values, changing environmental behaviors, and actions. Ecological literacy emphasizes the role of ecological knowledge in identifying the cause-effect relationships in order to enable more enlightened decision-making. More specifically, ecological literacy encompasses an understanding of our natural systems including the ecosystem, the local habitat, or the atmospheric sink, among others, and an appreciation of their ecological

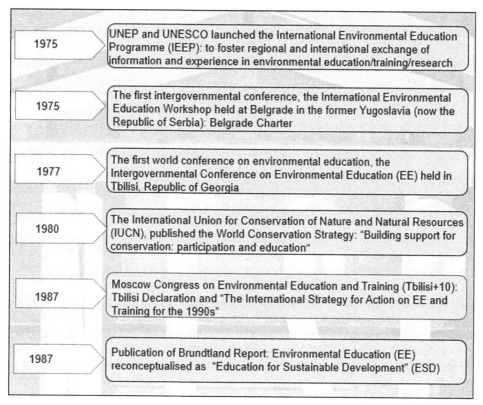

1975	UNEP and UNESCO launched the International Environmental Education Programme (IEEP): to foster regional and international exchange of information and experience in environmental education/training/research
1975	The first intergovernmental conference, the International Environmental Education Workshop held at Belgrade in the former Yugoslavia (now the Republic of Serbia): Belgrade Charter
1977	The first world conference on environmental education, the Intergovernmental Conference on Environmental Education (EE) held in Tbilisi, Republic of Georgia
1980	The International Union for Conservation of Nature and Natural Resources (IUCN), published the World Conservation Strategy: "Building support for conservation: participation and education"
1987	Moscow Congress on Environmental Education and Training (Tbilisi+10): Tbilisi Declaration and "The International Strategy for Action on EE and Training for the 1990s"
1987	Publication of Brundtland Report: Environmental Education (EE) reconceptualised as "Education for Sustainable Development" (ESD)

TIMELINE 6.1   The evolution of the United Nations/UNESCO environmental education structure (1970s and 1980s).

connectivity and human interaction with them, and for making informed decisions for their protection or conservation (Lewinsohn et al., 2015).

It thus follows that an ecologically literate person exhibits awareness about natural systems, their system dynamics and their links to global environmental issues, and understands the necessary ecological knowledge to make informed decisions. What may be said of this is that ecological literacy necessarily constitutes an indispensable component of environmental literacy. As discussed in the following subsection, (Timeline 6.1), the promotion of environmental literacy represents a primary goal of environmental education of the United Nations' agenda of environmental sustainability.

6.3.2 The United Nations' environmental education initiatives in the 1970s and 1980s

The importance of EE was reaffirmed at the Stockholm Conference held in 1972. Principle 19 of the Stockholm Declaration states that, "Education in environmental matters for the younger generation as well as adults...is essential in order to broaden the basis for an enlightened opinion and responsible conduct by individuals, enterprises and communities in protecting and improving the environment in its full human dimension..." (United Nations, 1972: 5). The

Conference recommended that environmental education be recognized and promoted in all countries. Following the recommendation of the Stockholm Conference, UNEP, and UNESCO launched the International Environmental Education Programme (IEEP) in 1975 (UNESCO, 2002). The Programme aimed at fostering regional and international exchange of information and experience in environmental education, training, and research. Here, it may be remarked that UNESCO plays a pivotal role in shaping, legitimizing, institutionalizing, and influencing formal and nonformal environmental education globally through a plethora of reports, agendas, programmes, and charters (Sauve, 1996; Gough, 1999; Kelsey, 2003).

As part of the follow-up of IEEP, UNESCO-UNEP organized an International Environmental Education Workshop which was held at Belgrade in the former Yugoslavia (now the Republic of Serbia) in 1975—the first intergovernmental conference convened especially for the promotion of EE (UNESCO, 1976). It produced a document known as the Belgrade Charter aimed at advancing the promotion of environmental literacy. In consonance with the IUCN's educational concept, the Charter defined environmental education as a process "to develop a world population that is aware of, and concerned about the environment and its associated problems, and which has knowledge, attitudes, motivations, commitment to work individually and collectively towards solutions of current problems and the prevention of new ones" (UNESCO, 1976: 2). The Charter also outlined the aims, objectives, key concepts, and guiding principles of an environmental programme with objectives that include the promotion of knowledge, awareness, attitude change, skill acquisition, evaluation ability, and participation.

The importance of the Belgrade Charter was further reaffirmed at the first world conference on environmental education, the Intergovernmental Conference on Environmental Education held in Tbilisi, Republic of Georgia in 1977. In his opening address to the conference, Mostafa K. Tolba, the Executive Director of UNEP stressed that, "...in the long run, nothing significant will happen to reduce local and international threats to the environment unless widespread public awareness is aroused concerning the essential links between environmental quality and the continued satisfaction of human needs" (UNESCO, 1978a: 61). He further stressed the importance of proper environmental education in raising environmental awareness. The Tbilisi Declaration was adopted at the conference (UNESCO, 1978b).

The goals of the Declaration coincide with the Belgrade Charter's objectives. These include:

1. fostering awareness and concern about the economic, social, political, and ecological interdependence in urban and rural areas;
2. promoting environmental knowledge, in particular in relation to the complexities of man-made and natural environment arising from their biological, social, economic, ethical, cultural, and political interactions;
3. practical skills acquisition in environmental problem-solving; and
4. the promotion of new patterns of environmental behavior among individuals, groups, and communities, among others.

It may be noted in light of the above that the Tbilisi Declaration already included the promotion of the fundamental elements of sustainable development namely, the social, economic, and environmental aspects of development, well ahead of the 1987 Brundtland Report. The Declaration also places great importance on environmental protection through value and behavioral change and skill acquisition.

6.3.3 United Nations' environmental education initiatives in the 1980s

As noted in Chapter 2, The United Nations' Journey to Global Environmental Sustainability since Stockholm: An Assessment, in 1980, the International Union for Conservation of Nature and Natural Resources (IUCN) published an important document on environmental conservation, the World Conservation Strategy. The document included a chapter on environmental education, "Building support for conservation: participation and education" (Chapter 12 of the UN document), which contains some of the philosophical and ethical principles discussed so far. More specifically, it explicitly states that achieving conservation objectives calls for the need for humans to embrace the ethical principles in their interactions with nature and to live in harmony with the natural world. It also stresses that the long-term goal of environmental education is to foster or reinforce environmental attitudes and behavior compatible with this new ethic.

Ten years after the Tbilisi Conference, the Moscow Congress on Environmental Education and Training (Tbilisi + 10) was held in 1987 to reaffirm the principles and goals of the Tbilisi Declaration amidst persisting environmental degradation in various countries (UNESCO, 1987, 2013). Some of the objectives of the conference were to review the progress and trends of environmental education since the Tbilisi Conference and the state of the environment and its educational and training implications, among others (UNESCO, 1987). In his opening address to the conference, Amadou-Mahtar M'Bow, the Director-General of UNESCO, spoke about the importance of humans living in harmony with the Earth system. He further stressed the role of environmental education in harnessing this objective (UNESCO, 1987).

The conference adopted "The International Strategy for Action on EE and Training for the 1990s" which serves as a basis for the development of national environmental education strategies in the member state countries and institutions (UNESCO-UNEP, 1988). The document is divided into two parts. Part one highlights the needs and priorities in the development of environmental education such as the repeated need to promote widespread environmental awareness through environmental education and the incorporation of an environmental dimension into the education system, among others. Part two touches on the narration of various environmental problems, the aims, principles, guidelines, and actions for international strategies for the 1990s.

6.3.4 The evolution of the United Nations/UNESCO Environmental Education discourse in the 1990s

As noted in Chapter 2, The United Nations' Journey to Global Environmental Sustainability since Stockholm: An Assessment, in 1987, the World Commission on Environment and Development (WCED) released its epoch-making document, the Brundtland Report, which popularized the concept of sustainable development. Since then, EE has adopted a new vision to focus on the concept of sustainable development. It was reconceptualized as "Education for Sustainable Development" (ESD) in order to promote education consistent with the principles of sustainable development (UNESCO, 2013). ESD has come to be seen "as a process of learning how to make decisions that consider the long-term future of the economy, ecology and social well-being of all communities" (UNESCO, 2002: 10) (Timeline 6.2).

1990	The World Conference on Education for All held at Jomtein in Thailand: "building awareness and practical knowledge in changing learner behaviour towards the goals of sustainable development"
1991	IUCN, UNEP and WWF jointly published the "Caring for the Earth. A Strategy for Sustainable Living": the importance of environmental education in promoting sustainable development
1992	Agenda 21: the importance of environmental education, public awareness and training in promoting sustainable development
1992	The World Congress for Education and Communication on Environment and Development (ECO-ED) was held in Toronto
1993	First International Congress on Population Education and Development held in Istanbul
1994	UNESCO launched a project, 'Education for a Sustainable Future' within the framework of the "Environment, Population and Development" (EPD)
1994	International Conference on Population and Development (ICPD) held in Cairo, Egypt and the launch of the "Environment and Population Education and Information for Human Development" project
1995	An inter-regional workshop on Re-orienting Environmental Education for Sustainable Development (MIO/ECSDE) was held
1997	International Conference on Environment and Society: Education and Public Awareness for Sustainability took place in Thessaloniki, Greece
1999	The Third UNESCO/Japan Seminar on Environmental Education in Asian-Pacific Region was held in Tokyo, Japan

TIMELINE 6.2   The evolution of the United Nations/UNESCO environmental education structure (1990s).

More specifically, ESD aimed to promote "changes in values and attitudes towards environment and development...and towards nurturing harmony between humanity and environment" (WCED, 1987: 111; see also, UNESCO, 2001). To achieve this, the Report called for the need for "vast campaigns of education, debate and public participation" (WCED, 1987: xiv). The concept of sustainable development has thus been explicitly associated with EE to promote development that meets present needs without compromising the ability of future generations to meet their own needs based on wise use of natural resources. It is claimed that from 1977 to 1987, the IEEP was associated with worldwide efforts to incorporate environmental education and training into the education systems (Mostafa et al., 1992).

The importance of environmental education/literacy in promoting sustainable development continued to gain international recognition after 1987. In 1989, for example, UNESCO-UNEP reaffirmed environmental literacy as the operating principle of environmental education in its newsletter entitled "Environmental Literacy for All" (UNESCO, 1989). The World Conference on Education for All held at Jomtein in Thailand in 1990 also placed substantial emphasis on the role of environmental education in promoting a more environmentally sound world through "building awareness and practical knowledge in changing learner behaviour towards the goals of sustainable development" (WCEFA Inter-Agency Commission, 1990: 25). In 1991, IUCN, UNEP, and WWF jointly published *Caring for the Earth: A Strategy for Sustainable Living*. The document reiterated the importance of environmental education and human behavioral change in promoting sustainable development (IUCN-UNEP-WWF, 1991).

Sustainable development is defined under the Strategy as "improving the quality of human life while living the carrying capacity of supporting ecosystems" (IUCN-UNEP-WWF, 1991: 11). The strategy also stressed the importance of formal and informal educational systems to disseminate information necessary to explain policies and actions needed for the survival and wellbeing of the world's societies (IUCN-UNEP-WWF, 1991: 10). The adoption of Agenda 21 at the Earth Summit held in 1992 also strongly emphasized the importance of environmental education, public awareness and training in promoting sustainable development as outlined in Chapter 36 of the Agenda. It explicitly states that, "Governments should strive to update or prepare strategies aimed at integrating environment and development as a cross-cutting issue into education at all levels within the next three years" (United Nations, 1992: Chapter 36). The Agenda also calls for the need to re-orient EE toward sustainability (UNESCO, 2013).

In 1992, the World Congress for Education and Communication on Environment and Development (ECO-ED) was held in Toronto as the first important follow-up of the 1992 Rio Earth Summit to explore in breadth and in depth how education can best be made to serve to achieve the goal of sustainable development (UNESCO, 1992a). The Congress placed sustainable development at the very core of EE. The Congress also emphasized that education must go beyond awareness building and become a basis for action by citizens, whose lifestyles and behavior may have an enormous cumulative impact on the environment (UNESCO, 1992a). The First International Congress on Population Education and Development held in Istanbul in 1993, fundamentally a population conference, also acknowledged the importance of EE as an integrated part of population education and international education (UNESCO-UNFPA, 1994).

Following up on the recommendations of Chapter 36 of Agenda 21 on Education, Awareness and Training and those of the 1993 International Congress on Population, Education and Development, UNESCO adopted a resolution at the 27th session of UNESCO's General Conference held in 1993 to re-orient EE towards fostering environmental and population education and communication. The focuses of the resolution were:

1. to refine knowledge-based and developing action frameworks to deal with environment, population, and development issues in an integrated manner;
2. to foster the development of new or reoriented education, training and information programmes and materials in order to strengthen capacity building; and
3. to mobilize the support of decision-makers and opinion leaders at international, regional and national levels on issues in relation to (1) and (2) above.

A transdisciplinary project known as the "Environment and Population and Information for Human Development (EPD)" was also approved at the 27th session of the UNESCO's General Conference. The title of the EPD may be simplified as 'EPD/UNESCO: Educating for a Sustainable Future'. The EDP project should encourage interdisciplinary and inter-agency cooperation in fostering a better understanding of interrelationships between environment, population, and development. It also aims to nurture "active and knowledgeable citizens and caring and informed decision makers capable of making the right choices about the complex and interrelated economic, social and environmental issues human society is facing" (UNESCO, 2002: 7).

The EDP project was launched following a critical assessment and based on the various recommendations on education, information and awareness adopted by the International Conference on Population and Development (ICPD) held in Cairo, Egypt in 1994 as well as several UN conferences as discussed earlier. The EPD project aims at "the development of education, training and information activities, designed to deal with the interwoven issues of population, environment and human development in an integrated manner" (UNESCO, 1994a: 2; UNPF, 2014). More particularly, the main objective of this integrated approach to education is to achieve people-centered, equitable, and sustainable development. It focused on:

1. sustainable development expressed in terms of improving the quality of life and of the environment;
2. the diversity of life and the balance between reasonable human activities and ecosystem preservation;
3. the impact of global environment and population change from local and global perspectives;
4. building human capacities, promoting people participation, and cooperation among people and institutions; and
5. re-orienting and improving the quality of education and the means to disseminate knowledge on human development (UNESCO, 1994b).

In 1995, UNESCO-UNEP International Environmental Education Programme (IEEP) and the Mediterranean Information Office for Environment, Culture and Sustainable Development, Athens, coorganized an inter-regional workshop on Re-orienting Environmental Education for Sustainable Development (MIO/ECSDE) in line with Chapter 36 of Agenda 21, "Reorienting Education Towards Sustainable Development" (UNESCO/UNEP- MIO/ECSDE, 1995). The main aim of the workshop was to re-orient the overall framework of IEEP toward sustainable development by taking into consideration the guiding principles of both EPD and IEEP. The workshop also sought to test new ideas and pilot EE approaches with the EPD focus based on diverse disciplines. The workshop produced four separate reports and a synthesis report of the working groups, "Re-orienting EE for Sustainable Development". The synthesis report unveiled a host of recommendations to promote ESD. These included the recognition of the importance and relevance of the fundamental elements of environmental education as contained in the Tbilisi Declaration and the critical dimensions of sustainable development of the 1992 Rio Declaration, and the incorporation of environment, population and development dimensions of the EPD approach, among others, into the EE system.

Increasingly, the expanded and refined meaning of EE, in particular the EPD, has come to serve as "a stimulus for transdisciplinary reflection and action" of the complex interrelationship of our socioeconomic system which involves cultural, political, religious, social, ethical, humanitarian, economic, and scientific and technological developments, among others (UNESCO, 1997a: 1). EPD has also become the main mechanism through which UNESCO addresses the recommendations of all United Nations conferences concerning education, information and public awareness related to sustainable development" (UNESCO, 1997a: 1). Through this development, EPD replaced IEE as the flagship programme in the field of environmental education in 1995 (UNESCO, 2002: 10).

Following the major findings of the conferences held in Tbilisi in 1977, Jomtein in 1990, Toronto in 1992, and Istanbul in 1993, a third conference on environmental education, the International Conference on Environment and Society: Education and Public Awareness for Sustainability took place in Thessaloniki, Greece in 1997. Its main focus was to highlight the role and the importance of environmental education and public awareness for achieving sustainability, and to provide inputs for further improvement in EE as stipulated in its background and main working document, "Educating for a Sustainable Future; Transdisciplinary Vision for Concerted Actions" (UNESCO, 1997a). The document was also prepared in response to the work programme of the United Nations Commission on Sustainable Development (CSD) "to refine the concept and key message of education for sustainable development, taking into account the experience of environmental education and integrating considerations pertaining to population, health, economics, social and human development, and peace and security" (UNESCO, 1997a: 2). The Conference discussed education for sustainability.

The culmination of this conference was the *Declaration of Thessaloniki* which constitutes the charter for future education for sustainability. The charter reaffirmed the importance of environmental education in promoting behavioral and lifestyle changes toward sustainability as emphasized in various international conferences discussed above. More specifically, point 11 of the Declaration reaffirms that, "Environment education, as developed within the framework of the Tbilisi recommendations and as it has evolved since then, addressing the entire range of global issues included in Agenda 21 and the major UN Conferences, has also been dealt with as education for sustainability. This allows that it may also be referred to as education for environment and sustainability" (UNESCO, 1997b: 2).

Thus the concept of EE as originally developed may also be referred to as "Education for Sustainability" (EfS) or Education for Environment and Sustainability (EfES). The Declaration also conveyed the strong and common message to the global community that Efs or EfEF constitutes "one of the pillars of sustainability together with legislation, economy and technology", and that it should be in the center of the international, regional and national agendas (UNESCO, 1997b: 1). More particularly, point 17 of the Declaration stressed that governments and international, regional and national financial institutions should "mobilize additional resources and increase investments in education and public awareness" (UNESCO, 1997b: 2).

After the Thessaloniki conference, a series of intergovernmental seminars and conferences were convened with the aim of reinforcing global momentum and commitment to education for sustainable development. For example, the Third UNESCO/Japan Seminar on Environmental Education in Asian-Pacific Region was held in Tokyo, Japan in 1999 to assess the shared common issues of teacher education for environmental education in the Asian Pacific region and to discuss possible solutions or recommendations (Japanese National Commission for UNESCO, 2000). In addition, the Fifth UNESCO-ACEID International Conference on Education was held in Bangkok in 1999 to undertake in-depth analysis into what reforms were required in learning, curriculum, and pedagogy in all sectors of education including environmental education especially in the Asian-Pacific region (UNESCO, 2000a).

6.3.5 The United Nations' environmental education initiatives in the 2000s: The emergence of the Decade of Education for Sustainable Development

In 2000, the World Education Forum was held in Dakar, Senegal. Acknowledging that education is "the key to sustainable development and peace and stability within and among countries", the forum called for a collective global commitment and action to achieve by 2015 the goals and targets of "The broad vision of Education for All" proclaimed at the Jomtien Conference (UNESCO, 2000b: 36, 9). In the same year, the United Nations adopted the UN Millennium Declaration which set out a series of targets including targets in relation to universal primary education and gender equality in education. (Timeline 6.3)

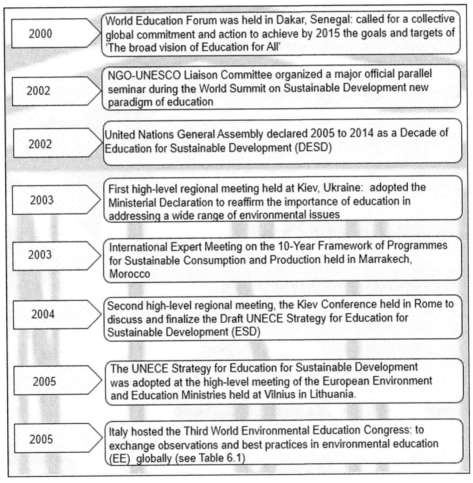

TIMELINE 6.3   The evolution of the United Nations/UNESCO environmental education structure (2000s and beyond).

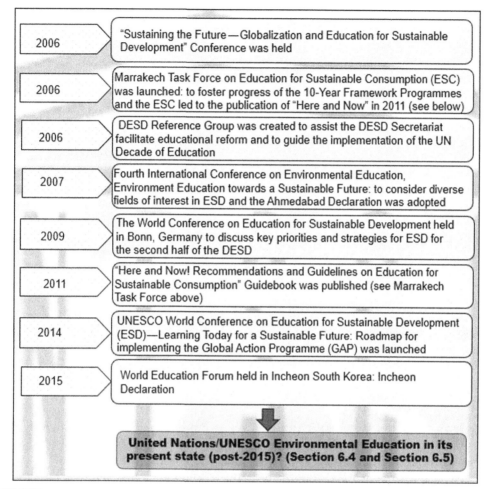

2006	"Sustaining the Future—Globalization and Education for Sustainable Development" Conference was held
2006	Marrakech Task Force on Education for Sustainable Consumption (ESC) was launched: to foster progress of the 10-Year Framework Programmes and the ESC led to the publication of "Here and Now" in 2011 (see below)
2006	DESD Reference Group was created to assist the DESD Secretariat facilitate educational reform and to guide the implementation of the UN Decade of Education
2007	Fourth International Conference on Environmental Education, Environment Education towards a Sustainable Future: to consider diverse fields of interest in ESD and the Ahmedabad Declaration was adopted
2009	The World Conference on Education for Sustainable Development held in Bonn, Germany to discuss key priorities and strategies for ESD for the second half of the DESD
2011	"Here and Now! Recommendations and Guidelines on Education for Sustainable Consumption" Guidebook was published (see Marrakech Task Force above)
2014	UNESCO World Conference on Education for Sustainable Development (ESD)—Learning Today for a Sustainable Future: Roadmap for implementing the Global Action Programme (GAP) was launched
2015	World Education Forum held in Incheon South Korea: Incheon Declaration

United Nations/UNESCO Environmental Education in its present state (post-2015)? (Section 6.4 and Section 6.5)

TIMELINE 6.3   *Continued*

In 2002, the World Summit on Sustainable Development (WSSD) was held in Johannesburg to assess and to reinforce global political commitment to sustainable development. UNESCO together with the South African Ministry of Education and in cooperation with the NGO-UNESCO Liaison Committee organized a major official parallel seminar during the WSSD. It aimed to review the lessons since the 1992 Rio Conference for a new paradigm of education which entailed different dimensions, range of perspectives, and pathways for sustainable development (UNESCO, 2004).

In the same year in 2002, the United Nations General Assembly through its Resolution 57/254 declared 2005–2014 as a Decade of Education for Sustainable Development (DESD) (UNESCO, 2007). The basic vision of DESD is "a world where everyone has the opportunity to benefit from quality education and learn the values, behaviour, and lifestyles required for a sustainable future and for positive societal transformation" (UNESCO, 2005: 6). The United Nations defined sustainable future in terms of "environmental integrity, economic viability, and a just society for present and future generations" (UNESCO, 2005: 6). The overall goal of DESD is "to integrate the principles, values and practices of sustainable development into all aspects of education and learning" (UNESCO, 2005: 6).

In 2003, the first high-level regional meeting to assess the progress in the implementation of environmental commitments of the region of the United Nations Economic Commission for Europe emanating from the 2002 WSSD was held at Kiev, Ukraine (UNESCO, 2003a). The Ministerial Declaration adopted at the conference reaffirmed the importance of education in addressing a wide range of environmental issues and called for all countries in the region to integrate sustainable development into education systems at all levels. The Kiev Conference was followed by a second regional meeting on education for sustainable development held in Rome in 2004 to discuss and finalize the Draft UNECE Strategy for ESD that would serve as a promotional contribution to the United Nations DESD led by UNESCO (UNESCO, 2004).

In the same year, 2003, at the International Expert Meeting on the 10-Year Framework of Programmes for Sustainable Consumption and Production (SCP) held in Marrakech, Morocco, the "Marrakech Process", an international collective effort to develop a 10-Year Framework of Programmes on SCP was launched. Its aim was to support the implementation of SCP at regional and national levels, to accelerate the shift toward SCP patterns and to decouple economic growth from environmental degradation. These initiatives were in response to the WSSD Johannesburg Plan of Implementation for SCP (UNEP, 2009). SCP may be defined as the consumptive use of goods and services to support basic needs and quality of life while minimizing the use of natural resources and toxic materials while reducing waste emissions and pollutants over the life cycle of the services or goods (UNEP, 2009: 8).

Among the aims of the Marrakech Process are to promote awareness and identify priorities and needs for SCP and to improve SCP tools and methodologies. The Marrakech Task Forces were also established. The Task Forces are voluntary initiatives led by governments and in cooperation with other institutions in both developed and developing countries to carry out a set of activities that serve to promote SCP (UNEP, 2009). Implementing the 10-Year Framework of Programmes on SCP is also stressed under Goal 12 of the 2030 Agenda for Sustainable Development Goals (United Nations, 2016). In this light, it is relevant to note that in its 32nd Plenary Session General Conference held in October 2003, UNESCO adopted a resolution (32 C/COM.III/DR.1) recognizing the Earth Charter as an important instrument to guide human efforts toward a sustainable future and in building a just, sustainable, and peaceful global society (UNESCO, 2003b).

In 2005, the UNECE Strategy for ESD was adopted at the high-level meeting of the European Environment and Education Ministries held at Vilnius in Lithuania. The Strategy acknowledged ESD as a human right and a prerequisite for sustainable development. It also recognizes that ESD helps to develop and strengthen "the capacity of individuals, groups, communities, organizations, and countries to make judgments and choices in favour of sustainable development" (United Nations, 2005: 1).

Following the first and second world congress on environmental education held in Espinho in (Portugal) and Rio de Janeiro (Brazil) in 2003 and 2004, respectively, Italy hosted the Third World Environmental Education Congress in 2005—the inaugural year of the United Nations DESD. The main objectives were to exchange observations and best practices in EE globally (see Table 6.1). A series of world congress on EE had also been held periodically to reinforce global commitment to promoting ESD (Table 6.1).

In response to the proclamation of DESD in 2005, an international conference, "Sustaining the Future—Globalization and Education for Sustainable Development", was convened jointly by UNESCO and the United Nations University in Nagoya, Japan in 2006. The objectives of the conference were to spearhead international coordinating and intersectoral efforts as well as programmatic actions to integrate sustainable development into educational processes, both formal and informal, to explore the interface between ESD and globalization including the role of higher education in this respect (UNESCO-UNU, 2006). Some of the visible outputs of the Conference included the Asia-Pacific Regional Launch of the DESD, which started from January 2005 to highlight the importance of education toward achieving a sustainable future, and the launch of the Regional Centres of Expertise (RCE) on ESD. The RCE aimed at contributing to the development of innovative ways of collaborating with various stakeholders including higher education institutes and local governments in the promotion of ESD (UNESCO-UNU, 2006).

In connection to the Marrakech Process as discussed above, at the 14th Session of the United Nations Commission on Sustainable Development held in 2006, the Italian Ministry of the Environment, Land and Sea launched the Marrakech Task Force on Education for Sustainable Consumption (ESC) (Ministry for the Environment Land and Sea of Italy, 2006). The main objective of the Task Force was to ensure ongoing contribution to the Marrakech Process and to foster progress of the 10-Year Framework Programmes (10YFPs) as noted above. It also aims to promote the integration of SCP issues into education processes especially formal learning processes with the aim of raising awareness and competencies in cultivating sustainable patterns of consumption and production. Education for Sustainable Consumption may be considered as a subset of ESD which contributes to its enhancement.

Together with UNEP and in collaboration with the United Nations Decade on Education for Sustainable Development and Hedmark University College in Norway, the Marrakech Task Force Education for Sustainable Consumption published the ESC guidebook, *Here and Now! Recommendations and Guidelines on Education for Sustainable Consumption*, in 2011 (UNEP, 2010). The publication, which is a valuable contribution to the UN Decade on Education for Sustainable Development, contains a series of recommendations and guidelines to enhance SCP behaviors among individuals or social groups. Accordingly, the basic learning outcomes of ESC can be defined as attitudes, knowledge, skills, and behavior leading to: (1) critical awareness; (2) ecological responsibility; (3) social responsibility; (4) action and involvement; and (5) global solidarity (UNEP, 2010: 24). It should be noted that responsible consumption and production which aims at "doing more and better with less" is also listed, as Goal 12, on the 2030 Agenda for Sustainable Development Goals (United Nations, 2016: 31).

In the same year, 2006, UNESCO created a Reference Group to assist the DESD Secretariat facilitate educational reform and to provide expert guidance to UNESCO in setting priorities and strategic direction for the implementation of the UN Decade of Education, and in the coordination and mobilization of a diverse set of partners and stakeholders (IAU, 2007; United Nations, 2010). The DESD Reference Group is also strategically involved in enhancing the linkages with other UN initiatives such as the Literacy Decade, Education for All

TABLE 6.1   World Environmental Education Congress (WEEC).

World Environmental Education Congress (WEEC)				
	Year	**Venue**	**Theme**	**Objectives**
1st WEEC	2003	Espinho (Portugal)	Strategies for sustainable future	To discuss worldwide key issues in EE in an integrated perspective.
2nd WEEC	2004	Rio de Janeiro (Brasil)	Building a possible future	To exchange experiences and research orientations in EE.
3rd WEEC	2005	Torino (Italy)	Educational paths toward sustainability	To exchange observations and best practices in EE at a worldwide level, to develop the main themes for the agenda on EE and to create a world community on environmental and sustainable education research and practice.
4th WEEC	2007	Durban (South Africa)	Learning in a changing world	To strengthen the diffusion of environmental and sustainability culture.
5th WEEC	2009	Montreal (Canada)	The Earth, our common home	(1) To promote the role of EE as a means of developing and enriching human sustainability identity, (2) to highlight the role of EE in social innovation and eco-development, (3) to emphasize the role of EE in public policy development, and (4) to examine how public policy can strengthen environmental education.
6th WEEC	2011	Brisbane (Australia)	Explore, experience, educate	To explore new trends in research and practice in EE including science impact on climate change in the Asia-Pacific, and nature and purpose of environmental education in the future, among others.
7th WEEC	2013	Marrakech (Morocco)	Environmental education in cities and rural areas: seeking greater harmony	To discuss a host of related EE issues which cover 11 thematic issues. These include: (1) Promoting EE and Networking, (2) research in environmental education, (3) greening education, and (4) pedagogy and learning, and (5) ethics, eco philosophy, human−nature relationships, among others.
8th WEEC	2015	Gothenburg (Sweden)	Planet and people— how they can develop together?	To discuss a series of global issues ranging from digital age to poverty reduction, green cities, climate change adaptation, and educational policy development for environment and sustainability.

EE, Environmental education.
Compiled by author from various sources.

Initiative, Global Initiative on Education and HIV-AIDS, and other major education initiatives, and the Millennium Development Goals (MDG) (Gadotti, 2009). The DESD Secretariat is supported by the Monitoring and Evaluation Expert Group, an ad hoc group created to provide guidance on the implementation of the global DESD Monitoring and Evaluation framework and to advise on DESD reporting progress (United Nations, 2010).

Three decades after Tbilisi, the Tbilisi + 30 or the Fourth International Conference on Environmental Education, Environment Education towards a Sustainable Future— Partners for the Decade of Education for Sustainable Development, organized by the Government of India in cooperation with UNESCO and UNEP, was held in India in 2007. The mission of the conference was to provide a platform to share diverse experience, best practices, and collective knowledge and ideas related to EE and ESD (CEE, 2007). The conference was intended to cover diverse fields of interest such as values and ethics, energy efficiency, ecosystem, disaster risk reduction, waste management, climate change, HIV-AIDS, culture, human rights, gender, peace, social justice, arts, technology, and other issues that have evolved over the last 30 years since Tbilisi. However, climate change was the cross-cutting theme of the conference, drawing important linkages between EE/ESD and global efforts in addressing the long-unresolved problem. The conference also touched on ESD in relation to biodiversity and Biosphere Reserve. The conference adopted the "Ahmedabad Declaration 2007: A Call to Action" which proclaimed that through education we can learn to prevent and resolve conflicts, respect cultural diversity, create a caring society and live in peace" (CEE, 2007: 20). It also reaffirmed the significance of ESD which serves to encourage "a shift from viewing education as a delivery mechanism to a lifelong, holistic and inclusive process" (CEE, 2007: 21). More importantly, the Declaration also called on the United Nations system and governments to support "Environmental Education and develop sound Education for Sustainable Development policy frameworks and commit to their implementation" (CEE, 2007: 21).

Following the Ahmedabad Declaration, the World Conference on Education for Sustainable Development was held in Bonn, Germany in 2009 to discuss key priorities and strategies for ESD for the second half of the DESD and to reinforce global engagement to ESD to address global challenges confronting the human race today (UNESCO-BMBF-German Commission for UNESCO, 2009). In consonance with the Ahmedabad Declaration, the Bonn Declaration adopted at the conference called for the need to put the experiences and knowledge accumulated in the first half of DESD into action. The Bonn conference discussed a host of environmental issues confronting the world, including climate change, sustainable lifestyles, AIDS, health and education, disaster risk management, food security, and biodiversity, among others, in relation to ESD. Insofar as it is relevant to the present work, two of the main workshop themes which are worth noting are strengthening educational response in addressing climate change and mainstreaming biodiversity in education and learning.

On the issue of climate change, the conference workshop sought to identify strategies and a practical plan of action toward scaling up the ESD to climate change. It further recognized the need for large-scale investments for a transformative education defined as "a critical values-based integrated participatory approach which enables empowered citizens to move from learning the facts towards taking action" (UNESCO-BMBF-German Commission for UNESCO, 2009: 48). The workshop participants, comprising mainly senior representatives from Government Ministries of education, environment and development, intergovernmental organizations and civil society from all over the world, called on UNESCO to develop a global strategy for ESD. This was articulated under point 16 of the Bonn Declaration adopted at the end of the conference which called on UNESCO to enhance its leadership and coordination role for the UN DESD on the International Implementation Scheme in cooperation with other UN agencies and programmes such as UNEP, UNU, the EFA convening agencies (UNICEF, UNDP, UNFPA, and the World Bank), among others. It also called on UNESCO "to incorporate ESD into 'one-UN' strategy at country level, particularly through UNDAF processes", and to "support member states and other partners in the implementation of the UN DESD particularly through upstream capacity-building and policy advice on the development of coherent national strategies, monitoring and evaluation...." (UNESCO-BMBF-German Commission for UNESCO, 2009: 121).

On the issue of ESD for biodiversity, the conference workshop acknowledged that "biodiversity (especially ecosystems) illustrates global interdependence, the consideration of which is vital to ESD" (UNESCO-BMBF-German Commission for UNESCO, 2009: 59). It also suggested to the global community to use the connected networks of knowledge, practices, and research emanating from various international events and processes such as the International Year of Biodiversity and the Multilateral Environmental Agreements dealing with biodiversity including those adopted at the Conference of Parties to promote the biodiversity-ESD nexus in a comprehensive concept.

In marking the end of the UN Decade of ESD, UNESCO and the Government of Japan coorganized the UNESCO World Conference on Education for Sustainable Development (ESD)—Learning Today for a Sustainable Future, in 2014. The main objectives of the conference were:

1. Celebrating a decade of action,
2. Reorienting education to build a better future for all,
3. Accelerating action for sustainable development,
4. Setting the agenda for ESD beyond 2014 (UNESCO, 2014a).

6.3.6 The Decade of Education for Sustainable Development and beyond

UNESCO launched its Roadmap for implementing the Global Action Programme (GAP) on ESD at the World Conference on Education for Sustainable Development as noted above. The Roadmap set a clear vision and direction for the future of ESD in line with DESD's vision for "a world where everybody has the opportunity to benefit from education and learn the values, behaviour and lifestyles required for a sustainable future and for positive societal transformation" (UNESCO, 2005: 6). The overarching goal of GAP is "to generate and scale up action in all levels and areas of education and learning to accelerate progress towards sustainable development" (UNESCO, 2014b: 4). The GAP has two fundamental objectives, namely, to re-orient education and learning, and to strengthen education and learning in all agendas, programmes, and activities in line with its overarching goal. The five priority action areas of GAP are (1) advancing policy, (2) transforming learning and training environment, (3) building capacities of educators and trainers, (4) empowering and mobilizing youth, and (5) accelerating sustainable solutions at local level (UNESCO, 2014b: 15). The Roadmap is also intended to contribute to the development of post-2015 agenda on ESD actions in all levels and areas of education, training, and learning.

The UNESCO Conference also adopted the Aichi-Nagoya Declaration on Education for Sustainable Development which reaffirmed UNESCO as the lead agency for ESD as emphasized in point 16 of the Bonn Declaration (UNESCO, 2014c). The Declaration also called upon all countries to implement the GAP on ESD, and to acknowledge that education is not just concerned about knowledge but also empowerment. The conference also launched the final report on the DESD "Shaping the Future We Want" which reaffirmed the significance of ESD for sustainable development. It also highlighted that "Leadership is essential for moving from policy commitments and demonstration projects to full implementation across the curriculum, teaching and operations, whether in formal systems or in non-formal learning and public awareness raising" (UNESCO, 2014c: 3). The 10YFP on Sustainable Lifestyles and Education (SLE) was also launched at the conference (UNESCO, 2014d). The Programme presses for the need to build a vision of sustainable lifestyles and to integrate sustainable lifestyle principles and practices across all sectors of society. This strategic move reflects the emphasis on the importance of sustainable lifestyle in the Third Plenary: Education: a game-changer for sustainable development, which states that "Education on its own is not enough...we must integrate education with sustainable lifestyles" (UNESCO, 2014a: 5).

In connection with this, it is noteworthy that Goal 4 of the 2030 Agenda for Sustainable Development Goals (SDG 4) also emphasizes the need to equip all learners with the necessary skills to promote sustainable development by 2030 (United Nations, 2016). More specifically, Target 4.7 of SDG 4 emphasizes that by 2030, "all learners acquire the knowledge and skills needed to promote sustainable development, including, among others, through education for sustainable development and sustainable lifestyles, human rights, gender equality, promotion of a culture of peace and non-violence, global citizenship, and appreciation of cultural diversity and of culture's contribution to sustainable development" (United Nations, 2016: 7). Goal 12 of the 2030 Agenda for Sustainable Development Goals (SDG 12) also calls for the need to raise awareness and to promote education on sustainable consumption and lifestyles among consumers in attaining the goal (United Nations, 2016).

In the post-2015 education agenda, the overarching goal of education as explicitly expressed in SDG 4 is to "Ensure inclusive and equitable quality education and promote lifelong learning opportunities for all" (United Nations, 2016). Following this, the World Education Forum held in Incheon in the Republic of Korea in May 2015 had the goal of reframing the global education agenda on how to ensure inclusive and equitable quality education and lifelong learning for all by 2030 (UNESCO, 2015). The conference adopted the Incheon Declaration reaffirming and supporting SDG 4. It also called for the urgent need to reinforce political will, policies, long-term commitment and bold and innovative measures to spearhead its achievement (UNESCO, 2015). The conference also acknowledged that education which lies at the heart of sustainable future for all, is a fundamental human right—a right that would enable the realization of all other economic, social, and cultural rights. The conference also emphasized that achieving sustainable development must necessarily begin from education that addresses the interdependence of environment, economy and society, and the need to bring about the fundamental change of mindsets to be inclined toward sustainable development practices.

6.4 The United Nations/UNESCO EE discourse: The changing face

Despite the United Nations/UNESCO's relentless effort over the past few decades at promoting EE by disseminating broad guiding principles to the global community for a sustainable future as discussed above, it is clear from Chapter 3, The United Nations' Journey to Global Environmental Sustainability Since Stockholm: The Paradox, that its EE discourse has been unable to prevent a host of human-induced environmental destructive practices, unprecedented environmental degradation, and biodiversity destruction. Even with so much information and knowledge about the environment in the advanced west, countries like Canada and the United States have been contributing to aggravating the ecological health of our global commons for the past few decades as discussed in Chapter 3, The United Nations' Journey to Global Environmental Sustainability Since Stockholm: The Paradox, and Chapter 4, Greening for a Sustainable Future: The Ethical Connection. Worse yet, as noted in Chapter 4, Greening for a Sustainable Future: The Ethical Connection, the Canadian government under Stephen Harper's administration even went to the extent of embarking on a chaotic drive to trash and destroy environmental documents and books in the name of "sustainable development" (read economic growth per se). Taking a similar antienvironmental stance as Stephen Harper, the environmental regulatory rollbacks of the Donald Trump administration as examined in Chapter 3, The United Nations' Journey to Global Environmental Sustainability Since Stockholm: The Paradox, is of no exception.

As noted above, the purported original goal of the United Nations/UNESCO EE initiative was the promotion of environmental literacy in order to raise public awareness. This was made clear in the Belgrade Charter and "Environmental Literacy for All". However, since then, with the changing themes of global development ideology, EE has evolved into a vast and diverse pedagogical landscape. More specifically, since the publication of the Brundtland Report in 1987, EE has taken a more defined stance with greater support for education for sustainability. One of the most valuable pointers of this evolving trend is the changing focus of the 1997 Thessaloniki Declaration. More specifically, instead of highlighting its original goal of promoting environmental education and public awareness for achieving environmental sustainability, the Declaration eventually shifted its focus to reaffirm UNESCO's preferred goals of "Education for Sustainable Development" and "Education for Sustainability".

It is increasingly clear that the concept of "sustainable development" instead of the environment itself has become the key orientating idea. Indeed, within the UNESCO education for sustainable development discourse, EE has changed into multifaceted domains. It is instructive to note that starting from the 1990s, in an attempt to reconcile the concept of sustainable development with educational reforms, UNESCO has moved EE out from its conventional goal of promoting environmental literacy. Instead, it now places great emphasis on promoting a sustainable future—a future which calls for the need to improve the quality of life of human beings based on wise use of natural resources as explicitly stated in the Brundtland Report or Agenda 21.

In view of the above development, the original version of EE has been facing a string of distinct currents of deflection. In hindsight, the First International Congress on Population Education and Development held in 1993 in Istanbul made EE concede a great deal by placing it within an interwoven tenet of population and international education. More specifically, the Congress acknowledged EE as an integral part of population and international education. It is along this line of thought that the International Conference on Population and Development held in Cairo in 1994 reshuffled EE by integrating it into population and human development issues. In 1997, EE was further rocked off its original philosophy of environmental literacy under point 11 of the Declaration of Thessaloniki. The Declaration relegated EE to education for environment and sustainability in order to allow it to address the entire range of global issues as revealed in Agenda 21 and at the major UN conferences.

Under the 2003 Fifth Ministerial Conference held in Kiev, in acknowledging that education for sustainable development was a much broader concept than EE, the legitimate acronym, "environmental education" which underpins the logical foundation of environmental literacy was obliterated by the general term "education" to be merged with "sustainable development". It further called on the global community to integrate sustainable development into education systems at all levels. Worse yet, the role of EE in promoting environmental literacy was further deflected by the UNECE Strategy for Education for Sustainable Development which acknowledged ESD as a human right and a prerequisite for sustainable development. The Ttbilisi + 30 also covered a wide range of issues which engendered not only values, ethics, climate change, ecosystems, and disaster risk reduction which are directly associated with environmental literacy but also covered other issues such as HIV-AIDS, gender, peace, arts, technology, and other diverse issues evolving over the past three decades—issues which may not be directly related to environmental literacy. The Bonn Declaration also took the same stance in acknowledging the same diverse approach to environmental education as discussed above.

All these changes succeeded so well that the concept of sustainable development with varying degrees of emphasis and objectives has dominated the United Nations/UNESCO EE discourse. More particularly, the original version of environmental education in the 1980s has become a subset of education for sustainable development. It is now being circumscribed by a mix of education for sustainability, education for sustainable development, education for sustainable consumption, education for a sustainable future, education for environment and sustainability, education for peace and sustainable future, education for sustainable development and global citizenship, inclusive quality education for children with disabilities, and education for a host of other things and issues. All these different versions of education are indeed overwhelming, covering a multitude of issues, which, apart from those mentioned above, also include, gender equity, cultural diversity, social values, empowerment, lifestyles, and ethics among others. This multidisciplinary inclination was designed to accommodate the recommendation of Agenda 21 and the Millennium Development Goals.

The evolution of education for sustainable development with incredibly broad perspectives and a seemingly endless series of objectives has blurred the goal of environmental literacy and environmental protection. For example, environmental literacy or environmental protection was hidden under the vision of the UNECE Strategy for Education for Sustainable Development (2005), which claims that education is a human right and a prerequisite for sustainable development (United Nations, 2005). It seems that sustainable development under its educational vision has been taken to mean "improving the quality of life" as explicitly stated—an anthropocentric view of sustainable development.

By the same token, the once commanding focus of environmental literacy suffered yet another blow at the Aichi-Nagoya Declaration on Education for Sustainable Development (2014) which anthropocentrically recognized that "people are at the centre of sustainable development". Its heavy anthropocentric resonance is reflected in its declaration that ESD offers an opportunity for both developed and developing countries to enhance poverty reduction and to promote social equity, people empowerment, environmental protection, and economic growth. Its opening plenary also set a heavy anthropocentric tone that framing ESD is a critically important social process which would ensure "good life for all" and facilitate "real empowerment and social transformation towards a more sustainable and just world order" (UNESCO, 2014a). More conspicuously, the Incheon Declaration proposed a new vision for education towards 2030—a vision which would ensure education for all. It was claimed that the new vision which captured the proposed SDG 4, was inspired by "a humanistic vision of education and development based on human rights and dignity; social justice; inclusion; protection; cultural, linguistic and ethnic diversity, and shared responsibility and accountability" (UNESCO, 2015: 67). The issue of environmental protection is glaringly absent from its preamble although it is mentioned scantily in the final report on the World Education Forum 2015 (UNESCO, 2015).

6.5 The United Nations/UNESCO environmental education initiatives: Some critical remarks

The evolution of the new concept of EE, education for sustainable development, into a diverse, complex and vast pedagogical landscape, does not necessarily help to raise environmental awareness for enhancing a sustainable world. On the contrary, it only reflects a fundamental disconnect from its original goal of promoting environmental education for environmental literacy, and hence an environmentally sustainable future. Moreover, excessive complexity of the UNESCO education milieu in terms of its multidisciplinary objectives has turned environmental education into a diffused and indistinctive form of educational framework.

To be sure, environmental education cannot be reflected with much precision under the general heading of ESD as no educational objectives as unveiled under its milieu cannot be called sustainable development goals. More to the point, each conspectus of educational objective embedded in different sustainable development tenets has its own special perspectives, nature, and characteristics which do not share equal environmental distinctiveness with environmental education. Although many of the ESD objectives are either mutually exclusive or directly or indirectly associated with the immediate contents of environmental education, they are, in the strict sense of the word, nonenvironmental. Hence, to envelop environmental education in multiple disciplines only clouds its original focus on environmental literacy, obscuring the specific function it aims to fulfil.

Also, faced with competing and, at times, conflicting epistemic descriptions over a host of sustainable development goals, some of which may be beyond one's own comprehension, individuals may encounter a great deal of confusion with regard to which disciplines, conceptions, or representations specifically contribute to the principal attribute of environmental literacy. More specifically, since we are encountering a consumed mass of educational objectives for a sustainable future, the gist of EE and the flow of environmental relationships cannot be immediately recognizable or identifiable. Conceivably, it no longer manifests simple educational means for conceptualizing environmental thought, especially for newcomers.

Still more important is the claim that it is unlikely that parents, teachers and other environmental educators, let alone the ordinary man on the street, could assimilate or cope with the profound depths of the United Nations/UNESCO educational foundations as well as the many mutually irreconcilable sustainable development goals, which are continuously under construction and refinement. Even educators may find it difficult to find a pragmatically clear path through this confusion and intellectual riddle in order to ascertain the direction that educational investigation should proceed to yield fruitful results.

Here, it is instructive to note that no individual carries a full mental image in his head of the world around him. As Jay Wright Forrester, a pioneer in computer engineering and systems science, rightly pointed out, an individual has only "selected concepts and relationships which he uses to represent the real system" (Forrester, 1971a: 14; see also Forrester, 1971b: 112; Forrester, 1961: 49–50). Hence, it is unlikely that individuals are capable of having a clear mental image of most of the concepts of ESD, let alone their interactions. Their inter-relationships and mutual interplay to produce the ultimate course of development ambitiously embraced by the United Nations/UNESCO are too complex to be conceived even by intellectuals. Moreover, if environmental education has no clear position or distinctive mark that can be easily conceived, its focus will become confused with other ESD specifics. As Jeff Myers and David Noebel elucidated, "the more information we have, the harder it is to figure out what to do with it all" (Myers and Noebel, 2015: 5). Similarly, Tom Berryman also pointed out that, "Too much emphasis on a planetary outlook can bring about global, disembodied, and placeless curriculum or programs" (Berryman, 1999: 53).

Furthermore, the consequence of the United Nations/UNESCO's unrealizable omnipotent vision of quality education on sustainable development in all nations, for citizens of all ages, curriculum at all levels, and in all social contexts, and for a host of objectives, many of which are full of broad statements of policy which may be mutually irreconcilable, is confusing. The United Nations/UNESCO's mess of multiple objectives is an outcome of its preference for multiple definitional spaces which rest on extraordinarily broad foundations and perspectives, namely, ethical, environmental, ecological, biological, anthropological, social, economic, pedagogical, educational, political, cultural, scientific and technological, among others. This is not only delimiting the sphere of the original goal of EE, but also blurring its focus on promoting environmental literacy—a prerequisite for environmental sustainability.

Metaphorically, the United Nations/UNESCO's broad and complex educational framework, for which there is no clear interpretative consensus, is caught in the Humpty Dumpty wonderland of conceptual chaos of interpretation. In Lewis Carroll's Through the Looking-Glass, Humpty Dumpty defends his own definitions of words thus: "When I use a word it means exactly what I want it to mean, neither more nor less" (Carroll, 2010: 88). The scenario for education for sustainable development seems to be a Humpty Dumpty world: bloated with fertile ideas, diverse conceptual foundations and aggressive scholarship, open to an almost endless definitional space and a wide range of interpretative possibilities. Thus when we pose the following simple question to a policy maker: "What is education for sustainable development?", he may respond in Humpty-Dumpty fashion that "it is a type of education that contributes to poverty alleviation and economic growth", which is very different from the general understanding that it is the long-term means to environmental conservation or protection.

This is an unfortunate way of saying that policy makers may efficaciously project this objective as a political vision into a programme of education which anthropocentrically favors development over environmental protection. This can subserve the tendency to encourage an anthropocentric view of nature—a view that condones environmentally destructive projects in the name of "sustainable development". Here sustainable development may be embraced as development that aims at poverty alleviation and the betterment of humankind. For this very reason, policy makers, in an attempt to fit willy-nilly their development bias into environmentally destructive projects, may take the educational position that the extension of logging concessions in the Heart of Borneo in Southeast Asia or the IndoMet Coal Project in Indonesia as discussed in Chapter 3, The United Nations' Journey to Global Environmental Sustainability Since Stockholm: The Paradox, is sustainable because the intention is to improve the life of the people. This anthropocentric position is indeed what the policy makers in this region have endorsed in unison. Consequently, all forms of education can be claimed as education for sustainable development insofar as they serve any of the MDG or SDG goals as postulated by the United Nations. The resultant impact is that the goal of environmental literacy simply pales in comparison with the collective nonenvironmental goals of MDG or SDG.

Furthermore, by blocking the prism of environmental thoughts with sustainable development ideas and concepts, it would be difficult for environmental education to see the light of day in our highly human-centric world. Indeed, in spite of the great progress the United Nations/UNESCO has made to promote ESD, policy makers tend to perceive EE as an appendage of ESD and from its most general angle, and hence accord it with lesser importance than given to other sustainable development objectives which are socioeconomic in nature. This apparent neglect is clearly reflected in the Asian developing countries where policies on higher education in the region tend to overtly place more weight on human resource and technological development than on environmental education (Ko and Osamu, 2015). In reality, it is no exaggeration to claim that environmental education has never been the key sustainable development concern for policy makers in the region for the past few decades since Stockholm.

As reflected in Fig. 6.1, there are various factors which constitute the barriers to environmental education mainstreaming. These include the lack of awareness of the relevance and importance of environmental education among educators or administrators in preparing students with the necessary knowledge needed for their career advancement (Lozanno, 2016). To compound the problem, the formal institutions of higher education such as the Ministry of Education often display a strong institutional bias toward mainstreaming environmental education into the national education systems. For, it is feared that promoting an eco-friendly or environmentally conscious society may obstruct government's efforts in launching or implementing various environmentally benign-development projects—a common phenomenon in the Asian developing countries.

To be sure, EE is not specifically about poverty alleviation, human rights issues, or a multitude of sustainable development objectives, although they may be related. Rather, it is about harnessing individual environmental obligations, duties, and restraints in association with our global commons. Here, moral education certainly has a crucial role to play to induce human beings toward harnessing an environmentally sustainable world. Nonetheless, it seems clear

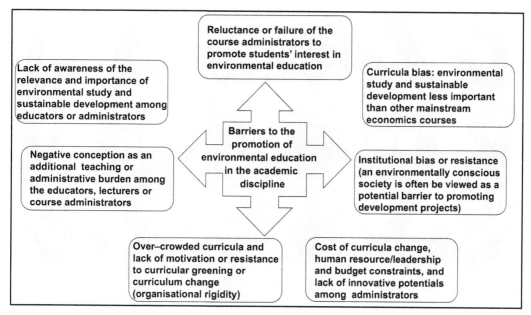

FIGURE 6.1 Environmental education mainstreaming—some entry barriers.

from the above discussion that the United Nations has been unable to give crucial weight in this highly significant discipline. Surely, this is not to deny the importance of other ESD goals. All other educational goals may be explicitly promoted under different education tenets for sustainable development in accordance with Sustainable Development Goals. Thus the United Nations or UNESCO's all-encompassing educational table should be turned. Section 6.8 below attempts to conceptually and critically revisit the United Nations/UNESCO environmental education framework based on Émile Durkheim's philosophy and suggest ways to circumvent the hitherto unresolved EE dilemmas.

6.6 The United Nations/UNESCO's global environmental education efforts: Some success stories

However, despite the above criticism, it is fair to remark that not all the United Nations' environmental education efforts ended in failure. There are also some clear success stories. Evidently, the United Nations/UNESCO's efforts at promoting a sustainable future through education have gathered force internationally as reflected by the implementation of regional and national strategies in various regions or countries across the globe to promote education for environmental protection and sustainable development. For example, in the area of exchange of information and experience, the quarterly newsletter *Connect* has continuously been published since 1976 first in English, French, and Spanish then in Arabic, Russian Chinese, Hindi, and Ukrainian. The newsletter was distributed to about 25,000 EE specialists, ministries, research centers, NGOs, private organizations and all levels of educational institutions, social and voluntary organizations in over 150 countries worldwide involved in the promotion of environmental education and training (UNESCO, 1984, 1995). The newsletter, which is the International Science, Technology and Environmental Education Newsletter of UNESCO-UNEP International Environmental Education Programme (IEEP), also sought to keep its readers, estimated at around 300,000 constantly informed about the state of EE around the world (UNESCO, 1995).

In the area of curriculum and educational materials development, the IEEP has produced an integrated "Environmental Education Series" comprising methodological guides, educational materials, textbooks, thematic modules, and manuals for general education and for the initial training and retraining of teachers. These have been published in Arabic, English, French, and Spanish and were distributed to the member state countries to guide them in their efforts to promote environmental education (UNESCO, 1990, 1995).

In the area of international cooperation, the IEEP launched a series of pilot projects, training workshops and seminars at the national, subregional, and international levels in different countries of various regions of the world to promote environmental education and to develop educational and informational materials suitable to local environmental situations and conditions (UNESCO, 1984, 1990, 1992b, 1995). Countries benefited from such projects include Mexico, Bolivia, Brazil, Chile, Colombia, Ecuador, Peru, Paraguay and Venezuela in Latin America, Bangladesh, Nepal, India, Bhutan, Pakistan, Sri Lanka, China, Indonesia, Malaysia, Mali, Mozambique, Afghanistan, Russia, Ukraine, Czechoslovakia, Poland, France, Portugal, Spain, United States and the United Kingdom, among others (UNESCO, 1984, 1990, 1995).

On a regional basis, a case in point is the launch of the Regional Strategy on Environmental Education in the Asia-Pacific by the Institute for Global Environmental Strategies (IGES), Japan, to foster environmental education in the Asian Pacific region. The Strategy provides a comprehensive framework of actions on environmental education for the Asian-Pacific region covering a host of countries including Bangladesh, Bhutan, India, Maldives, Nepal, Sri Lanka, Brunei Darussalam, Cambodia, Indonesia, Laos, Malaysia, Myanmar, Philippines, Singapore, Thailand, Vietnam, China, Japan, Korea, Mongolia, Australia, Fiji, Papua New Guinea, and Tonga, among others (IGES, 2002). The mission of the Strategy is to move the region toward "promoting, inspiring and fostering citizens to work towards achieving a sustainable future" (IGES, 2002: iv).

6.7 The United Nations' environmental education initiatives: Other success stories

Ineluctably, the United Nations' environmental education initiatives also offer an important push factor for higher educational institutions including universities for their intellectual quest to promote environmental education and sustainable development practices in accordance to the UN's goals of sustainability. Essentially, their intellectual innovations centered on waste management, resource optimization, curricular greening, and raising environmental awareness, among others (de Ciurana and Leal Filho, 2006). In reflection, a consortium of universities as shows in Table 6.2 below has brought to the surface an ACES Network in 2003 (curriculum greening of higher education, acronym in

TABLE 6.2 ACES Network University Consortium.

Country	University
Germany	Technical University Hamburg-Harburg
Argentina	National University of Cuyo
Argentina	National University of San Luis
Brazil	State University of Campinas
Brazil	State University of Paulista—Rio Claro
Brazil	Federal University of Sao Carlos
Cuba	University of Pinar del Rio
Spain	Autonomus University of Barcelona
Spain	University of Girona
Italy	Sannio's Studies University
Portugal	University of Aveiro

Spanish) with the aim to enhance curriculum greening and environmental research coordination.

Methodologically, the ACES Project is premised on cooperative efforts and participatory action research from a multi-disciplinary perspective. It aimed to generate and accumulate environmental knowledge from diverse disciplines and different fields of study. The project contributes strategically as an important coordinating platform in generating a cluster of knowledge that guides effective policy-making in addressing various environmental issues and problems. Equally important is the claim that the peojcet can be flexibly adjusted and applied to redesign curriculum greening in other universities to promote environmental awareness and responsible environmental behavior (de Ciurana and Leal Filho, 2006).

Another case study which is worth mentioning is the EE project launched by the Research and Transfer Centre (FTZ) "Applications of Life Sciences" (FTZ-ALS) at Hamburg University for the generation and implementation of innovative and ground-breaking ideas of sustainable development practices (Leal Filho, 2011). Some of its major achievements are shown in Table 6.3.

To tie thing up better, the FTZ-ALS also launched a series of topical and relevant projects as the seedbeds for germinating and promoting sustainable development education and practices. This is shown *in extenso* in Fig. 6.2

The projects as shown in the figure may be briefly discussed as follows:

A. The INSPIRE project

The INSPIRE project (2007–2009) was developed to analyze the potentials and needs of training courses for teachers. The more specific aims of the projects which are quite in line with the objectives of the UN DESD are listed in Table 6.4. The project contributes fruitfully to the development of various practically useful documents, reports and manuals in enhancing training courses for teachers (Table 6.4)

TABLE 6.3   FTZ-ALS—some major achievements.

Achievement	Remarks
Creation of the world's longest running book series on sustainability, "Environmental Education, Communication and Sustainable Development"	The series involved 300 authors from all over the world
Creation of the International Journal of Sustainable Development in Higher Education	The world's only journal focusing on sustainable development in institutions of higher education
Creation of the first "World Sustainable Development Teach-In Day"	To disseminate information as well as to raise awareness among university students on sustainable development
	To provide a platform for the discussions of the problems and the potentials in implementing sustainable development at the global, regional, and local levels
	To provide an opportunity for the introduction of sustainable development projects or initiatives at the regional and local level by schools, universities, government bodies, NGOs, and other stakeholders
	To encourage more networking and information exchange among participants for possible cooperation initiatives and new projects

Based on, Leal Filho, 2011. Applied sustainable development: a way forward in promoting sustainable development in higher education institutions. In: World Trends in Education for Sustainable Development. Peter Lang, Internationaler Verlag der Wissenschaften.

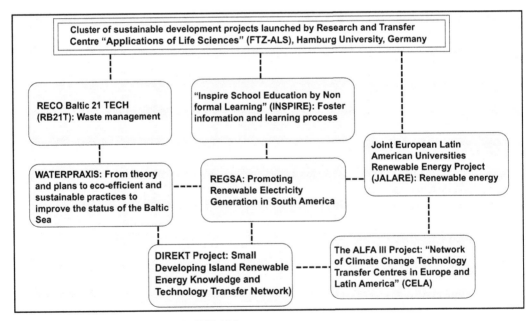

FIGURE 6.2   Projects launched under FTZ-ALS.

TABLE 6.4 INSPIRE project: Some basic information.

	Aims	Achievement
INSPIRE	To foster information and learning on renewable energy and climate change	Documentation and dissemination of the results of literature study and expert interviews
	To improve the quality and attractiveness of in-service teacher training by using extracurricular contexts and new learning places	(1) 4 project reports; (2) A lsit of out-of-school learning places and best practice examples; (3) Manuals and lecture notes for training courses (for teachers); (4) A manual for the project partners; and (5) A handbook on "Renewable Energy in Out-of-school Learning Places".
	To create synergies and links between out-of-school places of learning and curricular learning with the view of improving the knowledge European pupils sustainable development education	

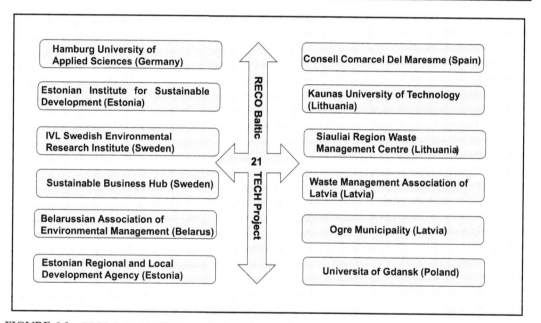

FIGURE 6.3 RECO Baltic 21 TECH: academic and research consortium. *Based on Keep.eu., 2017. Project—RECO Baltic 21 Tech. Keep.eu, European Union. <https://www.keep.eu/project/15692/reco-baltic-21-tech> (accessed 13.03.20.).*

B. The RECO Baltic 21 TECH project

The RECO Baltic 21 TECH programme was created to promote sustainable waste management and to explore the potentials of waste as a source of energy at local and regional levels (Keep.eu, 2017). The project involves the participation of a consortium of higher education establishments including the Hamburg University and other research institutions as shown in Fig. 6.3. Its main achievement is the development of a Joint BSR (Baltic Sea Region) Strategy for Sustainable Waste Management and Development and Investment Concept—both with the aim to assist the BSR to enhance waste management in line with the EU standards (Keep.eu, 2017).

C. The JELARE project

The JELARE project was established with the aim to improve academic quality of European and Latin American higher education institutions and, to strengthen their role in contributing to local economic development and social cohesion. The project also covered some of the following specific aims:

    i. to develop and implement labor market-oriented research and educational approaches in the field of renewable energy;

    ii. to increase the capacity of university staff to modernize their educational and research programmes and activities;

    iii. to strengthen the link between universities and the labor market, business and the public sector in the field of renewable energy;

    iv. to establish a long-term partnership and network between European and Latin American universities (The JELARE partner consortium, 2011).

D. The WATERPRAXIS project

The WATERPRAXIS project entitled "From theory and plans to eco-efficient and sustainable practices to improve the status of the Baltic Sea" was designed in 2007 to address the eutrophication problems facing the Baltic Sea. The project comprised the water management authorities and experts on environmental technology, economy, nature and social sciences and education from the following institutions:

    i. Hamburg University of Applied Sciences, Hamburg (Germany)

    ii. Finnish Environment Institute (Finland)

    iii. Centre for Economic Development, Transport and the Environment for North Ostrobothnia (Finland)

    iv. University of Aarhus, National Environmental Research Institute (Denmark)

    v. Municipality of Naestved (Denmark)

    vi. Environment Centre Nykøping F (Denmark)

    vii. Lodz Technical University (Poland)

    viii. Kaunas University of Technology (Lithuania)

    ix. Charity and Support Fund Sesupe Euroregion, Sakiai Office (Lithuania)

    x. Luleå University of Technology (Sweden)

    xi. Rezekne Higher Education Institution (Latvia)

The project covered some of the following objectives:

    i. To identify and suggest improvements to the existing water management practices,

    ii. To establish action plans based on River Basin Management Plans,

    iii. To prepare investment plans for water protection measures for pilot projects in selected sites, and

    iv. To disseminate information of best practices and measures of water management and offer education for planners of water management and environmental economy areas.

Some of the important achievements of the project include the realization of pilot scale investments for drinking water purification facilities in rural villages in the Lithuania. In addition, the main findings of the project have been compiled in the final project booklet and published in six Newsletter issues to serve as guides for promoting environmental knowledge and education (Interreg Baltic Sea Region, 2010).

E. CALESA project

The project, "Developing promising strategies using analogue locations in eastern and southern Africa" (CALSSA) was initiated in 2011 to investigate the impacts of climate change on agriculture in Africa and to develop sound adaptation strategies for future GHG-induced local warming using "analogue locations". The ultimate aim of the project is to enable the farmers in the semiarid tropics of subSaharan Africa, in particular Kenya and Zimbabwe, to adapt to progressive climate change through crop, sustainable soil and water management innovation, and appropriate crop genotype selection (De Trincheria et al., 2015). CALESA is considered as a research-oriented project for knowhow and technology creation, and development-oriented as well as activities-oriented undertaking for information sharing and capacity building (Leal Filho, 2011; De Trincheria et al., 2015). The project was led by International Crops Research Institute for the SemiArid Tropics (ICRISAT) in cooperation with the following partners:

i. Kenya Agricultural Research Institute (KARI), Kenya,
ii. Kenya Meteorological Department, Kenya,
iii. Zimbabwe Meteorological Department, Zimbabwe,
iv. Midlands State University, Zimbabwe, and
v. The Hamburg University of Applied Sciences (HAW), Germany.

The project has produced impressive results in promoting sustainable agricultural development in the targeted regions as mentioned above. Some of its concrete achievements are reflected in Table 6.5 below.

TABLE 6.5 CALESA's major achievements.

CALESA Project	Adaptation strategies focusing on varieties, plant populations, fertilizer use formulated
	Tools to analyze long-term climate data developed and, crop simulation model APSIM (Agricultural Production Systems Simulator) calibrated and validated
	Criteria for analogue locations developed and identified
	Necessary protocols for undertaking various activities developed and implemented
	Long-term climate data for all nine selected locations collected and analyzed (report prepared)
	Four locally relevant crops (two cereals and two legumes) identified
	Climate change scenarios for all the nine different locations in Kenya and Zimbabwe developed
	Potential management options including water conservation practices, adjustments to plant population, and fertility management identified
	Adaptation strategies based on existing available technologies identified

Based on, International Crops Research Institute for the SemiArid Tropics (ICRISAT), 2014. Adapting agriculture to climate change: developing promising strategies using analogue locations in eastern and southern Africa (CALESA). Final Report 1 January 2011–30 June 2014. International Crops Research Institute for the Semi-Arid Tropics (ICRISAT), Addis Ababa.

F. The CELA project

The ALFA III Project "Network of Climate Change Technology Transfer Centres in Europe and Latin America" (CELA) was initiated by FTZ-ALS to investigate the extent of environmental degradation of the ecosystems in Latin America. Its partnership comprised: (a) Hamburg University of Applied Sciences (Germany); (b) Tallinn Technical University (Estonia); (c) Universidad Galileo (Guatemala); (d) Universidad Católica Boliviana (Bolivia); (e) Universidad de Ciencias Comerciales (Nicaragua); and (f) Pontificia Universidad Católica del Perú (Peru). The project also aimed to promote joint research collaboration and exchanges of experiences between universities in Europe and Latin America in the field of climate change. Particularly, the core objectives of the project are as follows:

i. To improve the quality of research and technology transfer of Latin American Universities,
ii. To strengthen the role of Latin American Higher Education Institutions in sustainable socioeconomic development in the respective regions,
iii. To foster sustainable research and technology transfer cooperation between Higher Education Institutions in Latin America and the EU.

One of the important contributions of the project is the enhancement of environmental information and knowledge dissemination in the Latin American region and beyond. It also contributes to foster international technology transfer and capacity building in the climate change sector. More specifically, its main achievements are:

i. The conclusion of Transnational survey of Climate Change Information Needs,
ii. The establishment of the Network of Climate Change Transfer Centres in Bolivia, Estonia, Germany, Guatemala, Nicaragua, and Peru,
iii. The development of Transnational Climate Change Technology Transfer Strategy,
iv. Three pilot projects were completed: (a) Lifelong Learning Centre for sustainable forestry, Bolivia; (b) Monitoring of Water Quality at Napo River, Peru; and (c) Modelling the River Tamarindo, and Guatemala: Flood-risk management of River Coyolate, Nicaragua,
v. The convention of more than 24 Capacity-building seminars and more than 18 networking events, and
vi. The publication of recommendation report (FTS-ALS, 2015).

G. The DIREKT project

The Small Developing Island Renewable Energy Knowledge and Technology Transfer Network (DIREKT) is a cooperation scheme involving universities from Germany, Fiji, Mauritius, Barbados and Trinidad and Tobago as listed below:

i. The Hamburg University of Applied Sciences (HAW Hamburg), Germany (lead implementer),
ii. The University of the West Indies, Cave Hill, Barbados,
iii. The University of the South Pacific, Fiji,
iv. The University of Mauritius, Mauritius,
v. The University of the West Indies, St. Augustine, Trinidad & Tobago.

The project was implemented in 2009 with the aim of strengthening science and technology capacity in the field of renewable energy of a sample of ACP (Africa, Caribbean, Pacific) small island developing states based on technology transfer, information exchange, and networking. It also aimed to foster cooperation among ACP and EU institutions in the field of renewable energy. The project led to the production of a Transnational Recommendation Report which contains comprehensive and useful guides on the promotion of renewable energy. The Project also helps to enhance quality research capacity of the Small Developing Island States on matters pertaining to sustainable development (Leal Filho, 2011; Bijay et al., 2012; PRDR, 2015).

6.7.1 Environmental and sustainable development success stories—some remarks

The research and educational efforts undertaken by various institutions as discussed above are in accordance with the United Nations environmental education goals. In particular, the project-oriented, cooperative, and practical approaches undertaken by the FTZ-ALS in solving real world environmental problems while enhancing environmental course materials, ecological literacy, and environmental awareness are pragmatically compatible (Fig. 6.4). However, to render the project more fruitful and effective, it is necessary to turn the accumulated environmental knowledge into a flywheel of environmental education through organizing various educational programmes to raise ecological literacy and environmental awareness.

However, this is still not the end of the game. It is instructive to note that environmental awareness does not in itself constitute an action; it is a preparation for it, that is, it precedes environmental action. Indeed, environmental awareness is a poor judge of

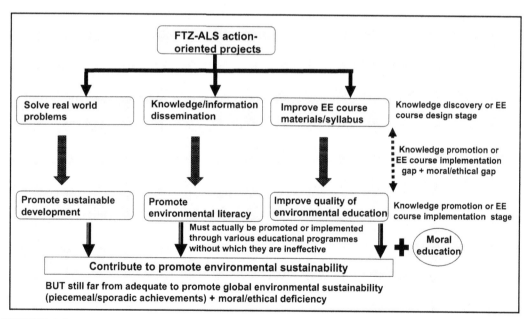

FIGURE 6.4 Action-oriented programme and environmental education—some practical applications.
Note: Moral education as shown in the figure constitutes the ethical nuts and bolts for cementing an environmentally inclined moral society.

environmental action because it does not penetrate deep enough into our inner-self to morally motivate and ethically command us to orient our conduct and actions sustainably. To illustrate, in the United States, polls conducted on green behavior consistently show a high level of environmental awareness in the country. However, as it turned out, increasing environmental awareness frequently does not translate into action (Ritchie, 2017).

Two gaps exist in the above situations. One is the value-action gap which is defined as "the observed disparity between people's reported concerns about key environmental, social, economic or ethical concerns and the lifestyle or purchasing decisions that they make in practice" (SDC, 2006: 63). Another gap is the knowledge-attitude-practice (KAP) gap. This refers to a situation where despite an increase in the individual understanding or knowledge of a given environmental problem due to widespread environmental information and the subsequent changing attitudes in favor of environmental protection, such a positive relation does not, in reality, lead to the application of knowledge nor translate it into real action in practice (Kifle et al., 2014).

In other words, people may claim that they are concerned about, say, carbon emissions and global warming, but generally, they may be unwilling to take real mitigating actions such as reducing energy consumption or changing lifestyles or refraining from high consumerist practices if such actions necessarily involved tradeoffs, high perceived costs, or sacrifices that would inconvenience them even slightly (Flynn et al., 2009). Technically speaking, such expressions of environmental intention are merely a judgement of existence or of reality which defines a person's relations with certain objects. In other words, it does not confer value on the objects but merely affirms the state of the subject. Thus an individual who discloses a judgment of reality may or may not be obliged to fulfil his duty in line with his moral sentiments. Judgment of reality is, as a matter of fact, an individual sentimental reaction, personal attitude, or private judgement that lacks the authoritativeness or moral force to guide real action.

Here, it is relevant to distinguish judgement of existence from value judgement. Conceptually, when an individual expresses the worth of an object in relation to himself, he is not only affirming its existence but also, psychologically demonstrates its "validity" by supporting it with "impersonal argument", that is, conferring a value on it. This may be understood as an inner experience stemming from an individual's understanding of how things are related in the universe. The inner experience is the cornerstone for internal moral incentive to act in an environmentally responsible manner. It is this moral element that represents a stable and enduring motivational force which categorically justifies the adoption of environmentally responsible behavior and action. Thus value judgments may be viewed as external stimuli with authoritative force that morally motivate individuals into collective action.

Returning to the environmental education domain of analysis, it is beyond doubt that universities or higher education institutions (HEIs) can contribute immensely to promote ecological literacy and environmental awareness. However, as noted above, environmental education, compared to mainstream economics, is less attractive to students who generally exhibit a strong preference for the latter which is commonly believed to offer better career prospects. This biased trend of thought is especially prominent in the Asian developing countries. To be sure, there are also numerous success stories across the world especially in Europe where some universities have become the prime mover for environmental or sustainable development education (EE for short). However, many of those efforts were confined to addressing only one or two of the sustainability domains.

Furthermore, educational establishments, public and private alike, are often constrained by limited funding and resources. To compound the problem, the promotion of environmental education in HEIs is further hampered by the ignorance or environmental knowledge gap prevailing among the course administrators, or resistance to change among educators, among others (Leal Filho et al., 2017, Figure 6.1). For example, Verhulst and Lambrechts (2015) have identified two types of resistance related to financial and structural support and resistance related to empowerment and personal support.

That said, the mainstreaming of environmental education into HEIs is mostly thwarted by human barriers characterized by the deeply engrossed biased views and regressive minds of educationists and students alike concerning the importance and relevance of EE education compared to mainstream economics courses. To break this human impasse, it is necessary to convincingly portray EE as a multidisciplinary and interdisciplinary course that adequately adapts to the rapidly changing global economic, social, and environmental conditions today. More pertinently, EE may be introduced as a fertile educational seedbed which brings about a process of intellectual efflorescence that enhances an individual's scholarship, critical global thinking, and cooperative learning, and the development and sharpening of problem-solving skills. Promoted as such, EE may find its way into the mainstream of education as an indispensable component.

To complete the analysis, let us turn our attention to the role of the UN in promoting environmental literacy and awareness. To start with, the UN, as the authoritative global organization, must display an overriding consideration, with a clear-cut objective, for promoting not only environmental education but more importantly, the moral discipline of environmentalism. The focus is on activating and reinforcing the moral spirit of environmentalism in shaping economic behavior, environmental attitudes, and moral values primordially and lastingly—the prerequisite for global environmental sustainability.

The *raison d'être* for this line of thought is that without promoting an environmentally conscious global community and ethically responsible global society, it is practically impossible to bend the will of the global society to embrace a more sustainable mode of human practice. As argued in Section 6.4, environmental education per se is unlikely to acquire a sweeping and imperative significance to drive human will to restrain and contain environmentally destructive behaviors and action without concomitantly bringing ethical and moral disciplines into its fold—an indispensable discipline which invokes the moral value of environmentalism (Section 6.15 of this chapter).

The UN-led environmental and moral education may be promoted through a series of cooperative programmes with local political and educational institutions, government bodies, NGOS, and civil society with the overarching goal of enhancing an environmentally conscious society. These programmes may include, for example, the action-oriented environmental projects as launched by FTZ-ALS discussed above and other educational initiatives such as the establishment and design of eco-schools, outdoor environmental activities, enhancing environmental education mainstreaming and environmental course streamlining, among others. These may be oriented at some of most important issues facing humanity today, such as climate change and biodiversity loss, nature and moral education, including ecosystem services and human–nature relationships, and outdoor programmes, among others (Fig. 6.5).

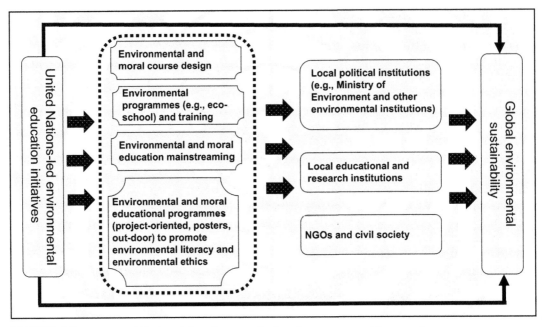

FIGURE 6.5   The United Nations-led environmental education—a proposed framework.

Also, no less important is the need for the United Nations to make concerted efforts at promoting and encouraging the mainstreaming of environmental and moral education in the educational system in countries across the world. However, global environmental education efforts must be persistent and continuous rather than conducted sporadically or on a piecemeal basis, as the latter is unlikely to be effective in establishing an enduring system of ethical relationship between human and nature—the touchstone of global environmental sustainability.

6.8 The United Nations/UNESCO education for sustainable development revisited: The return of Émile Durkheim's philosophical principle of simplicity

Following from the above discussion, we may now turn to Émile Durkheim for some clues to broaden our understanding of the United Nations/UNESCO's environmental education dilemmas, and to suggest ways for improvement. Durkheim, a French sociologist, is often cited as the founding father of classical sociology and the principal architect of modern social science along with Karl Marx, and Max Weber (Coffey, 2004; Calhoun et al., 2012; Dahnke and Dreher, 2016). Durkheim's work is fundamentally associated with the analysis and exploration of social study and morality and includes published influential works on a number of sociological and moral topics (see, for example, Durkheim, 1973a,b). Despite their social and moral epistemological positions, they provide invaluable insights into various conceptual underpinnings which allows us to understand the current United Nations/UNESCO EE fiascos and challenges. This in turn enables us to ratify its shortcomings by developing mechanisms that serve to improve environmental education structure and standards applicable to most other environmental settings.

For a start, in *The Elementary Forms of Religious Life*, Durkheim contends that "Every time that we undertake to explain something human, taken at a given moment in history—be it religious belief, a moral concept, a legal principle, an aesthetic style or an economic system—it is necessary to commence by going back to its most primitive and simple form..." (Durkheim, 1915: 3). A religious system is said to be "most primitive" when: (1) it is found in a society whose organization is surpassed by no others in simplicity; and (2) it is possible to explain this system without using any element borrowed from a previous religion (Durkheim, 1915: 1). According to Durkheim, primitive religions are privileged cases, because they are simple cases which "do not merely aid us in disengaging the constituent elements of religion; they also have the great advantage that they facilitate the explanation of it" (Durkheim, 1915: 7). He further reveals that, "since the facts there are simpler, the relations between them are more apparent" (Durkheim, 1915: 7). It thus follows that concepts and perspectives introduced to explain certain things, such as environmental issues must not be ambitiously multiplied beyond necessity. The key word here is simplicity, which offers one of the best means to explain issues at hand. In other words, people especially the uninitiated will be able to appreciate something if it turns out to be simple enough to comprehend.

Durkheim emphasizes that, "If we have taken primitive religion as the subject of our research, it is because it has seemed to us better adapted than any other to lead to an understanding of the religious nature of man, that is to say, to show us an essential and permanent aspect of humanity" (Durkheim, 1915: 1–2). More telling is his argument that "at the foundation of all systems of belief and all cults, there ought necessarily to be a certain number of fundamental representations or conceptions and of ritual attitudes which, in spite of the diversity of forms which they have taken, have the same objective significance and fulfil the same functions everywhere. These are the permanent elements which...form all the objective contents of the idea which is expressed when one speaks of religion in general" (Durkheim, 1915: 5).

In *The Elementary Forms of Religious Life* (1915), Durkheim unveils a theory of functionalism in association with religion which provides additional important clues for the design and structure of the contemporary environmental education system. According to Durkheim, religion is something eminently social. It serves to "excite, maintain or recreate certain mental states" that assemble people together (Durkheim, 1915: 10). It is worth emphasis that according to Durkheim, religion is a "collective thing" and "a unified system of beliefs and practices relative to sacred things" which symbolizes the same objective significance and ideas, and serves the same unifying function everywhere (Durkheim, 1915: 47, see also, p. 5 and p. 47). Durkheim further contends that religion plays an important function in providing cohesion and norms in a society. This line of argument represents the basis of the functionalist perspective, or functionalism in sociology. Within Durkheim functionalist theory, sometimes also referred to as structural-functionalism, consensus or equilibrium theory, a system or an institution exists because it serves its vital role in the function of society, as in the case of religion which serves its function in unifying society as mentioned above (Ballantine and Hammack, 2012: 11).

What may be inferred from the above discussion is that to render environmental education more effective, it is necessary to avoid complexity. In other words, environmental education, like Durkheim's work on primitive religions, should be simply made clear and distinct in order to allow it to serve its function to establish a mental image concerning our relationship with the natural system. This in turn will allow it to serve its functional role to "excite, maintain or recreate certain mental states" in society toward a sustainable future. In addition, environmental education, just like religion, may be treated as a "collective thing" and "a unified system of beliefs and practices" which symbolize the same objective significance and ideas in association with our ethical relationship with the natural environment, and it must serve this "same unifying function everywhere".

In fulfilling these conditions, it is necessary to allow environmental education to stand on its own as an authentic rigorous educational framework rather than blended with a plethora of mutually exclusive educational aims. The question here is which of the alternative models of educational system would best allow individuals to have a clear mental image of various environmental concepts and our relationship with the natural system—a prerequisite for raising environmental awareness. Thus following Durkheim's sociological logic, it is increasingly clear that if environmental education is to be an effective means of promoting a sustainable future, it must follow a simple course of promoting environmental literacy/environmental awareness as the "objective significance" fulfilling "the same functions everywhere". These are the "permanent elements" when one speaks about environmental education in general.

6.9 The foundation of environmental education

6.9.1 The first ring of predominance mechanism

In connection to the above, it is relevant to examine the question of how the "permanent elements" of environmental education may be crafted in a simple and intelligible way to serve the "same unifying function everywhere". Again, we may turn to Durkheim for his discussion concerning René Descartes' first ring principle of predominance in the chain of scientific truths (Durkheim, 1915: 4). Descartes (1596–1650), a famous French philosopher and mathematician, is considered by many scholars as the father of modern philosophy (Byrne, 1996; Ariew et al., 2010). Durkheim's conception of Descartes' first ring principle rests on the premise that the first ring, on which the rest of the rings are anchored, manifests the fundamental foundation of a progressively ascending building block for an unbreakable chain of knowledge. If the first ring is weak and breaks off, the whole chain of knowledge will fall apart.

According to Descartes, the first ring, which constitutes a keystone of knowledge or a firm building block toward human perfection and indubitable judgment "must be not only clear, but also distinct" (Descartes, 1913: 152). Descartes calls it the "principal property of every substance, which constitutes its nature or essence, and upon which all the others depend" (Descartes, 1913: 157). He further emphasizes that it is an attribute of a substance "without which the substance is unintelligible" (Descartes, 1985: 214). The nature or the essence of a clear and distinct substance in Descartes may also be portrayed as the "first principles", that is the basis from which a certain subject matter of thought is known.

To avoid confusion, it may be reiterated that a substance is just a self-subsisting subject matter of thought. If it ceases to endure, it will also cease to exist (Descartes, 1913: 163). In other words, a substance cannot exist without the principal attribute. Moreover, a principal attribute is a distinct property of a substance that determines what that self-subsisting substance or subject matter of thought is. The principal attribute is the indispensable building block or the most essential first ring for the expansion of an ascending chain of knowledge. In fact, the principal attribute is akin to "essence" in Descartes. It may be deduced that "nature or essence" in Descartes is akin to "permanent elements" in Durkheim.

Descartes expresses the first principles along with its elements as aforementioned as follows:

> I should have desired, in the first place, to explain in it what philosophy is, by commencing with the most common matters, as, for example, that the word philosophy signifies the study of wisdom, and that by wisdom is to be understood not merely prudence in the management of affairs, but a perfect knowledge of all that man can know, as well for the conduct of his life as for the preservation of his health and the discovery of all the arts, and that knowledge to subserve these ends must necessarily be deduced from first causes; so that in order to study the acquisition of it (which is properly called philosophizing), we must commence with the investigation of those first causes which are called Principles. Now these principles must possess Two Conditions: in the first place, they must be so clear and evident that the human mind, when it attentively considers them, cannot doubt of their truth; in the second place, the knowledge of other things must be so dependent on them as that though the principles themselves may indeed be known apart from what depends on them, the latter cannot nevertheless be known apart from the former. It will accordingly be necessary thereafter to endeavour so as to deduce from those principles the knowledge of the things that depend on them, as that there may be nothing in the whole series of deductions which is not perfectly manifest (Descartes, 1913: 107–108).

It may further be revealed that according to Descartes, things should be explained with the most common matters which are clear and distinct for human conception. This is the condition which underpins Durkheim's logic of simplicity as discussed above. Furthermore, these most common matters are the initial distinct properties of the substance or the principal attributes constituting the essential and firm foundation for human gradual acquisition of a chain of related subject matter of thought and other substance. This systematic orientation of knowledge acquisition foundation allows us to discover its long chain of mutually connected reasoning.

Implicitly, viewed from the environmental education perspectives, its tenets must also commence from first principles which should be simple, clear, and distinct. More specifically, environmental education, exhibiting as the "principal attribute" for promoting environmental literacy, should commence from the investigation of the most common issue on the inextricable link between human and nature and the infallible truth of nature connectedness that underpins the integrity of our Earth systems and human civilization. Connectedness refers to the extent which individuals belief that they are part of the Earth systems as argued by Aldo Leopold. This initial building block of the first ring of predominance constitutes the nature and essence from which a chain of related substances may be built and extended in length, breadth, and depth in order to allow us to intellectually reach the "highest degree of wisdom" concerning the natural systems around us, in Descartes' words (Descartes, 1913: 115).

6.9.2 The impact-oriented maxim

In designing an effective environmental education model, apart from the principle of the first ring of predominance as discussed above, it is also necessary to consider the impact-oriented ways which have the effect to "excite, maintain or recreate certain disciplined thought" concerning our place in nature and our moral relationship with it. Here, we may again turn to Durkheim for edification.

According to Durkheim, no knowledge of the world is possible without humanity in some way representing it. More specifically, "we can only become attached to things through the impressions or images we have of them...Not only must we repeat this representation, but in repeating it, give the idea enough colour, form and life to stimulate action. It must warm the heart and set the will in motion. The point here is not to enrich the mind with some theoretical notion, a speculative conception...In other words, the representation must have something emotional; it must have the characteristics of a sentiment more than of a conception" (Durkheim, 2002: 229). Durkheim also emphasizes the need to broaden and infuse the greatest possible numbers of other ideas and feelings into education as vividly and as forcefully as possible to reinforce one's moral consciousness (Durkheim, 2002: 229–230).

Drawing from Durkheim's insights, and based on my several years of environmental education teaching experiences, one of the most effective ways to emotionally "excite, maintain or recreate certain mental states" concerning our natural environment is through impactful power point slides, vivid illustrations, and visual aids based on academic research with relevant case studies, and environmental field trips. This is followed by a gradual and progressive demonstration in an ascending order on how the various basic and simple environmental concepts are related to these studies.

Here, it needs to be cautioned that while the promotion of environmental literacy and awareness involves complex discussions of a wide range of environmental and scientific concepts and issues as well as complex interconnected ecological processes, the first step toward achieving this goal is not to fill the individuals' minds with complex theoretical constructs or environmental concepts. Rather it should aim at stimulating lasting impression on our relationship with nature based on a clear and distinct environmental theme in order to "excite, maintain or recreate certain mental states" in our moral relationship with nature. This will be demonstrated in the later sections of the chapter (Sections 6.10 and 6.11).

6.10 From the first ring of environmental predominance to the second ring of moral essentiality

Commencing from the first ring of knowledge, environmental education may be progressively developed by admitting other environmental substances into it in order to enable us to discover and grasp their relationships. The gradual admission of the subordinate rings of knowledge into the first ring of postulates will connect the intellectual chain of our understanding on the inextricable relationship between human beings and the natural environment, and our place and role in nature in the fullest sense of the word. This culminates in the self-realization of our ethical responsibility toward nature. This first ring of environmental predominance necessarily leads to the second most important part of environmental education analysis—moral or environmental ethics education, or the second ring of moral essentiality. This will be dealt with in the last section of the chapter (Section 6.15 onwards). The following section is an attempt to examine the first ring of environmental education with the view of reflecting how the United Nations' environmental framework may be progressively developed from a simple framework to a complex structure.

6.10.1 The first ring of environmental education: Where do we start from here?

Environmental literacy may be enhanced based on a systematically developed educational framework of the environment comprising various types of environmental content such as nature studies, ecology, conservation, environmental issues, environmental concepts, and ecological principles. These closely interrelated environmental academic perspectives constitute the first ring of environmental education which allows individuals to gradually recognize the complexity of nature and our dependence on its capacity to serve as a life-support system for continued human existence. To achieve this, it is necessary to avoid embracing a large number of concepts at the initial stage as this could be confusing.

Bearing this in mind, and avoiding multiplying its contents beyond necessity, we first lay the foundation in an ascending order for the environmental education framework by attempting to answer the following simple, clear, distinct, and practical question: "Why should we care about our environment?" This is the "most primitive and simple" question which constitutes the fundamental representation of our relationship with nature. It is also the principal attribute or the foundation of a comprehensive system of environmental belief from which other substances or subject matters of thought ("essence" in Descartes) are facilitated. In Durkheim's terms, it represents "functionalism" in environmentalism which leads us to understand the essential and permanent aspect of human–nature relationship. It also affects how humans view our natural environment and advances public thinking about environmental values and environmental ethics, which in turn contributes to inducing strong public commitment to environmental conservation. More specifically, it gives us a shared set of clear ideas on the interconnectedness of human beings and nonhuman natural environment, our indispensable duty in environmental protection, and the likely cumulative environmental consequences of human inactions. Logically articulated, this line of analysis universally serves the "same unifying function everywhere" in promoting environmental literacy and proenvironmental behavior.

Returning to the above question, we should care for the environment because human long-term existence hinges on the ecological integrity of our Earth systems, and our relationship with nature is that of interaction and interdependence. We depend on the Earth for a stream of goods and services for survival. These include provisioning services such as the provision of food, water, raw material, medicinal resources and genetic resources, regulating services such as air quality regulation, climate regulation and moderation of extreme events, supporting services such as nutrient cycling, and GHG assimilation, and cultural and amenity services such as recreational benefits. This will be further demonstrated in the following section based on impact-oriented illustrations (Section 6.10).

6.10.2 The first ring of education: Weak anthropocentric inducement

Conceivably, the first ring of environmental education as elucidated above commenced from the anthropocentric mark of environmentalism in that it is concerned about the choice of acting in one's self-interest vis-à-vis collective interest in protecting the instrumental values of nature which underpin each individual's socioeconomic sustenance. However, it must be clarified that this is not an assertion or endorsement about the exclusive importance of nature's instrumental value in motivating individuals to protect the Earth system as a whole. Rather, its main aim is to firstly provide the empirical and moral reasons for human dependence on nature (weak anthropocentric inducement) and the ethical imperative for environmental protection.

Furthermore, in our daily interaction with the natural environment, there exists a real life decision-making problem called the social dilemmas which may generally be defined as the conflicts between individual and collective interests. For the former, an individual judges his environmental actions based on perceived personal cost and benefit, and may act ethically when he believes that his position is threatened due to environmental changes. On the other hand, an individual in pursuit of collective interest will not only consider his own self-interest but also, the interests of his fellow members, nonhuman species and the ecosystems. In this case, it is said that collective interest is prioritized over self-interest.

Thus the pendulum of social dilemma which swings between anthropocentrism and ecocentrism is inherently a moral issue, that is, a question of value orientation which can best be dealt with by moral education (see Section 6.15). In addition, as revealed in Chapter 4, Greening for a Sustainable Future: The Ethical Connection, our main purpose is to opt for a pragmatic approach to solve practical problems rather than conceptual conflicts or theoretical differences. Thus approaching environmental issues premised on the conceptual polarization between an anthropocentric view of nature and nonanthropocentric philosophy of environmental conservation is unhelpful in tackling practical questions in a real world system.

Viewed from a realist perspective, it will be impractical to reject anthropocentrism totally as humanity imperatively depends on the Earth system for socioeconomic sustenance. Furthermore, as examined in Chapter 4, Greening for a Sustainable Future: The Ethical Connection, even Aldo Leopold and Albert Schweitzer's environmental philosophies, which constitute the gist of environmental ethics are not exclusively non-anthropocentric in nature. They clearly acknowledge human need for the instrumental use of nature. In other words, they are anthropo-bioecocentric in construction (Sections 4.3 and 4.4). That said, in a real world system, it is impractical and nonfeasible to suggest ethical commitment to protect all intrinsically valuable beings or natural entities.

Arguably, given that by and large, human beings are in general motivated by short-term self-interest rather than long-term collective interest as reflected in Chapter 3, The United Nations' Journey to Global Environmental Sustainability Since Stockholm: The Paradox, and Chapter 4, Greening for a Sustainable Future: The Ethical Connection, individuals are likely to extend moral consideration to nature and protect the environment if motivated by the rationale to do so. In particular, individuals who hold a weak anthropocentric view of nature are likely to be induced to behave and act environmentally if given an appreciable dose of environmental and moral education. This will be further discussed in Section 6.15 in relation to moral education.

6.11 Environmental education in practice: An illustration

6.11.1 Nature study and conservation practice

To build up the structure of environmental education in an ascending order from the first ring of predominance as discussed above, we may add more environmental substance to its principal attributes. However, not every environmental substance can be added here due to space constraints. For the purpose of demonstration, we may begin our educational excursion based on practical knowledge of ecological processes resting on the clearest and most distinct environmental theme concerning food supply and human survival which is inextricably linked to the principal environmental question raised in the preceding section. This theme, which constitutes one of the most common and fundamental themes of environmental education, is distinctly clear in that it is the basic belief we are acquainted with through our daily experience with the natural world. Basic belief is true, incontestable, and noncontroversial by virtue of its self-evident rationality or its empirical confirmation (Kirk, 2007: 233).

The dissection of this theme provides the fundamental knowledge and understanding about how the ecosystem functions and supports life, and creates one of the essential foundations for environmental literacy. It also provides an intelligible understanding on the question of "Why should we care about our environment?" as raised in the previous section. More specifically, the construction of knowledge based on this educational theme not only enriches our mind with a distinct understanding of our indispensable reliance on nature for continued existence but also, enables us to see and relate things as functions of each other. Apart from lending weight to the importance of biodiversity conservation as discussed in Chapter 3, The United Nations' Journey to Global Environmental Sustainability Since Stockholm: The Paradox, and Chapter 4, Greening for a Sustainable Future: The Ethical Connection, the theme also reflects obligatory elements that allow us to interpret our relationships with things in nature and provide them with meaning and value. This may warm our heart and set the will in motion for our ethical responsibility toward our natural systems. Imparting knowledge along this educational theme also equips us with a clear understanding of various basic and principal ecological concepts underlying the functions of our ecosystems.

6.11.2 Environmental education in practice: A step-by-step approach in nature study and conservation practice

To provide a vantage point for the above educational theme, we may safely turn to Albert Einstein's view on the interdependence between humans and nature which illuminates a collective representation of a number of essential ideas governing human−nature relationship (see Section 6.16.2 for the explanation of collective representation). We could begin with what Albert Einstein wrote in 1941: "Remove the bee from the earth and at the same stroke you remove at least one hundred thousand plants that will not survive" (Vujica, 2013). This seven decades' old incredibly powerful ecological thought, called here "the parable of the bees", is the collective representation of the multifunctional dynamics of ecosystems which are inextricably linked to our wellbeing and long-term survival (Fig. 6.6). It poses the following pertinent questions to the readers:

Why does this matter? What are the relationships between bees and their natural surroundings, and the implications for the food supply system that underpins human long-term existence? In short, why should we care about our environment?

To start with, Fig. 6.6 provides an infinitely clear understanding of human dependence on the ecosystem for its life-supporting services, namely, provisioning services, regulating services, supporting services, and cultural and amenity services, for long-term existence. It

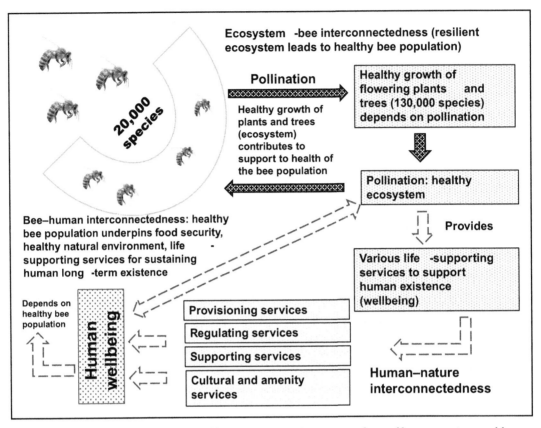

FIGURE 6.6 The parable of the bees and human existence: interconnectedness of bees, ecosystem, and human beings.
Notes: The ecosystem provides various services, namely, provisioning service, regulating service, supporting service, and cultural and amenity services to support human long-term existence. The plants and trees or natural environment, the bees and humans interact with one another to ensure an ecologically sound ecosystem which underpins the long-term and healthy existence of the human and nonhuman living organisms.

also reveals clearly that the ecological health of the ecosystem depends on pollinators especially the bees. It may well be that out of 250,000 species of flowering plants and trees, 130,000 of them rely on insects, mostly 20,000 species of bees, to ensure reproduction through pollination (Schacker, 2008). Thus they are the players in pollination and the key drivers in the maintenance of biodiversity and ecosystem health, helping to ensure crop plants produce full harvests. The bees also help to maintain the genetic diversity of the plant community in the ecosystem. Hence, their population degeneration (bee-kill) may result in the destruction of hundreds of thousands of plants.

Furthermore, cross-pollination by bees helps at least 30% of the world's crops and 90% of the wild plants that thrive in the United States (NRDC, 2011; Grossman, 2013). In addition, 70 out of the top 100 human food crops are pollinated by bees—this amounts to the supply of about 90% of the world's nutrition (Weyler, 2013). Remarkably, a single bee colony can pollinate 300 million flowers each day. The European Commissioner for Health and Consumer Policy, Tonio Borg has calculated that bees contribute more than $30 billion annually to European agriculture. Worldwide, it has been estimated that bees pollinate human food valued at more than $350 billion (Weyler, 2013). Bees and other insects also pollinate 95% of the canopy and 75% of the shade-tolerant subcanopy (Buchmann and Nabhan, 1996). In comparison, only less than 3% of all tropical lowland forest plants of the Americas rely on the wind for pollination while vertebrates pollinate 5% of the canopy and 20 to 25% of the subcanopy (Buchmann and Nabhan, 1996).

It is thus increasingly clear that bees play an important role in sustaining the health of the global ecosystem as well as the continued existence of human beings in terms of food security. More specifically, bees constitute the crucial lifeblood to global agricultural systems. They may be considered as keystone species in our natural system. The term "keystone species" refer to a species that plays a disproportionately large and unique role relative to the rest of the community in supporting the healthy functions of an ecosystem (Paine, 1969). Without the keystone species, other species in the habitat would also disappear or become extinct; causing the ecosystem to be dramatically different from its original position or causing it to cease functioning altogether (Mills et al., 1993). The removal of a keystone species will result in a host of destructive impacts across the system, and this could destabilize the system as a whole.

Upon gaining the basic knowledge about human—nature—bee relationship based on a distinctively clear narration as elucidated above, it is possible to go deeper into subject matter by revealing how human unsustainable environmental practices such as uncontrolled use of pesticides may destabilize the human—nature—bee relationship to the detriment of human long-term survival. Demonstrably, extensive use of pesticides in the agricultural sector tends to pollute the natural environment (abiotic factor).

When the pollutants interact with the natural environment, it creates adverse environmental conditions (abiotic factor) which are detrimental to the continued existence of various living organisms such as the bees (biotic component). For instance, the common use of neonicotinoids including its subclasses, nitroguanidines, and cyanoamidines in the agricultural sector is particularly harmful to the bees. More particularly, nitroguanidines, which include imidacloprid, clothianidin, thiamethoxam and dinotefuran, are acutely toxic to honeybees, and their oral toxicity is extremely high at 4—5 ng/individual bee (Greenpeace, 2013). These insecticides, which have been used for the past two decades, help to control a range of pests such as sap-feeding insects and root-feeding grubs, by blocking neural pathways in the insect's nervous system (Long, 2014).

To convince the prospective learners, it may further be revealed that the extensive use of these pesticides in the United States has in part resulted in the decline of the honeybee population from about 6 million hives in 1947 to 2.4 million hives in 2008, a 60% reduction (Weyler, 2013). Extensive and uncontrolled use of pesticides also worsen the strong immune system of bee colonies caused by immunotoxic effects (Goulson et al., 2015; Sánchez-Bayo et al., 2016; Brandt et al., 2016). This renders the bees more susceptible to disease and more vulnerable to the harmful effects of pesticides. Habitat loss, fungicide exposure and the introduction of exotic species are also some of the factors that have led to the decline of the bee population (NRDC, 2011; Bernauer et al., 2015; Stanley and Raine, 2016; Stanley et al., 2016).

Thus starting from the first and simple basis from which some self-evident environmental idea such as human dependence on nature or the bees for that matter, for long-term existence which is known or clearer to us, it is possible to work all the way backward by breaking them down into parts until we get the core building blocks on which those ideas are built on. This will allow us to grasp the dynamics, complexity and multiple, or causal connections of those environmental ideas. In this way, we will be able to see environmental problems from a different perspective.

In more specific terms, the narration of the parable of the bees extended from the first ring of predominance, comprising a series of progressive environmental and ecological explanations, aids us in understanding a chain of scientific truths in association with the complex interactions and interconnectedness among various biotic components (bees, flora, and humans) and abiotic factors (environmental pollution) within an ecosystem, and its importance in underpinning global food security. This allows us to see how unique humans are bound together into a collective system of human–nature relations. This collectivism informs our sense of belonging and our place in nature. We may experience these collective relationships as obligations, duties, and restraints on our behavior when interacting with the natural environment.

To round up, the parable of the bees provides the basic educational theme for incorporating broader concepts of environmental literacy and global environmental issues in a gradual manner, allowing individuals to have a greater understanding and practical knowledge of how our natural system functions, and our dependence on its ecological health to sustain human long-term existence. This serves to enhance individuals' ability to apply environmental knowledge, ecological disciplines, and moral principles for making informed sustainable and moral decisions in bridging the human–nature gap. This in turn leads to the second ring of moral thought which instrumentally encourages us to interpret and order our position in the natural world and our ethical responsibility toward it.

Although there is no one single factor that contributes to the overall global decline in bee population, two of the major factors causing their decline are the uncontrolled use of pesticides and habitat destruction in the agricultural sector. In other words, bee decline is essentially an ethical issue when farmers unethically and violently disrupt their way of life in the utilitarian quest for profit maximization. These unsustainable practices are basically caused by the lack of moral attributes in the use of natural environment; attributes such as those moral principles embedded in Schweitzer's "reverence for life" or Leopold's "love and respect" for the natural environment. In reversing the bees' decline and restoring their abundance, it is necessary to foster a better moral relationship between human beings and the natural environment, and to forge right patterns of conduct premised on Leopold-Schweitzer's environmental philosophy or anthropo-bioecocentric ethics as discussed in Chapter 4, Greening for a Sustainable Future: The Ethical Connection, toward the ecosystem, including the bees, through limiting the use of pesticides and other chemicals, or by making a moral shift toward using least-toxic pesticides and chemicals. The ethical behavior in biodiversity conservation may be enhanced through the second ring of moral education as discussed in Section 6.15.

6.12 Environmental education: Outdoor education programmes

In extension, environmental literacy can be more effectively developed and enhanced through direct engagement with the natural world based on outdoor educational programmes. The rationale behind this approach is that placing individuals in an actual environmental setting where some of the ecological processes take place allows them to appreciate and discover personally the essence of ecological truths governing human survival in a real world system. It connects real ecological processes to a series of discoveries, a wealth of insights and a clear image of the dynamic behavior of the ecological system and our relationship with it. This will have a considerable influence on individuals' moral discipline in Durkheim's sense.

To demonstrate, as an extension of the "the parable of the bees" narrative mentioned earlier, we may take on yet another educational theme with the same objective significance. This can be done by laying open the functional properties of one of the ecosystem services which affects human long-term existence, namely, ecosystem regulating and supporting services in association with GHG emissions. This educational theme also serves to enable a better understanding of the global GHG emission problems as discussed in Chapter 3, The United Nations' Journey to Global Environmental Sustainability Since Stockholm: The Paradox, and Chapter 4, Greening for a Sustainable Future: The Ethical Connection. As will be made clear in the subsequent section, this environmental educational theme, when developed progressively, sheds light into a series of interlocking relationships between the functional properties of nature and human activities. The theme, initially taking its simple form but gradually developing in depth with additional related information, also reveals the distinct and concrete reality of the looming environmental crisis confronting human civilization.

The curriculum-linked outdoor learning and environmental education in association with GHG emissions and global warming may be conducted progressively in the ocean beach surrounding area and the atmosphere which serve as a platform for environmental outreach for the prospective learners. Placing the learners into this natural setting allows them to have a broad scope of mental information that connects human activities to environmental issues. To put it bluntly, it has the impact "to excite, maintain or recreate certain mental states" in individuals concerning their moral duty toward the natural environment or the atmospheric carbon sink, for that matter.

To start with, to many people, the image of the ocean is that it is "beautiful, sublime, and glorious and, mild, majestic, foaming, free", constituting an "image of eternity", to borrow the lines from a poem by Bernard Barton (1784−1849), *The Sea* (Barton, 1981: 377). However, behind the veil of its "image of eternity", the global oceans undertake a critical life-support function in sustaining human long-term existence. For example, prospective learners may be enlightened by the fact that the global oceans which cover about 71% of the Earth's surface, about 360 million square kilometers, and with a volume of about 1370 million cubic kilometers, constitute a critical part of the climate system.

More specifically, the global oceans (the Pacific, Atlantic, Indian, Southern, and the Antarctic Ocean) play a crucially important role as a global climate control system in regulating climate change and global warming through atmosphere-ocean surface exchange of the heat trapping carbon dioxide (CO_2). This is due to its large capacity to absorb CO_2, hence reducing or keeping the amount of CO_2 concentration in the atmosphere and maintaining its concentration in a relatively stable state. More than 25% of the anthropogenic CO_2 emission has been absorbed by the oceans since the industrial revolution (Ocean and Climate, 2015; Wold et al., 2013).

To emphasize, in a process called photosynthesis, phytoplankton use the dominant abiotic components, namely, sunlight, CO_2 and water to convert into proteins, fats, and carbohydrates, which they use for food, and to make their cells. CO_2 is an extensive source of inorganic carbon found in the atmosphere. When CO_2 dissolves in seawater through ocean-surface exchange as shown in Fig. 6.7, it can exist in various dissolved forms including aqueous carbon dioxide, (H_2CO_3) (1%), bicarbonate (HCO_3^-) ($\sim$90%), carbonate (CO_3^{2-}) ($\sim$9%), and the minor form, carbonic acid which constitutes less than 0.3% of aqueous CO_2 concentration (Dickson, 2010; Sen Gupta and McNeil, 2012; Zeebe, 2012; Fig. 6.7).

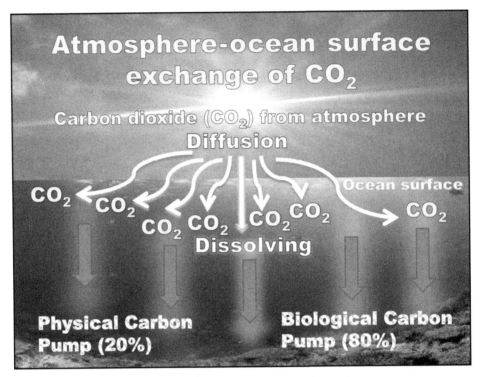

FIGURE 6.7 Ocean's CO_2 absorption through atmosphere-ocean surface exchange.
Note: CO_2 moves between the atmosphere and the ocean by molecular diffusion when there is a difference in partial pressure of CO_2 (pCO_2) between the atmosphere and oceans. For example, when the atmospheric pCO_2 is higher than the surface ocean, CO_2 diffuses across the air-sea boundary into the sea water (Sabine et al., 2004). It is then taken up by phytoplankton as one of the basic ingredients (inorganic carbon) which is converted to organic matter through photosynthesis. The biological process of transporting the organic matter to the deep ocean is known as the biological carbon pump (biological carbon export). The biological carbon export, technically termed as the vertical gradient in dissolved inorganic carbon is fueled by a network of marine phytoplankton or "photoautotrophs" (photosynthesizing organisms). The biological carbon pump contributes to the bulk of the total carbon transported (about 80%) from the ocean surface to the oceanic interior. The physical carbon pump operates through the upward and downward movement of water (upwelling and downwelling) where the dense cold waters sink from the mixed layer in the ocean surface carrying with it the CO_2 absorbed from the atmosphere. This CO_2-laden water may take centuries to millennia to return to the ocean surface. In passing, the physical carbon pump accounts for about 20% of the total carbon export (Choy, 2019). *From, Choy, Y. K., 2019. The Ocean Carbon Sink and Climate Change: A Scientific and Ethical Assessment. International Journal of Environmental Science and Development, 10 (8): 246−251.*

Bicarbonate is a form of carbon that does not escape the ocean easily. The sum of the concentrations of all these dissolved forms of CO_2 is the inorganic carbon concentration (Dodds and Whiles, 2010). Once dissolved, inorganic carbon is converted to organic matter through photosynthesis. This is known as primary production, that is, the synthesis of organic compounds from atmospheric CO_2. A small fraction of this organic carbon is subsequently transferred to the deep ocean through gravitational sinking which is generally termed as carbon export where it is sequestered from the atmosphere on time scales of months to millennia (Feely et al., 2001; Ocean and Climate, 2015; Kämpf and Chapman, 2016; Guidi et al., 2016; Honjo et al., 2014; Siegel et al., 2016, Fig. 6.7). The physical carbon pump or the solubility carbon pump removes atmospheric CO_2 by sinking dense cold water where CO_2 is sequestered from the atmosphere in the form of dissolved organic carbon (Sundquist and Visser, 2005; Sen Gupta and McNeil, 2012; Fig. 6.7).

Once absorbed into the oceans, CO_2 no longer traps heat. Over the past 200 years, the oceans have taken up 500 $GtCO_2$ from the atmosphere out of 1300 $GtCO_2$ total anthropogenic carbon emissions (IPCC, 2005). A recent estimate by the IPCC (Intergovernmental Panel on Climate Change) indicates that the total carbon emission from 1750 to 2011 was 545 GtC or 2000 $GtCO_2$. Of this, 44% (240 GtC or 880.8 $GtCO_2$) had accumulated in the atmosphere while 28% (155 GtC or 568.85 $GtCO_2$) had been taken up in the oceans. Another 28% (150 GtC or 550.5 $GtCO_2$) had accumulated in the terrestrial ecosystems (IPCC, 2013). It is thus clear that the ocean is an enormous carbon sink (ocean's supporting function). To note in passing, the factor used to convert carbon (C) to (CO_2) is 1 GtC = 3.67Gt CO_2. In addition, while carbon sink removes carbon from the atmosphere, carbon reservoir stores carbon previously removed from the atmosphere (Wold et al., 2013).

Even areas hundreds of miles away from any coastline are still largely influenced by the global ocean CO_2 regulating system (NOAA, 2006). Most of the carbon in the oceans is stored in deep waters and sediments of the ocean floor. The oceans contain 50 times more carbon than the atmosphere in the form of dissolved inorganic carbon (Ocean and Climate, 2015). The oceans are indeed the largest inorganic carbon reservoir in exchange with CO_2 in the atmosphere (World Ocean Review, 2010). It is no exaggeration to claim that the oceans are also the principal "blue lung" of the planet that serves to regulate our global climate system (Ocean and Climate, 2015).

6.13 Environmental education for children

Insofar as environmental education for children is concerned, a predominantly scientific approach or the use of rich, complex, and vast pedagogical environmental educational themes should be avoided. The focus is to allow them to understand and think about the natural world, and to relate things with one another and to establish relations between them based on various outdoor environmental activities or place-based outdoor learning. As David Sobel suggested, simple outdoor moments of childhood in keenly remembered wild or semiwild places constitute an important factor in motivating their environmental engagement in future adulthood (Sobel, 2008).

It may well be that simply teaching children prematurely at the age of eight or nine about overwhelming environmental problems or nightmarish global environmental issues beyond their understanding and control will not make them become environmentally responsible adults (Sobel, 1996, 1998). Instead, this may cut off the possible sources of their potential environmental conservation attitudes and strength. It may also result in an "eco-phobia" syndrome among them—a fear of ecological problems and the natural world such as the fear of rainforest destruction, whale hunting, and acid rain or the fear of just being out in the natural world (Sobel, 1996, 1998). As a result, instead of loving the natural world, the children may learn to distrust it and dislike it (Berryman, 1999).

What is important is that "children have an opportunity to bond with the natural world, to learn to love it, before being asked to heal its wounds" (Sobel 1996: 10; see, also, Sobel, 1998). Indeed, children's environmental education should aim at developing a sense of place in children by connecting them with nature with pleasurable experiences and memories, to get them started on a lifetime of loving and wanting to protect and conserve the natural environment (Sobel, 2012; Soule, 1998). This is certainly in consistent with the indigenous environmental perspective on the natural world as discussed in Chapter 5, The Nexus of Environmental Ethics and Environmental Sustainability: An Empirical Assessment. It is suggested along this line of argument that in order to encourage children to grow up to actively care about the environment, it is necessary to provide them with plenty of time to play in the "wild" and to increase the nature observation skills, and develop their sense of nature wonder before they are 11 years old. These wild nature activities include camping, fishing, hunting, or playing in the woods (Wells and Lekies, 2006). Some of these experiences may leave the children with vivid memories and can affect their behavior in establishing a lifetime bond with nature (Soule, 1998).

6.14 Environmental education: Some remarks

The environmental education theme above premised on a vast network of related environmental substance, ideas and concepts help us to instrumentally see things as functions of each other, and to understand logically how nature operates as a critical life-support system of the human race. They also allow us to order and interpret our place in nature and our moral consideration and ethical responsibility toward it. It is also apparently clear from the above educational theme that environmental education or environmental literacy for that matter is concerned with the conceptual connections, knowledge and understanding as well as general awareness of the relationship between human and natural systems and processes. The above educational theme also demonstrates succinctly that environmental education building from the first ring predominance moves in an ascending-progressing order of an unbreakable chain of knowledge in different aspects of environmental, biological, and human interconnectedness.

In contrast, the United Nations' attempt to capture our attention with an overload of information inputs and sheer volume of ideas on sustainable development issues under the single grand vision of education for a sustainable future has only resulted in a mass mental blurriness on environmental literacy. This has impeded our ability to sustain concentration on serious environmental problems that touch the core of human existence, in turn resulting in deficits in core mental skills in our connection with the natural system and our place in it. Consequently, we may become unaware of what is happening around us and are therefore clueless about how to interact with pressing environmental issues confronting us.

Addressing this predicament inevitably calls for the need to bring back the single focus on environmental education that is capable of allowing us to holistically connect and conceptualize the overall environmental story, the interconnectedness between humans and nature and its functional properties, and to be involved in soul-searching discussions on urgent environmental issues facing humanity. The upshot is that creating a scientifically informed society requires a concerted, systematic approach to promote environmental literacy through environmental education as a stand-alone environmental education framework.

6.15 Moral education

Admittedly, environmental education as discussed above alone is less likely to lead to effective and meaningful change of environmentally behaviors, attitudes, and actions without having in place a moral education system to promote environmental philosophies as comprehensively discussed in Chapter 4, Greening for a Sustainable Future: The Ethical Connection. As unveiled in Chapter 4, Greening for a Sustainable Future: The Ethical Connection, recognizing one's relationship with and dependence on nature may or may not trigger an ethical sense of environmental behavior over self-interested attitudes or a moral commitment to environmentalism. Also, as noted in Sections 4.21 and 6.9.2 above, a weak anthropocentric individual may subscribe to proenvironmental value orientation. However, such orientation is unstable in that it may or may not be activated in reality because the perception of proenvironmental behavior originally stems from self-interest with little intuition of social, cultural, or environmental needs and spiritual growth. Hence, when a weak anthropocentric individual is caught in a social dilemma, his proenvironmental behavior may revert to antienvironmental actions when his economic welfare or material progress is threatened.

That being said, in an attempt to genuinely trigger human constraint of self-interested behavior or unsustainable resource use practices, it is necessary to go a step further, transcending education about the environment to education for the environment. This may be achieved by nurturing a sense of collective environmental conscience through moral education. In other words, environmental education is more than enhancing the individual's ability to understand the environment. It also involves developing an individual's moral sense of care or the environment premised on the environmental philosophies as examined in Chapter 4, Greening for a Sustainable Future: The Ethical Connection. This will be systematically examined below.

6.16 Why moral education can lead to proenvironmental behavior and actions?

Although Chapter 4, Greening for a Sustainable Future: The Ethical Connection, has dealt in detail on the relationship between moral or ethical principles and human proenvironmental behavior, it refrained from embarking on a more profound psychoanalysis on the facets of human environmental behavior development in order to avoid intellectual confusion. In essence, if our analysis excludes this intellectual tenet, we will not be able to see on what conceptual basis the Leopold-Schweitzer environmental philosophy or other ethical principles as discussed in Chapter 4, Greening for a Sustainable Future: The Ethical Connection, is practically tenable and intelligibly feasible in a real world system.

More specifically, without putting in place a psychoanalytic framework of analysis, we will not be able to conceive the factors, motivations, or disciplines that impact on the inner workings of the human mind in transcending its anthropocentric view of nature to eco-centric percepts of our global environment. We will also remain unclear on the principal tenets, conceptual underpinnings, and subtle psychological processes on which the ultimate cause of changing human ethical behaviors are premised.

With this view in mind, this section aims to unveil the concrete conceptual foundations on which these moral principles are premised in order to render them pragmatically clear and intelligible. This analytical approach allows us to discover that the moral frame of analysis as expounded in Chapter 4, Greening for a Sustainable Future: The Ethical Connection, constitutes a universally valid and tenable philosophical discipline *par excellence* which fulfils the same functions everywhere in promoting proenvironmental behavior, if effectively disseminated through moral education.

Within the present context, psychoanalysis refers to a conceptual analysis of a matrix of relationships between human mind, nature, and society (Dodds, 2011). More particularly, it is primarily concerned with the "dynamics of mental processes and individual experience that are influenced by biological, social, and environmental contributions" (Marans et al., 2002: 382). Simply put, it deals with the complex relationship between the body and the mind and the role of emotions in environmental behavior. This will be discussed in the following subsections.

6.16.1 Social facts: The moving force behind Leopold-Schweitzer environmental philosophy

As discussed in Chapter 4, Greening for a Sustainable Future: The Ethical Connection, morality deals with the ethical principle of doing the right thing as expounded by Aldo Leopold and reinforced by Albert Schweitzer. However, conceptually, what are the motivational forces that guide and induce individuals to do so? How do these motivational forces operate in a real world system? To answer this question, we may begin our investigation based on Durkheim's concept of "social facts", a concept defined as the "manners of acting or thinking, distinguishable through special characteristics of being capable of exercising a coercive influence on the consciousness of individuals" (Durkheim, 1982: 43). These special characteristics, defined externally to the individuals and imbued with "a compelling and coercive power", may comprise morality, customs, religious beliefs and practices, among others (Durkheim, 1982: 51).

Individuals experience them as moral obligations and duties or restraints on their behavior, which operate outside their will or consciousness (Durkheim, 1982: 47−51). Moreover, these intrinsic characteristics assert themselves when individuals attempt to resist. As Durkheim contended "If I attempt to violate the rules of law, they react against me so as to forestall my action to conform to the norm..." (Durkheim, 1982: 51). Durkheim further pointed out that, "We accept and adopt them because, since...they are invested with a special authority that our education has taught us to recognize and respect. For example, as Durkheim demonstrates, "When I perform my duties...and carry out the commitments I have entered into, I fulfil obligations which are defined in law and custom and which are external to myself and my actions" (Durkheim, 1982: 50).

Taking an analogy from Durkheim's philosophical finesse, the motivational forces that guide and induce individuals in their dealings with the external world are built on the power of external coercion. This in turn translates into individuals' self-imposed constraints which may be interpreted as "a limitation on freedom of action in the struggle for existence", according to Leopold, when interacting with the natural systems in order to preserve "the integrity, stability, and beauty of the biotic community."

It may further be remarked that the external power of coercion or Leopold's environmental limit for that matter is deducible from Durkheim's concept of *"représentations collectives"* or collective representations. Collective representations may be defined as a body of collective beliefs, sentiments, values, and experience which reflect things in reality, and which have the impact to "excite, maintain, or recreate certain mental states" among members of society (Durkheim, 1915: 10). In other words, collective representations are instrumental in helping humans to order and make sense of their place in the world surrounding them and to decide how the external world should be represented.

It thus follows that the perception of our universe, or our Earth system for that matter, rests on the way we represent it based on a shared set of environmental attitudes, moral norms, and ethical beliefs which have evolved through human interactions with the natural system. It is precisely this shared system of beliefs that creates a collective sense of environmental and ethical consciousness called here the collective conscience (see the following section) that morally binds humanity in its interaction with nature. This set of shared ethical and environmental beliefs (collective conscience) which may be promoted through moral education informs us about our position in the Earth system and exerts coercive moral constraints upon us in our interaction with natural systems.

How is it possible to cultivate this collective conscience through moral education among members of a given society that come from different backgrounds and with different environmental ideologies, moral attitudes and ethical behaviors especially those anthropocentric individuals? More specifically, why and how can moral education serve its function in changing anthropocentric inclined beings into ecocentric-skewed individuals in a complex and heterogeneous society? This will be the subject of interest in the following subsection.

6.16.2 Collective conscience, duality of humans and moral education: The nexus

In answering the above questions, we may again turn to Durkheim for intellectual clues. According to Durkheim, "man is double" (Durkheim, 1915: 16). Durkheim uses the term *homo duplex* to depict this human dualism. Accordingly, humans are rooted in two opposing, yet interacting, aspects of their being: insatiable individualistic desires and appetites which are, in turn, constrained by socially generated moral norms and ethical beliefs—a dilemma of individualism and holism (Fish, 2013; Šubrt, 2017). Manifestly, the concept of *homo duplex* as promulgated by Durkheim corresponds to the social dilemma discipline as briefly discussed in Section 6.9.2

To wit, according to Durkheim, there are, in each of us, two consciences: "one which is common to our group in its entirety...the other, in contrast, represents that in us which is personal and distinct, that which makes us an individual" (Durkheim, 1893: 129). The former, known as collective conscience which is common to our group, and which represents "the highest reality in the intellectual and moral order" (Durkheim, 1915: 16, see also, Durkheim, 1893: 105, 129). More specifically, collective conscience is "the totality of beliefs and sentiments common to average citizens of the same society...It is...independent of the particular conditions in which individuals are placed. It is the same in the North and in the South, in great cities and in small ones, in different professions. Moreover, it does not change with each generation, but on the contrary, it connects successive generations with one another..." (Durkheim, 1893: 79−80).

In other words, collective conscience exhibits the will of society as a whole and constitutes the basis of social systems of moral representation and action about various aspects of the external world. Furthermore, collective conscience is external to, and cohesive of, individuals in a given society (Gurbuz, 2008). This is what makes us social beings, or moral and ethical social beings, for that matter. Viewed from the environmental sustainability perspective, collective conscience has an important role to play in creating a shared way of understanding and interpreting the order of the Earth system on which humanity exists which may be premised on Leopold-Schweitzer's moral principles and which are in line with Durkheim's interpretation of morality. Morality in Durkheim refers to the impersonal orientation of self-serving interests toward collective interest. Put differently, the rule of morality "is always behaviour in pursuit of impersonal ends" and to regularize and constrain our desire within bounds (Durkheim, 1961: 58).

The other conscience is that of an individual being, or conscience that contains states which are personal and distinct, that which makes us an individual. As distinct from collective conscience, individual conscience represents the will of an individual rather than that of society as a whole. More particularly, it is is egoistically aligned to individual self-interest and insatiable desires (Durkheim, 1893; 1973a,b,c). As Durkheim pointed out clearly, human habit is such that "the more one has, the more one wants, since satisfactions received only stimulate instead of filling needs" (Durkheim, 1951: 248). Under this circumstance, individual conscience overwhelms collective conscience, and society or social norms based on shared sentiments and responsibilities will be weakened, and individuals will not be subjected to some kind of moral restraint.

Thus the functional role of collective conscience in regulating human moral behavior declines, that is, it can no longer regulate human affairs effectively. This eventually leads to a breakdown in social norms regulating behaviour and individuals will become more impersonal in social life (Durkheim, 1951). Hence, as Durkheim pointed out tellingly, "To the extent the individual is left to his own devices and freed from all social constraint, he is unfettered too by all moral constraint" (Durkheim, 1958: 7). Viewed from the present analytical perspective, this will result in irreparable damage to the environment out of anthropocentric quest for economic progress or material wellbeing. However, Durkheim further contends that when an individual moves away from self-interested position, he will transcend himself in both his thought and act morally and intellectually (Durkheim, 1915: 16). Thus he calls for the need to restrain our passions, desires, and conduct based on constraint, which may be manifested in the forms of law, morality, beliefs, or customs (see, for example, Durkheim, 1973a,b).

This is precisely the role of moral education which has the "immediate aim of guiding conduct", in Durkheim's words. As Durkheim points out, while education theories, or environmental educational frameworks, for that matter, "do not constitute action in themselves, they are the preparation for it...Their *raison d'être* is in action" (Durkheim, 1961: 2). This reflects the significance of moral education as a collective body of ethical representation in guiding and influencing human beings to order and interpret the natural environment surrounding them. This collective body of ethical representation necessarily involves the philosophical and ethical concepts as expounded by Albert Schweitzer, Aldo Leopold and others as discussed in Chapter 4, Greening for a Sustainable Future: The Ethical Connection. More particularly, these philosophical or ethical concepts serve an important role in influencing individuals anthropocentrically guided by self-interest to become public-selves guided by morality (ecocentrism).

Durkheim's philosophical thought on human behavior sheds immense light on the *raison d'être* of environmental ethics and moral education in promoting environmentally sound behaviors, attitudes, and actions as examined in Chapter 4, Greening for a Sustainable Future: The Ethical Connection. It also necessarily links to the implications for an ethical value system in constraining human unsustainable practices. This is further discussed in Chapter 7, Summary and Conclusion. It may further be remarked that the United Nations' aim at promoting environmental awareness through environmental education alone is far from adequate to direct this functionalist perspective of environmental education. That is to say, environmentally sustainable behavior cannot be logically manifested without the second ring of moral education which provides clear ethical guidelines to stimulate and bind people collectively toward ecocentric behavioral patterns and practices. This line of approach to environmental education—from the first principal ring of environmental attributes building upward to the second ring of ethical substance provides a remarkable *tour de force* to excite, maintain or recreate certain environmental states that collectively bring people together toward a progressive march to environmental sustainability.

6.17 Environmental education framework: A suggested model

The logical and necessary way to proceed in this volume is to optimally exploit its epistemological edifice and analytical insights to propose a pedagogical framework to promote environmental sustainability. Environmental education will develop the faculty of ecological literacy and instill the instinctive protective love for environmental sustainability. EE provides a practical means of inculcating values that shape our conduct when interacting with nature. It lays the moral foundation for practical prescriptions to address the idiosyncratic impulses of unsustainable human behavior and action. Precisely, what kinds of EE systems do we need?

First, an effective EE framework which is capable of exerting a constraining influence upon individuals must necessarily be interdisciplinary and multidisciplinary in perspective (de Ciurana and Leal Filho, 2006). Sustainable development, as reflected by the analyses expounded so far, is intrinsically a complex discipline. It covers wide ranging issues: environmental considerations on the one hand, and economic matters, development issues, social, cultural and moral events and institutional and political spheres on the other (Leal Filho, 2011). Implicitly, no single discipline or even a cluster of a few related disciplines is capable of allowing us to come to grips with the complex process of causation that has given rise to the environmental impairment humanity is enmeshed in today.

It thus follows that an intelligible EE framework must contain adequate perspectives and concepts in order to allow us to see clearly the causal chain of events leading to environmental disruptions. Granted, the essential pedagogical components that are capable of meeting these conditions must essentially but not exclusively cover some of the following (Fig. 6.8)

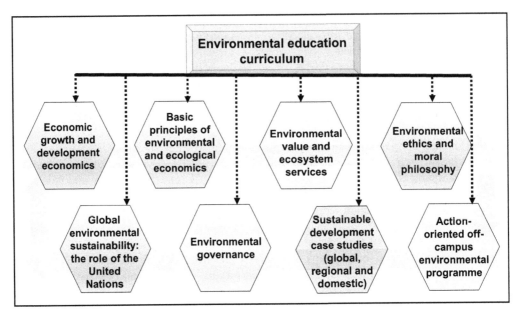

FIGURE: 6.8   A suggested environmental education framework.

1. Economic growth and development economics: the focus is on the discussions of the human economy (production and consumption processes) which are closely associated with environmental exploitation, resource depletion, and environmental decline.
2. Environmental and ecological economics provides the essential concepts and principles to allow students to understand environmental events, behaviors and/or situations in a systematic way. However, a high degree of theoretical abstraction and modelling which may be beyond the intellectual reach of the uninitiated must be avoided. For one thing, apart from dampening interest, complex mathematical modelling represented by continuous differentiable functions lacks the explanatory power of world reality and of systems relations that govern the environmental dimension of human affairs. Thus the educational game can be lost even before the first move is made as it defeats the purpose of enhancing environmental literacy.
3. As reflected in Sections 6.10.2 and 6.11 above, the subject on ecosystem services helps to incite emotional mental images of human−nature relations. It also provides a clear understanding of human dependency on nature for continued existence. This raises the logical need for environmental protection which is related to the study of environmental value and ethics and values as mentioned below (Simon, 2002; Santos and Leal Filho, 2005).
4. As reflected in Chapter 4, Greening for a Sustainable Future: The Ethical Connection, environmental ethics, if properly and intelligibly harnessed, is an effective ethical means to regulate and constrain a broad sweep of human environmental affairs that ultimately lead to the fruition of environmental sustainability. In addition, our interaction with the natural environment is expressed in terms of values which play a

crucial role in determining the ways we ought to behave and act when dealing with nature. Hence, the necessity for this module. However, rather than simply constructing purely intellectual moral principles and abstract ethical doctrines, the subject would be rich with practical relevance based on field studies conducted with the forest-dwelling indigenous people as described in Chapter 5, The Nexus of Environmental Ethics and Environmental Sustainability: An Empirical Assessment. This is also relevant in the action-oriented programme as explained below. This approach brings theory to life and allows students to grasp intelligibly the degree of validity between moral theory and ethical practice in a real-world system (Chapter 5: The Nexus of Environmental Ethics and Environmental Sustainability: An Empirical Assessment; see also, e.g., Simon, 2002).

5. As reflected in Chapter 3, The United Nations' Journey to Global Environmental Sustainability Since Stockholm: The Paradox, the world is awash with a profusion of well-formulated development plans and environmental policies crafted in accord with the UN environmental initiatives. Yet, they do not guarantee, nor in any way constitute, actual implementation. In other words, as discussed in Section 4.7, there always exists the capacity of policy formulation and the political will of implementation. Granted, the subject on environmental governance must also be considered in the EE framework as this allows students to evaluate the ways in which a more sustainable mode of environmental governance may be structured. It may be taught concomitantly with institutional economics.

6. The UN's role in global environmental sustainability as discussed in Chapter 2, The United Nations' Journey to Global Environmental Sustainability Since Stockholm: An Assessment, is also of great relevance in that it provides greater understanding of the policy formulation—implementation gap as mentioned above in reality when students are exposed to a case-study module as described below. This helps to deepen a student's critical inquiry into examining the conditions and modalities that are capable of mitigating the major failings of policy implementation.

7. Sustainable development case studies: Global environmental sustainability cannot be intelligibly grasped without empirical study. This subject is also associated with (5) and (6) above. This module also helps to raise awareness on a range of environmental problems and the magnitude of ecological threat facing humanity today. This has the impact of motivating students to examine the underlying causes of environmental impairments and to evaluate the necessary mitigating measures.

8. An action-oriented programme serves to enhance students' motivation, awareness, capacity building, and empowerment in solving real world problems. It connects environmental information acquired in class to real activities and provides a practical platform for students to apply theories into practice. It also provides the venue for students to acquire and accumulate diverse fields of practical knowledge from various experts and complement their studies in the classroom (Simon, 2002; de Ciurana and Leal Filho, 2006; Leal Filho et al., 2015).

It is also instructive to note that to provide a motivating environment to inspire the learning process, classes should be conducted based on a mixture of pedagogical approaches and interactive and collaborative learning processes. While the former involves imparting knowledge of discipline specific content, theory and concepts to students, the latter engenders an active learning process. One of the recommended methods for the latter is the Case Method Programme pioneered by the Harvard Business School which, based on the author's teaching experiences at Kyoto University and Keio University, is very effective in stimulating students' quest for knowledge acquisition and active class participation. It also helps students to develop critical thinking, analytical reasoning, and communication as well as in articulation of thoughts (Simon, 2002; Leal Filho et al., 2015). In addition, outdoor education programmes as discussed in Section 6.11 may also be planned to complement the above teaching efforts.

It is now beyond any doubt that environmental and moral education provides a seedbed and external stimuli for germinating our moral will and ethical imperative toward the goal of environmental sustainability. The EE framework as proposed in this section is by no means exhaustive. Nonetheless, it serves to unify the diverse relevant and essential aspects of environmentalism into a dynamic whole as demonstrated in (Fig. 6.9 and Box 6.1)

In other words, "explaining part through part, and the whole through the parts", to borrow Durkheim's phrase (Durkehim, 1953: xvii), leads us to waves of thought and new insights that allow us to connect the dots of the inextricably complex human–nature relationship and the interwoven causal connections between human perceptions of nature, the ethics of sustainability, and environmental sustainability as extensively discussed in this volume. All nations across the globe, especially in the developing Asian countries should devote every concerted effort to mainstream environmental and moral education in their education systems in order to lead humanity toward a Future We Want, for now, and for the indefinite future.

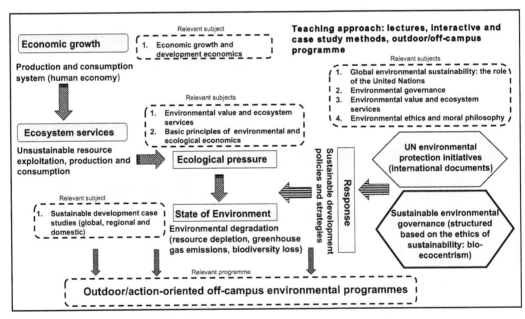

FIGURE 6.9 Environmental Education in practice—a diagrammatic representation.
Note: (a) Environmental governance may be studied in the context of institutional economics; (b) The United Nations sustainable development framework may cover the major environmental efforts undertaken by the United Nations in promoting global environmental sustainability.

BOX 6.1

EE in practice—some notes.

The pursuit of economic growth necessarily involves the extraction of natural resources as production input (Ecosystem Services). However, unsustainable resource exploitation, consumption, and production processes tend to lead to a range of ecological disturbances such as increasing greenhouse gases emissions and atmospheric concentrations (increasing "Ecological Pressure"). This serves to alter the quantity as well as quality of the natural environment/resources in the term of, for example, overly saturated carbon and waste sinks (worsening "State of Environment"). The "State" refers to the quantity and quality of the natural capital which is affected by ecological disturbances or pressures arising from a range of human economic activities. This calls for mitigation measures ("Response") to address the problems arise. These causal chains of events enrich our minds (state and nonstate actors alike) on the environmentally destructive impacts of human unrestrained economic activities. It raises environmental awareness and ecological literacy in association human–nature relationship and our dependence on nature for continued existence. This may incite our emotional sentiments of fear and of the desire of environmental mitigation (environmental ethic, environmental governance, UN international documents). Outdoor and off-campus study programmes may be organized to expose students to the environmental reality facing humanity today. Overall, the proposed environmental education structure serves as a basis for unifying the diverse aspects of causals events into a dynamic whole and provides the students with clear perspectives of the complex causal mechanisms governing the study of sustainable development.

6.18 Concluding remarks

The current looming environmental crisis reflects in part a failure in environmental education. Admittedly, environmental education is not a panacea for every environmental problem, but it is nonetheless an indispensable means to help humans to order and interpret our natural world, and to ensure that individuals have a holistic understanding of our global ecosystems and their life-supporting functional properties which underpin human long-term existence. This would in turn help to create a sense of urgency for the need to show greater respect and moral consideration for the natural environment.

It is important to understand that there is essentially a great difference between environmental education for environmental literacy and ESD as embraced by the United Nations/UNESCO. In the former, the environment is at the center of concern in its educational realm. It functionally aids us in understanding our natural systems as well as our relationship and interactions with our natural environment. However, in the latter, its educational framework touches on any issue that contributes to a sustainable future. It is concerned about a multitude of objectives and an all-encompassing agenda with greater emphasis being placed on development and poverty alleviation, inclusive growth or other

SDGs. Despite the fact that environmental protection represents a fragment of its essential traits, it is overall a human-centric form of education.

Although ESD helps to broaden our knowledge as well as enlighten us on a wide range of sustainable development issues, it cannot serve as the first ring of principle for unveiling the long chain of mutually connected reasoning for the protection of our environment. Indeed, adding too many elements into the ESD framework with different focuses and under different names weakens the focus or reformative strength, the level of consolidation and precision and also detracts from the perceptible effects of the original message of environmental protection.

Furthermore, burdening the environmental education framework with voluminous perspectives and objectives tends to lead to difficulty in comparative pedagogic deduction and induction of ideas and concepts adapted to environmental literacy. This will negate the functionalist role of environmental education as the prime mover for the evolution of a lasting trend of environmental thought and ethical behavior. It may well be that environmental education is what determines and guides our environmentally sustainable course of behavior and actions. Consequently, when alienating environmental education to a substratum of the larger ESD educational framework, policy makers, in fulfilling their global commitment, are inclined to make a cursory enquiry into the most general views of environmental education. Consequently, they will not be drawn into a collective drive in crafting policies that prioritize the overarching objective of environmental education in promoting environmental literacy and environmental awareness—the prerequisite for a sustainable future.

Clearly, if environmental education is to serve as a means to awaken environmental consciousness, it has to be explained and consolidated in its own terms—the first ring principle of predominance in the chain of environmental literacy. This very first ring is the fundamental foundation of and an ascending-progressing building block for an unbreakable chain of environmental literacy. It thus follows that environmental education must stand on its own as an authentic and rigorous educational framework to serve its function "to excite, maintain or recreate certain mental states" in society toward a sustainable future.

Thus environmental education should be distinguished from the rest of the educational disciplines through its special role of being capable of not only exercising a coercive influence on our consciousness but also imposing moral beliefs and practices upon us with regard to our ethical relationship with the natural system. In addition, environmental education goes hand in glove with moral education, which may be conceived as the second ring of the larger environmental education framework in helping to guide and regulate human engagement with nature based on the moral philosophies as discussed in Chapter 4, Greening for a Sustainable Future: The Ethical Connection. The idea is to create a collective sense of environmental consciousness and ethical responsibility among individuals in their engagement with nature.

Furthermore, given the present dire global environmental conditions, there is little time left for the United Nations/UNESCO to continuously serve as a "ritual and symbolic club" for leaders and people around the world to talk at each global educational conference with hardly any real and concrete actions. It cannot be denied that the long list of high-profile conferences and mega meetings on environmental education for the past few decades since the Stockholm Conference (1972) may reflect the United Nations/ UNESCO's determination in promoting environmental literacy across the globe. The disturbing truth, however, is that they have been unable to jolt the global leaders, especially those in the developing countries, into responding with lasting conviction to reorient their educational programmes/curricula to foster a better understanding of our natural environment and our ethical responsibility toward it. Also, for all the talk at each mega conference, the United Nations/UNESCO has not been able to effectively galvanize the developing countries to develop strategies to integrate environmental ideas, beliefs, and practices into education and training programmes to promote a sustainable consumption lifestyle among the citizens or for better transition to green economies and societies.

Indeed, even moral education, let alone environmental education, has not been given high consideration in the mainstream education system in the developing regions. Environmental or moral education has also not been proven popular in those regions with a crowded and rigorous school or university curriculum focusing on the traditional academic disciplines such as science, mathematics and economic studies, or the acquisition of technical and vocational skills. Furthermore, whatever environmental education embraced in school is predominantly concerned about the descriptive information of the environment rather than environmental preservation or harnessing an understanding of natural processes and moral care or ethical consideration for nature. Environmental education devoid of critical perspectives has also been rhetorically introduced in the developing countries to serve as an instrumental means to lend support for various government development projects rather than for the promotion of an environmentally literate society. Globally, the spread of environmental education is also hampered by people spending endless hours engrossed in a virtual world of tweets through their electronic gadgets.

There is no more time for global rhetorical conversation. We need real and concrete action and less talk, and we need it now. As it has turned out, the United Nations/UNESCO's piecemeal efforts in promoting issue-specific environmental education in various developing countries and on a short-term basis is unlikely to make concrete contributions to permanently excite, maintain, or recreate certain environmentally benign mental states among global citizens. It is time for the United Nations/UNESCO to concentrate more of its energy and effort on launching solid educational partnerships with representatives of higher institutions of learning or other relevant environmental authorities across the globe to motivate, support, and guide the design and integration of environmental components into national education strategies and action plans for sustainable societies. The United Nations/UNESCO also needs to exert real and concrete universal effort on the ground to launch diplomatic missions to governments of the developing countries to identify viable approaches to effectively implement environmental education programmes and curriculum, community environmental oriented projects, extracurricular environmental activities, and environmental education of children for the promotion of environmental literacy in the region.

To heal the planet, it is necessary to heal ourselves at the very outset through environmental education for environmental literacy and moral education for environmental ethics—the *condition sine qua non* for global environmental sustainability.

7

Summary and conclusion

7.1 The United Nations striving for global environmental sustainability since Stockholm

As discussed in Chapter 4, Greening for a Sustainable Future: The Ethical Connection, the ways human beings view nature determines the way we manage our relationship with the Earth system. Treating nature instrumentally, for instance, tends to induce an anthropocentric view and exploitative use of our natural world. This human-centered perception of nature was the very authoritative attitude of society during the Industrial Revolution. After more than 150 years since the Industrial Revolution began in Great Britain in the late 1700s and early 1800s, humans have slowly and gradually ascended to the biospheric supremacy as the dominant geological force that is capable of inflicting massive or irreversible alteration of our planetary system. Man's dominance over nature has exhibited itself as a threat to the ecological resilience of our planetary system in incidents such as the London Killer Fog, the Liberian Torrey Canyon supertanker oil spill, the California Santa Barbara Channel's oil spill, and the Japanese "Big-4" industrial pollution-related diseases (Minamata Disease, Niigata Disease, Itai-Itai Disease, and Yokkaichi Asthma), among others.

At the forefront of the challenges, confronting the West is a controlling framework to arrest the adverse impacts of industrialization. In response to this challenge, the United Nations since Stockholm has convened a host of international environmental meetings, summits, and conferences to address a myriad of environmental problems which are of critical global concerns. Through these meetings, hundreds of environmental treaties, agreements, declarations, action plans, environmental documents, and regulations were signed, adopted, or endorsed by the member states of the United Nations.

Some of the important documents which sought to guide the United Nations' journey to global environmental sustainability include the Stockholm Declaration, the Convention on Biological Diversity (CBD), the United Nations Framework Convention on Climate Change (UNFCCC), and the World Conservation Strategy, among others. Particularly worth mentioning is the World Charter for Nature which places great emphasis on the importance of environmental ethics in guiding our relationship with nature. At the same time, in acknowledging the importance of environmental literacy in promoting environmental awareness for a sustainable future, the United Nations or

UNESCO for that matter, also wisely made concerted drives to promote environmental education through the adoption of the Belgrade Charter (1975), Tbilisi Declaration (1977), and the Tbilisi + 10 Declaration (1987).

On the national fronts, countries across the world have devoted much intellectual effort and innovation in crafting out in breadth and depth the global sustainability agendas and environmental protection laws or regulations based on international environmental declarations and agreements adopted. They have also set up environmental ministries and agencies to reinforce and streamline environmental policy implementation processes.

This series of events reflects the determination of the United Nations and the global community at large to protect our Earth system from large scale, potentially catastrophic, or irreversible environmental degradation.

7.2 From the United Nations' journey to global environmental sustainability to the emergence of an anthropocentric world of environmental destruction

The analysis in Chapter 3, The United Nations' Journey to Global Environmental Sustainability Since Stockholm: The Paradox, and Chapter 4, Greening for a Sustainable Future: The Ethical Connection, however, provides incontrovertible evidence that the United Nations' journey to a sustainable future has ended as a dismal failure. While sustainable development is originally understood as a balancing act between economic growth, environmental sustainability, and social equity, in the contemporary human-centered world, it is anthropocentrically embraced within the economic framework of progress. Every nation across the globe was set in motion to compete with each other in the egoistic race of sustainable development expressed in term of sustainable economic growth. To put it bluntly, it has turned into a Donald Trump *anthropocentrism proclivity sui generis* phenomenon in our dealings with nature as discussed in Chapter 3, The United Nations' Journey to Global Environmental Sustainability Since Stockholm: The Paradox.

The situation we have now is that sustainable development has taken a new twist of sustainability orientation determined by utilitarian concern for economic growth, and the moral concern for nature has been overwhelmingly removed from its ethical imperative of environmentalism as emphasized under the World Charter for Nature. Sustainable development, as it turned out, has perceptually and paradigmatically shifted from sustaining the ecological integrity of the environment to sustaining the economic health of material progress. This raises the following pertinent question: What is sustainable development, really? Casual observation suggests the perception that sustainable development is economic growth or socioeconomic progress *ad infinitum*. Thus an environmentally sustainable world has failed to take shape in the global society as it has inexorably been replaced by "sustainable economic growth system."

As empirically demonstrated in Chapter 3, The United Nations' Journey to Global Environmental Sustainability Since Stockholm: The Paradox, and Chapter 4, Greening for a Sustainable Future: The Ethical Connection, for more than 45 years, sustainable development has resulted in creating a world of profligate resource exploitation and consumption, widespread environmental degradation, extensive deforestation, and biological impoverishment across the globe especially in the Asian developing region including, for example,

Africa, the Southeast Asian nations, South America and in the world's most powerful nation—the United States. Human economic activities have been so pervasive that they are endangering our Earth system's capacity to sustain long-term human survival of the present and future generations. One of the most salient and perilous transgressions of our planetary system which is worth reiterating is the Canadian "war against nature and humanity" and libricide launched by the former conservative government led by Stephen Harper in the name of "sustainable development."

In addition, despite putting in place the legally binding CBD, no country in the world has fulfilled the targets to significantly reduce the rate of biodiversity loss by 2010. Indeed, Chapter 3, The United Nations' Journey to Global Environmental Sustainability Since Stockholm: The Paradox, provides compelling evidence that international efforts to meet targets to stem the loss of wildlife and habitats ended as a broken promise. Worse yet, the overall global or regional environment have increasingly been put under serious pressure by the rapid growth of economic activities and intensified exploitation of natural resources. The UN Global Assessment Report on Biodiversity and Ecosystem Services (Summary for Policy Makers) released in May 2019 further confirmed the persistent trends of this crisis-ridden environmental episode across the globe which are unprecedented in human history—with around 1 million species of plant and animal facing the risk of extinction, many within decades.

What is more, all past climate talks have failed to halt the ever-expanding CO_2 emission and concentration trends. For example, the 2020 Climate Change Performance Index (CCPI) presented at the 25th session of the Conference of the Parties (COP 25) held in Madrid in December 2019 reveals that no country is yet on a path compatible with the Paris climate targets. Worse yet, COP 25 was ended in "utter failure" as deal is stripped of ambition, leaving all the major issues including a key decision on global carbon markets for the next summit (COP 26) scheduled to be convened in Glasgow, Scotland, in 2021.

The extensive evidence of wide spread and unprecedented human-induced environmental change reflects that the monumental promise of a sustainable world since Stockholm has predominantly turned into a false hope of environmentally bleak future. Implicitly, for more than 45 years, the United Nations' journey to global environmental sustainability since Stockholm has only resulted in the ascendance of humans to biospheric supremacy as the dominant force capable of instituting large scale and irreversible anthropogenic transformation of our Earth system. This has placed our natural world at a precarious and potentially catastrophic tipping point which warrants our immediate mitigating actions.

7.3 The United Nations ethical dilemmas

The earlier-mentioned human-induced environmental degradation illustrates the structural inability of the United Nations' global environmental controlling mechanisms comprising megaconferences, environmental regulations, action plans, declarations, protocols, and treaties to safeguard our Earth system. These documents signed or adopted at each conference are the products of compromises after rancorous debates, and long and tedious processes of multilateral consultations, assessments, negotiations between governments and intergovernmental organizations behind the scenes. Thus the stakeholders were assumed to be well aware of the worrisome global environmental conditions and their environmental obligations. The past UN conferences have also resulted in the rapid growth of environmental awareness of the devastating consequences of man's unsustainable environmental practices.

Furthermore, more scientific information on global pressing issues such as climate change and biodiversity loss has become readily available through a continuous series of environmental publications such as the Intergovernmental Panel on Climate Change Assessment reports and the Millennium Ecosystem Assessment report. which deals with the critical relationships between ecosystem services and human social economic well-being.

Viewed from this perspective, the United Nations has contributed immensely to cementing international accords on environmental protection. The United Nations has also, through its relentless efforts, contributed significantly to guide global direction toward environmental sustainability through new environmental information, knowledge transfer, new norms and modalities, action plans, and guidelines. Through the dissemination of new information and knowledge, it has helped the global community to promote a common understanding and concern for the state of our environment and generate political will for environmental protection. By the same token, the revelation of new modalities, and the like has helped the global community to redefine problems in line with the existing changing environmental status quo and to shape global environmental direction in more definite forms.

Thus as briefly revealed earlier, showing high aspirations, enlightened awareness, strong commitments and high hopes, national environmental controlling policies or measures and comprehensive institutional and legal mechanisms to address global-scale environmental degradation have flourished across the globe since Stockholm. A myriad of international cooperation and action plans have also been initiated to address various environmental problems especially those with transboundary or global implications since Stockholm. There has also been an upsurge in global concern over the treacherous state of our environment such as biodiversity extinction and global warming which demand our urgent attention.

Despite these, nonetheless, global leaders, once back home, generally display lukewarm responses and actions for environmental conservation or protection. It seems that the legal regimes established have mostly been reactive to emerging environmental problems instead of displaying serious concern to tackle them. Governments, especially in the developing countries, for the most part are reluctant to allocate sufficient funds for environmental causes. Enforcement of environmental laws or policies at the national level is lax especially in the South. As it turned out, the South continues to focus on GDP growth centered on capital accumulation and material progress. To the South, the socioeconomic pillars supporting the framework of development matter most in filling the development gap between the North and the South. The world, utterly divided on how to fight climate change, is further polarized by Donald Trump's anticlimate and antienvironmental sentiments.

To compound the problems, most of the environmental documents or regulatory frameworks unveiled by the United Nations, many of which are repetitive and contradictory, have no coercive value. That is to say, they are not legally binding on the member states. Thus the United Nations does not hold the institutional capacity or mandate to compel member states to cooperate internationally on collective environmental concerns such as climate change or biodiversity loss. Any attempt to push through this interstate cooperative framework will be viewed as amounting to the infringement of state sovereignty by the stakeholders. Without meaningful sanctions for noncompliance, it is hard to see how the political commitments for environmental protection or conservation can be translated into real actions. It is also noteworthy that the World Charter for Nature advanced nothing more serious and practical than proposing and elevating the concept of sustainable development and the importance of the ethical formulation of environmentalism.

Conceivably, it appears that the United Nations' series of conferences fundamentally serve as a venue for the formal adoption or endorsement of eloquent rhetorical UN documents. More particularly, they seem more like a "mega talk shop" than the delivery of genuine political commitment for real action by the stakeholders. A specter is thus haunting the United Nations: the specter of ineffective implementation and lack of positive behavioral changes amidst progressive development of multilateral cooperation and institution building across the globe since Stockholm. Indeed, as revealed succinctly in Chapter 3, The United Nations' Journey to Global Environmental Sustainability Since Stockholm: The Paradox, environmental law and regulations, especially those enacted, revised, or refined in response to the United Nations' environmental sustainability initiatives were hardly enforced effectively on the ground. The Baiji dilemma in China, the South American Big-4 deforestation scenarios, the complete and irreversible destruction of the legally gazetted Bikam Permanent Forest Reserve in Perak and the wholesale destruction of rainforests in Kelantan in Malaysia, and the environmentally destructive IndoMet Coal Project in Indonesia, and not the least, the Canadian "anthropocentric conquest of nature" à la Francis Bacon's philosophy of "Dominion of Man over the Universe" mentioned earlier, and Donald Trump's anticlimate and antienvironmental stance, are some of the telling indictments of the inability of the United Nations to achieve global bargains in arresting the fast deteriorating environmental degradation across the globe.

In a nutshell, the United Nations has been less than successful in mobilizing political actions in curtailing global environmental degradations. In the final analysis, the sovereign state still reigns supreme, and enforcement of sustainable development policies attuned with the United Nations' aspirations rests entirely on the political will of the member states. Thus political commitments, as reflected in extensive and varied case studies revealed in Chapter 3, The United Nations' Journey to Global Environmental Sustainability Since Stockholm: The Paradox, do not necessarily translate into real actions in the practical world.

Obviously, global environmental problems involve economic, political, and social issues. However, what is less obvious is that they also entail ethical questions for the human race. Since Stockholm, it seems that the glaringly obvious and important ethical aspects of environmental problems have been sidelined from discussion. To be sure, adoption or endorsement of international environmental treaties and global environmental pledges or commitment per se alone will not endure without strong and genuine commitment to protect our Earth system.

In more specific terms, getting world leaders or the global community to transform their commitment into real action cannot happen in a vacuum. Rather, it must necessarily be motivated from something which is distinctly capable of exerting a certain form of coercive influence on human mental representation of nature based on shared moral sentiments and ethical responsibilities. Implicitly, without emphasizing human moral causes for environmentally destructive practices and ethical imperatives for environmental protection, the international documents signed or endorsed at each UN conference or national environmental law enacted in each member state will not produce real or lasting effect.

In rounding up, environmental problems cannot be effectively circumvented without first breaking the ethical impasse of environmentalism. The ethical reasoning for environmentalism is particularly important in serving as a more effective motivation for environmental obligatory actions given that there has been a lack of authoritative UN controlling and monitoring apparatus or mechanisms to legitimately coerce or scrutinize actual political commitments to environmental protection or conservation. Hence, the need for an ethical framework and moral justification to command, regulate, and constrain human environmental attitudes, behaviors, and actions within the planetary boundary limit of growth or material progress, is a Kantian categorical imperative. What is fundamentally important now is not more celebrated agenda settings but a critical reflection on how global summits can be used to revive the moral discipline of environmentalism to enhance global sustainability governance and environmentally friendly international community.

7.4 Environmental ethics

To break the ethical impasse, as discussed in Chapter 4, Greening for a Sustainable Future: The Ethical Connection, an important step is to embrace environmental ethics as a unifying means to excite, maintain, or recreate our mental representation of nature and the moral beliefs about the rightness and wrongness of our actions as eloquently argued in Aldo Leopold's land ethic, Albert Schweitzer's reverence for life philosophy, and Immanuel Kant's deontological principle of morality. In addition, the practical importance of environmental ethics in promoting environmental sustainability is well reflected in Émile Durkheim's philosophical thoughts on the functionalist perspective in sociology as discussed in Chapter 6, The United Nations Environmental Education Initiatives: The Green Education Failure and the Way Forward. Viewed from Durkheim's perspective, morality, or environmental ethics for that matter, is invested with a special authority and unifying function that cements society together within the moral purview of environmentalism. Individuals experience these moral value orientations as "obligations, duties, and restraints on their behavior" in their interactions with nature. These tendencies have been empirically verified based on evidence gathered from extensive field research with the forest-dwelling indigenous people in Malaysia as examined in Chapter 5, The Nexus of Environmental Ethics and Environmental Sustainability: An Empirical Assessment.

The upshot is that the unprecedented global environmental problems and pervasive ecological destruction to a large extent are moral crises. They can only be mitigated or controlled if the global community at large shakes off its enduring anthropocentric skin and abides by the ethical code of conduct in its dealing with the Earth system. That being said, to translate political commitment into real action requires a more environmentally responsible institutional realignment and a much stronger social ethical consensus with obligatory elements which provide the basis for moral value orientations.

The thesis is that ethical reasoning for environmental sustainability works as a more effective motivation to induce world leaders and global society into embracing sustainable agendas where the fecund years of United Nations environmental efforts have largely failed. The voluminous UN environmental documents and national law and policy which were mostly targeted at addressing the symptoms of environmental problems, will not be effective or produce lasting impact unless environmental ethics or more specifically, the bioecocentric ethics as examined in Chapter 4, Greening for a Sustainable Future: The Ethical Connection, is brought down to earth to address the moral causes of unsustainable environmental practices at the roots—an indispensable condition which seems to have been missed by the United Nations for the past few decades.

7.5 United Nations environmental education framework: The changing face

Mere acknowledgment of the importance of environmental ethics will lead us to nowhere. For one thing, the wheels of environmental ethics cannot move in a vacuum. They must be activated and propelled through the promotion of environmental and moral education. As discussed in Chapter 6, The United Nations Environmental Education Initiatives: The Green Education Failure and the Way Forward, the purpose of environmental education is to promote environmental literacy to gain an understanding of the environment surrounding us including the functional properties of ecosystems, and the moral insights into the relationships between man and nature. It also aims to raise practical awareness of environmental problems or issues and the ways to deal with them as well as sustainable living and consumerism within bounds, among others. This is the way in which we realize our place in nature—that is, we are a part of the whole.

To be sure, the importance of environmental education in promoting an environmentally sustainable world was reaffirmed at the Stockholm Conference held in 1972. Since then, the United Nations, or UNESCO for that matter, has progressively convened or organized a series of conferences or workshops to streamline environmental education frameworks to promote environmental literacy, and sustainable environmental attitudes and actions such as sustainable living and consumerism within bounds.

However, since the publication of the Brundtland Report in 1987 which popularized the concept of sustainable development, there was a major shift in the degree of importance in its original goal of raising environmental literacy and sustainable environmental actions. Through a series of subsequent conferences, environmental education has overwhelmingly been restructured with new fields of emphasis and called by different names such as "educating for a sustainable future," "education for sustainability," "education for all," and education for a host of other objectives. Worse yet, since Brundtland, the seedbed of environmental education has been broadened to such an extent that it covers any issues which are deemed to come under the purview of sustainable development. These include HIV-AIDS, culture, human rights, gender, peace, social justice, arts, technology, and other issues that have evolved over the past few decades since 1987. In the end, environmental education has transformed into a substratum of the UN supreme and multidisciplinary sustainable development educational framework.

Although the issue of environmental sustainability was never severed from the UN omnipotent global education framework, its incredibly expansive perspectives and a boundless series of objectives has blurred the original goal of environmental education in raising environmental literacy. What is more, each educational objective embedded in the

UN supreme sustainable development education framework has its own special focus and specific characteristics which do not share equal environmental distinctiveness with environmental education. Also, faced with competing and, at times, conflicting epistemic descriptions and overwhelmingly rich pedagogical landscape of sustainability visions or goals, some of which may be beyond one's own comprehension, individuals may be unclear as to which disciplines, conceptions, or representations specifically contribute to the principal attribute of environmental literacy

Thus the diffused form of environmental education will not be able to attain a level of consolidation and precision to render it effective in our human-centric world. To be sure, environmental education does not appertain to the sustainability order of poverty alleviation, human rights or a multitude of sustainable development objectives, although they may be related. Rather, it is about harnessing individual obligations, moral duties, and self-restraints in our interactions with nature. Thus to distill the nonenvironmental attributes into the domain of environmental education will only turn it into a complex, indeterminate and elusive discipline which cannot be intelligibly understood and hence cannot gear to some useful end or fruitful results.

7.6 The United Nations' environmental education programs: Where do we go from here?

Following from the above discussion, it is of critical importance for the United Nations to obtain clearer insights into the approaches to assist the public to become more environmentally literate and to be more inclined to display proenvironmental behaviors based on various environmental philosophical principles. However, environmental issues abound and are diverse, and hence it is not instructive to cover every aspect at the very outset as this can be confusing especially to the uninitiated. Thus we may start from the fundamental principle of the first ring of predominance in constructing environmental education themes which are simple, clear, and distinct and with the objective significance of fulfilling the same function everywhere, that is, "to excite, maintain, or recreate certain mental states" in society. The mental states are collectively associated with a unified system of ethical beliefs and moral practices in our relationship with the Earth System.

René Descartes' first ring principle of predominance in the chain of scientific truths offer an amenable solution. Accordingly, the first ring which constitutes a firm building block toward human acquisition of environmental knowledge must be clear and distinct. In other words, environmental education should firstly be explained based on the keystone of knowledge, that is, with the most common matters which are clear, simple, and distinct for human conception. Once this foundation or the first ring of knowledge is firmly established, a system of integrated knowledge can be developed gradually and intelligibly in an ascending order to the frontier of knowledge. It may be commenced in an ascending progression by admitting other environmental substances into its keystone foundation to enable us to discover and grasp their complex relationships. This may cover such issues as the inextricable link between human and nature and the infallible truth of human dependence on the environment for continued existence, among others, as identified in Chapter 6, The United Nations Environmental Education Initiatives: The Green Education Failure and the Way Forward.

To render environmental education clear and distinct, two demonstrative themes have been developed based on basic true beliefs which cannot be doubted. These are important and notable environmental problems which are in line with the analysis in Chapter 3, The United Nations' Journey to Global Environmental Sustainability Since Stockholm: The Paradox: (1) "The parable of the bees"—the first ring of collective representation of the multifunctional dynamics of ecosystems, and (2) "The collective representation of the ocean biological system"—an outdoor program aimed at exemplifying the inextricable connection between the ecological integrity of ocean and climate change. This case study approach pioneered by Harvard Business School is built on descriptive information and empirical and scientific studies, and it offers an effective means and adequate way of achieving the objective significance of environmental education. It draws upon ecological and biological concepts and scientific knowledge from different disciplines in environmental studies where skill development in tackling environmental issues may be induced. The nonenvironmental information such as poverty issues, human rights, and gender, among others as embraced under the UN omnipotent education for sustainable development framework are all excluded from the environmental education structure. The fact is that their content of discipline cannot be intelligibly evinced and integrated into its environmental education setting. To do so will violate or infringe its environmental educational bound, losing its quality of particularity.

Building up from their first rings of predominance in an ascending order little by little to cover the scientific facts and concepts underlying their frames of analysis, it provides the basis for unifying the dynamics, complex, and diverse aspects of the functional properties of ecosystems and their inextricable relationship to human long-term existence. More importantly, it aids us to connect real human activities to a treasure trove of environmental information and shed light into such inquiries as to what constitutes an ecosystem and life-support system, why it needs human ethical concern, to whom the ethical concern is directed, and who decides, among others. It inspirationally evokes the moral imperative of humans to embrace the priceless philosophical thoughts of Aldo Leopold and Albert Schweitzer in our dealings with the already distressed and fragile natural world. Correspondingly, there will be linear cause—effect relations in that a human-induced environmental problem is perceived, an external moral force of constraint is generated, and an environmental value orientation is instituted to guide against the erratic, brutal, and undisciplined force of environmental impairment.

However, environmental education for children at the early stage must necessarily be crafted under a different setting to avoid the "ecophobia" syndrome. It is best to start using various outdoor environmental activities or place-based outdoor learning to firstly nurture their love for nature with pleasurable experiences and memories. This will build the bridges between environmental awareness and moral proenvironmental behaviors and to inspire them to embark on a lifetime of loving and wanting to keenly respect, protect, and preserve the natural environment from anthropocentric afflictions. In this way, children are slowly and gradually inculcated into the indispensable self-constraint of moral discipline, understanding that humans cannot transcend the biophysical limits of nature with in rebellion without placing themselves at odds with the Earth system.

Furthermore, in another aspect of the United Nations' environmental education methodology, as noted in Chapter 6, The United Nations Environmental Education Initiatives: The Green Education Failure and the Way Forward, it took the initiative to organize various environmental education programs in various developing countries to raise environmental literacy. But these programs were sketchily presented and sporadically organized on a piecemeal or a one-time project basis. They were far from being effectively developed and have yet to yield much fruitful results corresponding to the efforts expended. It may well be that individuals, especially those uninitiated, cannot adjust instantaneously to the new environmental visions and practices within a short period of time. Above all, the programs also conveyed, ipso facto the idea of the basic and specific aspects of environmental issues without expounding the moral aspects of social life in releasing us from immoral and amoral forces besetting the individual–collective duality of our minds.

Thus everything hinges on the imagination of individuals to learn greater self-control based on their personal vision to grasp the meaning of moral values, the full significance of impersonal end and collective interest of environmental protection. Consequently, the chain of environmental education is upset, weakened, or broken through the loss of the moral import of environmentalism. For this reason, there has been little motivation for individuals to discover and ascertain moral precepts to induce the spirit of abnegation, that would guide them to delimit themselves within the planetary boundary and to live in harmony with nature. The momentary faith of moral value orientation can be easily attenuated or shaken especially in the prevailing global economic system engulfed by the waves of anthropocentrism.

Furthermore, environmental education is not a one-time endeavor or a piecemeal project-specific undertaking. Rather, it is a systematic, ongoing, continuous, and revitalizing process involving the establishment of the Durkheimian discipline of collective conscience in humanity based on a shared set of environmental beliefs and moral values which dictates and commands the moral will of society to live in harmony with nature. Indisputably, this collective conscience of environmentalism and morality must be continuously and everlastingly maintained with eternal equanimity in our anthropocentric economic system if we are to rescue our Earth system from descending further into the bottomless abyss of human annihilation.

The *raison d'être* of this claim is that despite the fact that society is continually evolving concomitantly with socioeconomic progress and science and technological advancement, and notwithstanding that social life across the globe differs from one place to another under different socioeconomic conditions, customs, or traditions, there are few self-evident truths that are undeniable: environmental ethics constitutes an ethical agent par excellence for all moral change in our dealings with the Earth system and it takes the same function and common characteristics everywhere irrespective of time, space, and location. It also manifests the ultimate force in addressing the roots of global environmental disruptions.

Thus there is an urgent need for the United Nations to restore the past to its normal state by reconsidering its environmental education strategy. The vantage point is to construct a solid foundation via environmental and moral education for enhancing an environmentally responsible global community. To do this, as argued in Chapter 6, The United Nations Environmental Education Initiatives: The Green Education Failure and the

Way Forward, it is necessary to bring back environmental education as the first ring of predominance in raising environmental literacy and awareness as originally envisaged. Furthermore, it must be integrated into the second chain of moral education to arrive at a complete domain of a genuine and forceful environmental education framework.

The current precarious environmental status quo does not permit putting old wine into new bottles. Expansive past conferencing and voluminous international documents adopted are more than sufficient to provide insights, information and concepts to systematically guide the establishment or reconstruction of improved educational frameworks, the design of supporting curricula and textbooks as well as effective environmental education policy-making. The United Nations needs to put in place a team of experts comprising environmental education advisory committees and coordinators and go to the ground to liaise with the relevant authorities to promote environmental education and to guide curriculum development and implementation. In particular, attention needs to be focused on promoting and developing effective environmental education models in the South where environmental education has hitherto failed to gain popularity or is substantially marginalized in the curricula of most public schools and universities, and in public media.

7.7 Environmental sustainability: The current state and future outlook

7.7.1 The world as it is

In closing, we reflect again on the time when potently strong hurricanes swept across the globe in 2017 and 2018, bringing about catastrophic effects on humans as well as to the natural and physical environment at incalculable social and economic costs. Equally to the point, in 2019, typhoons, storms, bushfires, floods, and heatwaves swept across the planet, affecting millions worldwide. As a case in point, Super Typhoon Hagibis, the strongest and the most devastating typhoon ever to hit Japan in 61 years, paralyzed the capital city of Tokyo, slamming it at maximum wind speeds of 216 km/h, unleashing torrential rain and howling winds, causing rivers to overflow, triggering landslides and floods that damaged thousands of homes.

Furthermore, the destruction of tropical rainforests in South America, Africa, and Southeast Asia has continued unrelentlessly. In the Brazilian Amazon, for example, 1 million ha of forests were unscrupulously decimated for commercial use in 2019—the highest level in a decade. In Indonesia, by the end of September 2019, more than 850,000 ha of forests had been burned mostly for agricultural development, contributing to widespread ecological destruction and acute greenhouse emissions. Furthermore, between September 2019 and January 2020, around 5.8 million ha of forest in Australia were destroyed in fires induced by drought linked to climate change, significantly modifying its natural landscapes. It is estimated that nearly 3 billion mammals, birds, and reptiles have been killed or displaced by the bushfires, and their diverse ecosystems acutely affected.

Elsewhere, poaching surged to an unprecedented rate in Botswana with at least 46 rhinos (protected species) slaughtered in 10 months, while in its neighboring country, South Africa, 594 rhinos were poached in the same year. In the United States, the species extinction crisis worsened, attributable to its failure to protect 241 plant and animal species listed under the

Endangered Species Act. To compound the problem, the list of protected species under the Endangered Species List was reduced to only 21 species, easing the way for the government to encroach further into the protected federal lands including the Arctic National Wildlife Refuge in Alaska for oil and gas drilling.

As we welcomed the new decade in 2020, one tipping point after another continued to follow. The most serious crisis is that human-induced global warming has not shown any sign of retreat as reflected by broken heat records with the world experiencing its hottest January in recorded history triggered by increasing atmospheric carbon concentration and accelerating global land and ocean warming. Global land and ocean temperatures in the same month exceeded all temperatures recorded in the past 141 years of data at 1.13°C (2.05°F) above the 20th century average. What is more. Siberia, one of the coldest places on Earth became one of the warmest with its temperature soared to 38°C (100.4°F) in June 2020—the highest temperature ever recorded north of the Arctic Circle. The record-breaking heatwave which is said to link to oil spills, smoldering wildfires, thawing permafrost and a plague of tree-eating moths, is undoubtedly alarming as it can have severe environmental consequences for the rest of the planet. Also, the world is still a long way from containing climate emergency as evidenced by the almost persistent increasing trends of atmospheric carbon concentration since Stockholm, reaching 416.18 ppm in April 2020 compared to 413.52 ppm in the same month of the previous year. Some of the regions adversely impacted by this warming effect with record temperatures included parts of Central and South America, Asia, Scandinavia, the Indian and Atlantic Oceans and the central and western Pacific Ocean.

The persistent warming effects are also causing massive loss of ice in Greenland and the Antarctica, melting 6 times faster than in the 1990s with the East Antarctica's Denman glacier retreating nearly 3 miles in the last 22 years. Together, these two regions contributed to an aggregate loss of 6.4 trillion tonnes of ice in three decades with Greenland losing 600 billion tonnes of ice in its exceptionally warm summer in 2019 alone. It is also instructive to note that a recent study conducted by scientists from Edinburgh and Leeds universities and University College London reveal that the Earth has lost 28 trillion tonnes of ice since 1994 because of global warming triggered by rising greenhouse gas emissions. Without taking drastic measures to thwart further glacier retreat, ice losses in both regions are likely to accelerate in the coming decades or even centuries, aggravating future sea level rise to unprecedented chronic levels, leading to massive flooding in all the low-lying Asian, American and European coastal cities.

Ironically, almost all countries across the world missed the February 9 deadline to revise and submit their plans to fight climate change by cutting their emissions by 50%–55% below 1990 levels by 2030 from the previous target of at least 40%. This starkly reveals the lack of real political will to earnestly respond to the Paris Agreement. Given the present trend of development, limiting global warming to 1.5°C as proposed under the Paris Agreement is almost certainly not going to materialize. On the contrary, it is likely to reach further to 2°C or worse yet, 3°C above the preindustrial levels if no serious political will or genuine societal efforts are taken immediately to reduce warming effects.

In the same year, 2020, tropical rainforests across the globe continued to be razed. Forest depletion in the Brazilian Amazon, for example, was more than doubled in January compared with the previous year with more than 28,000 ha cleared in that month alone—an area 83 times the size of New York's Central Park. Furthermore, the satellite imagery captured by the European Space Agency reveals that during the coronavirus pandemic, logging and mining operations in the Brazilian Amazon have accelerated the destruction of its rainforest extensively. To put matter into perspective, according to the Brazilian National Institute for Space Research (INPE), in the first 6 months of 2020, deforestation of the Brazilian Amazon rose 25% to 306,600 ha compared to the same period last year. If the current rate of deforestation remains unmitigated, the Amazon forest could become a net carbon emitter in as little as 15 years. To compound the problem, the government in Congo has announced a plan to drill one of the largest carbon sinks on Earth, Congo's Cuvette Centrale peatlands, for their oil deposits, threatening to release 30 billion tonnes of carbon into the atmosphere. In addition, in Indonesia, the government, motivated by its anthropocentric current of thought, has approved a road project to a coal company to allow it to transport the black rock from its coal mine. However, the infrastructure project can cause violent destruction to the Sundaic dry lowland rainforest (Harapan forest) which is not only among the most biologically diverse on earth, but also one of the most threatened forests in the world. Undoubtedly, the environmentally destructive project is detrimental to the continued survival of some 1,350 of animal and plant species. These include 133 globally threatened species such as the legally protected Sumatran tiger and Sumatran elephant and the Dipterocarp trees Hopea mengerawan, Hopea sangal, Shorea acuminata, and Syzygium ampliflorum which are listed as Critically Endangered on the International Union for Conservation of Nature (IUCN) Red List.

If human insatiable exploitation of nature remains unrestrained, a massive collapse of the ecosystem the size of the Amazon forests could occur in approximately 50 years once the tipping point is reached. Worse yet, with the tropical forest in serious trouble as a result of man's persistent and uncontrolled exploitation, the looming sixth mass extinction is accelerating faster than scientists previously thought. In a nutshell, the world is currently facing not only a real and looming ecological crisis attributed to massive anthropogenic transformation of the global landscapes but also, a crisis of the same magnitude as that of the Covid-19 pandemic.

To digress a little, it is relevant to note that many wild animals serve as hosts that harbor pathogens, that is, infectious agents such as viruses or bacteria, providing them nourishment and shelter. However, widespread land clearing, deforestation and habitat destruction tend to generate high levels of stress among wild animals, upsetting this host—parasite relationship. This destabilization effect renders wild animals to become immunosuppressed, causing them to stimulate the excretion of pathogenic organisms which may be transmissible to humans. This may result in widespread contagious diseases, just like the disease caused by the Ebola virus in the past and the Covid-19 pandemic we are witnessing today.

Over time, when the wild animals become immune, they take on a new function as reservoirs (living hosts) for the pathogen without being harmed by it. Viewed from this perspective, wild animals, just like our carbon sink, serve as a critically important storehouse for deadly infectious agents. Hence, humans should leave the wild animals alone in the wild: since 75% of all emerging contagious diseases originate from wildlife, humans are at risk for many more pandemics in years to come.

The above illustrations are by no means exhaustive, but they suffice to convince us that, overall, our environmental future is bleak. Our Earth system is in an unprecedented freefall driven by unrestrained human activities. If the current status quo continues, many of the extreme weather events we have witnessed will not only become commonplace but also become even more oppressive and life threatening. A storm is certainly brewing and the scientific warning signs are encyclopedically clear and depressingly hard to ignore. Humanity is now at a crossroads in the wake of an impending environmental crisis. This raises the most pertinent question: can we still save the world? This will be discussed in the following section as the conclusion to this volume.

7.7.2 Future outlook: Mapping the way forward—the Covid-19 way

Certainly, thanks to modern social media, human beings in general are well aware of the various environmental problems, especially climate change, or biodiversity loss facing humanity today. They may also display their environmental sentiments of fear, despair, or hope. However, to a striking extent, these feelings are often overshadowed by the lack of conscious recognition of the causal rings of events that give rise to the current environmental disruptions and their multiple causal connections to their practical life. Thus it may be reiterated that environmental awareness which gives rise to certain environmental sentiments is a poor judge of action.

As expounded in Chapter 6, The United Nations Environmental Education Initiatives: The Green Education Failure and the Way Forward, environmental awareness is not a shared set of environmental beliefs (collective consciousness). These are but personal moods synonymous with judgments of existence or of reality which, unlike value judgements, do not entail regulative power to guide, direct, and propel human environmental behavior and moral action aligned with the ethics of sustainability or bioecocentrism for the matter. The turning point from this moral impasse is to put in place effective environmental and moral education programs, as has been extensively examined in this volume. The main aim here is to harness a shared set of environmental beliefs, ethical principles, and moral values—the collective consciousness which constitutes the final manifestation of the will of society.

The last point is an important one. In mapping for an environmentally sustainable future, the need for a collective conscientious global community should never be questioned. For one thing, this is the *raison d'être* in the shaping of environmental ideas, economic behaviors, and moral values of individuals in a global society collectively. It informs the sense of their belonging and identity in the natural world within which they exist. Only through harnessing a shared way of understanding and behaving in the natural world will it be possible to address in a lasting manner the environmental problems confronting us today.

The Covid-19 or coronavirus pandemic currently facing the world today has perfectly demonstrated that it is not only possible but also clearly desirable for a global society to stand together for a common cause. The current unprecedented existential threat of the Covid-19 pandemic has remarkably created an aggressive force of social cohesion across the world. Particularly, it has transcended into a binding force joining individuals with diverse interests and from different socioeconomic and cultural backgrounds across the Planet together into a collective global society with a shared overarching goal—to fight a common enemy, the new coronavirus named Covid-19. It is no exaggeration to claim that the global collective efforts which take the forms of massive mobilization and deployment of the world's resources are unparalleled in the history of humankind.

Why and how is this possible? To answer this question, we need to conceptualize the causal setting from which it evolved. We need to trace back to the earlier circumstances or causes that led to the final emergence of collective consciousness. The starting point was when the grim march of Covid-19 had ascended into a weighty force propelling individuals across the world to behave and act collectively (final causal effect) because of the collective conscious recognition of the disturbing reality of its life-threatening property (initial emotional cause) and its abysmal devastation of the global economy (initial economic cause). In the subsequent stage, these initial causes and final effects together metamorphosed further into atom-like Covid-19 forces along the causality chain, sparking off the highest level of alert and urgency in practically all individuals across the globe, binding them to act without fail (final effect) (Fig. 7.1).

More particularly, the binding forces manifest as the centrifugal moral imperative to unify and command individuals around the world to collectively join forces in unison to fight against a common enemy—the Covid-19. It is precisely in this setting that the coronavirus became ascendant and universal as an ineluctable and authoritative moral force

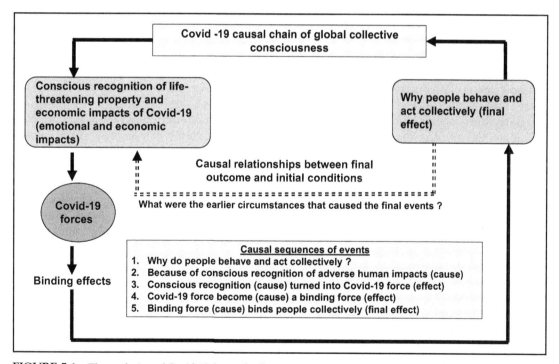

FIGURE 7.1   The evolution of Covid-19 force of collective consciousness.

sui generis enunciating the conditions of collective consciousness to guide and direct humankind toward a common cause of action in confronting the viral pandemic.

To reflect on and interpret the Covid-19 effect in an environmental perspective, the dark cataclysmic events of the coronavirus pandemic causing humankind to transcend self-serving individualism to collectivism may yet bring hope for a sustainable environmental future. Conceivably, the Covid-19 crisis evoked our faith, instinct, and drive in the present and future moral reasoning on our relationship with nature. What is needed is to project the environmental light of morality similar to that of the Covid-19 light of collectivism into the consciousness of collective human conscience. When moral collectivism gains its urgency and breath from humanity, it will work like the force of moral consolidation in response to Covid-19 in driving the human race into collective action toward a common goal—The Future We Want.

To date, the similitude prism of environmental moralism has not been able to penetrate deeply into man's consciousness in the manner that the Covid-19 pandemic has ignited collectivism because, unlike the pandemic, the impacts of the existential environmental problems we are facing today are not immediately evident, tangibly dramatic, and concretely deep rooting. Thus the disconnect caused by the lack of immediacy and urgency of fact and idea of the dystopian realities of life and collective fears, unlike the Covid-19 world of havoc. Hence, the environmental light of moral urgency becoming lost in the penumbra.

The turning point may come when people finally confront the stark impact in reality of human inaction toward dreadful existential environmental problems such as climate change. The environmental reality or global environmental emergency has to be conveyed with distinct urgency and in utmost clarity with data, pictures, digital images, videos, or other visual aids in the manner that the Covid-19 pandemic provoked humankind to urgent action. The purpose is to create a common collective spirit—akin to that evoked in our fight against Covid-19—of environmental awareness, and to sustain the spirit of urgency in conscious recognition of the cumulative effects of environmental impairment.

When the intuitive grasp of this environmental representation is incontrovertibly settled and deeply embedded into the mainstream of society, it will warm the hearts of individuals who will collectively set their moral will in motion to act and live sustainably. Here, the importance of moral education as discussed in Chapter 6, The United Nations Environmental Education Initiatives: The Green Education Failure and the Way Forward, and further elaborated earlier can be seen metaphorically as an engine of moral impulses to propel the global community to rationally retreat from its egoistic anthropocentric practices to the moral domain of bioecocentrism in optimizing the economic use of nature. Nothing is more certain that this moral order of environmentalism is the only primordially lasting key to the door of global environmental sustainability.

Furthermore, it is instructive to note that environmental problems are a global concern which involves all radically heterogenous individuals across the world with diverse interests. Granted, it is practically impossible for all of them to act collectively in response to the United Nations' call for action based on the all-pervasive reason that man's unsustainable practices are jeopardizing the Earth's natural life-support systems to the detriment of human long-term existence. This may not evoke the moral sense of immediate urgency, thus negating the hope for a weighty response for real action. Equally to the point is the fact that, as reflected in this volume, it is all too easy for global leaders to pledge or commit to the UN environmental agreements for a global common cause but without a genuine inner moral drive to real action. That said, it is worth reiterating that without enunciating the fundamental conditions of collective consciousness that governs social moral solidarity, it would be immensely difficult if not impossible to drive the global community toward the goal of The Future We Want, and our environmental future would remain as bleak as ever.

We may thus conclude that the United Nations must awake to the looming reality that unless efforts are marshaled to mark a milestone in establishing impacts of environmentalism of the "Covid-19 response magnitude" through incessant streams of life-long environmental and moral education as propounded in this volume, the mission of achieving global environmental sustainability may continue to be beyond our reach. The purpose here is not to recreate a Covid-19 fear but to rekindle a shared set of environmental ideas, moral beliefs, and ethical values among individuals based on environmental and moral education as extensively discussed in the milieu. The ultimate aim is to establish a collective environmental consciousness that binds the human race on the victory march of global environmental sustainability.

7.8 Concluding thoughts

Evidently, the current unfolding discourse reveals that no number of global environmental conferences, international environmental treaties or agreements, or any volume of national law and environmental policy enactments would be able to effectively and perpetually address the pervasive worldwide environmental degradation. The concrete reality is that we cannot solely pin all our hopes on The Future We Want based solely on these international and national environmental initiatives although they do have a considerable influence in pulling the global community into the perspective of environmentally sustainable development discourse as reflected by various positive achievements in mitigating environmental ills across the world.

The stark truth is that for more than 45 years since Stockholm, we have paradoxically failed to bring our Earth system back into the orbit of environmental sustainability. We are now presented with a different natural world—a world that is perilously transgressed with unprecedented human inflictions and irreversible anthropogenic transformations. This impels us to discern more succinctly that there is no way to heal our Earth system unless we heal ourselves morally and ethically. In essence, there is no other way to save our planet unless and until environmental and moral education becomes accessible to guide human direction toward an environmentally sustainable development. The self-evident truth is that for more than 45 years since Stockholm, many global leaders across the world seem to lack the real political will to transform environmental commitments into reality—the United States under Donald Trump's administration is an obvious case in point. What is still lacking in their leadership is the holistic environmental ethical systems of thought and the guiding rules of moral engagement to transform political commitments into effective policies and real actions.

The denouement is that they have lost their grip on the racetrack in the deterministic march to global environmental sustainability. To be sure, environmental sustainability, by and large, cannot be authoritatively legislated in the current state of international relations. It is unfortunate that for more than 45 years since Stockholm, the strength of moral and ethical force has not always been fully recognized as a cornerstone for environmental sustainability. In this sense, to bring our Earth system in closer accord with environmental sustainability, the global community, and especially the world leaders, must analyze their actions and introspect on their ecological integrity, on what is demanded of them to rectify their mistakes. In doing so, we will be able to unite to commit genuinely to a new integrated ethical vision of environmentally sustainable development.

We live in a world which has evolved over the past 4.5 billion years through natural causes. However, we have had no part in its creation: we arrived in the natural world in the blink of an eye in geologic time, 3.5 billion years ago when life first emerged on earth. Hence, our Earth system does not play by human rules and we, constituting only an infinitesimally small part of it, by no means rule supreme over it. Hence, declaring the Baconian war against nature (read uncontrolled environmental exploitation) is akin to declaring war on ourselves. It is high time for us to reconsider the classical concept of sustainable development as defined in the Brundtland Report which is an ideal that can never be realized in the current human-centered world. Sustainable development must be redefined as development that promotes a continual process of environmental improvement which underpins long-term human survival of present and future generations. The concept is that the current decisions in the economic exploitation of nature should not impair the ecological integrity or Holling sustainability of the Earth system which underscores the indefinite existence of human civilization. Without environmental sustainability, there will be no sustainable development, no matter how one defines the concept.

The years of disasters from 2017 to 2019 as elucidated in this volume, clearly affirms this logical line of thought. We have already observed that when the potently strong hurricanes swept across the Atlantic, they brought with them catastrophic effects on humans, and on the natural and physical environment at incalculable economic costs. What has been constructed by humans for years can be destroyed and wiped out by the brutal force of hurricanes instantaneously. Manifestly, there is no way humans can deploy science and technology effectively to avert the destructive force of a fully developed hurricane with its explosive heat energy potential equivalent to a 10-megatonne nuclear bomb exploding every 20 minutes.

Can we still save the world? Yes, there is still hope on the horizon because the future of the world is within our ability to manage. Clearly, the only way is to transcend the individualistic utilitarian force of our anthropocentrically stimulated "invisible hands" and move into the prism of bioecocentrically driven hands of ethical care. In a nutshell, there is no way of effectively halting the current environmental predicament no matter how many more global environmental conferences are convened, or how expansive the international environmental treaties or agreements adopted may be, or how voluminous the national law and regulations enacted in the future. The only way is by drawing all the communities worldwide under the dual mandates of moral will and ethical imperative and to unite in the race to set our Earth system back onto the elliptical path of environmental sustainability.

Bibliography

Chapter 1

Boon, E.K., Eyong, C.T., 2009. History and civilizations: impacts on sustainable development in Africa. In: Boon, E.K. (Ed.), Area Studies—Regional Sustainable Development: Africa Review, vol. 2. Eolss Publisher Co. Ltd., Oxford, pp. 241–275.

Marong, A.B.M., 2003. From Rio to Johannesburg: reflections on the role of international legal norms in sustainable development. Georget. Int. Environ. Law Rev. 16 (21), 21–76.

May, J.R., 1998. Of Development, daVinci and Domestic Legislation: the prospects for sustainable development in Asia and its untapped potential in the United States. Widener Law Symposium J. 3, 197–212.

Rolston, H., 1999. Environmental ethics: values in and duties to the natural world. In: Baker, E., Richardson, M. (Eds.), Ethics Applied, second ed. Simon and Shuster, New York, pp. 407–437.

WCED (World Commission on Environment and Development), 1987. Our Common Future. Oxford University Press, Oxford, New York.

WTO, 2014. Trade and Development: Recent Trends and the Role of the WTO. World Trade Report 2014. World Trade Organization (WTO), Geneva.

Chapter 2

ABS, 2003. 2003 Year Book Australia. Australian Bureau of Statistics (ABS), Canberra, ACT.

Adams, D., 2002. The Kyoto Protocol in 2002—Opportunities for Coal. International Energy Agency (IEA) Clean Coal Centre, United Kingdom.

Ahmad, I.M., et al., 2010. A report of the High-level Plenary Meeting of the 65th Session of the UN General Assembly on the Millennium Development Goals (MDGs). MDG Summit Bull. 153 (9), 1–15.

Alimonda, H. (Ed.), 2011. La Naturaleza Colonizada - Ecología Política y Minería en América Latina. CLACSO/Ciccus, Buenos Aires.

Ambrose, J., 2020. Cop26 Climate Talks in Glasgow Postponed Until 2021. The Guardian, United Kingdom. <https://www.theguardian.com/environment/2020/apr/01/uk-likely-to-postpone-cop26-un-climate-talks-glasgow-coronavirus> (accessed 02.04.20.).

Baker, S., 2006. Sustainable Development. Routledge, London, New York.

Bowman, M., 2010. Environmental protection and the concept of common concern of mankind. In: Fitzmaurice, M., Ong, D.M., Merkouris, P. (Eds.), Research Handbook on International Environmental Law. Edward Elgar, Cheltenham; Northampton, MA, pp. 493–518.

Brunnée, J., 2004. The United States and International Environmental Law: living with an elephant. Eur. J. Int. Law 15 (4), 617–664.

Cai, S.Q., Voigts, M., 1993. The Development of China's Environmental Policy. Pac. Rim Law Policy Assoc. 3 (special ed.), pp. S-17–S42.

Caponera, D.A., 1972. Towards a new methodological approach in environmental law. Nat. Resour. J. 12 (2), 133–152.

Carpenter, C., et al., 1998. Summary of the Nineteenth United Nations General Assembly Special Session to Review Implementation of Agenda 21: 23–27 June 1997. Earth Negotiations Bull. 5 (88), 1–14.

Carson, R., 1962. Silent Spring. Fawcett Publications, Greenwich, CT.

CBD Secretariat, 2005. Working Together for Biodiversity: Regional and International Initiatives Contributing to Achieving and Measuring Progress Towards the 2010 Target. Abstracts of Poster Presentations at the Tenth Meeting of the Subsidiary Body on Scientific, Technical and Technological Advice of the Convention on Biological Diversity. CBD Technical Series No. 17, Montreal, QC.

CBD Secretariat, 2010a. Global Biodiversity Outlook 3. Secretariat of the Convention on Biological Diversity (CBD), Montreal, QC.

CBD Secretariat, 2010b. Decision Adopted by the Conference of the Parties to the Convention on biological Diversity at Its Tenth Meeting. (UNEP/CBD/COP/DEC/X/2), Secretariat of the Convention on Biological Diversity (CBD), Montreal, QC.

CBD Secretariat, 2011a. Nagoya Protocol on Access to Genetic Resources and the Fair and Equitable Sharing of Benefits Arising From Their Utilization to the Convention on Biological Diversity. Secretariat of the Convention on Biological Diversity (CBD), United Nations Environmental Programme (UNEP), Montreal, QC.

CBD Secretariat. 2011b. Aichi Targets: Assisting Parties to Implement the Strategic Plan for Biodiversity 2011–2020. Secretariat of the Convention on Biological Diversity (CBD) News Letter, vol. 1 (1), Montreal, QC, pp. 1–16.

CBD Secretariat, 2016a. Press Release: Recent Moves by China, Finland and Zambia Edge Nagoya Protocol Closer to Goal of 100 Ratifications. Secretariat of the Convention on Biological Diversity (CBD), United Nations Environment Programme (UNEP), Montreal, QC.

CBD Secretariat, 2016b. Report of the Conference of the Parties to the Convention on Biological Diversity on Its Thirteenth Meeting. Secretariat of the Convention on Biological Diversity (CBD), Montreal, QC.

CBD Secretariat, 2018. Latest NBSAPs. Secretariat of the Convention on Biological Diversity (CBD), Montréal, QC.

CBD Secretariat, 2019. Post-2020 Global Biodiversity Framework: Discussion Paper. Montréal, QC.

Chandra, M., 2015. Environmental concerns in India: problems and solutions. J. Int. Bus. Law 5 (1), 1–13.

Chechi, A., 2016. Risks relating to the protection of cultural heritage: from climate change to disasters. In: Mišćenić, E., Raccah, A. (Eds.), Legal Risks in EU Law: Interdisciplinary Studies on Legal Risk Management and Better Regulation in Europe. Springer, Switzerland, pp. 199–224.

Choy, Y.K., 2015a. From Stockholm to Rio + 20: the ASEAN environmental paradox, environmental sustainability and environmental ethics. Int. J. Environ. Sustain. 12 (1), 1–25.

Choy, Y.K., 2015b. Sustainable resource management and ecological conservation of mega-biodiversity: the Southeast Asian Big-3 reality. J. Environ. Sci. Dev. 6 (11), 876–882.

Choy, Y.K., 2015c. 28 years into "our common future": sustainable development in the post-Brundtland world. In: Brebbia, C.A. (Ed.), Sustainable Development, vol. II. WIT Press, Southampton, pp. 1197–1211.

Choy, Y.K., 2016. Economic growth, sustainable development and ecological conservation in the Asian developing countries: the way forward. In: Das, I., Alek Tuen, A. (Eds.), Naturalists, Explorers and Field Scientists in South-East Asia and Australasia. Topics in Biodiversity and Conservation Series, vol. 15. Springer, Cham, Heidelberg, New York, Dordrecht, London, pp. 239–283.

Choy, Y.K., 2018. Sustainable development and environmental stewardship: the Heart of Borneo paradox and its implications on green economic transformations in Asia. In: Hsu, S. (Ed.), Routledge Handbook of Sustainable Development in Asia. Routledge, Oxon, New York, pp. 532–549.

Choy, Y.K., 2019. Containing the world's environmental problems: an interdisciplinary approach applied to Malaysia. In: Ecological Economics and Social Ecological Movements: Science, Policy and Challenges to Global Processes in a Troubled World. Autonomous Metropolitan University, Mexico, pp. 445–467.

CITES Secretariat, 2010. Activity Report of the CITES Secretariat. Convention on International Trade in Endangered Species of Wild Fauna and Flora 2008–2009. The Convention on International Trade in Endangered Species of Wild Fauna and Flora (CITES) Secretariat, Switzerland.

CPSG, 2019. CPSG Annual Meeting 2019: "Engaging Governments in Species Conservation Planning" October 31–November 3, 2019 Buenos Aires, Argentina Briefing Book. Conservation Planning Specialist Group (CPSG), United States.

Dalal-Clayton, B., Sadler, B., 2014. Sustainability Appraisal: A Sourcebook and Reference Guide to International Experience. Routledge, London, New York.

Delang, C.O., 2016. China's Water Pollution Problems. Routledge, Oxon, New York.

Dembowski, H., 2001. Taking the State to Court: Public Interest Litigation and the Public Sphere in Metropolitan India (Law in India). Oxford University Press, Oxford.

Dine, M.K., 2012. Forests: does state sovereignty hinder their protection at the international level? In: Sancin, V., Dine, M.K. (Eds.), International Environmental Law: Contemporary Concern and Challenge. Papers presented at the First Contemporary Challenges of International Environmental Law Conference, Ljubljana, June 28–29, 2012, GV Založba, Ljubljana, pp. 109–128.

Doran, P., et al., 2012. Summary of the United Nations Conference on Sustainable Development 13–22, June 2012. Earth Negot. Bull. 27 (51), 1–24.

Dresner, S., 2008. The Principles of Sustainability, second ed. Earthscan, London, Sterling, VA.

Drexhage, J., Murphy, D., 2010. Sustainable Development: From Brundtland to Rio 2012. Background Paper Prepared for Consideration by the High Level Panel on Global Sustainability at Its First Meeting, 19 September 2010, United Nations Headquarters, New York.

ECOSOC, 1969. Problems of the Human Environment. Report of the Secretary General. Economic and Social Council (ECOSOC), United Nations, New York.

Edwards, G., Roberts, J.T., 2015. A Fragmented Continent: Latin America and the Global Politics of Climate Change. The MIT Press, Cambridge, MA.

Egelston, A.E., 2013. Sustainable Development. A History. Springer, Dordrecht, Heidelberg, New York, London.

Ellison, K., 2014. Rio + 20: how the tension between developing and developed countries influenced sustainable development. Pac. McGeorge Global Bus. Dev. Law J. 27 (1), 107–129.

Engfeldt, L-G., 2009. From Stockholm to Johannesburg and Beyond. The Evolution of the International System for Sustainable Development and Its Implications. Ministry for Foreign Affairs. Government Offices of Sweden, Stockholm.

European Communities, 1997. Agenda 21—The First 5 Years. Implementation of Agenda 21 in the European Community. European Communities, Luxembourg.

European Communities, 1999. Communication From the Commission—Europe's Environment: What Directions for the Future? The Global Assessment of the European Community Programme of Policy and Action in Relation to the Environment and Sustainable Development, "Towards Sustainability". European Commission, Luxembourg.

French, D., 2016. Common concern, common heritage and other global(-ising) concepts: rhetorical devices, legal principles or a fundamental challenge? In: Bowman, M., Davies, P., Goodwin, E. (Eds.), Research Handbook in Biodiversity and Law. Edward Elgar, Cheltenham, Northampton, MA, pp. 334–358.

Gaines, S.E., 1997. Rethinking environmental protection, competitiveness, and international trade. Univ. Chic. Leg. Forum 1997 (1), 231–292. Article 9.

GFCS, 2009a. World Climate Conference-3. Conference Statement. Summary of the Expert Segment. Global Framework for Climate Services (GFCS), Geneva. Available: <https://gfcs.wmo.int/wwc_3>.

GFCS, 2009b. WCC-3 High Level Declaration. Global Framework for Climate Services (GFCS), Geneva. Available: <https://gfcs.wmo.int/wwc_3>.

Glowka, L., et al., 1994. A Guide to the Convention on Biological Diversity. IUCN (International Union for Conservation of Nature), Gland, Cambridge.

Goldsmith, E., et al., 1972. A blueprint for survival. Ecologist 2 (1), 1–43.

Hale, T., 2003. Managing the Disaggregation of Development. How the Johannesburg "Type II" Partnerships Can Be Made Effective? Woodrow Wilson School of Public and International Affairs, Princeton University.

Hardin, G., 1968. Tragedy of the commons. Science 162, 1243–1248.

Haq, G., Paul, A., 2012. Environmentalism Since 1945. Routledge, London, New York.

Hazlewood, P., Mock, G., WRI, 2010. Ecosystems, Climate Change and the Millennium Development Goals (MDGs): Scaling Up Local Solutions. A Framework for Action. Working Paper Prepared for the UN MDG Summit September 2010. United Nations Development Programme (UNDP), New York; World Resources Institute (WRI), Washington, DC.

Hecht, S., Cockburn, A., 1992. Rhetoric and reality in Rio. The Nation 254 (24), 848–854.

Hey, C., 2005. EU environmental policies: a short history of the policy strategies. In: Scheuer, S. (Ed.), EU Environmental Policy Handbook. A Critical Analysis of EU Environmental Legislation. European Environmental Bureau (EEB), Brussels, pp. 17–30.

Hironaka, A., 2014. Greening the Globe. Cambridge University Press, New York.

Holling, C.S., 1973. Resilience and stability of ecological systems. Annu. Rev. Ecol. Syst. 4, 1–23.

Holling, C.S., 1986. The resilience of terrestrial ecosystems; local surprise and global change. In: Clark, W.C., Munn, R.E. (Eds.), Sustainable Development of the Biosphere. Cambridge University Press, Cambridge, pp. 292–317.

Holling, C.S., Gunderson, L.H., 2002. Resilience and adaptive cycles. In: Gunderson, L.H., Holling, C.S. (Eds.), Panarchy: Understanding Transformations in Human and Natural Systems. Island Press, Washington, DC, pp. 25–62.

Horn, L., 2004. The implications of the concept of common concern of a human kind on a human right to a healthy environment. Macquarie J. Int. Comp. Environ. Law 1, 233–269.

House of Commons London, 2013. Thirty-Ninth Report of Session 2012–13. House of Commons European Scrutiny Committee, London.

Howe, J.P., 2014. Behind the Curve: Science and the Politics of Global Warming. University of Washington Press, Seattle, London.

ICSU (International Council for Science) and ISSC (International Social Science Council), 2015. Review of the Sustainable Development Goals: The Science Perspective. International Council for Science (ICSU), Paris.

ITDP, 2012. World's Largest Development Banks Pledge $175 Billion for the Creation of More Sustainable Transport. Institute of Transportation and Development Policy (ITDP), New York.

IUCN, 2010. Draft International Covenant on Environment and Development, fourth ed. IUCN (International Union for Conservation of Nature and Natural Resources), Gland, International Council of Environmental Law (ICEL), Bonn, Germany.

IUCN, UNEP, WWF, 1980. World Conservation Strategy. Living Resources for Sustainable Development. The International Union for Conservation of Nature and Natural Resources (IUCN), Gland, United Nations Environment Programme (UNEP), Washington, DC; World Wide Fund for Nature (WWF), Washington, DC.

Ivanova, M., 2005. Environment: the path of global environmental governance—form and function in historical perspective. In: Ayre, G., Callway, R. (Eds.), Governance for Sustainable Development. A Foundation for the Future. Earthscan, London, Sterling, VA, pp. 45–72.

Jackson, P., 2007. From Stockholm to Kyoto: A Brief History of Climate Change. UN Chronicle, XLIV (2), United Nations, New York.

Jacobsen, S., 1973. Stockholm—a year later. Sci. Public. Aff. XXIX (6), 35–40.

Kapur, N., 2015. Asia Pacific relations and the globalization of the environment. In: Johnson, R.D. (Ed.), Asia Pacific in the Age of Globalization. Palgrave Macmillan, United Kingdom, pp. 13–23.

Klemmensen, B., et al., 2007. Environmental Policy—Legal and Economic Instruments. Baltic University Press, Sweden.

Koh, K.L., Robinson, N.A., 2002. Regional environmental governance: examining the Association of Southeast Asian Nations (ASEAN) model. In: Esty, D.C., Ivanova, M.H. (Eds.), Global Environmental Governance: Options and Opportunities. Yale School of Forestry and Environmental Studies, New Haven, CT, pp. 101–120.

Kumar, A., Kumar, A., 2015. Environmental Studies (As per Latest VTU Syllabus), third ed. New Age International Pvt Ltd Publishers, New Delhi.

Kutter, A., 2009. The United Nations Convention to Combat Desertification: policies and programs for implementation. In: Verheye, W.H. (Ed.), Encyclopedia of Land Use, land Cover and Soil Science, vol. V. Eolss Publisher Co. Ltd., Oxford, pp. 97–115.

Larsson, M.-L., 1999. The Law of Environmental Damage: Liability and Reparation. Kluwer Law International, The Hague, London, Boston, MA.

La Viña, A., et al., 2003. Making Participation Work: Lessons From Civil Society Engagement in the WSSD. World Resources Institute, Washington, DC.

Leggett, J.A, Carter, N.T., 2012. Rio + 20: The United Nations Conference on Sustainable Development, June 2012. CRS Report for Congress. Congressional Research Service, Washington, DC.

Marong, A.B.M., 2003. From Rio to Johannesburg: reflections on the role of international legal norms in sustainable development. Georget. Int. Environ. Law Rev. 16 (21), 21–76.

Martinez-Alier, J., Baud, M., Sejenovich, H., 2016. Origins and perspectives of Latin America environmentalism. In: De Castro, F., Barbara, H.B., Baud, M. (Eds.), Environmental Governance in Latin America. Palgrave Macmillan, United Kingdom, pp. 29–57.

Matthews, W.H., Kellogg, W.W., Robinson, G.D. (Eds.), 1971. Man's Impact on the Climate. The MIT Press, Cambridge, MA.

Meadows, D.H., Meadows, D.L., Randers, J., Behrens III, W.W., 1972. The Limits to Growth. A Report for the Club of Rome's Project on the Predicament of Mankind. Universe Books, New York.

M'Gonigle, R.M., Zacher, M.W., 1981. Pollution, Politics and International Law. Tankers at Sea. University of California Press, Berkeley, Los Angeles, London.

Mol, A.P.J., Carter, N.T., 2007. China Environmental Governance. In: Carter, N.T., Mol, A.P.J. (Eds.), Environmental Governance in China. Routledge, Oxon, pp. 1–22.

Mu, Z., Bu, S., Xue, B., 2014. Environmental Legislation in China: achievements, challenges and trends. Sustainability 6, 8967–8979.

Najam, A., 2005. Developing countries and global environmental governance: from contestation to participation to engagement. Int. Environ. Agreem. 5 (3), 303–321.

NILOS, 1998. (The Netherlands Institute for the Law of the Seas). In: Kwiatkowska, B., Molenaar, E., Elferink, A. O., Soons, A. (Eds.), International Organizations and the Law of the Sea: Documentary Yearbook 1996, vol. 12.. Kluwer Law International, The Hague, The Netherlands.

OHCHR, OHRLLS, UNDESA, UNEP, UNFPA, 2013. UN System Task Team on the Post-2015 UN Development Agenda. Global Governance and Governance of the Global Commons in the Global Partnership for Development Beyond 2015. Office of the High Commission for Human Rights (OHCHR), Switzerland, Countries and Small Island Developing States (OHRLLS), New York; United Nations Department of Economic and Social Affairs (UNDESA), New York; United Nations Environment Programme (UNEP), New York; United Nations Population Fund (UNFPA), New York.

Osborn, D., Bigg, T., 1998. Earth Summit II: Outcomes and Analysis. Earthscan, London.

Pintasilgo, Md.L., 1992. Crisis and Change in Latin America. High-Level Expert Group Meeting 28–29 February 1992. InterAction Council, Washington, DC.

Pisano, M., Endl, A., Berger, G., 2012. The Rio + 20 Conference 2012: Objectives, Processes and Outcomes. The European Sustainable Development Network, Vienna, ESDN Quarterly Report No 25, pp. 3–52.

Poulden, G., 2013. Launch of Partnership for Action on Green Economy (PAGE). International Cooperation and Development. European Commission.

PSCD, 2012. EPA 21. Philippine Council for Sustainable Development (PSCD), Pasig City.

Rajamani, L., 2000. The principle of common but differentiated responsibility and the balance of commitments under the climate regime. Rev. Eur. Commun. Int. Environ. Law 9 (2), 120–131.

Ramesh, J., 2012. Foreword. In: Dubash, N. (Ed.), Handbook of Climate Change and India: Development, Politics and Governance. Earthscan, Abingdon, Oxon, pp. xix–xxii.

Roberts, J.T., Newell, P., 2017. Editors' Introduction: The Globalization and Environmental Debate. In: Newell, P., Roberts, J.T. (Eds.), The Globalization and Environment Reader. John Wiley & Sons, The Atrium, Southern Gate, Chichester, West Sussex.

Rosen, A.M., 2015. The wrong solution at the right time: the failure of the Kyoto Protocol on climate change. Polit. Policy 43 (1), 30–58.

Rowland, W., 1973. The Plot to Save the World. The Life and Times of the Stockholm Conference on the Human Environment. Clark, Irvin and Company, Toronto.

Sands, P., Peel, J., Fabra, A., MacKenzie, R., 2012. Principles of International Environmental Law, third ed. Cambridge University Press, United Kingdom.

Sasaki, T., et al., 2015. Perspectives for ecosystem management based on ecosystem resilience and ecological thresholds against multiple and stochastic disturbances. Ecol. Indic. 57, 398–408.

Sawyer, D.R., 1998. The society and its environment. In: Hudson, R.A. (Ed.), Brazil: A Country Study, fifth ed. Federal Research Division, Library of Congress, Washington, DC, pp. 87–156.

SCEP (Study of Critical Environmental Problems), 1970. Man's Impact on the Global Environment. Assessment and Recommendation for Action. The MIT Press, Cambridge, MA.

Schumacher, E.F., 1973. Small is Beautiful. Blond & Briggs, London.

Schunz, S., Belis, D., 2011. China, India and Global Environmental Governance: The Case of Climate Change. Leuven Centre for Global Governance Studies, Belgium, Policy Brief No. 15.

SCOPE, 1971. Global Environmental Monitoring. A Report Submitted to the United Nations Conference on the Human Environment, Stockholm 1972. Commission on Monitoring of the Scientific Committee on Problem of the Environment (SCOPE) of the International Council of Scientific Union (ICSU), Sweden.

Selin, H., Linnér, B.-O., 2005. The Quest for Global Sustainability: International Efforts on Linking Environment and Development. Science, Environment and Development Group, Center for International Development, Harvard University, Cambridge, MA, Working Paper No. 5.

Shapiro, J., 2001. Mao's war against nature: legacy and lessons. J. East Asian Stud. 1 (2), 93–119.

Sodhi, N.S., et al., 2004. Southeast Asian biodiversity: an impending disaster. Trends Ecol. Evol. 19 (12), 654–660.

Sohn, L.B., 1973. The Stockholm declaration on the environment. Harv. Int. Law J. 14 (3), 422–515.

Srebotnjak, T., Polzin, C., Giljum, S., Herbert, S., Lutter, S., 2010. Establishing Environmental Sustainability Thresholds and Indicators Final Report. Final Report to the European Commission's DG Environment. Ecologic Institute, EU and Sustainable Europe Research Institute (SERI), Vienna.

Stokke, O., 2009. The UN and Development: From Aid to Cooperation. Indiana University Press, Bloomington, Indianapolis.

Taneja, P., 1998. Entrepreneurs. In: Mackerras, C., McMillen, D.H., Watson, A. (Eds.), Dictionary of the Politics of the People's Republic of China. Routledge, London, New York, pp. 91–93.

Tinker, J., 1975. Cocoyo revisited. New Sci. 480–483 (August 28).

Tollefson, J., Gilbert, N., 2012. Rio report card. Nature 486, 21–23.

Tsioumani, S., et al., 2018. Summary of the UN Biodiversity Conference 13–29 November 2018, Sharm El-Sheikh, Egypt. Earth Negotiations Bull. 9 (725), 1–27.

Tsioumani, S., et al., 2019. 11th Meeting of the Ad Hoc Open-Ended Working Group on Article 8(j) and Related Provisions and 23rd Meeting of the Subsidiary Body on Scientific, Technical and Technological Advice of the Convention on Biological Diversity: 20–22 and 25–29 November 2019. Earth Negotiations Bull., 9 (741), 1–2.

UNDP, 2009. India's National Capacity Needs Self-Assessment Report & Action Plan. United Nations Development Programme (UNDP), India.

UNDP, 2013a. Global MDG Conference—Making the MDGs Work. United Nation Development Programme (UNDP), New York.

UNDP, 2013b. 2013 Global MDG Conference Working Paper Series. United Nation Development Programme (UNDP), New York.

UNDP, 2013c. Accelerating Progress Sustaining Results: The MDGs to 2015 and Beyond. United Nation Development Programme (UNDP), New York.

UNEP, 2006. GEO Year-Book: An Overview of Our Changing Environment. United Nations Environmental Programme (UNEP), New York.

UNEP, 2010. Driving a Green Economy Through Public Finance and Fiscal Policy Reform. Working Paper v.1.0: 1–33. United Nations Environment Programme (UNEP), New York.

UNEP, 2011. Towards a Green Economy. Pathways to Sustainable Development and Poverty Eradication. A Synthesis for Policy Makers. United Nations Environment Programme (UNEP), New York.

UNEP and UNCTAD, 1974. The Cocoyoc Declaration. Adopted at Symposium on Patterns of Resource Use, Environment and Development Strategies, Cocoyoc, Mexico, 8–12 October. Reprinted in: International Organization, 29 (3) (summer, 1975), pp. 893–901.

UNEP-WCMC, IUCN, 2016. Protected Planet Report 2016. United Nation Environmental Programme (UNEP), World Conservation Monitoring Centre (WCMC) International Union for Conservation of Nature (IUCN), Cambridge, Gland.

UNFCCC, 2006. United Nations Framework Convention on Climate Change: Handbook. Climate Change Secretariat, Bonn.

UNFCCC, 2010. Report of the Conference of the Parties on Its Fifteenth Session, Held in Copenhagen From 7 to 19 December 2009 (FCCC/CP/2009/11/Add.1). United Nations Framework Convention on Climate Change (UNFCCC), Bonn.

UNFCCC, 2014. Lima Call for Climate Action. United Nations Framework Convention on Climate Change (UNFCCC), Bonn (Decision-/CP.20).

UNFCCC, 2015. Adoption of the Paris Agreement (FCCC/CP/2015/L.9/Rev.1). United Nations Framework Convention on Climate Change (UNFCCC), Bonn.

UNFCCC, 2018a. Report of the Conference of the Parties on Its Twenty-Third Session, Held in Bonn From 6 to 18 November 2017 (FCCC/CP/2017/11/Add.1). United Nations Framework Convention on Climate Change (UNFCCC), Bonn.

UNFCCC, 2018b. Conference of the Parties Twenty-Fourth Session Katowice, 2–14 December: COP 24 Agenda as Adopted. United Nations Framework Convention on Climate Change (UNFCCC), Bonn.

United Nations, 1950. Proceedings of the United Nations Scientific Conference on the Conservation and Utilization of Resources, 17 August–6 September 1949. United Nations, Lake Success, New York.

United Nations, 1968. 2398 (XXIII). Problems of the Human Environment (3 December 1968). Resolutions Adopted by the General Assembly During Its Twenty-Third Session. United Nations, New York.

United Nations, 1969. 2581 (XXIV): United Nations Conference on the Human Environment (15 December 1969). Resolutions Adopted by the General Assembly During Its Twenty-Fourth Session. United Nations, New York.

United Nations, 1971. Development and Environment. Report Submitted by a Panel of Experts Convened by the Secretary-General of the United Nations, Conference on Human Environment. Founex, 4–12 June 1971 (UN Doc. A/CONF.48/10 Annex I). United Nations, New York.

United Nations, 1972. Report of the United Nations Conference on the Human Environment, Stockholm, June 1972 (UN.Doc. A/CONF/48/14/REV.1). United Nations, New York.

United Nations, 1974. Declaration on the Establishment of a New International Economic Order. Resolutions Adopted by the General Assembly During Its Sixth Special Session (Document RES/S-6/3201). United Nations, New York.

United Nations, 1982. World Charter for Nature Adopted During the General Assembly During the 48th plenary meeting (28 October 1982) (Document A/RES/37/7). United Nations, New York.

United Nations, 1983. Process of Preparation of the Environmental Perspective to the Year 2000 and Beyond Adopted at Meeting no. 102 on 19 December 1983 (A/RES/38/161). United Nations, New York.

United Nations, 1988. Protection of Global Climate for Present and Future Generations of Mankind. Resolution A/RES/43/53 Adopted at 70th Plenary Meeting on 6 December 1988. United Nations, New York.

United Nations, 1992a. Rio Declaration on Environment and Development. (Document A/CONF.151/26, vol. I). United Nations, New York.

United Nations, 1992b. Intergovernmental Negotiating Committee for a Framework Convention on Climate Change Adopted During Fifth Session (9 May 1992) (Document A/AC.237/18). United Nations, New York.

United Nations, 1992c. Convention on Biological Resources: Preamble. United Nations, New York.

United Nations, 1997a. United Nations Conference on Environment and Development (1992). United Nations Department of Public Information, New York.

United Nations, 1997b. Programme for the Further Implementation of Agenda 21 Adopted During Nineteenth Special Session (19 September 1997) (Document A/S-19/29). United Nations, New York.

United Nations, 2000a. We the Peoples—the Role of the United Nations in the 21st Century. United Nations Department of Public Information, New York.

United Nations, 2000b. United Nations Millennium Declaration Adopted During Fifty-Fifth Session of General Assembly on 18 September 2000 (Document A/RES/55/2). United Nations, New York.

United Nations, 2001. Ten-Year Review of Progress Achieved in the Implementation of the Outcome of the United Nations Conference on Environment and Development Adopted During Fifty-Fifth Session of the General Assembly on 5 February 2001 (Document A/RES/55/199). United Nations, New York.

United Nations, 2002a. Report of the World Summit on Sustainable Development. United Nations Department of Public Information, New York, Johannesburg, South Africa, 26 August–4 September 2002 (Document A/CONF.199/20).

United Nations, 2002b. Guidance in Preparing National Sustainable Development Strategies: Managing Sustainable Development in the New Millennium. United Nations Department of Economic and Social Affairs, New York, Background Paper No. 13. (Document DESA/DSD/PC2/BP13).

United Nations, 2002c. Summit Secretary-General Says Completed Negotiations Provide a Solid Foundation for Action. United Nations, New York.

United Nations, 2002d. Partnership Initiatives Announced at Sustainable Development Summit in Johannesburg. United Nations Department of Public Information, New York. (Document ENV/DEV/J/10).

United Nations, 2002e. Report by the Directors-General on the Status of Preparation for the World Summit on Sustainable Development and Its Expected Outcomes Adopted During the Hundred and Sixty-Fourth Session on 3 May 2002 (Document 164 EX/45). United Nations Educational, Scientific and Cultural Organization, New York.

United Nations, 2002f. Johannesburg Summit 2002: Key Outcomes of the Summit. Department of Economic and Social Affairs (DESA), New York.

United Nations, 2005. Resolution Adopted by the General Assembly on 16 September 2005: 2005 World Summit Outcome (Document A/RES/60/1). United Nations, New York.

United Nations, 2010a. The Millennium Development Goals Report 2010. United Nations, New York.

United Nations, 2010b. High-Level Plenary Meeting on the Millennium Development Goals. United Nations, New York, Conference Room Paper.

United Nations, 2010c. Keeping the Promise: United to Achieve the Millennium Development Goals Adopted During the Sixty-Fifth Session of the General Assembly on 17 September 2010 (Document (A/65/L.1.)). United Nations, New York.

United Nations, 2012a. Report of the United Nations Conference on Sustainable Development Rio de Janeiro, Brazil 20–22 June 2012 (Document A/CONF.216/16). United Nations, New York.

United Nations, 2012b. The Future We Want. Outcome Document of the United Nations Conference on Sustainable Development. United Nations, New York.

United Nations, 2012c. Over 700 Commitments, $513 Billion Pledged at Rio + 20. Integrated Implementation Framework, United Nations, New York.

United Nations, 2012d. UN Senior Officials Highlight Rio + 20 Achievements. United Nations, New York.

United Nations, 2013a. MDG Success: Accelerating Action and Partnering for Impact. Overview. United Nation, New York.

United Nations, 2013b. MDG Success: Accelerating Action and Partnering for Impact. United Nation, New York, Remarks by H.E. Ambassador John W. Ashe President of the 68th Session of the United Nations General Assembly (Closing Session).

United Nations, 2013c. MDG Success: Accelerating Action and Partnering for Impact. United Nation, New York, Press Release (Update).

United Nations, 2015a. Transforming Our World: The 2030 Agenda for Sustainable Development Adopted During the Seventieth Session of the General Assembly on 21 October 2015 (Document (A//RES/70/1)). United Nations, New York.

United Nations, 2015b. UN Adopts New Global Goals, Charting Sustainable Development for People and Planet by 2030. UN News Centre, United Nations, New York.

United Nations, 2016a. UN Climate Conference to Continue Momentum After Paris Agreement Comes Into Force. United Nations, New York.

United Nations, 2016b. Marrakech: 'The Eyes of the World are Upon Us,' Chair of UN Conference Says as New Round of Climate Talks Opens. United Nations, New York.

United Nations, 2018. Chapter XXVII: Environment-Convention on Biological Diversity. United Nations, New York, United Nations Treaty Collection.

United Nations, 2019a. SDG Summit, 24–25 September 2019, New York: Summary of the President of the General Assembly. United Nations, New York.

United Nations, 2019b. Resolution Adopted by the General Assembly on 15 October 2019 (A/RES/74/4). United Nations, New York.

United Nations, 2019c. The Future is Now: Science for Achieving Sustainable Development. United Nations, New York.

US Department of Commerce, 2002. China Environmental Technologies Export Market Plan. International Trade Administration, Washington, DC.

US Department of State, 1995. Global Environmental Issues: Fact Sheet. US Department of States, Washington, DC.

van Dever, S.D., 2006. European politics with a scientific face: framing, asymmetrical participation, and capacity in LRTAP. In: Farrell, A.E., Jäger, J. (Eds.), Assessment of Regional and Global Environmental Risks. Designing Process for the Effective Use of Science in Decision-making. Resource for the Future, Washington, DC, pp. 25–63.

Wang, J.S., 1989. Water pollution and water shortage problems in China. J. Appl. Ecol. 26, 851–857.

WCED (World Commission on Environment and Development), 1987. Our Common Future. Oxford University Press, Oxford, New York.

World Bank, 1998. Five Years After Rio: Innovations in Environmental Policy. The World Bank, Washington, DC.

World Bank, 2015. Joint Statement by the Multilateral Development Banks on Sustainable Transport and Climate Change. The World Bank, Washington, DC.

World Development, 1975. The Declaration of Cocoyoc. World Development 3 (2 & 3), 141–143.

WWF, 2015. What Are the MDGs? World Wildlife Fund (WWF), Gland.

Zillman, J.W., 2009. A history of climate activities. WMO Bull. 58 (3), 141–150.

Chapter 3

ACB, 2010. ASEAN Biodiversity Outlook. ASEAN Centre for Biodiversity (ACB), The Philippines.

ADP, 2013. Heart of Borneo: Saving Forests in Southeast Asia. Asian Development Bank (ADP), The Philippines, <https://www.adb.org/news/features/heart-borneo-saving-forests-southeast-asia> (accessed 12.12.19.).

AFP, 2020. Nearly 50 Rhino Killed in Botswana in 10 Months as Poaching Surges. AFP. <https://m.news24.com/Africa/News/nearly-50-rhino-killed-in-botswana-in-10-months-as-poaching-surges-20200224> (accessed 25.02.20.).

Afrizal, J., 2018. Lost Habitats Push Sumatran Tiger Out of Forests. Jakarta Post, Indonesia, <http://www.theja-kartapost.com/news/2018/03/28/lost-habitats-push-sumatran-tiger-out-of-forests.html> (accessed 11.12.19.).

Ahlenius, H., 2006. Extent of Deforestation in Borneo 1950–2005, and Projection Towards 2020. UNEP/GRID-Arendal. Available from: <http://www.grida.no/resources/8324>.

Alave, K.L., 2011. Hottest of Biodiversity Hot Spots Found in PH. Philippines Daily Inquirer, <http://newsinfo.inquirer.net/10996/hottest-of-biodiversity-hot-spots-found-in-ph#ixzz2caPpT3ME> (accessed 11.12.19.).

Aldred, J., 2016. Poaching Drives Huge 30% Decline in Africa's Savannah Elephants. The Guardian, United Kingdom, <https://www.theguardian.com/environment/2016/aug/31/poaching-drives-huge-30-decline-in-africas-savannah-elephants> (accessed 10.12.19.).

Allan, J.R., et al., 2019. Hotspots of human impact on threatened terrestrial vertebrates. PLoS Biol. 17 (3), 1–18. e3000158. Available from: <https://doi.org/10.1371/journal.pbio.3000158>.

Allsopp, M., et al., 2007. Oceans in peril: protecting marine biodiversity. In: Mastn, L. (Ed.), Worldwatch Report 174. Worldwatch Institute, Washington, DC.

Amahowé, I.O., et al., 2013. Transboundary protected areas management: experiences from W-Arly-Pendjari Parks in West Africa. Parks 19 (2), 95–105.

Andrew, R.M., 2018. Global CO_2 emissions from cement production. Earth Syst. Sci. Data 10, 195–217.

Angelo, C., 2012. Brazil's Atlantic forests lose key species. Nature. Available from: <http://www.nature.com/news/brazil-s-atlantic-forests-lose-key-species-1.11175>.

Anstee, S., 2019. Gurney's Pitta: lost and found—and lost again? BirdLife Int <https://www.birdlife.org/world-wide/news/gurneys-pitta-lost-and-found-and-lost-again> (accessed 10.12.19.).

Anup, K.C., 2017. Community Forestry Management and its role in biodiversity conservation in Nepal. In: Lameed, G.S.A. (Ed.), Global Exposition of Wildlife Management. IntechOpen Limited, London, pp. 51–72.

Aratrakorn, S., Thunhikorn, S., Donald, P.F., 2006. Changes in bird communities following conversion of lowland forest to oil palm and rubber plantations in Southern Thailand. Bird Conserv. Int. 16 (1), 71–82.

Arbi, I.A., 2019. Helmeted Hornbill Poached to Brink of Extinction. The Jakarta Post, Indonesia, <https://www.thejakartapost.com/news/2019/03/28/helmeted-hornbill-poached-brink-extinction.html> (accessed 11.12.19).

Ardiansyah, F., Marthen, A.A., Amalia, N., 2015. Forest and Land-Use Governance in a Decentralized Indonesia. A Legal and Policy Review. Center for International Forestry Research (CIFOR), Bogor, Occasional Paper 132.

Ariain, N., 2013. Focus Group Discussion on Heart of Borneo Green Economy in Indonesia. World Wildlife Fund (WWF), Washington, DC.

Arnold, M., et al., 2011. Editorial: forests, biodiversity and food security. Int. Forest. Rev. 13 (3), 259–264.

ASEAN Secretariat, 2009. Fourth ASEAN State of the Environment Report 2009. ASEAN Secretariat, Jakarta.

ASEAN Secretariat, 2017. Fifth ASEAN State of the Environment Report. ASEAN Secretariat, Jakarta.

Associated Press, 2020. Trump Rollback of Mileage Standards Guts Climate Change Push. Associated Press. <https://www.nbcnews.com/politics/donald-trump/trump-rollback-mileage-standards-guts-climate-change-push-n1173026> (accessed 02.04.20.).

AU-IBAR, 2012. Project site—The W-Arly-Pendjari Parks Complex. African Union-Intra-African Bureau for Animal Resources (AU-IBAR). <http://www.au-ibar.org/l4lp-project-overview/99-en/programmes-and-pro-jects/completed-programmes-and-projects/l4lp/project-overview/9-project-site-the-w-arly-pendjari-parks-complex> (accessed 10.12.19.).

Bacon, F., 1901. Preface to the Novum Organum. In: Elliot, C.W. (Ed.), Preface and Prologues to Famous Book. P.F Collier & Son Corporation. The Harvard Classics, United States, pp. 143–147.

Bacon, F., 1964. The masculine birth of time, or the great instauration of the dominion of man over the universe. In: Farrington, B. (Ed.), The Philosophy of Francis Bacon: An Essay on Its Development From 1603 to 1609. Liverpool University Press, United Kingdom, pp. 59–62.

Bacon, F., 2011. The works of Francis Bacon. In: Spedding, J., Ellis, R.L., Heath, D.D. (Eds.), Translations of the Philosophical Works. 1, vol. 4. Cambridge University Press, Cambridge, New York.

Baguinon, N.T., Quimado, M.O., Francisco, G.J., 2005. Country report on forest invasive species in the Philippines. In: McKenzie, P., et al., (Eds.), Unwanted Guests: Proceedings in Asia-Pacific Forest Invasive Species Conference; 17–22 August 2003. United Nations Food and Agriculture Organization (UNFAO), Kunming, pp. 108–113.

Baillie, J.E.M., Hilton-Taylor, C., Simon, S.N. (Eds.), 2004. 2004 IUCN Red List of Threatened Species. A Global Species Assessment. International Union for Conservation of Nature and Natural Resources (IUCN), Gland, Cambridge.

Baker, J., 2019. All Eyes on China as National Carbon Market Plan Emerges From Haze. Ethical Corporation, England and Wales, <http://www.ethicalcorp.com/all-eyes-china-national-carbon-market-plan-emerges-haze> (accessed 12.12.19.).

Baldwin, A., et al., 2005. Summary of the Fifth Session of the United Nations Forum on forests: May 16–27, 2005. Earth Negotiation Bull. 13 (133), 1–16.

Barlow, J., et al., 2016. Anthropogenic disturbance in tropical forests can double biodiversity loss from deforestation. Nature 535 (7610), 144–147.

Barney, K., 2005. Central Plans and Global Exports: Tracking Vietnam's Forestry Commodity Chains and Export Links to China. Forest Trend. Center for International Forestry Research (CIFOR), Washington, DC, Bogor, Indonesia; York Centre for Asian Research (YCAR), Toronto, Ontario.

Bauch, S., et al., 2009. Forest policy reform in Brazil. J. For., 107(3), 132–138.

Bauer, H., et al., 2015. Lion (Panthera leo) populations are declining rapidly across Africa, except in intensively managed areas. PNAS (Proc. Natl Acad. Sci.) 112 (48), 14894–14899.

BBC, 2018. Amazon Rainforest Deforestation 'Worst in 10 Years', Says Brazil. British Broadcasting Corporation (BBC), London, <https://www.bbc.com/news/world-latin-america-46327634?fbclid = IwAR29X_C79_WDBK20MzXeqdY2N_1wr8b6qIMXUQseLXQClxspEntzfCekFWY> (accessed 07.12.19.).

Beastall, C., et al., 2016. Trade in the Helmeted Hornbill Rhinoplax vigil: the 'ivory hornbill'. Bird Conserv. Int. 26 (2), 137–146.

Beitsch, R., 2020a. With Polar Bear Study Open for Comments, Critics See Effort to Push Drilling in ANWR. The Hill, <https://thehill.com/policy/energy-environment/483482-with-polar-bear-study-open-for-comments-critics-see-effort-to-push> (accessed 08.03.20.).

Beitsch, R., 2020b. Trump Budget Calls for Slashing Funds to Climate Science Centers. The Hill, <https://thehill.com/policy/energy-environment/483689-trump-budget-calls-for-slashing-funds-to-climate-science-centers> (accessed 08.03.20.).

Beitsch, R., 2020c. Overnight Energy: EPA Suspends Enforcement of Environmental Laws Amid Coronavirus. The Hill, <https://thehill.com/policy/energy-environment/overnights/489764-overnight-energy-epa-suspends-enforcement-of> (accessed 30.03.20.).

Beitsch, R., Frazin, R., 2020. Trump Budget Slashes EPA Funding, Environmental Programs. The Hill, <https://thehill.com/policy/energy-environment/482352-trump-budget-slashes-funding-for-epa-environmental-programs> (accessed 08.03.20.).

Bennington, W., et al., 2014. The Green Shock Doctrine. Global Ecology Project, Buffalo, New York.

Berlinger, J., 2018. World's Last Male Northern White Rhino Dies. CNN. <https://edition.cnn.com/2018/03/20/africa/last-male-white-rhino-dies-intl/index.html> (accessed 10.12.19.).

Bhandari, P.K.C., et al., 2019. Importance of Community Forestry Funds for rural development in Nepal. Resources 8 (85), 1–14. Available from: https://doi.org/10.3390/resources8020085.

BirdLife International, 2019 Species Factsheet: Rhinoplax vigil. <http://datazone.birdlife.org/species/factsheet/22682464> (accessed 11.12.19.).

Bittel, J., 2019. Last male Sumatran rhino in Malaysia dies. National Geographic. <https://www.nationalgeographic.com/animals/2019/05/last-sumatran-rhino-malaysia-dies/> (accessed 11.12.19.).

Boden, T.A., Marland, G., Andres, R.J., 2016. Global, Regional, and National Fossil-Fuel CO_2 Emissions. Carbon Dioxide Information Analysis Center, Oak Ridge National Laboratory, U.S. Department of Energy, Oak Ridge, TN, <https://doi.org/10.3334/CDIAC/00001_V2016>.

Borger, J., 2001. Bush Kills Global Warming Treaty. The Guardian, United Kingdom, <https://www.theguardian.com/environment/2001/mar/29/globalwarming.usnews> (accessed 12.12.19.).

Boucher, D., Roquemore, S., Fitzhugh, E., 2013. Brazil's success in reducing deforestation. Trop. Conserv. Sci 6 (3, special issue), 426–445.

BP, 2011–2019. BP Statistical Review of World Energy (June 2011 to June 2019 issues). British Petroleum (BP), United Kingdom.

BP, 2018. BP Statistical Review of World Energy. British Petroleum (BP), United Kingdom.

Bradbury, I., Kirkby, R., 1996. China's agenda 21. A critique. Appl. Geogr. 16 (2), 97–107.

Braimoh, A.K., et al., 2011. Climate and Human-Related Drivers of Biodiversity Decline in Southeast Asia. UNU-IAS Policy Report. The United Nations University Institute of Advanced Studies (UNU-IAS), Yokohama.

Breitburg, et al., 2018. Declining oxygen in the global ocean and coastal waters. Science 359, 1–11.

Brito, B., Barreto, P., Rothman, J., 2005. Brazil's New Environmental Crimes Law: An Analysis of Its Effectiveness in Protecting the Amazon Forests. Instituto do Homem e Meio Ambiente da Amazônia (IMAZON), Belém.

Brown, E., Jacobson, M.F., 2005. Cruel Oil. How Palm Oil Harms Health, Rainforest & Wildlife. Center for Science in the Public Interest, Washington, DC. Available from: <http://www.cspinet.org/palmoilreport/PalmOilReport.pdf>.

Brown, G., 2017. World's largest tropical peatland found in Congo basin. Guardian. <https://www.theguardian.com/environment/2017/jan/11/worlds-largest-peatland-vast-carbon-storage-capacity-found-congo> (accessed 16.07.20.).

Brown, T., et al., 2012. Reducing CO_2 emissions from heavy industry: a review of technologies and considerations for policy makers. Grantham Institute for Climate Change Briefing Paper No.7, Imperial College London, pp. 1–29.

Bruno Manser Fonds, 2019. The Mulu Land Grab Results of a Fact-Finding Mission to Sarawak (Malaysia) on Palm Oil-Related Deforestation and a Land Conflict Near the UNESCO-Protected Gunung Mulu World Heritage Site. Bruno Manser Fonds, Basel.

Buckley, T., Nicholas, S., Brown, M., 2018. China 2017 Review World's Second-Biggest Economy Continues to Drive Global Trends in Energy Investment. The Institute for Energy Economics and Financial Analysis (IEEFA).

Bugayong, L.A., 2006. Effectiveness of logging ban policies in protecting the remaining natural forests of the Philippines. In: Paper Presented at the 2006 Berlin Conference on Human Dimensions of Global Environmental Change—Resource Policies: Effectiveness, Efficiency, and Equity, held at Freie University, Berlin, Germany on 17–18 November 2006.

Busby, J., 2018. 4 Things to Know About the Trump Budget's Environmental Cuts. The Washington Post. <https://www.washingtonpost.com/news/monkey-cage/wp/2018/02/15/4-things-to-know-about-the-trump-budgets-environmental-cuts/> (accessed 13.12.19.).

Butchart, et al., 2010. Global biodiversity: indicators of recent declines. Science 328, 1164–1168.

Butler, R.A., 2006a. Indonesia. Mongabay.com. <http://rainforests.mongabay.com/20indonesia.htm> (accessed 09.12.19.).

Butler, R.B., 2006b. Rainforest Diversity—Origins and Implications. Mongabay.com. <http://rainforests.mongabay.com/0301.htm> (accessed 10.12.19.).

Butler, R.A., 2010. Orangutans vs Palm Oil in Malaysia: Setting the Record Straight. Mongay.com. <https://news.mongabay.com/2010/01/orangutans-vs-palm-oil-in-malaysia-setting-the-record-straight/> (accessed 12.12.19.).

Butler, R.A., 2012a. Industrial Logging Leaves a Poor Legacy in Borneo's Rainforests. Mongay.com. <http://news.mongabay.com/2012/07/industrial-logging-leaves-a-poor-legacy-in-borneos-rainforests/> (accessed 12.12.19.).

Butler, R.A., 2012b. In Pictures: Rainforests to Palm Oil. Mongay.com (July 2). <http://news.mongabay.com/2012/07/in-pictures-rainforests-to-palm-oil/>.

Butler, R.A., 2013a. Palm Oil Now Biggest Cause of Deforestation in Indonesia. Mongabay.com. <http://news.mongabay.com/2013/0903-rspo-vs-greenpeace-palm-oil-deforestation.html#LQuXJ2f0KsKQH1i3.99> (accessed 09.12.19.).

Butler, R.A., 2013b. Malaysia Has the World's Highest Deforestation Rate, Reveals Google Forest map. Mongabay.com. <https://news.mongabay.com/2013/11/malaysia-has-the-worlds-highest-deforestation-rate-reveals-google-forest-map/> (accessed 09.12.19.).

Butler, R.A., 2013c. Countries With the Highest Biological Diversity. Mongabay.com. <http://rainforests.mongabay.com/03highest_biodiversity.htm> (accessed 10.12.19.).

Butler, R.A., 2014. Tradeoff: Sabah Banks on Palm Oil to Boost Forest Protection. Mongay.com. <http://news.mongabay.com/2014/12/tradeoff-sabah-banks-on-palm-oil-to-boost-forest-protection/> (accessed 12.12.19.).

Butler, R.A., 2016a. Brazil: Deforestation in the Amazon Increased 29% Over Last Year. Mongabay.com. <https://news.mongabay.com/2016/11/brazil-deforestation-in-the-amazon-increased-29-over-last-year/> (accessed 07.12.19.).

Butler, R.A., 2016b. The Top 10 Most Biodiverse Countries. Mongabay.com. <https://news.mongabay.com/2016/05/top-10-biodiverse-countries/> (accessed 10.12.19.).

Butler, R.A., 2019. Deforestation. Mogabay.com. <https://rainforests.mongabay.com/08-deforestation.html> (accessed 09.12.19.).

Byrne, J., 2018. New Report Documents Soy-Linked Deforestation in Argentina and Paraguay. William Reed's Feed Navigator. <https://www.feednavigator.com/Article/2018/03/28/New-report-documents-soy-linked-deforestation-in-Argentina-and-Paraguay> (accessed 09.12.19.).

Cabico, G.K., 2018. Recovering the Philippines' Forest Cover. Philstar Global, The Philippines, <https://www.philstar.com/headlines/2018/03/04/1793446/recovering-philippines-forest-cover> (accessed 09.12.19.).

CAFI, 2018. FAQ: What Is the Trend in Deforestation in DRC? Central African Forest Initiative (CAFI). United Nations Development Programme, Switzerland. Available from: <https://www.cafi.org/content/cafi/en/home/all-news/drc-forests---frequently-asked-questions/what-is-the-trend-in-deforestation-in-drc-.html>.

Carbon Brief, 2019. Guest Post: Why China's CO_2 Emissions Grew 4% During First Half of 2019. Carbon Brief, United Kingdom, <https://www.carbonbrief.org/guest-post-why-chinas-co2-emissions-grew-4-during-first-half-of-2019> (accessed 12.12.19.).

C4, 2015. An Insight to the Kelantan Timber Industry. (A Need for Better Forest Governance). Malaysia: C4 (C4 is a local NGO in Malaysia which represents Clean, Conscious, Competent and Credible). Available from: <https://c4center.org/sites/default/files/KT%20Final%20final.pdf>.

Cardona, W.C., Claros, M.Z., Cordoba, A.L., 2014. Forest regulation flexibility, livelihoods, and community forest management in northern Bolivian Amazon. In: Katila, P., Galloway, G., de Jong, W., Pacheco, P., Mery, G. (Eds.), Forests Under Pressure—Local Responses to Global Issues. IUFRO World Series, vol. 32. International Union of Forest Research Organizations (IUFRO), Vienna, pp. 97–111.

Carew-Reid, J., Kempinski, J., Clausen, A., 2010. Biodiversity and Development of the Hydropower Sector: Lessons From the Vietnamese Experience Volume I: Review of the Effects of Hydropower Development on Biodiversity in Vietnam. ICEM International Centre for Environmental Management, Prepared for the Critical Ecosystem Partnership Fund, Hanoi.

Casado, L., Londoño, E., 2019. Under Brazil's Far-Right Leader, Amazon Protections Slashed and Forests Fall. New York Times. <https://www.nytimes.com/2019/07/28/world/americas/brazil-deforestation-amazon-bolsonaro.html> (accessed 07.12.19.).

Castrén, T., 1999. Timber Trade and Wood Flow Study—Vietnam. Regional Environmental Technical Assistance 5771 Poverty Reduction & Environmental Management in Remote Greater Mekong Subregion (GMS) Watersheds Project (Phase I). Available from: <http://www.mekonginfo.org/assets/midocs/0002787-environment-timber-trade-and-wood-flow-study-viet-nam.pdf>.

Catherine Yap, C., Kong, F.Y., Lin, S.L., 2010. Pollution across Chinese provinces. In: Eichengreen, B., Gupta, P., Kumar, R. (Eds.), Emerging Giants: China and India in the World Economy. Oxford University Press, Oxford, pp. 281–306.

Catterson, T.M., Fragano, F.V., 2004. Tropical Forestry and Biodiversity Conservation in Paraguay: Final Report of a Section 118/119 Assessment EPIQ II Task Order No.1. United States Agency for International Development (USAID).

CBD Secretariat, 2011. NBSAP Training Modules Version 2.1—Module 4 Setting National Biodiversity Targets in Line With the Framework of the Strategic Plan for Biodiversity 2011–2020, Including Aichi Biodiversity Targets. Secretariat of the Convention on Biological Diversity (CBD), Montreal, QC.

CBD Secretariat, 2012. Report of the High-Level Panel on Global Assessment of Resources for Implementing the Strategic Plan for Biodiversity 2011–2020 (Document UNEP/CBD/COP/11/INF/20). Secretariat of the Convention on Biological Diversity (CBD), Montreal, QC.

CBD Secretariat, 2014. Global Biodiversity Outlook 4. Secretariat of the Convention on Biological Diversity (CBD), Montreal, QC.

CBD Secretariat, 2018. Status of Development of National Biodiversity Strategies and Action Plans or Equivalent Instruments (NBSAPs) at 7 May 2018. Secretariat of the Convention on Biological Diversity (CBD), Montreal, Canada. <https://www.cbd.int/nbsap/introduction.shtml> (accessed 09.12.19.).

CBD Secretariat, 2019. Latest NBSAPs. Secretariat of the Convention on Biological Diversity (CBD), Montreal, QC, <https://www.cbd.int/nbsap/about/latest/> (accessed 09.12.19.).

Center for Biological Diversity, 2016. Pacific Bluefin Tuna One Step Closer to Endangered Species Act Protection. Center for Biological Diversity, Tucson, AZ, <https://www.biologicaldiversity.org/news/press_releases/2016/bluefin-tuna-10-07-2016.html> (accessed 09.12.19.).

Center for Sustainable Systems, 2016a. US Energy System Factsheet. University of Michigan, Pub. No. CSS03-11.

Center for Sustainable Systems, 2016b. US Food System Factsheet. University of Michigan, Pub. No. CSS01-06.

Center for Sustainable Systems, 2016c. US Environmental Footprint. University of Michigan, Pub. No. CSS08-08.

CEPF, 2005. Atlantic Forest Hotspot: Brazil Briefing Book. Ecosystem Partnership Fund (CEPF). Conservation International, Washington, DC. Available from: <http://lerf.eco.br/img/publicacoes/final.atlanticforest.brazil.briefingbook.pdf>.

CEPF, 2014. Ecosystem Profile. Madagascar and Indian Ocean Islands. Arlington: Critical Ecosystem Partnership Fund (CEPF). Conservation International, Washington, DC. Available from: <http://www.cepf.net/SiteCollectionDocuments/madagascar/EcosystemProfile_Madagascar_EN.pdf>.

Chase, M.J., et al., 2016. Continent-wide survey reveals massive decline in African savannah elephants. PeerJ4 e2354. Available from: https://doi.org/10.7717/peerj.2354.

Chaytor, B., 2001. The Development of Global Forest Policy: Overview of Legal and Institutional Frameworks. International Institute for Environment and Development (IIED), World Business Council for Sustainable Development (WBCSD).

Chazdon, R.L., et al., 2016. When is a forest a forest? Forest concepts and definitions in the era of forest and landscape restoration. Ambio 45, 538–550.

Chemnick, J., 2017. Trump Drops Climate Threats From National Security Strategy. Scientific American. <https://www.scientificamerican.com/article/trump-drops-climate-threats-from-national-security-strategy/> (accessed 13.12.19.).

Chen, H., 2017. The Road From Paris: China' Progress Toward Its Climate Pledge. National Development and Reform Commission (NDRC), People's Republic of China.

Chen, M., 2020. Trump Wants to Gut Crucial Environmental Regulation. The Nation. <https://www.thenation.com/article/environment/national-environmental-policy-act-nepa/> (accessed 25.03.20.).

Cheung, E., 2020. Up to 23 Feet Long, the Chinese Paddlefish Was the Giant of the Yangtze. And We Killed It. CNN. <https://edition.cnn.com/2020/01/07/asia/chinese-paddlefish-extinct-study-intl-hnk-scli/index.html> (accessed 1.11.20.).

Che Yeom, F.B., Chandrasekharan, C., 2002. Achieving Sustainable Forest Management in Indonesia. IITO Tropical Forest Update, pp. 10–14.

Chia, J., 2015. Sarawak to Have More, Bigger TPAs Under HoB—Adenan, Borneo Post. <http://www.theborneopost.com/2015/11/04/sarawak-to-have-more-bigger-tpas-under-hob-adenan/> (accessed 12.12.19.).

Chiavari, J., Lopes, C., 2015. Brazil's New Forest Code: How to Navigate the Complexity. Climate Policy Initiative, San Francisco, London, Rio de Janeiro, New Delhi, Jakarta.

Chien, T.L., 2019. Key Water Policies 2018–19. China Water Risk, Hong Kong, <http://www.chinawaterrisk.org/resources/analysis-reviews/key-water-policies-2018-2019/> (accessed 12.12.19.).

Chisleanschi, R., 2019. Gran Chaco: South America's Second-Largest Forest at Risk of Collapsing (S. Engel, Trans.). Mongabay.com. <https://news.mongabay.com/2019/09/gran-chaco-south-americas-second-largest-forest-at-risk-of-collapsing/> (accessed 09.12.19.).

Chong, R., 2015. Sabah Doubles Protected Areas to 1.55 mln. Borneo Post. <http://www.theborneopost.com/2015/08/12/sabah-doubles-protected-areas-to-155-mln/> (accessed 12.12.19.).

Chow, C.C., 2004. Economic reform and growth in China. Ann. Econ. Financ. 5, 127–152.

Chow, G.C., 2008. China's energy and environmental problems and policies. Asia-Pacific J. Account. Econ. 15 (1), 57–70.

Chow, D., 2014. 2013 Was Record Year for Rhino Poaching in South Africa. Live Science, New York, <http://www.livescience.com/42776-rhino-poaching-south-africa.html> (accessed 10.12.19.).

Choy, Y.K., 2015a. Sustainable resource management and ecological conservation of mega-biodiversity: the Southeast Asian Big-3 reality. Int. J. Environ. Sci. Dev. 6 (11), 876–882.

Choy, Y.K., 2015b. From Stockholm to Rio + 20: the ASEAN environmental paradox, environmental sustainability and environmental ethics. Int. J. Environ. Sustain. 12, 1–25.

Choy, Y.K., 2016a. Economic growth, sustainable development and ecological conservation in the Asian develop-ing countries: the way forward. In: Indraneil Das, I., Tuen, A.A. (Eds.), Naturalists, Explorers and Field Scientists in South-East Asia and Australasia. Topics in Biodiversity and Conservation Series, vol. 15. Springer, Cham, Heidelberg, New York, Dordrecht, London, pp. 239−283.

Choy, Y.K., 2016b. Ecosystem health, human existence, and bio-capacity deficit: the ethical relationship. Int. J. Sustain. Dev. Plan. 11 (6), 1004−1016.

Choy, Y.K., 2018. Sustainable development and environmental stewardship: the Heart of Borneo paradox and its implications on green economic transformations in Asia. In: Hsu, S. (Ed.), Routledge Handbook of Sustainable Development in Asia. Routledge, London, New York, pp. 532−549.

Christophersen, T., 2010. The Convention on Biological Diversity and the Post-2010 Target. Secretariat of the Convention on Biological Diversity. Available from: <http://ec.europa.eu/environment/archives/green-week2010/sites/default/files/speeches_presentations/christophersen_17.pdf>.

CIFOR, 2013. Use Hansen High-Res Forest Cover Maps Wisely, Experts Say. Center for International Forestry Research (CIFOR), Bogor. Available from: <https://forestsnews.cifor.org/27435/use-hansen-high-res-forest-cover-maps-wisely-experts-say?fnl = en>.

CITES, 2014. Elephant Poaching and Ivory Smuggling Figures Released Today. The Convention on International Trade in Endangered Species of Wild Fauna and Flora (CITES), Geneva, <https://cites.org/eng//elephant_poaching_and_ivory_smuggling_figures_for_2013_released> (accessed 10.12.19.).

CITES, 2016. African Elephants Still in Decline due to High Levels of Poaching. The Convention on International Trade in Endangered Species of Wild Fauna and Flora (CITES), Geneva, <https://cites.org/eng/news/pr/african_elephants_still_in_decline_due_to_high_levels_of_poaching_03032016>   (accessed 10.12.19.).

CITES Secretariat, 2013. List of Contracting Parties. Convention on International Trade in Endangered Species of Wild Fauna and Flora (CITES). The Convention on International Trade in Endangered Species of Wild Fauna and Flora (CITES), Geneva. Available from: <https://www.cites.org/eng/disc/parties/chronolo.php> (accessed 09.12.19.).

CITES Secretariat, 2017. CITES: African Elephant Poaching Down, Ivory Seizures Up and Hit Record High. The Convention on International Trade in Endangered Species of Wild Fauna and Flora (CITES), Geneva. Available from: <https://cites.org/eng/news/pr/African_elephant_poaching_down_ivory_seizures_up_and_hit_record_high_24102017> (accessed 10.12.19.).

Clark, P., 2017. Introduction. In: Clark, P., Niemi, M., Catharina Nolin, C. (Eds.), Green Landscapes in the European City, 1750−2010. Routledge, London, New York, pp. 1−15.

Climate Focus, 2015. Progress on the New York Declaration on Forests—An Assessment Framework and Initial Report. Prepared by Climate Focus, in collaboration with Environmental Defense Fund, Forest Trends, The Global Alliance for Clean Cookstoves, and The Global Canopy Program. Climate Focus, Amsterdam, Netherlands.

CMD, 2014. IndoMet Coal Project. Center for Media and Democracy (CMD), Madison, WI.

CMEA, 2011. Masterplan for Acceleration and Expansion of Indonesia Economic Development 2011−2025. Coordinating Ministry for Economic Affairs(SMEA), Republic of Indonesia.

CNRES (Center for Natural Resources and Environmental Studies of the Vietnam National University, Hanoi), 2000. Vietnam: North and Central Highlands. In: Wood, A., Stedman-Edwards, P., Mang, J. (Eds.), The Root Causes of Biodiversity Loss. Earthscan, London, pp. 337−370.

Cole, S., 2015. NASA, USGS Begin Work on Landsat 9 to Continue Land Imaging Legacy (Updated 2017). National Aeronautics and Space Administration (NASA), Washington, DC. Available from: <https://www.nasa.gov/press/2015/april/nasa-usgs-begin-work-on-landsat-9-to-continue-land-imaging-legacy>.

Colman, Z., Waldman, S., E&E News reporters, 2018. Trump Admin Sees Grim Climate Outcome in Car Rule. E&E News. <https://www.eenews.net/stories/1060092895> (accessed 13.12.19.).

Conniff, R., 2009. The Growing Specter of Africa Without Wildlife. Yale Environment 360. Yale School of Forestry & Environmental Studies. <http://e360.yale.edu/content/feature.msp?id = 2183> (accessed 10.12.19.).

Conservation International, 2010. Biodiversity Hotspots—Atlantic Forest (January, 2010). Conservation International, Arlington, VA. Available from: <http://www.biodiversityhotspots.org/xp/Hotspots/atlantic_forest/Pages/default.aspx>.

Conservation International, 2011. The World's 10 Most Threatened Forest Hotspots. Conservation International, Arlington, VA. Available from: <http://www.conservation.org/newsroom/pressreleases/Pages/The-Worlds-10-Most-Threatened-Forest-Hotspots.aspx>.

Conservation International, 2012. New Assessment Finds Madagascar's Lemurs to Be the Most Threatened Mammal Species in the World. Conservation International, Arlington, VA, <http://www.conservation.org/NewsRoom/pressreleases/Pages/New-Finding-Lemurs-Most-Threatened-Mammal-Species.aspx> (accessed 10.12.19.).

Contreras-Hermosilla, A., Fay, C., 2005. Strengthening Forest Management in Indonesia Through Land Tenure Reform: Issues and Framework for Action. Forest Trends, Washington, DC.

Cotovio, V. 2019. Amazon Destruction Accelerates 60% to One and a Half Soccer Fields Every Minute. CNN. <https://edition.cnn.com/2019/07/02/americas/amazon-brazil-bolsonaro-deforestation-scli-intl/index.html> (accessed 19.12.19.).

Council on Ethics for the Government Pension Fund Global, 2013. Annual Report 2013. Council on Ethics for the Government Pension Fund Global, Norway.

Counsell, S., 2006. Forest Governance in the Democratic Republic of Congo. An NGO Perspective. FERN, The Netherlands.

CPF Secretariat, 2013. Promoting the Sustainable Management of All Types of Forests. The United Nations Forum on Forests Secretariat, New York.

Crippa, et al., 2019. Fossil CO_2 and GHG Emissions of All World Countries. 2019 Report. European Commission, Joint Research Centre. Publications Office of the European Union, Luxembourg.

D'Angelo, C., 2020a. Trump Is the Most Anti-Conservation President in History, Analysis Finds. HuffPost, <https://www.huffpost.com/entry/analysis-trump-public-lands-rollbacks_n_5ec59af1c5b6df8b159bf287> (accessed 16.07.20.).

D'Angelo, C., 2020b. Activists Slam Trump's Plan to Cut Public Out of Environmental Review Process. Huffpost. <https://www.huffpost.com/entry/trump-nepa-environmental-justice-rule-change_n_5e557ac0c5b63b9c9-ce498f4?guccounter = 2> (accessed 08.03.20.).

D'Angelo, C., Kaufman, A.C., 2020. Not Even a Pandemic Can Stop Trump From Pushing Fossil Fuels. Huffpost. <https://www.huffpost.com/entry/trump-pandemic-fossil-fuels_n_5e7288efc5b63c3b6489c2c2> (accessed 28.03.20.).

Daly, M., 2019. Trump Issues New Permit for Stalled Keystone XL Pipeline. Associated Press, <https://apnews.com/d62e3aaa796c402cbe29bcece6cee009> (accessed 13.12.19.).

Daly, N. 2020. Two rare white giraffes killed in Kenya. National Geogrpahic. <https://www.nationalgeographic.com/animals/2020/03/rare-white-giraffes-poached/> (accessed 16.07.20.).

da Matta, M.I.M., 2015. The Brazilian Forest Code. Three Years Later. Earth Day Network, Washington, DC. Available from: <http://www.earthday.org/2015/06/06/the-brazilian-forest-code-three-years-later/>.

Dargie, G.C., et al., 2017. Age, extent and carbon storage of the central Congo Basin peatland complex. Nature 542, 86–90.

Dauvergne, P., 2001. Loggers and Degradation in the Asia-Pacific. Cambridge University Press, Cambridge.

Davenport, C., Rubin, A.J., 2017. Trump Signs Executive Order Unwinding Obama Climate Policies. The New York Times, <https://www.nytimes.com/2017/03/28/climate/trump-executive-order-climate-change.html> (accessed 13.12.19.).

de Bolle, M., 2019. The Amazon Is a Carbon Bomb: How Can Brazil and the World Work Together to Avoid Setting It Off? PIIE Policy Brief 19-15. Peterson Institute for International Economics (PIIE), Washington, DC. Available from: <https://www.piie.com/sites/default/files/documents/pb19-15.pdf>.

Debroux, L., et al., (Eds.), 2007. Forests in Post-Conflict Democratic Republic of Congo. Analysis of a Priority Agenda. Center for International Forestry Research (CIFOR), Jakarta, The World Bank, Washington, DC; Centre International de Recherche Agronomique pour le Développement (CIRAD), Paris, France.

de Jong, W., Do, D.S., Trieu, V.H., 2006. Forest Rehabilitation in Vietnam: Histories, Realities and Future. The Center for International Forestry Research (CIFOR), Bogor.

Deng, Y.Q., 2018. Swimming Upstream. Beijing Review. <http://www.bjreview.com/Nation/201804/t20180428_800128189.html> (accessed 12.12.19.).

DENR, 1998. The First Philippine National Report to the Convention on Biological Diversity. Protected Areas and Wildlife Bureau. Department of Environment and Natural Resources (DENR), Republic of the Philippines.

DENR, 2012. Communities in Nature: State of Protected Areas Management in the Philippines. Protected Areas and Wildlife Bureau (PAWB). Department of Environment and natural Resource (DENR), The Philippines.

DENR, 2016. Philippine Biodiversity Strategy and Action Plan 2015–2018. Biodiversity Management Bureau. Department of Environment and Natural Resources, Republic of the Philippines.

DENR, 2018. Over 125,000 Hectares of Forests Rehabilitated in 2018—DENR. Department of Environment and Natural Resources (DENR), The Philippines. Available from: <https://www.denr.gov.ph/index.php/news-events/press-releases/738-over-125-000-hectares-of-forests-rehabilitated-in-2018-denr>.

DENR-UNEP, 1997. Philippine Biodiversity. An Assessment and Action Plan. Department of Environment and Natural Resources (DENR). United Nations Environment Programme (UNEP), The Philippines.

Denton, J., 2014. A Hunger for Coal Threatens the Heart of Borneo. The Jakarta Post, Indonesia, <http://www.thejakartapost.com/news/2014/05/20/a-hunger-coal-threatens-the-heart-borneo.html> (accessed 09.12.19.).

DFRS, 2015. State of Nepal's Forests. Forest Resource Assessment (FRA) Nepal. Department of Forest Research and Survey (DFRS), Kathmandu.

Dhiraj, A.B., 2017. Top 20 Most Polluting Rivers in the World: China's Yangtze and India's Ganges Tops. Geo World Magazine. <https://ceoworld.biz/2017/06/09/top-20-most-polluting-rivers-in-the-world-chinas-yangtze-and-indias-ganges-tops/> (accessed 12.12.19.).

Diamond, J., Kaufman, E., 2018. EPA Rolls Back Obama-Era Coal Pollution Rules as Trump Heads to West Virginia. CNN. <https://edition.cnn.com/2018/08/21/politics/epa-climate-power-plants-trump-west-virginia/index.html> (accessed 13.12.19.).

Diaz, R.J., Rosenberg, R., 2008. Spreading dead zones and consequences for marine ecosystems. Science 321, 921–929.

Díaz, et al., 2019. Summary for Policymakers of the Global Assessment Report on Biodiversity and Ecosystem Services of the Intergovernmental Science-Policy Platform on Biodiversity and Ecosystem Services. Intergovernmental Science-Policy Platform on Biodiversity and Ecosystem Services (IPBES), Bonn.

Diela, T., 2019. Indonesia Has Just Made Its Moratorium on Forest Clearance Permanent. World Economic Forum. <https://www.weforum.org/agenda/2019/08/indonesia-president-makes-moratorium-on-forest-clearance-permanent/> (accessed 09.12.19.).

Dimagiba, L., 2014. Too Many Exotic Species' in Philippine Greening Plan. SciDev.Net. <https://www.scidev.net/asia-pacific/forestry/news/too-many-exotic-species-in-philippine-greening-plan.html> (accessed 09.12.19.).

Di Minin, E., et al., 2015. Identification of policies for a sustainable legal trade in rhinoceros horn based on population projection and socioeconomic models. Conserv. Biol. 29, 545–555.

Ding, W., et al., 2006. Conservation of the baiji: no simple solution. Conserv. Biol. 20, 623–625.

Dipby Wells Environmental, 2015a. Herpetological Study for Feronia, Yaligimba Oil Palm Plantation. High Conservation Value Assessment. Digby Wells and Associates (South Africa) (Pty) Ltd. Available from: <https://www.feronia.com/uploads/2018-02-08/v3-herpetological-study-yaligimba-final80046.pdf>.

Dipby Wells Environmental, 2015b. Herpetological Study for Feronia, Boteka Oil Palm Plantation. High Conservation Value Assessment. Digby Wells and Associates (South Africa) (Pty) Ltd. Available from: <https://www.feronia.com/uploads/2018-02-08/v3-herpetological-study-boteka-final39414.pdf>.

Dlugokencky, Ed, Tans, P., 2016. Trends in Atmospheric Carbon Dioxide. National Oceanic and Atmospheric Administration (NOAA), U.S. Department of Commerce.

do Valle, R.S.T., 2018. Forests Specialist Group Reports on Outcomes of the 13th Session of the UN Forum on Forests. International Union for Conservation of Nature (ICUN), Gland, <https://www.iucn.org/news/world-commission-environmental-law/201807/forests-specialist-group-reports-outcomes-13th-session-un-forum-forests>.

Drollette, D., 2013a. A Plague of Deforestation Sweeps Across Southeast Asia. Yale Environment 360. Available from: <http://e360.yale.edu/feature/a_plague_of_deforestation_sweeps_across_southeast_asia/2652/>.

Drollette, D., 2013b. Gold Rush in the Jungle: The Race to Discover and Defend the Rarest Animals of Vietnam's "Lost World". Crown Publisher, New York.

Duckworth, J.W., et al., 2012. Why South-East Asia should be the world's priority for averting imminent species extinctions, and a call to join a developing cross-institutional programme to tackle this urgent issue. S.A.P.I.E. N.S Surv. Perspect. Integrat. Environ. Soc. 5 (2), 77–95.

Dunnivant, F.M., Anders, E., 2019. Pollutant Fate and Transport in Environmental Multimedia. John Wiley & Son, Hoboken, NJ.

Early, C., 2019. Can the Global Cement Industry Cut Its Carbon Emissions? Eco Business. <https://www.eco-business.com/news/can-the-global-cement-industry-cut-its-carbon-emissions/> (accessed 12.12.19.).

Earth Policy Institute, 2015. Climate, Energy, and Transportation: Carbon Dioxide Emissions From Fossil Fuel Burning in Top Ten Countries, 1950–2012. Earth Policy Institute, Rutgers University. Available from: <http://www.earth-policy.org/data_center/C23>.

Economist, The, 2019. On Thin ice COP25, the UN Climate Talks in Madrid, ENDS in a Sad Splutter. The Economist. <https://www.economist.com/science-and-technology/2019/12/15/cop25-the-un-climate-talks-in-madrid-ends-in-a-sad-splutter> (accessed 17.12.19.).

Edwards, P., 2015. The Rise and Potential Peak of Cement Demand in the Urbanized World. Cornerstone. Available from: <http://cornerstonemag.net/the-rise-and-potential-peak-of-cement-demand-in-the-urbanized-world/>.

EIA, 2016. Illegal Trade Seizures: Helmeted Hornbills. Mapping the Crimes. Environmental Investigation Agency (EIA), London. Available: <https://eia-international.org/illegal-trade-seizures-helmeted-hornbills>.

Elliott, S., Kuaraksa, C., 2008. Producing framework tree species for restoring forest ecosystem in northern Thailand. Small-Scale Forestry 7 (3), 403–415.

Emslie, R.H., et al., 2016. African and Asian Rhinoceroses—Status, Conservation and Trade. A Report From the IUCN Species Survival Commission (IUCN SSC) African and Asian Rhino Specialist Groups and TRAFFIC to the CITES Secretariat pursuant to Resolution Conf. 9.14 (Rev. CoP15).

EPA, 2019. Global Greenhouse Gas Emissions Data. United States Environmental Protection Agency (EPA). <https://www.epa.gov/ghgemissions/global-greenhouse-gas-emissions-data> (accessed 12.12.19.).

Erickson-Davis, M., 2018. The Importance of Keeping Forests Intact. Pacific Standard. <https://psmag.com/environment/preserving-the-worlds-forests> (accessed 09.12.19.).

ESCAP, 2011. Statistical Yearbook for Asia and the Pacific 2011. Economic and Social Commission for Asia and the Pacific (ESCAP), United Nations, Thailand.

European Environmental Agency, 2018. Increase in the Number and Size of Nationally Designated Protected Areas, 1838–2017. European Union. Available from:<https://www.eea.europa.eu/data-and-maps/daviz/growth-of-the-nationally-designated-3#tab-chart_4>.

Evans, K., 2013. How Much Credit Can Brazil Take for Slowing Amazon Deforestation—And How Low Can It Go? Forest News, Center for International Forestry Research (CIFOR), Bogor, <https://forestsnews.cifor.org/13491/how-much-credit-can-brazil-take-for-slowing-amazon-deforestation-and-how-low-can-it-go?fnl = en> (accessed 07.12.19.).

FAO, 2000. On Definitions of Forest and Forest Change. Food and Agriculture Organization of the United Nations (FAO), Rome.

FAO, 2001. Global Forest Resources Assessment 2000. Food and Agriculture Organization of the United Nations (FAO), Rome, FAO Forestry Paper 140.

FAO, 2007. The State of World Fisheries and Aquaculture 2006. Food and Agriculture Organization of the United Nations (FAO), Rome.

FAO, 2011. Global Forest Resources Assessment 2010. Trends in Extent of Primary Forest. FAO Forestry Paper 169. Food and Agriculture Organization of the United Nations (FAO), Rome.

FAO, 2012a. NFPs in Practice. Food and Agriculture Organization of the United Nations (FAO), Rome.

FAO, 2012b. FRA 2015 Terms and Definitions. Forest Resources Assessment Working Paper 180. Food and Agricultural Organization of the United Nations (FAO), Rome.

FAO, 2013. Forests Challenge Badge. Food and Agriculture Organization of the United Nations (FAO), Rome.

FAO, 2015. Global Forest Resources Assessment 2015. How Are the World Forest Changing? second ed. Food and Agriculture Organization of the United Nations (FAO), Rome.

FAO, 2016. The State of World Fisheries and Aquaculture. Contributing to Food Security and Nutrition for all. Food and Agriculture Organization of the United Nations (FAO), Rome.

FAO, 2017. National Forest Programme (NFP). Food and Agriculture Organization of the United Nations (FAO), Rome.

FAO, 2018. FAO Regional Office for Latin America and the Caribbean. Food and Agriculture Organization of the United Nations (FAO), Rome.

FAO, UNDP, and UNEP, 2010. National Programme Document—Paraguay. UN-REDD Programme 5th Policy Board Meeting 4–5 November 2010, Washington, DC. Food and Agriculture Organization (FAO), Rome; United Nations Development Programme (UNDP), New York; United Nations Environment Programme (UNEP), Nairobi.

Farand, C., 2019. Irreconcilable Rift Cripples UN Climate Talks as Majority Stand Against Polluters. Climate Home News Ltd., <https://www.climatechangenews.com/2019/12/15/irreconcilable-rift-dominates-un-climate-talks-majority-stand-polluters/> (accessed 17.12.19.).

Felbab-Brown, V., 2011. The Disappearing Act. The Illicit Trade in Wildlife in Asia. Foreign Policy at Brooking, Working Paper 6. Available from: <https://www.brookings.edu/wp-content/uploads/2016/06/06_illegal_wildlife_trade_felbabbrown.pdf>.

Fideler, D., 2014. Restoring the Soul of the World: Our Living Bond With Nature's Intelligence. Inner Traditions, Rochester, Vermont.

Fondation Ensemble, 2014. Mauritania: Biodiversity Conservation in the Diawling National Park Through Sustainable and Participatory Management. Fondation Ensemble. <https://www.fondationensemble.org/en/projet/conservation-de-la-biodiversite-du-parc-national-du-diawling-par-la-gestion-durable-et-participative/> (accessed 27.2.20.).

Franca, F., et al., 2016. Do space-for-time assessments underestimate the impacts of logging on tropical biodiversity? An Amazonian case study using dung beetles. J. Appl. Ecol. 53, 1098−1105.

Frank, T., 2020. Report Detailing U.S. Threats Ignores Climate Change. E&E News. <https://www.scientificamerican.com/article/report-detailing-u-s-threats-ignores-climate-change/> (accessed 20.02.20.).

Frazin, R., 2020. EPA Proposes Additional Rollback to Obama-Era Coal Ash Regulation. The Hill. <https://thehill.com/policy/energy-environment/483757-epa-proposes-additional-rollback-to-obama-era-coal-ash-regulation> (accessed 08.03.20.).

Fredericksen, T.S., et al., 2003. Sustainable Forestry in Bolivia. J. Forest. 101, 37−40.

Friedman, L., 2019a. U.S. Significantly Weakens Endangered Species Act. New York Times. <https://www.nytimes.com/2019/08/12/climate/endangered-species-act-changes.html> (accessed 16.07.20.).

Friedman, L., 2019b. Trump Serves Notice to Quit Paris Climate Agreement. The New York Times. <https://www.nytimes.com/2019/11/04/climate/trump-paris-agreement-climate.html> (accessed 13.12.19.).

Friedman, L., Plumer, B., 2017. E.P.A. Announces Repeal of Major Obama-Era Carbon Emissions Rule The New York Times. <https://www.nytimes.com/2017/10/09/climate/clean-power-plan.html> (accessed 13.12.19.).

Fry, I., et al., 1999. Summary of the third session of the Intergovernmental Forum on forests: 3−14 May 1999. Earth Negotiation Bull. 13 (55), 1−14.

FSIV (Forest Science Institute of Vietnam), 2009. Vietnam Forestry Outlook Study. Working Paper No. APFSOS II/WP/2009/09. FAO Regional Office for Asia and the Pacific, Bangkok. Available from: <http://www.fao.org/3/am254e/am254e00.pdf>.

Fu, C.Z., et al., 2003. Freshwater fish biodiversity in the Yangtze River basin of China: patterns, threats and conservation. Biodivers. Conserv. 12, 1649−1685.

FWI−GFW, 2002. The State of Forest: Indonesia. Forest Watch Indonesia (FWI), and Global Forest Watch (GFW), Bogor, Washington, DC.

Gabay, M., Dolter, S., Sa, M., 2014. Model forests in Argentina: creating place and time for participatory sustainable forest management. In: Katila, P., Galloway, G., de Jong, W., Pacheco, P., Mery, G. (Eds.), Forests Under Pressure—Local Responses to Global Issues. International Union of Forest Research Organizations (IUFRO), Vienna, pp. 49−69.

Galindo-Leal, C., de Gusmão Câmara, I. (Eds.), 2003. Atlantic forest hotspot status: an overview. The Atlantic Forest of South America: Biodiversity Status, Threats, and Outlook. Center for Applied Biodiversity Science. Conservation International and Island Press, Washington, DC, pp. 3−11.

Garbow, A., 2020. Trump Administration Is Rushing to Gut Environmental Protections. CNN. <https://edition.cnn.com/2020/04/10/opinions/trump-rushing-to-rollback-environmental-protections-during-pandemic-garbow/index.html> (accessed 15.04.20.).

Gates, V., 2013. Pacific Leatherback Turtle Faces Extinction in 20 Years. Reuters. <http://www.reuters.com/article/us-turtles-leatherback-idUSBRE91Q0VA20130227> (accessed 11.12.19.).

Gaworecki, M., 2017. New Carbon Map Will Help Protect the DRC's Rainforests. Mongabay.com. <https://news.mongabay.com/2017/06/new-carbon-map-will-help-protect-the-drcs-rainforests/> (accessed 09.12.19.).

Gelletly, L.A., 2014. Africa Progress and Problems: Ecological Issues. Mason Crest, Broomall, PA.

Germanos, A., 2020. We Can't Let Trump Roll Back 50 Years of Environmental Progress. EcoWatch. <https://www.ecowatch.com/fracking-colorado-federal--land-2645703280.html?rebelltitem = 1#rebelltitem1> (accessed 01.05.20.).

GJEP, 2017. 'Tsunamis From the Sky' Result of Deforestation in Argentina. Global Justice Ecology Project (GJEP), Buffalo, NY. Available from: <https://globaljusticeecology.org/tsunamis-from-the-sky-result-of-deforestation-in-argentina/>.

GLAD, 2018. Global—Forest Change 2000 to 2016—Global Land Analysis & Discovery. Global Land Analysis & Discovery Laboratory, University of Maryland. Available from: <https://oasishub.co/dataset/global-forest-change-2000-to-2016-global-land-analysis-discovery>.

Global Cement, 2013. China: First in Cement. Global Cement. <http://www.globalcement.com/magazine/articles/796-china-first-in-cement> (accessed 12.12.19.).

Global Footprint Network, 2019. Humanity's Ecological Footprint Contracted Between 2014–2016. Global Footprint Network. Available from: <https://www.footprintnetwork.org/2019/04/24/humanitys-ecological-footprint-contracted-between-2014-and-2016/>.

Global Forest Watch, 2019. Thailand. Dashboards. Global Forest Watch. <https://www.globalforestwatch.org> (accessed 09.12.19.).

Global Oxygen Network, 2018. The ocean is losing its breath: Declining oxygen in the world's ocean and coastal waters. In: Denise, B., Marilaure, G., Isensee, K. (Eds.), IOC-UNESCO, IOC Technical Series, No. 137 (IOC/2018/TS/137). The Intergovernmental Oceanographic Commission of UNESCO (IOC-UNESCO), France. Available from: <https://unesdoc.unesco.org/ark:/48223/pf0000265196>.

Global Tiger Initiative Secretariat, 2011. Global Tiger Recovery Program 2010–2022. The World Bank, Washington, DC.

Global Witness, 2015. The Threat to a Sustainable 2020 Tokyo Olympics Games Posed by Illegal and Unsustainable Logging in Sarawak, Malaysia. Global Witness, United Kingdom.

Goldenberg, S., 2013. 2013 in Review: Obama Talks Climate Change—But Pushes Fracking. The Guardian, United Kingdom. <https://www.theguardian.com/environment/2013/dec/20/2013-climate-change-review-obama-fracking> (accessed 12.12.19.).

Gomez, J., Karmini, N., 2019. Haze From Indonesian Fires Now Affecting Philippines. AP News. <https://apnews.com/ec862cbc41d84e008d37ed05137d02d2> (accessed 09.12.19.).

Gore, A., 2011. Al Gore: Climate of Denial. Can Science and the Truth Withstand the Merchants of Poison? Rolling Stone. <https://www.rollingstone.com/politics/politics-news/al-gore-climate-of-denial-244124/> (accessed 12.12.19.).

Green Climate Fund, 2019. Enhancing the Resilience to Climate Change of Livelihoods and Food Security of Agro-Sylvo-Pastoral Communities in Southwestern Mauritania. Green Climate Fund, Songdo, Incheon City, Republic of Korea.

Greenpeace, 2004. Malaysia Biodiversity Under Threat. Greenpeace International. Available from: <http://www.greenpeace.org/international/Global/international/planet-2/report/2004/1/malaysia-s-mega-diversity-unde.pdf>.

Greenpeace, 2007. Carving out the Congo. Greenpeace International., Amsterdam.

Greenpeace, 2009. Slaughtering the Amazon. Greenpeace International, United States.

Greenpeace, 2010. Swimming in Poison—An Analysis of Hazardous Chemicals in Yangtze River Fish. Greenpeace, Beijing. Available: <https://www.greenpeace.org/eastasia/Global/eastasia/publications/reports/toxics/2010/swimming-in-poison-yangtze-fish.pdf>.

Greenpeace, 2017. Cut From Congo. Industrial Logging and the Loss of Intact Forest Landscapes in the Congo Basin. Greenpeace International, Africa.

Greenpeace, 2019. Burning Down the House How Unilever and Other Global Brands Continue to Fuel Indonesia's Fires. Greenpeace International, Amsterdam.

Gregg, J.S., Andres, R.J., Marland, G., 2008. China: emissions pattern of the world leader in CO_2 emissions from fossil fuel consumption and cement production. Geogr. Res. Lett. 35, 1–5.

Greshko, M., 2017. What You Need to Know About Trump's Proposed Climate Cuts. National Geographic Society. <http://news.nationalgeographic.com/2017/03/trump-cuts-epa-noaa-environmental-science-climate-change-impacts/?utm_source = Facebook&utm_medium = Social&utm_content = link_fb20170310news-EPAimpact&utm_campaign = Content&sf61479938 = 1> (accessed 10.12.19.).

Greshko, M., Parker, L., Howard, B.C., 2018. A Running List of How Trump Is Changing the Environment. National Geographic Society. <https://news.nationalgeographic.com/2017/03/how-trump-is-changing-science-environment/> (accessed 12.12.19.).

GRID-Arendal, 2015. How, and How Much, Tropical Forests Absorb and Store Carbon. GRID-Arendal, Norway. Available from: <http://www.grida.no/graphicslib/detail/how-and-how-much-tropical-forests-absorb-and-store-carbon_7690>.

Groves, S., Schaefer, B., Loris, N., 2016. The U.S. Should Withdraw From the United Nations Framework Convention on Climate Change. The Heritage Foundation. Available from: <https://www.heritage.org/environment/report/the-us-should-withdraw-the-united-nations-framework-convention-climate-change>.

Guidi, R., 2016. Seven Million Hectares of Forests Have Been Lost in Argentina Over the Past 20 Years. Mongabay.com. <https://news.mongabay.com/2016/02/seven-million-hectares-of-forests-have-been-lost-in-argentina-in-the-past-20-years/> (accessed 08.12.19.).

Guillén, G., Lefebvre, B., 2019. Trump Administration to Seek Rollback of Methane Pollution Rule. Politico. <https://www.politico.com/story/2019/08/29/methane-pollution-rule-trump-administration-1692262> (accessed 13.12.19.).

Gunarso, P., et al., 2013. Oil palm and land use change in Indonesia, Malaysia and Papua New Guinea. In: Killeen, T.J., Goon, J. (Eds.), Reports From the Technical Panels of the 2nd Greenhouse Gas Working Group of the Roundtable on Sustainable Palm Oil (RSPO). Roundtable on Sustainable Palm Oil (RSPO), pp. 29–63.

Guynup, S., 2014. Illegal Tiger Trade: Why Tigers Are Walking Gold. National Geographic. <https://blog.nationalgeographic.org/2014/02/12/illegal-tiger-trade-why-tigers-are-walking-gold/> (accessed 25.2.20.).

Guzder-Williams, B., 2017. FORMA250: Global Forest Watch's Original Real-Time System Gets an Upgrade. Global Forest Watch, Washington, DC. Available from: <https://blog.globalforestwatch.org/data/forma250-global-forest-watchs-original-near-real-time-alert-system-gets-an-upgrade.html>.

Hall, J.E., et al., 2014. Model forests in Argentina: creating place and time for participatory sustainable forest management. In: Katila, P., Galloway, G., de Jong, W., Pacheco, P., Mery, G. (Eds.), Forests Under Pressure—Local Responses to Global Issues. IUFRO World Series, vol. 32. International Union of Forest Research Organizations (IUFRO), Vienna, pp. 49–69.

Hamilton, D., 1987. African elephants: population trends and their causes. Oryx 21 (1), 11–24.

Hance, J., 2009. Emotional Call for Palm Oil Industry to Address Environmental Problems. Mongay.com. <http://news.mongabay.com/2009/10/emotional-call-for-palm-oil-industry-to-address-environmental-problems> (accessed 12.12.19.).

Hance, J., 2012. Over 30 Yangtze Porpoises Found Dead in China as Population Nears Extinction. Mongabay.com. <http://news.mongabay.com/2012/0501-hance-yangtze-porpoise.html#ixzz1tdJ4hzme> (accessed 12.12.19.).

Hance, J., 2013. Yangtze Finless Porpoise Drops to Critically Endangered. Mongabay.com. <http://news.mongabay.com/2013/0707-hance-finless-porpoise-ce.html> (accessed 12.12.19.).

Hand, C., 2016. Dead Zones: Why Earth's Waters Are Losing Oxygen. Twenty-first Century Books, Minneapolis, MN.

Hanel, R., 2009. Tigers. Creative Publication, St. Paul, MN.

Hansen, M.C., et al., 2013a. High-resolution global maps of 21st-century forest cover change. Science 342 (6160), 850–853.

Hansen, M.C., et al., 2013b. Tree Cover Loss and Gain Area. University of Maryland, Google, USGS, and NASA. Accessed Through Global Forest Watch on [May 2018]. Available from: <https://www.globalforestwatch.org/countries/overview>.

Hargrave, J., 2012. Evaluation of the Action Plan to Prevent and Control the Deforestation in the Brazilian Amazon (2007–2010). (co-authors: Gomez, J.J., Maia, H., and Röper, M.). First European Environmental Evaluators' Network (EEEN) Forum held at HIVA-K.U. Leuven, Belgium, 9–10 February, 2012.

Harris, N., et al., 2016. Insider: Global Forest Watch and the Forest Resources Assessment, Explained in 5 Graphics. World Resources Institute (WRI), Washington, DC. Available from: <http://www.wri.org/blog/2016/08/insider-global-forest-watch-and-forest-resources-assessment-explained-5-graphics>.

Harris, et al., 2017. Using spatial statistics to identify emerging hot spots of forest loss. Environ. Res. Lett. 12, 1–13.

Harris, N.C., et al., 2019. First camera survey in Burkina Faso and Niger reveals human pressures on mammal communities within the largest protected area complex in West Africa. Conserv. Lett. 1–8. Available from: https://doi.org/10.1111/conl.12667.

Hayes, D.J., 2020. Trump Is Aggressively Pushing His Anti-Environment Agenda Amid a Pandemic. It's Inexcusable. Washington Post. <https://www.washingtonpost.com/opinions/2020/04/01/trump-is-aggressively-pushing-his-anti-environment-agenda-amid-pandemic-its-inexcusable/> (accessed 03.04.20.).

Hays, B., 2018. ESA Satellite Image Highlights Deforestation in Bolivia. United Press International, <https://www.upi.com/ESA-satellite-image-highlights-deforestation-in-Bolivia/5541521724170/> (accessed 09.12.19.).

He, J.K., 2016. Global low-carbon transition and China's response strategies. Adv. Clim. Res. 7, 204–212.

He, J.S., Xie, Z.G., 1995. The impact of the Three Gorges Hydroelectric Project on and the preservation strategies for the biodiversity in the affected region. Chin. Biodivers. 3, 63–72.

Hickey, V., et al., 2004. Crouching Tiger, Hidden Langur: World Bank Support to Biodiversity Conservation in East Asia and the Pacific, A Portfolio Review, 2004 (November). The World Bank, Washington, DC.

Hilton-Taylor, C., et al., 2009. State of the world's species. In: Vié, J.-C., Hilton-Taylor, C., Simon, N., Stuart, S.N. (Eds.), Wildlife in a Changing World. An Analysis of the 2008 IUCN Red List of Threatened Species. International Union for Conservation of Nature and Natural Resource (ICUN), Gland, pp. 5–42.

Hoang, T.Q., Thanh, P.N., Lan, L.V., 2017. Forest Governance in Vietnam: A Literature Review. UK Department for International Development (DFID) and the European Union.

Holden, E., 2019a. War on Science: Trump Administration Muzzles Climate Experts, Critics Say. The Guardian, United Kingdom. <https://www.theguardian.com/us-news/2019/jul/26/war-on-science-trump-administration-muzzles-climate-experts-critics-say> (accessed 13.12.19.).

Holden, M., 2019b. Trump Opens Protected Alaskan Arctic Refuge to Oil Drillers. The Guardian, United Kingdom. <https://www.theguardian.com/us-news/2019/sep/12/trump-arctic-national-wildlife-refuge-oil-gas-drilling> (accessed 13.12.19.).

Hu, A., 2014. China: Innovative Green Development. Springer, Heidelberg, New York, Dordrecht, London.

Hu, J.J., et al., 2009. Malformations of the endangered Chinese sturgeon, Acipenser sinensis, and its causal agent. PNAS (Proc. Natl. Acad. Sci. USA) 106 (23), 9339–9344.

Hu, F., Tan, D., Xu, Y.C., 2019. Yangtze Water Risks, Hotspots & Growth. China Water Risk, Hong Kong.

Humphrey, C., 2018. Tracing the Safeguards Against Illegal Logging in Vietnam. Mongabay.com. <https://news.mongabay.com/2018/12/tracing-the-safeguards-against-illegal-logging-in-vietnam/> (accessed 09.12.19.).

Humphrey, C., 2019. Indigenous Communities, Nat'l Parks Suffer as Malaysia Razes Its Reserves. Mongabay.com. <https://news.mongabay.com/2019/08/indigenous-communities-natl-parks-suffer-as-malaysia-razes-its-reserves/> (accessed 10.12.19.).

Hutchison, S., Aquino, L., 2011. Making a Pact to Tackle Deforestation in Paraguay. World Wildlife Fund (WWF), Paraguay.

Hovi, J., Sprinz, D.F., Bang, G., 2010. Why the United States did not become a party to the Kyoto Protocol: German, Norwegian and US perspectives. Eur. J. Int. Relat. 18 (1), 129–150.

Howard, J., 2019. Dead Zones, Explained. National Geographic. <https://www.nationalgeographic.com/environment/oceans/dead-zones/> (accessed 09.12.19.).

IBP, Inc, 2014. Brazil Ecology, Nature Protection Laws and Regulation Handbook, vol. 1. International Business Publication, Washington, DC.

IBRD, World Bank, 2011. Socialist Republic of Vietnam. Forest Law Enforcement and Governance. International Bank for Reconstruction and Development (IBRD), Washington, DC, The World Bank, Washington, DC.

ICAP, 2019. China National ETS. International Carbon Action Partnership (ICAP), Berlin.

ICSU and ISSC, 2015. Review of the Sustainable Development Goals: The Science Perspective. 2015. International Council for Science (ICSU). Paris. International Social Science Council (ISSC), France.

IEA, 2018. Global Energy & CO$_2$ Status Report 2017. International Energy Agency (IEA), OECD.

IEA, 2019. Global Energy & CO$_2$ Status Report. The Latest Trends in Energy and Emissions in 2018. International Energy Agency (IEA), OECD.

IMF, 2011. Islamic Republic of Mauritania: Poverty Reduction Strategy Paper, vol. I: PRSP 2006–2010. Post Implementation Review. International Monetary Fund, Washington, DC.

IMF, 2013. Report on Implementation of the Third PRSP Action Plan. Islamic Republic of Mauritania: Poverty Reduction Strategy Paper. International Monetary Fund, Washington, DC.

IMF, 2018. Islamic Republic of Mauritania. Economic Development Documents. International Monetary Fund, Washington, DC.

INCLUDEPROJECT, 2017. The Legal Framework for the Protection of Native Forests in Salta. Indigenous Communities Land Use and Tropical Deforestation (INCLUDE), Argentina. Available from: <https://includeproject.wordpress.com/author/includeproject/>.

INTERPOL, 2011. Call Issued to Save Wild Tigers From Extinction. INTERPOL, Lyon, <https://www.interpol.int/News-and-Events/News/2011/Call-issued-to-save-wild-tigers-from-extinction> (accessed 25.2.20.).

INTERPOL, 2013. Nepal Police Seize Tiger Parts and Arrest Seven During Intelligence-Led Actions. INTERPOL, Lyon, <https://www.interpol.int/News-and-Events/News/2013/Nepal-police-seize-tiger-parts-and-arrest-seven-during-intelligence-led-actions> (accessed 25.2.20.).

INTERPOL, 2018a. Global Wildlife Enforcement. Strengthening Law Enforcement Cooperation Against Wildlife Crime. INTERPOL, Lyon, France.

INTERPOL, 2018b. INTERPOL Progress Report on Asian Big Cat Trade. INTERPOL, Lyon. Available from: <https://cites.org/sites/default/files/eng/com/sc/70/E-SC70-51-A5.pdf>.

IPCC, 2014. Climate change 2014: mitigation of climate change. In: Edenhofer, O., et al., (Eds.), Contribution of Working Group III to the Fifth Assessment Report of the Intergovernmental Panel on Climate Change. Cambridge University Press, Cambridge, United Kingdom, New York, NY.

Iracambi, 2016. The Atlantic Rainforests. The Iracambi Research Center, Iracamni, Minas Gerais. <https://en.iracambi.com/about-us/where-we-are/the-atlantic-rainforest> (accessed 10.12.19.).

Isaacson, A., 2011. A New Species Bonanza in the Philippines. Science & Nature, Smithsonian Magazine. <http://www.smithsonianmag.com/science-nature/A-New-Species-Bonanza-in-the-Philippines.html> (accessed 11.12.19.).

ISC, 2012. ISC Released the First Guide for the Development of Low Carbon Zones in China. Institute for Sustainable Communities (ISC), Stone Cutters Way, Montpelier, VT.

IUCN, 1985. Vietnam. National Conservation Strategy. Prepared by the Committee for Rational Utilisation of Natural Resources and Environmental Protection. International Union for Conservation of Nature and Natural Resources (IUCN), Gland.

IUCN, 2005. Benefits Beyond Boundaries Proceedings of the Vth IUCN World Parks Congress. Durban, South Africa, 8–17 September 2003. International Union for Conservation of Nature and Natural Resources (IUCN), Gland.

IUCN, 2010. 50 Years of Working for Protected Areas. A Brief History of IUCN World Commission on Protected Areas. International Union for Conservation of Nature and Natural Resources (IUCN), Gland.

IUCN, 2012. Facts and Figures on Forests. The International Union for Conservation of Nature (IUCN), Gland.

IUCN, 2013a. Elephant Database: Continental Totals for Africa. International Union for Conservation of Nature and Natural Resources (IUCN), Gland. Available from: <http://www.elephantdatabase.org/report/2007/Africa>.

IUCN, 2013b. Rhinos in Crisis—Poaching and Illegal Trade Reach Highest Levels in Over 20 Years. The International Union for Conservation of Nature and Natural Resources (IUCN), Gland, <https://www.iucn.org/content/rhinos-crisis-%E2%80%93-poaching-and-illegal-trade-reach-highest-levels-over-20-years> (accessed 10.12.19.).

IUCN, 2015a. A Review of Thailand's Proposed Mae Wong Dam. International Union for Conservation of Nature and Natural Resources (IUCN), Gland.

IUCN, 2015b. The IUCN Red List of Threatened Species. The International Union for Conservation of Nature (IUCN), Gland. Available from: <https://cmsdocs.s3.amazonaws.com/keydocuments/IUCN_Red_List_Brochure_2015_LOW.pdf>.

IUCN, 2016a. New Bird Species and Giraffe Under Threat—IUCN Red List (Press Release) (December 8). International Union for Conservation of Nature and Natural Resources (IUCN), Gland, Switzerland. <http://www.iucnredlist.org/news/new-bird-species-and-giraffe-under-threat-iucn-red-list> (accessed 10.12.19.).

IUCN, 2016b. Four Out of Six Great Apes One Step Away From Extinction—IUCN Red List (September 4). International Union for Conservation of Nature and Natural Resources (IUCN), Gland. <https://www.iucn.org/news/four-out-six-great-apes-one-step-away-extinction-%E2%80%93-iucn-red-list> (accessed 10.12.19.).

IUCN, 2016c. Rhinoplax vigil. The IUCN Red List of Threatened Species. International Union for Conservation of Nature and Natural Resources (IUCN), Gland, <http://www.iucnredlist.org/details/22682464/0> (accessed 10.12.19.).

IUCN, 2016d. Pithecophaga jefferyi—IUCN Red List. International Union for Conservation of Nature and Natural Resources (IUCN), Gland, <http://www.iucnredlist.org/details/22696012/0> (accessed 11.12.19.).

IUCN, 2017. The IUCN Red List of Threatened Species. Version 2017-3. International Union for Conservation of Nature and Natural Resources (IUCN), Gland.

IUCN World Commission on Protected Areas, 2016. Parks. The International Journal of Protected Areas and Conservation, vol. 22.1. International Union for Conservation of Nature and Natural Resources (IUCN), Gland.

Jackson, P., Nowell, K., 2008. *Panthera tigris* ssp. balica. The IUCN Red List of Threatened Species 2008. International Union for Conservation of Nature and Natural Resources (IUCN), Gland. Available from: <https://doi.org/10.2305/IUCN.UK.2008.RLTS.T41682A10510320.en>.

Jackson, R.B., et al., 2018. Global energy growth is outpacing decarbonization. Environ. Res. Lett. 13 (12), 1−7.

Jain, et al., 2018. Securing safe havens for the Helmeted Hornbill *Rhinoplax vigil*. BirdingASIA 30, 26−32.

Jardeleza, J.M., Gotangco, C.K., Guzman, M.A.L., 2019. Simulating national-scale deforestation in the Philippines using land cover change models. Philippine J. Sci. 148 (4), 597−608.

JG, 2018. The Plight of the Sumatran Elephant. Jakarta Globe (JG), Indonesia, <https://jakartaglobe.id/vision/plight-sumatran-elephant/> (accessed 11.12.19.).

Jhala, Y.V., Qureshi, Q., Nayak, A.K. (Eds.), 2019. Status of Tigers, Co-Predators and Prey in India 2018. National Tiger Conservation Authority, Government of India, New Delhi & Wildlife Institute of India, Dehradun, Summary Report.

Jiang, H., 2014. The laws of climate change in China. Environ. Pract. 16 (3), 205−229.

Jiang, X.Y., 2015. Climate change and energy law. In: Qin, T.B. (Ed.), Research Handbook on Chinese Environmental Law. Edward Elgar, United Kingdom, pp. 162−195.

Kahfi, K., 2019. Look at the Amazon, Not Us: Indonesia Claims Handling Forest Fires Better Than Other Nations. The Jakarta Post, Indonesia, <https://www.thejakartapost.com/news/2019/12/07/look-at-the-amazon-not-us-indonesia-claims-handling-forest-fires-better-than-other-nations.html> (accessed 09.12.19.).

Kann, D., 2020. Australia Is Burning. The Arctic Is Melting. Yet Trump Keeps Gutting Climate Change Regulations. CNN. <https://edition.cnn.com/2020/01/10/politics/trump-climate-change-environmental-policy-rollbacks/index.html> (accessed 20.02.20.).

Kaufman, A.C., 2020. States Quietly Pass Laws Criminalizing Fossil Fuel Protests Amid Coronavirus Chaos. Huffpost. <https://www.huffpost.com/entry/pipeline-protest-laws-coronavirus_n_5e7e7570c5b6256a7a2aab41> (accessed 30.03.20.).

Keenan, R.J., et al., 2015. Dynamics of global forest area: results from the FAO Global Forest Resources Assessment 2015. For. Ecol. Manag. 352, 9−20.

Ker, T., 2011. Javan Rhino Extinct in Mainland Asia. National Geographic News. <http://news.nationalgeographic.com/news/2011/10/111028-vietnam-javan-rhinos-extinct-species-science-animals/> (accessed 11.12.19.).

Kiprop, J., 2017. 5 Countries With the Largest Rainforest Coverage. Worldatlas.com. Available from: <https://www.worldatlas.com/articles/5-countries-with-the-largest-rainforest-area.html>.

Klar, R., 2020. Trump Official Inserted Debunked Climate Change Language Into Scientific Documents: Report. The Hill. <https://www.vanityfair.com/news/2020/03/the-trump-administration-is-just-flat-out-lying-about-climate-change> (accessed 25.03.20.).

Knickmeyer, E., 2019. Trump Administration Sues to Block California's Climate Change Agreement With Quebec. Associated Press, <https://www.newscentermaine.com/article/news/nation-world/trump-vs-california-climate-emissions-lawsuit/507-05061f61-b34f-4380-809d-0de894c099da> (accessed 13.12.19.).

Knoema, 2020. Nepal—Gross Domestic Product per Capita Based on Purchasing-Power-Parity in Current Prices. Knoema. <https://knoema.com/atlas/Nepal/GDP-per-capita-based-on-PPP> (accessed 30.03.20.).

Kolbert, E., 2020. An Earth Day Reminder of How the Republicans Have Forsaken the Environment. The New Yorker. <https://www.newyorker.com/news/daily-comment/an-earth-day-reminder-of-how-the-republicans-have-forsaken-the-environment> (accessed 01.05.20.).

Kostka, G., 2016. Command without control: the case of China's environmental target system. Regul. Gov. 10, 8−74.

Kovacevic, M., Galindo, G.R., 2012. Brazil Forges Forward on Path to Sustainable Forest Development. Forest News, Center for International Forestry Research (CIFOR), Bogor, <https://forestsnews.cifor.org/8196/brazil-forges-forward-on-path-to-sustainable-forest-development?fnl = en> (accessed 07.12.19.).

Krishnasamy, K., Leupen, B., Or, O.C., 2016. Observations of the Helmeted Hornbill Trade in Lao PDR.TRAFFIC. Southeast Asia Regional Office, Selangor.

Krugman, P., 2019. Trump and His Party of Pollution Environmental destruction May Be Their Biggest Legacy. The New York Times. <https://www.nytimes.com/2019/11/14/opinion/trump-republicans-pollution.html> (accessed 13.12.19.).

Kumar, U., et al., 2019. Do conservation strategies that increase tiger populations have consequences for other wild carnivores like leopards? Nature 9, 1−8.

Kumari, K., 1995. Is Malaysian forest policy and legislation conducive to multiple-use forest management? Trade and Marketing of Forest Products. Unasylva No 183. Food and Agriculture Organization of the United Nations (FAO), Rome, pp. 51–56.

Laman, T., 2018. Poached for Its Horn, This Rare Bird Struggles to Survive. National Geographic. <https://www.nationalgeographic.com/magazine/2018/09/helmeted-hornbill-bird-ivory-illegal-wildlife-trade/> (accessed 11.12.19.).

Lang, C., 2001. Deforestation in Vietnam, Laos and Cambodia. In: Vajpeyi, D.K. (Ed.), Deforestation, Environment, and Sustainable Development: A Comparative Analysis. Praeger, Westport, CT, London, pp. 111–137.

Lavelle, M. 2016. Obama's Climate Legacy Marked by Triumphs and Lost Opportunities. Inside Climate News. <https://insideclimatenews.org/news/23122016/obama-climate-change-legacy-trump-policies> (accessed 12.12.19.).

Lawson, K., Vine, A., 2014. Global Impacts of the Illegal Wildlife Trade: The Costs of Crime, Insecurity and Institutional Erosion. The Royal Institute of International Affair, Chatham House, London. Available from: <https://www.chathamhouse.org/publications/papers/view/197367#>.

Leão, T.C.C., et al., 2014. Predicting extinction risk of Brazilian Atlantic forest angiosperms. Conserv. Biol. 28 (5), 1349–1359.

Leblond, J.-P., 2014. Thai forest debates and the unequal appropriation of spatial knowledge tools. Conserv. Soc. 12 (4), 425–436.

Lee, H.S., et al., 2002. The 52-Hectare Forest Research Plot at Lambir Hills, Sarawak, Malaysia. Tree Distribution Maps, Diameter Tables and Species Documentation. Forest Department Sarawak, Smithsonian Tropical Research Institute, Sarawak, Washington, DC. Available from: <http://www.ctfs.si.edu/data////Lambir_HIlls.pdf>.

Lehmann, E., et al., 2016. Trump's Election Could Threaten Global Climate Agreement. Scientific American. Available from: <https://www.scientificamerican.com/article/trump-s-election-could-threaten-global-climate-agreement/>.

Lewis, M., 2020. 3 States Pass Anti-Fossil-Fuel Protest Bills in 3 Weeks. It's Not Coincidental. Electrek. <https://electrek.co/2020/03/31/three-states-pass-anti-fossil-fuel-protest-bills-three-weeks/> (accessed 30.03.20.).

Li, R., Leung, G.C.K., 2012. Coal consumption and economic growth in China. Energy Policy 40, 438–443.

Liang, W., 2010. Changing climate? China new interest in global climate change negotiations. In: Kassiola, J.J., Guo, S. (Eds.), Global Political Impacts and Responses. Palgrave Macmillan, New York, pp. 61–84.

Liao, D., 2017. China's Hi-Tech Economy Is Taking Shape, Thanks to Its Entrepreneurs and Innovators. South Morning Post. <https://www.scmp.com/comment/insight-opinion/article/2095771/chinas-hi-tech-economy-taking-shape-thanks-its-entrepreneurs> (accessed 12.12.19.).

Linder, J.M., Palkovltz, R.E., 2016. The threat of oil palm expansion to primates and their habitats. In: Waller, M.T. (Ed.), Ethnoprimatology: Primate Conservation in the 21st Century. Springer International Publishing, Switzerland, pp. 21–45.

Liu, Z., 2015. China's Carbon Emissions Report 2015. Cambridge, MA: Report for Sustainability Science Program, Mossavar-Rahmani Center for Business and Government, Harvard Kennedy School; Energy Technology Innovation Policy Research Group, Belfer Center for Science and International Affairs, Harvard Kennedy School.

Liu, Z., et al., 2015. Reduced carbon emission estimates from fossil fuel combustion and cement production in China. Nature 524, 335–346.

Londoño, E., 2018. As Brazil's Far Right Leader Threatens the Amazon, One Tribe Pushes Back. New York Times. <https://www.nytimes.com/2018/11/10/world/americas/brazil-indigenous-mining-bolsonaro.html?module=inline> (accessed 07.12.19.).

López-Pujol, J., Zhang, F.M., Ge, S., 2006. Plant biodiversity in China: richly varied, endangered and in need of conservation. Biodivers. Conserv. 15, 3983–4026.

López-Pujol, J., Hua-Feng, W., Zhi-Yong, Z., 2011. Conservation of Chinese plant diversity: an overview. In: Pavlinov, I. (Ed.), Research in Biodiversity—Models and Applications. InTech Publication. Available from: <http://www.intechopen.com/books/research-in-biodiversity-models-and-applications/conservation-of-chinese-plant-diversity-an-over>.

Loris, N., 2016. Top 5 Reasons Congress Should Reject Obama's Climate Change Treaty. The Daily Signal. <https://www.dailysignal.com/2016/04/19/top-5-reasons-congress-should-reject-obamas-climate-change-treaty> (accessed 12.12.19.).

Lovgren, S., 2007. World's Largest River Fish Feared Extinct. National Geographic News. <http://news.national-geographic.com/news/2007/07/070726-china-fish.html> (accessed 12.12.19.).

Lund, H.G., 2014. What is a forest? Definitions do make difference. An example from Turkey. Avrasya Terim Dergisi 2 (1), 1—8.

Lutz, E., 2020. The Trump Administration Is Just Flat-Out Lying About Climate Change. Vanityfair. <https://www.thedailybeast.com/trump-official-keeps-adding-climate-denial-into-scientific-reports-says-report> (accessed 25.03.20.).

Ly, O.K., Zein, S.A.O.M., 2006. Building Viet Nam's Protected Areas System—Policy and Institutional Innovations Required for Progress: Policy Brief. United Nations Development Programme (UNDP), Hanoi.

Maala, C.P., 2001. Endangered Philippine wildlife species with special reference to the Philippine Eagle (*Pithecophaga jefferyi*) and Tamaraw (*Bubalus mindorensis*). J. Int. Dev. Coop. 8 (1), 1—17.

MAAP, 2019. MAAP #110: Major Finding—Many Brazilian Amazon Fires follow 2019. Monitoring of the Andean Amazon Project, Brazil. Available from: <https://maaproject.org/2019/amazon-fires-deforestation/>.

MacKinnon, K., et al., 1996. The Ecology of Kalimantan: Indonesian Borneo. Periplus Editions (HK) Ltd, Hong Kong.

Maguire, R., 2013. Global Forest Governance: Legal Concepts and Policy Trends. Edward Elgar, Cheltenham, Northampton, MA.

Maisels, et al., 2013. Devastating decline of forest elephants in Central Africa. PLoS One 8 (3), 1—13.

Malaysiakini, 2019. NGO Calls for Halt on Logging Near Mulu National Park. Malaysiakini. <https://www.malaysiakini.com/news/463906> (accessed 09.12.19.).

Maloney, M.P., Ward, M.P., 1973. Ecology. Let's hear from the people. Am. Psychol. 28, 583—586.

Manila Times, 2017. Why 'Rainforestation' Brings Back Healthier Forests. The Manila Times. <https://www.manilatimes.net/2017/08/31/business/green-business/rainforestation-brings-back-healthier-forests/347736/> (accessed 09.12.19.).

Mansourian, S., et al., 2014. A comparison of governance challenges in forest restoration in Paraguay's privately-owned forests and Madagascar's co-managed state forests. Forests 5 (4), 763—783.

Marco Elías, C.T., 2017. The Impact of Public Polices on Deforestation in the Brazilian Amazon. Dissertation. der Landwirtchaftlichen Fakultät der Rheinischen Friedrich-Wilhelms-Universität, Bonn.

Margono, B.A., et al., 2014. Primary forest cover loss in Indonesia over 2000—2012. Nat. Clim. Chang. 4, 730—735.

Mark, Y.A., Wei, L., Dennis Tao, Y., 2001. China's Great Leap: Forward or Backward? Anatomy of a Central Planning Disaster. Centre for Economic Policy Research, United Kingdom, Discussion Paper 2824.

Mason, J., 2017. China to Boost Non-Fossil Fuel use to 20 Percent by 2030: State Planner. Reuters. <https://www.reuters.com/article/us-china-energy/china-to-boost-non-fossil-fuel-use-to-20-percent-by-2030-state-planner-idUSKBN17R0QK> (accessed 12.12.19.).

Mayberry, K., 2018. Malayan Tiger in Crisis as Poaching Threatens to Wipe Out Big Cat. Aljazeera. <https://www.aljazeera.com/news/2018/12/malayan-tiger-crisis-poaching-threatens-wipe-big-cat-181214042412554.html> (accessed 11.12.19.).

Mayuga, J.L., 2019. Unique Species on the Brink of Extinction. Business Mirror, The Philippines. <https://businessmirror.com.ph/2019/03/11/unique-species-in-the-brink-of-extinction/> (accessed 11.12.19.).

McBeath, G.A., et al., 2014. Environmental Education in China. Edward Elgar, Cheltenham.

McCay, K., 2017. Argentina Named 9th Worst Country in Deforestation by the UN. The Bubble, Buenos Aires, <http://www.thebubble.com/argentina-named-9th-worst-country-in-deforestation-by-the-un/> (accessed 08.12.19.).

McGinley, M., 2013. Biological Diversity in the Philippines. Conservation International, Arlington, VA. Available from: <http://www.eoearth.org/view/article/150648>.

McGrath, M., 2017a. Paris Climate Deal: Trump Pulls US Out of 2015 Accord. BBC, London. <https://www.bbc.com/news/world-us-canada-40127326> (accessed 13.12.19.).

McGrath, M., 2017b. Trump Administration Approves Keystone XL Pipeline. BBC, London. <https://www.bbc.com/news/world-us-canada-39381324> (accessed 13.12.19.).

MEA (Millennium Ecosystem Assessment), 2005. Ecosystems and Human Well-being: Synthesis. Island Press, Washington, DC.

Mee, L., 2006. Reviving dead zones, Scientific American, 295, pp. 78–85. Available from: <http://faculty.bennington.edu/~sherman/the%20ocean%20project/reviving%20dead%20zones.pdf>.

Meijaard, et al., 2011. Quantifying killing of orangutans and human-orangutan conflict in Kalimantan, Indonesia. PLoS One 6 (11), 1–10.

Mendes, K., 2019. Deforestation Drops in Brazil's Atlantic Forest, but Risks Remain: Experts. Mongabay.com. <https://news.mongabay.com/2019/07/deforestation-drops-in-brazils-atlantic-forest-but-risks-remain-experts/> (accessed 10.12.19.).

Mendonça, E., 2019. Bolsonaro's Brazil Unlikely to Achieve Paris Agreement Goals: Experts. Mongabay.com. <https://news.mongabay.com/2019/09/bolsonaros-brazil-unlikely-to-achieve-paris-agreement-goals-experts/> (accessed 07.12.19.).

Milliken, T., et al., 2016. Addendum to the Elephant Trade Information System (ETIS) and the Illicit Trade in Ivory: A Report to the 17th Meeting of the Conference of the Parties to CITES [CoP17Doc.57.6 (Rev. 1)]. Convention on the International Trade in Endangered Species of Wild Fauna and Flora (CITES), Geneva, Switzerland.

Milner-Gulland, E.J., Beddington, J.R., 1993. The exploitation of elephants for the ivory trade: an historical perspective. Proc. R. Soc. Lond. 252, 29–37.

Ministry of Agriculture, Beijing, 1995. China: Country Report to the FAO International Technical Conference on Plant Genetic Resources. Ministry of Agriculture, Beijing.

Ministry of Development, Brunei, 2015. Brunei Darussalam's Intended Nationally Determined Contribution (INDC). Ministry of Development, Brunei Darussalam. Available from: <http://www4.unfccc.int/submissions/INDC/Published%20Documents/Brunei/1/Brunei%20Darussalam%20INDC_FINAL_30%20November%202015.pdf>.

Ministry of Ecology and Environment, 2017. The State Council Rolls Out a Three-Year Action Plan for Clean Air. Ministry of Ecology and Environment of the People's Republic of China.

Ministry of Ecology and Environment, China, 2018. China's Policies and Actions for Addressing Climate Change 2018. Ministry of Ecology and Environment, The People's Republic of China.

Ministry of Environment, Brazil, 1998. First National Report for the Convention on Biological Diversity. Ministry of Environment, Brazil.

Ministry of Environment, Brazil, 2004. Second National Report to the Convention on Biological Diversity. Ministry of Environment, Brazil.

Ministry of Environment, Brazil, 2005. Report: Brazil and the Conservation of the Amazon Forest. Ministry of Environment, Brazil.

Ministry of Environment, Brazil, 2010. Fourth National Report to the Convention on Biological Diversity. Ministry of Environment, Brazil.

Ministry of Environment, Brazil, 2015. Fifth National Report to the CBD. Ministry of Environment, Brazil.

Ministry of Environment, Brazil, 2016. National Biodiversity Strategy and Action Plan. Secretariat of Biodiversity and Forests, Ministry of Environment, Brazil.

Ministry of Environment and Forestry of Indonesia, 2014. The Fifth National Report of Indonesia to the Convention on Biological Diversity. Ministry of Environment and Forestry of Indonesia.

Ministry of National Development Planning, Indonesia, 1993. Biodiversity Action Plan for Indonesia. National Development Planning Agency. Ministry of National Development Planning, Indonesia.

Ministry of Natural Resources and Environment, Malaysia, 2005a. Second National Report to the Convention on Biological Diversity. Conservation and Environmental Management Division. Ministry of Natural Resources and Environment.

Ministry of Natural Resources and Environment, Malaysia, 2005b. Third National Report to the Convention on Biological Diversity. Conservation and Environmental Management Division. Ministry of Natural Resources and Environment, Malaysia.

Ministry of Natural Resources and Environment, Malaysia, 2006. Biodiversity in Malaysia. Ministry of Natural Resources and Environment., Malaysia.

Ministry of Natural Resources and Environment, Malaysia, 2009. 4th National Report to the Convention on Biological Diversity. Ministry of Natural Resources and Environment, Malaysia.

Ministry of Natural Resources and Environment, Malaysia, 2014. The Fifth National Report to Convention on Biological Resources. Biodiversity and Forestry Management Division, Ministry of Natural Resources and Environment.

Ministry of Natural Resources and Environment, Malaysia, 2016. Biodiversity and Forestry Management Division. Ministry of Natural Resources and Environment, Malaysia.

Ministry of Natural Resources and Environment, Thailand, 2006. Thailand. National Report on the Implementation of the Convention on Biology Diversity. Office of Natural and Environmental Policy and Planning. Ministry of Natural Resources and Environment, Thailand.

Ministry of Natural Resources and Environment, Vietnam, 2007. Socialist Republic of Vietnam. Third National Report. Ministry of Natural Resources and Environment.

Ministry of Natural Resources and Environment, Vietnam, 2008. 4th Country Report. Vietnam's Implementation of the Biodiversity Convention. Vietnam Environment Administration. Ministry of Natural Resources and Environment.

Ministry of Natural Resources and Environment, Vietnam, 2014. Vietnam's Fifth National Report to the United Nations Convention on Biological Diversity. Ministry of Natural Resources and Environment.

Ministry of Resources and Environment, Thailand, 2009. National Report on the Implementation of Convention on Biological Diversity. Office of Natural Resources and Environmental Policy and Planning, Ministry of Natural Resources and Environment, Thailand.

Ministry of Science, Technology and the Environment, Malaysia, 1998. Malaysia. First National Report to the Conference of the Parties of the Convention on Biological Diversity. Ministry of Science, Technology and the Environment, Malaysia.

Milman, O., 2017. Trump Budget Would Gut EPA Programs Tackling Climate Change and Pollution. The Guardian, United Kingdom. <https://www.theguardian.com/environment/2017/mar/16/trump-budget-cuts-climate-change-clean-up-programs-epa> (accessed 13.12.19.).

Mol Arthur, P.J., Carter, N.T., 2007. China environmental governance. In: Carter, N.T., Mol Arthur, P.J. (Eds.), Environmental Governance in China. Routledge, Oxon, pp. 1–22.

Monga, V., 2019. Trump Moves Again to Clear Path for Keystone XL Pipeline. Wall Street J. <https://www.wsj.com/articles/trump-issues-new-permit-for-keystone-xl-pipeline-11553894429> (accessed 13.12.19.).

Mongabay, 2009. Deforestation Jumps 55% in Vietnam Province. Mongabay.com. <http://news.mongabay.com/2009/0218-vietnam.html> (accessed 10.12.19.).

Mongabay, 2011. Philippines Forest Information and Data. Mongabay.com. <http://rainforests.mongabay.com/deforestation/2000/Philippines.htm> (accessed 11.12.19.).

Mongabay, 2012. Deforestation Accounts for 10 Percent of Global Carbon Emissions, Argues New Study. <https://news.mongabay.com/2012/06/deforestation-accounts-for-10-percent-of-global-carbon-emissions-argues-new-study/> (accessed 06.12.19.).

Mongabay, 2014. NASA Detects Surge in Deforestation in Malaysia, Bolivia During First Quarter of 2014. Mongabay.com. <https://news.mongabay.com/2014/04/nasa-detects-surge-in-deforestation-in-malaysia-bolivia-during-first-quarter-of-2014/> (accessed 09.12.19.).

Mongabay, 2018. Indonesian President Signs 3-Year Freeze on New Oil Palm Licenses. Mongabay.com. <https://news.mongabay.com/2018/09/indonesian-president-signs-3-year-freeze-on-new-oil-palm-licenses/> (accessed 09.12.19.).

Mongabay, 2019. Three Pangolin Species Closer to Extinction: IUCN. Monhabay.com. <https://news.mongabay.com/2019/12/three-pangolin-species-closer-to-extinction-iucn/> (accessed 20.12.19.).

Morton, K., 2006. Surviving an environmental crisis: can China adapt? Brown J. World Aff. 13 (1), 63–75.

Mpoyi, A.M., et al., 2013. The Context of REDD + in the Democratic Republic of Congo: Drivers, Agents and Institutions. Center for International Forestry Research (CIFOR), Bogor, Occasional Paper 94.

Mu, Z.L., Bu, S.C., Xue, B., 2014. Environmental legislation in China: achievements, challenges and trends. Sustainability 6, 8967–8979.

Muller, R., Pacheco, P., Carlos Montero, J., 2014. The Context of Deforestation and Forest Degradation in Bolivia. Drivers, Agents and Institutions. Center for International Forestry Research (CIFOR), Bogor, CIFOR Occasional Paper 108.

Munthe, B.C., Nangoy, F., 2019. Area burned in 2019 forest fires in Indonesia exceeds 2018—official. In: Schmollinger, C., (Ed.) Reuters. <https://www.reuters.com/article/us-southeast-asia-haze/area-burned-in-2019-forest-fires-in-indonesia-exceeds-2018-official-idUSKBN1X00VU> (accessed 09.12.19.).

Nabhitabhata, J., Chan-ard, T., 2005. Thailand Red Data: Mammals, Reptiles and Amphibians. Office of Natural Resources and Environmental Planning, Thailand.

Nalang, V.S., 2003. Indonesia Biodiversity Action Plan 2003–2020. Ministry of Environment, Indonesia.

NASA, 2013. NASA-USGS Landsat Data Yield Best View to Date of Global Forest Losses, Gains. National Aeronautics and Space Administration (NASDA), Washington, DC.

NASA Earth Observatory, 2015. Smoke Blankets Indonesia. NASA Earth Observatory, Washington, DC. Available from: <https://earthobservatory.nasa.gov/images/86681/smoke-blankets-indonesia>.

Nash, S., 2019. Vietnam's Empty Forest. New York Times. <https://www.nytimes.com/2019/04/01/travel/vietnam-wildlife-species-ecotravel-tourism.html> (accessed 09.12.19.).

National Environmental Agency, Vietnam, 2001. Vietnam: Second National Report. National Environmental Agency, Vietnam.

National Planning Commission, Nepal, 2011. Nepal Status Paper United Nations Conference on Sustainable Development 2012 (Rio + 20). National Planning Commission Government of Nepal Singhadurbar, Kathmandu.

Naughton, B., 1993. Deng Xiaoping: the economist. China Q. 135, 491–514.

NDRC, 2012. The People Republic of China National Report on Sustainable Development. National Development and Reform Commission (NDRC), The People Republic of China.

NDRM, 2009. China's Policies and Actions for Addressing Climate Change 2009. National Development and Reform Commission (NDRM), Beijing.

NESDB, 2017. The Twelfth National Economic and Social Development Plan (2017–2021). Office of National Economic and Social Development Board (NESDB), Prime Minister Office, Bangkok. Available from: <http://www.nesdb.go.th/nesdb_en/ewt_w3c/ewt_dl_link.php?nid = 4345>.

Newburger, E., 2020. Trump weakens environmental law to speed up permits for pipelines and other infrastructure. CNBC. <https://www.cnbc.com/2020/07/15/trump-to-weaken-national-environmental-policy-act.html> (accessed 16.07.20.).

Nguyen, Q.T., 2005. Trends in forest ownership, forest resources tenure and institutional arrangements. Are they contributing to better forest management and poverty reduction? Case study from Viet Nam. Understanding Forest Tenure in South and Southeast Asia. Forestry Policy and Institutions Working Paper. Food and Agriculture Organization (FAO), Rome, pp. 355–407.

Nguyen, Q.T., 2008. The household economy and decentralization of forest management in Vietnam. In: Pierce Colfer, C.J., Dahal, G.R., Doris Capistrano, D. (Eds.), Lessons From Forest Decentralization: Money, Justice, and the Quest for Good Governance in Asia-Pacific Justice and the Quest for Good. Earthscan, United Kingdom, United States, pp. 187–209.

Nicholas Jong, H., 2018. Indonesia Launches Bid to Restore National Park That's Home to Tigers, Elephants. Mongabay.com. <https://news.mongabay.com/2018/03/indonesia-launches-bid-to-restore-national-park-thats-home-to-tigers-elephants/> (accessed 10.12.19.).

Nicholas Jong, H., 2019a. Indonesian Court Fines Palm Oil Firm $18.5 m Over Forest Fires in 2015. Mongabay.com. <https://news.mongabay.com/2019/10/palm-oil-indonesia-arjuna-utama-sawit-musim-mas-forest-fires/> (accessed 09.12.19.).

Nicholas Jong, H., 2019b. Indonesia Forest-Clearing Ban Criticized as 'Government Propaganda'. Mongabay.com. <https://www.ecowatch.com/indonesia-deforestation-ban-propaganda-2639853596.html> (accessed 09.12.19.).

Nicholas Jong, H., 2019c. Indonesia Fires Emitted Double the Carbon of Amazon Fires, Research Shows. Mongabay.com. <https://news.mongabay.com/2019/11/indonesia-fires-amazon-carbon-emissions-peatland/> (accessed 09.12.19.).

Nicholas Jong, H., 2019d. RSPO Questions Effectiveness of Indonesian Palm Plantation Moratorium. Mongabay.com. <https://news.mongabay.com/2019/11/rspo-indonesia-palm-oil-plantations-moratorium/> (accessed 09.12.19.).

Nilsson, K., 2001. The Proposals for Action Submitted by the Intergovernmental Panel on Forests (IPF) and the Intergovernmental Forum on Forests (IFF)—in the Swedish Context. The National Board of Forestry, Sweden.

Niu, C.H., 1958. China Will Overtake Britain. Foreign Language Press, Peking.

NOAA, 2015. Gulf of Mexico Dead Zone 'Above Average'. National Oceanic and Atmospheric Administration (NOAA). United States Department of Commerce, Washington, DC, <http://www.noaanews.noaa.gov/stories2015/080415-gulf-of-mexico-dead-zone-above-average.html> (accessed 10.12.19.).

Nuberg, I.K., Shrestha, K.K., Bartlett, A.G., 2019. Pathways to forest wealth in Nepal in Nepal. Australian Forestry 82 (supp1), 106–120. Available from: https://doi.org/10.1080/00049158.2019.1614805.

Nuccitelli, D., 2018. Trump's Disbelief Won't Stop Dangerous Climate Change. The Guardian, United Kingdom. <https://www.theguardian.com/environment/2018/dec/05/trumps-disbelief-wont-stop-dangerous-climate-change> (accessed 23.12.19.).

NYDF Assessment Partners, 2019. Protecting and Restoring Forests: A Story of Large Commitments yet Limited Progress. New York Declaration on Forests Five-Year Assessment Report. Climate Focus (Coordinator and Editor). Available from: <forestdeclaration.org>.

Nyingi, D.W., Lebădă, A.-M., Ripley, K., 2019. Summary of the fourteenth session of the united nations forum on forests: 6−10 May 2019. Earth Negotiation Bull. 13 (215), 1−15.

Obidzinski, K., Andrianto, A., Wijaya, C., 2007. Cross-border timber trade in Indonesia: critical or overstated problem? Forest governance lessons from Kalimantan. Int. Forest. Rev. 9 (1), 526−535.

Ochoa-Quintero, J.M., et al., 2015. Threshold of species loss in Amazonian deforestation frontier landscapes. Conserv. Biol. 29 (2), 440−451.

OECD, 2015. Material Resources, Productivity and the Environment. OECD Publishing, Paris, <https://doi.org/10.1787/9789264190504-en>.

OEPP, Thailand, 2000. Biodiversity Conservation in Thailand: A National Report. Implementation of Article 6 of the Convention on Biological Resources. Biological Resources Section. Natural Resources and Environmental Management Division. Office of Environmental Policy and Planning (OEPP), Thailand.

OEPP, Thailand, 2002. National Report on the Implementation on Biological Diversity. Office of Environmental Policy and Planning (OEPP). Ministry of Science, Technology and Environment (present is Ministry of Natural Resources and Environment), Thailand.

Ohlheiser, A., 2016. Trump Didn't Delete His Tweet Calling Global Warming a Chinese Hoax. The Washington Post. <https://www.washingtonpost.com/news/the-intersect/wp/2016/09/27/trump-didnt-delete-his-tweet-calling-global-warming-a-chinese-hoax/?utm_term = .884b6ec1f772> (accessed 12.12.19.).

Oliver, W.L.R., 2006. Philippines biodiversity conservation programme: integrating institutional partnerships and practical conservation measures in some of the world's highest priority areas. In: Hiddinga, B. (Ed.), Proceedings of the EAZA Conference 2005. EAZA Executive Office, Amsterdam, pp. 249−263.

Olivier, J.G.J., Peters, J.A.H.W., 2018. Trends in Global CO_2 and Total Greenhouse Gas Emissions 2018 Report. PBL Netherlands Environmental Assessment Agency, The Hague.

Olivier, J.G.J., et al., 2016. Trends in Global CO_2 Emissions: 2016 Report. PBL Netherlands Environmental Assessment Agency, The Hague.

Olivier, J.G.J., et al., 2017. Trends in Global CO_2 and Total Greenhouse Gas Emissions 2017 Report. PBL Netherlands Environmental Assessment Agency, The Hague.

O'Neill, T., 2013. Why African Rhinos Are Facing a Crisis. National Geographic Society. <https://news.national-geographic.com/news/2013/02/130227-rhino-horns-poaching-south-africa-iucn/> (accessed 10.12.19.).

ONEP, 2015. Master Plan for Integrated Biodiversity Management B.E.2558−2564 (2015−2021). Office of Natural Resources and Environmental Policy and Planning (ONEP), Natural Resources and Environmental Management Division, Thailand. Available from: <https://www.cbd.int/doc/world/th/th-nbsap-v4-en.pdf>.

ONEP, 2019. Thailand's Sixth National Report on the Implementation of the Convention on Biological Diversity. Office of Natural Resources and Environmental Policy and Planning (ONEP), Natural Resources and Environmental Management Division, Thailand. Available from: <https://www.cbd.int/doc/nr/nr-06/th-nr-06-en.pdf>.

ONEP, Thailand, 2014. Thailand Fifth National Report on the Implementation of the Convention on Biological Diversity. Office of Natural Resources and Environmental Policy and Planning (ONEP), Ministry of Natural Resources and Environment. Available from: <https://www.cbd.int/doc/world/th/th-nr-05-en.pdf>.

Ongprasert, P., 2011. Forest Management in Thailand. International Forestry Cooperation Office Royal Forest Department. Ministry of Natural Resources and Environment, Thailand.

Overly, S., Eilperin, J., 2017. President Trump to Reopen Review of Obama-Era Fuel Economy Standards. The New York Times. <https://www.washingtonpost.com/news/innovations/wp/2017/03/13/president-trump-to-reopen-review-of-obama-era-fuel-economy-standards/> (accessed 13.12.19.).

Pacheco, P., 2017. Decoding Deforestation in Brazil and Bolivia. CIFOR Forest News. Center for International Forestry Research (CIFOR), Bogor, <https://forestsnews.cifor.org/49057/decoding-deforestation-in-brazil-and-bolivia?fnl = en> (accessed 07.12.19.).

Palma, R.A., 2016. Philippine Biodiversity: Issues, Challenges, and Initiatives. Department of Environment and Natural Resources. Biodiversity Management Bureau (BMB).

Partono, S., 2011. Action Plan for Implementing the Convention on Biological Diversity's Programme of Work on Protected Areas. Ministry of Forestry, Jakarta. Available from: <https://www.cbd.int/doc/meetings/mar/cbwsoi-seasi-01/other/cbwsoi-seasi-01-indonesia-en.pdf>.

Peltonen, M., 1992. Politics and science: Francis Bacon and the true greatness of states. Hist. J. 35 (2), 279−305.

Petersen, R., 2016. Global Forest Watch: Transforming Big Data Into Big Decisions for Forests. Global Forest Watch, Washington, DC. Available from: <https://www.rnrf.org/2016cong/Petersen.pdf>.

Petersen, R. et al. 2015. Satellites Uncover 5 Surprising Hotspots for Tree Cover Loss. Global Forest Watch. Available from: <http://blog.globalforestwatch.org/data/2014-tree-cover-loss-2.html>.

Petersen, R., et al., 2016. Mapping Tree Plantations With Multispectral Imagery: Preliminary Results for Seven Tropical Countries. Technical Note. World Resources Institute, Washington, DC. Available from: <https://www.wri.org/sites/default/files/Mapping_Tree_Plantations_with_Multispectral_Imagery_-_Preliminary_Results_for_Seven_Tropical_Countries.pdf>.

Petras, J., Veltmeyer, H., 2014. The New Extractivism: A Post-Neoliberal Development Model or Imperialism of the Twenty-First Century? Zed Books, London, New York.

Perkins, D.H., 1967. Economic growth in China and the cultural revolution (1960−April 1967). China Q. 30, 33−48.

Perkins, D.H., 1991. China's economic policy and performance. In: MacFarquhar, R., Fairbank, J.K. (Eds.), The Cambridge History of China. Cambridge University Press, Cambridge, pp. 473−539.

Pham, M.T., 2013. REDD + Benefit Sharing in Vietnam. Background Paper for Field Dialogue on REDD + Benefit Sharing 24−27 September, 2013. The Forest Dialogue. <http://theforestsdialogue.org/dialogue/field-dialogue-redd-benefit-sharing-vietnam> (accessed 09.12.19.).

Pham, T.T., et al., 2012. The Context of REDD + in Vietnam. Drivers, Agents and Institutions. Center for International Forestry Research (CIFOR), Bogor, Occasional Paper 75.

Phillips, T., Harvey, F., Yuhas, A., 2016. Breakthrough as US and China Agree to Ratify Paris Climate Deal. The Guardian, United Kingdom. <https://www.theguardian.com/environment/2016/sep/03/breakthrough-us-china-agree-ratify-paris-climate-change-deal> (accessed 13.12.19.).

Phys.org, 2019. Deforestation in Brazil's Amazon Up by More Than Double: Data. Phys.org. <https://phys.org/news/2019-12-deforestation-brazil-amazon.html> (accessed 19.12.19.).

Piesse, M., 2019. Chinese Air and Water Pollution Levels Continue to Decline. Future Directions International, <http://www.futuredirections.org.au/publication/chinese-air-and-water-pollution-levels-continue-to-decline/> (accessed 12.12.19.).

Pillai, V., 2017.Orang Asli Under Threat as Loggers 'Skirt' Court Ruling. Malaysiakini. <https://www.malaysiakini.com/news/387833> (accessed 09.12.19.).

Platt, J.R., 2015. Ring-Tailed Lemurs Threatened by Illegal Pet Trade. Scientific American, <https://blogs.scientificamerican.com/extinction-countdown/ring-tailed-lemurs-pet-trade/> (accessed 10.12.19.).

Potapov, P.V., et al., 2012. Quantifying forest cover loss in Democratic Republic of the Congo, 2000−2010, with Landsat ETM + data. Remote Sens. Environ. 122, 106−116.

Popescu, A., 2018. No More Elephants? Poaching Crisis Takes Its Toll in the Central African Republic. Mongabay.com. <https://news.mongabay.com/2018/01/no-more-elephants-poaching-crisis-takes-its-toll-in-the-central-african-republic/> (accessed 10.12.19.).

PRB, 2006. Making the Link in the Philippines. Population, Health, and the Environment. Population Reference Bureau (PRB), Washington, DC. Available from: <http://www.prb.org/pdf06/05MakingtheLinkPhilippines.pdf>.

Pritam, S., 2017. Climate Change in Trump Times. Alternatives International, Montreal. <http://www.alterinter.org/spip.php?article4560> (accessed 12.12.19.).

Qin, T.B., 2014. Challenges for sustainable development and its response in China: a perspective for social transformation. Sustainability 6, 5075−5106.

Qiu, J., 2012. Yangtze Finless Porpoises in Peril. Nature. Available from: <http://www.nature.com/news/yangtze-finless-porpoises-in-peril-1.12125>.

Rainey, J., 2018. The Trump Administration Scrubs Climate Change Info From Websites. These Two Have Survived. NBC News. <https://www.nbcnews.com/news/us-news/two-government-websites-climate-change-survive-trump-era-n891806> (accessed 13.12.19.).

Rakyat Post, 2014. DAP: Why Was Oil Palm Project Approved in Forest Reserve in Sabah? Rakyat Post, Kuala Lumpur. <http://www.therakyatpost.com/news/2014/11/27/dap-oil-palm-project-approved-forest-reserve-sabah/> (accessed 12.12.19.).

Raman, M., 2002. An analysis of Malaysia's Implementation of the Convention on Biological Diversity With a Focus on Forests. Fern, United Kingdom.

Raufer, R., Wang, S.J., 2003. Navigating the policy path for support of wind power in China. China Environ. Ser. 6, 37–54.

Rebugio, L.L., et al., 2007. Forest restoration and rehabilitation in the Philippines. In: Don, K.L. (Ed.), Keep Asia Green, vol. 1. Southeast Asia. Austria, IUFRO World Series, Vienna.

REDD-Monitor, 2016. Democratic Republic of Congo Threatens to Open Forests to Industrial Logging. <https://redd-monitor.org/2016/03/02/democratic-republic-of-congo-threatens-to-open-forests-to-industrial-logging/> (accessed 09.12.19.).

Refkin, A., Cray, S., 2013. Conducting Business in the Land of the Dragon: What Every Businessperson Needs to Know About China. iUniverse, Bloomington, IN.

Reilly, S., E&E News, 2018. Trump's EPA Scraps Air Pollution Science Review Panels. Science. <https://www.sciencemag.org/news/2018/10/trump-s-epa-scraps-air-pollution-science-review-panels> (accessed 13.12.19.).

Republic of Indonesia, 1999. Law of the Republic of Indonesia Number 41 of 1999 Regarding Forestry. Republic of Indonesia. Available from: <http://theredddesk.org/sites/default/files/uu41_99_en.pdf>.

Reuters, 2018. China Prohibits Glass, Cement Capacity Expansion in 2018. Reuters. <https://www.reuters.com/article/us-china-commodities-cement/china-prohibits-glass-cement-capacity-expansion-in-2018-idUSKBN1FW066> (accessed 12.12.19.).

Reuters, 2019. China's Renewable Power Capacity Up 9.5% Year-on-Year in June. Reuters. <https://www.reuters.com/article/us-china-renewables/chinas-renewable-power-capacity-up-9-5-year-on-year-in-june-idUSKCN1UK1MF> (accessed 12.12.19.).

Reuters, 2020. White House Unveils Plan for Major Projects to Bypass Environmental Review. The Guardian, United Kingdom. <https://www.theguardian.com/environment/2020/jan/09/white-house-projects-permits-climate-impact-plan> (accessed 20.02.20.).

Revesz, R.L., 2020. Trump Shows His Cards on Environmental Protections—or a Lack Thereof. The Hill. <https://thehill.com/opinion/energy-environment/495457-trump-shows-his-cards-on-environmental-protections-or-lack-thereof> (accessed 02.05.20.).

Richter, W., 2019. Crude Steel Production: China Knocks the Socks Off Rest of the World. Wolf Street. <https://wolfstreet.com/2019/06/21/crude-steel-production-china-v-the-rest-of-the-world/> (accessed 12.12.19.).

Ripley, K., et al., 2018. Thirteenth Session of the United Nations Forum on Forests: 7–11 May 2018. Earth Negotiation Bull. 13 (214), 1–14.

Riskin, C., 1987. China's Political Economy. Oxford University Press, Oxford.

Ritchie, H.R., Roser, M. 2019. CO_2 and Greenhouse Gas Emissions. Our World in Data. Available from: <https://ourworldindata.org/co2-and-other-greenhouse-gas-emissions>.

Rock, M.T., Toman, M.A., 2015. China's Technological Catch-Up Strategy: Industrial Development, Energy Efficiency, and CO_2 Emissions. Oxford University Press, New York, Oxford.

Rodgers, L., 2018. Climate Change: The Massive CO_2 Emitter You May Not Know About. BBC. <https://www.bbc.com/news/science-environment-46455844> (accessed 12.12.19.).

Rosen, T., 2011. INTERPOL, World Bank Launch Project Predator. International Institute for Sustainable Development (IISD), Canada, <https://sdg.iisd.org/news/interpol-world-bank-launch-project-predator/> (accessed 24.2.20.).

Ross, J., 2020. Trump Official Keeps Adding Climate Denial Into Scientific Reports, Says Report. The Daily Beast. <https://www.thedailybeast.com/trump-official-keeps-adding-climate-denial-into-scientific-reports-says-report> (accessed 25.03.20.).

Ross, L., Silk, M.A., 1987. Environmental Law and Policy in the People's Republic of China Quorum Books, New York, London.

RSPB, 2009. Gurney's Pitta Research and Conservation in Thailand and Myanmar. Final Report of Darwin Project 162/13/030. The Royal Society for the Protection of Birds (RSPB), United Kingdom. Available from: <https://www.darwininitiative.org.uk/documents/13030/14067/13-030%20FR%20-%20edited.pdf>.

RTE, 2019. Deforestation in Brazil's Amazon Up by 104%—Data. Raidió Teilifís Éireann (RTE), Ireland. <https://www.rte.ie/news/2019/1214/1099159-deforestation-amazon-brazil/> (accessed 19.12.19.).

Rudqvist, A., Woodford-Berger, P., 1996. Evaluation and Participation: Some Lessons. SIDA Studies in Evaluation 96/1. SIDA's Department for Evaluation and Internal Audit, Swedish International Development Cooperation Agency (SIDA), Stockholm.

Russell, K., 2017. Landsat 9 Satellite Progressing on Schedule, Says Orbital ATK. Via Satellite, Rockville, MD. Available from: <https://www.satellitetoday.com/government-military/2017/08/09/landsat-9-satellite-pro-gressing-schedule-says-orbital-atk/>.

Saatchi, et al., 2017. Carbon Map of DRC. High Resolution Carbon Distribution in Forests of Democratic Republic of Congo. A Summary Report of UCLA Institute of Environment & Sustainability. UCLA Institute of Environment & Sustainability, Los Angeles, CA.

Said, M.Y., et al., 1995. African Elephant Database 1995. International Union for Conservation of Nature (IUCN), Gland.

Salim, E., Ullsten, O., 1999. Our Forests, Our Future. Report of the World Commission on Forests and Sustainable Development. Cambridge University Press, Cambridge.

Sample, I., 2012. Amazon's Doomed Species Set to Pay Deforestation's 'Extinction Debt'. The Guardian, United Kingdom. <https://www.theguardian.com/environment/2012/jul/12/amazon-deforestation-species-extinc-tion-debt> (accessed 10.12.19.).

Sandalow, D., 2018. Guide to Chinese Climate Policy 2018. Center on Global Energy Policy, Columbia.

Sanger, D.E., 2001. Bush Will Continue to Oppose Kyoto Pact on Global Warming. New York Times. <https://www.nytimes.com/2001/06/12/world/bush-will-continue-to-oppose-kyoto-pact-on-global-warming.html> (accessed 12.12.19.).

Sarawak Forestry Corporation, 2019. National Parks and Natural Reserves. Sarawak Forestry Corporation, Sarawak, <https://www.sarawakforestry.com/national parks/> (accessed 12.12.19.).

Sarawak Report, 2018. Kelantan—Creed or Greed? Sarawak Report. Available from: <http://www.sarawakre-port.org/2018/08/kelantan-creed-or-greed/>.

Save the Rhino, 2019. Poaching Statistics. Save the Rhino International, London. <https://www.savetherhino.org/rhino_info/poaching_statistics> (accessed 10.12.19.).

Savransky, R., 2018. FEMA Eliminates Mentions of Climate Change From Strategic Planning Document. The Hill. <https://thehill.com/policy/energy-environment/378747-fema-eliminates-mentions-of-climate-change-from-strategic-planning> (accessed 13.12.19.).

Schiffman, R., 2015. Brazil's Deforestation Rates Are on the Rise Again. Newsweek. <http://www.newsweek.com/2015/04/03/brazils-deforestation-rates-are-rise-again-315648.html> (accessed 07.12.19.).

Schochet, J., 2018. Rainforest Primer: Biodiversity—How Much Biodiversity Is Found in Tropical Rainforests? Rainforest Conservation Fund, Chicago, IL.

Schoene, D., et al., 2007. Forests and Climate Change Working Paper 5. Definitional Issues Related to Reducing Emissions From Deforestation in Developing Countries. Food and Agriculture Organization of the United Nations (FAO), Rome.

Schrier-Uijl, A.P., et al., 2013. Environmental and social impacts of oil palm cultivation on tropical peat. A scientific review. In: Killeen, T.J., Goon, J. (Eds.), Reports From the Technical Panels of the 2nd Greenhouse Gas Working Group of the Roundtable on Sustainable Palm Oil (RSPO). Roundtable on Sustainable Palm Oil (RSPO), pp. 131—168.

Schwartz, J., 2019. Major Climate Change Rules the Trump Administration Is Reversing. The New York Times. <https://www.nytimes.com/2019/08/29/climate/climate-rule-trump-reversing.html> (accessed 13.12.19.).

Schwitzer, C., et al., 2014. Averting lemur extinctions amid Madagascar's political crisis. Science 343 (6173), 842—843.

Scott, J., 2016. Ring-Tailed Lemurs of Madagascar: Going, Going, Gone? University or Colorado Boulder, <https://www.colorado.edu/today/2016/12/19/ring-tailed-lemurs-madagascar-going-going-gone> (accessed 10.12.19.).

Scott, D., Willits, F.K., 1994. Environmental attitudes and behavior. A Pennsylvania survey. Environ. Behav. 26 (2), 239—260.

SDGF, 2017. Mauritania: Mainstreaming Local Environmental Management in the Planning Process. Sustainable Development Goal Fund (SDGF), United Nations.

Sengupta, S., 2018. What Jair Bolsonaro's Victory Could Mean for the Amazon, and the Planet. New York Times. <https://www.nytimes.com/2018/10/17/climate/brazil-election-amazon-environment.html?module = inline> (accessed 07.12.19.).

SEPO, 2015. Philippine Forests at a Glance. Senate Economic Planning Unit (SEPO), The Philippines. Available from: <https://www.senate.gov.ph/publications/SEPO/AAG%20on%20Philippine%20Forest_Final.pdf>.

Sepp, C., Mansur, E., 2006. National forest programmes—a comprehensive framework for participatory planning. National Forest Programmes Unasylva, 225 (57), 2006/3. Food and Agriculture Organization of the United Nations (FAO), Rome, pp. 6−12.

Seyler, J.R., et al., 2010. Democratic Republic of Congo: Biodiversity and Tropical Forestry Assessment (118/119). Final Report. United States Agency for International Development (USAID), Washington, DC.

Shan, Y.L., et al., 2019. Peak cement-related CO_2 emissions and the changes in drivers in China. J. Ind. Ecol. 23, 959−971.

Shen, Y., 2013. Environmental policies concerning climate change in China: a contemporary and holistic view. Environ. Law Report. 43 (12), 11086−11097.

Shrestha, B., 1998. Changing Forest Policies and Institutional Innovations: Users Group Approach in Community Forestry of Nepal. International Workshop on Community-Based Natural Resource Management (CBNRM), May 10−14, Washington, DC.

Shrestha, U.B., Shrestha, B.B., Shrestha, S., 2010. Biodiversity conservation in community forests of Nepal: rhetoric and reality. Int. J. Biodivers. Conserv. 2 (5), 98−104.

Sikor, T., 1998. Forest policy reform: from state to household forestry. In: Poffenberger, M. (Ed.), Stewards of Vietnam's Upland Forests. Asia Forest Network, Berkeley, CA, pp. 118−138.

Sizer, N., Hansen, M., Moore, R., 2013. New High-Resolution Forest Maps Reveal World Loses 50 Soccer Fields of Trees per Minute. World Watch Institute, Washington, DC. Available from: <http://www.wri.org/blog/2013/11/new-high-resolution-forest-maps-reveal-world-loses-50-soccer-fields-trees-minute>.

Smil, V., 2004. China's Past, China's Future. Energy, Food, Environment. Routledge, New York, London.

Smithsonian, 2014. Dead Zones: Mussels on Beach. Smithsonian Newsdesk, Washington, DC, <http://newsdesk.si.edu/photos/dead-zones-mussels-beach> (accessed 09.12.19.).

Socialist Republic of Vietnam, 1995. Biodiversity Action Plan for Vietnam. The Government. Socialist Republic of Vietnam. Available from: <https://www.thegef.org/sites/default/files/project_documents/Ntl%2520BD%2520Strategy%2520n%2520Action%2520Plan%2520-%2520Pt%25201_3.pdf>.

Southerland, E., 2020. We Can't let Trump Roll Back 50 Years of Environmental Progress. The Guardian. <https://www.theguardian.com/commentisfree/2020/apr/22/earth-day-50-years-anniversary-environment-trump> (accessed 25.04.20.).

Stanway, D., 2019. China's Water Quality Improvements 'Imbalanced' in First Quarter: Xinhua. Reuters. <https://www.reuters.com/article/us-china-pollution/chinas-water-quality-improvements-imbalanced-in-first-quarter-xinhua-idUSKCN1SO096> (accessed 12.12.19.).

State Council of the PRC, 2008. China's Policies and Actions for Addressing Climate Change 2008. Information Office of State Council of the People's Republic of China, Beijing.

Steffen, W., et al., 2011. The anthropocene: conceptual and historical perspectives. Philos. Trans. R. Soc. 369, 842−867.

Sterling, E.J., Hurley, M.M., Duc Minh, L., 2006. Vietnam: A Natural History. Yale University Press, New Haven, CT.

Steyn, P., 2016. African Elephant Numbers Plummet 30 Percent, Landmark Survey Finds. National Geographic Society. <http://news.nationalgeographic.com/2016/08/wildlife-african-elephants-population-decrease-great-elephant-census/> (accessed 10.12.19.).

Stibig, H.-J., et al., 2007. Forest Cover Change in Southeast Asia. The Regional Pattern. European Commission Joint Research Centre, Italy.

Stoddard, E., 2015. African Vultures Targeted by Poachers, Headed for Extinction—Report. Reuters, Cape Town. <http://af.reuters.com/article/kenyaNews/idAFL8N12S44J20151029> (accessed 10.12.19.).

Stokes, M., 2017. As Thailand Ramps Up Its Palm Oil Sector, Peat Forests Feel the Pressure. Mongabay.com. <https://news.mongabay.com/2017/03/as-thailand-ramps-up-its-palm-oil-sector-peat-forests-feel-the-pressure/> (accessed 09.12.19.).

Stolle, F., Gingold, B., 2011. Indonesia's Ambitious Forest Moratorium Moves Forward. World Resource Institute, Washington, DC. Available from: <https://www.wri.org/blog/2011/06/indonesia-s-ambitious-forest-moratorium-moves-forward>.

Stracqualursi, V., Wallace, G., 2019. EPA Proposes Rule Easing Regulation of Methane Emissions. CNN. <https://edition.cnn.com/2019/08/29/politics/methane-emissions-regulations-epa-rollback/index.html> (accessed 13.12.19.).

Sullivan, Z., 2017. Mining Activity Causing Nearly 10 Percent of Amazon Deforestation. Monabay.com. <https://news.mongabay.com/2017/11/mining-activity-causing-nearly-10-percent-of-amazon-deforestation/> (accessed 07.12.19.).

Sunderlin, W.D., Huynh, T.B., 2005. Poverty Alleviation and Forests in Vietnam. The Center for International Forestry Research (CIFOR), Bogor.

Suntikul, W., Butler, R., Airey, D., 2010. Implications of political change on national park operations: *doi moi* and tourism to Vietnam's national parks. J. Ecotourism 9 (3), 201–218.

Superville, D., Freking, K., 2019. Trump Signs Orders Making It Harder to Block Pipelines. Associated Press (US News) <https://www.usnews.com/news/business/articles/2019-04-10/trump-order-would-make-it-harder-to-block-pipelines> (accessed 13.12.19.).

Tabarelli, M., et al., 2005. Challenges and opportunities for biodiversity conservation in the Brazilian Atlantic Forest. Conserv. Biol. 19 (3), 695–700.

Tabuchi, H., Rigby, C., White, J., 2017. Amazon Deforestation, Once Tamed, Comes Roaring Back. New York Times. <https://www.nytimes.com/2017/02/24/business/energy-environment/deforestation-brazil-bolivia-south-america.html?hp&action = click&pgtype = Homepage&clickSource = story-heading&module = photo-spot-region&region = top-news&WT.nav = top-news&_r = 1> (accessed 07.12.19.).

Tangwisutijit, N., 2018. Saving Forests Must Remain a Focus for Slowing Climate Change. The Nation, Thailand, <https://www.nationthailand.com/national/30355892> (accessed 09.12.19.).

Tans, P., Keeling, R., 2018. NOAA ESRL Data. Scripps Institution of Oceanography. National Oceanic and Atmospheric Administration Earth System Research Laboratory Global Monitoring Division.

The ASEAN Post Team, 2018. Environment Under Threat in Thailand. The ASEAN Post. <https://theaseanpost.com/article/environment-under-threat-thailand> (accessed 09.12.9.).

The Nations, 2016a. Mae Wong Dam Project a Test for Checks and Balances. The Nations, Thailand. <https://www.nationthailand.com/politics/30301708> (accessed 09.12.19.).

The Nations, 2016b. Activists Fume Over Plan to use Article 44 for dam. The Nations, Thailand. <https://www.nationthailand.com/national/30294469> (accessed 09.12.19.).

The Nations, 2016c. Court Delivers Blow to Govt Plans for Mae Wong Dam. The Nations, Thailand. <https://www.nationthailand.com/national/30300791> (accessed 09.12.19.).

Tian, Z.Q., et al., 2007. Plant diversity and its conservation strategy in the inundation and resettlement districts of the Yangtze Three Gorges, China. Acta Ecol. Sin. 27 (8), 3110–3118.

Ting, M., 2011. Yangtze River Pollution Concerns. Global Times, Beijing, <http://en.people.cn/90882/7644184.html> (accessed 12.12.19.).

TNC, 2014. Applicability of the Hansen Global Forest Data to REDD + Policy Decisions. The Nature Conservancy (TNC), Arlington, VA. Available from: <https://www.conservationgateway.org/ConservationPractices/ClimateChange/ForestCarbon/Documents/tnc_REDD + _Hansen.pdf>.

To, X.P., Sikor, T., 2008. The Politics of Illegal Logging in Vietnam. DEV Working Paper 05. The School of Development Studies. University of East Anglia, United Kingdom.

Tollefson, J., 2016. Obama's science legacy: climate (policy) hots up. Nature 536, 387. Available from: <https://www.nature.com/news/polopoly_fs/1.20468!/menu/main/topColumns/topLeftColumn/pdf/536387a.pdf>.

Tollefson, J., 2020. Five ways that Trump is undermining environmental protections under the cover of coronavirus. Nature <https://www.nature.com/articles/d41586-020-01261-4> (accessed 02.05.20.).

Tordoff, A.W., et al., 2012. Indo-Burma Biodiversity Hotspot. Critical Ecosystem Partnership Fund, Conservation International, Arlington, TX. Available from: <http://www.cepf.net/SiteCollectionDocuments/working_group/Draft_IndoBurma_Ecosystem_Profile.pdf>.

Trading Economics, 2019a. China GDP. Trading Economics. <https://tradingeconomics.com/china/gdp> (accessed 12.12.19.).

Trading Economics, 2019b. United States GDP. Trading Economics. <https://tradingeconomics.com/united-states/gdp> (accessed 12.12.19.).

Trisurat, Y., Alkemade, R., Verburg, P.H., 2011. Modeling land use and biodiversity in Northern Thailand. In: Trisurat, Y., Prasad Shrestha, R., Alkemade, R. (Eds.), Land Use, Climate Change and Biodiversity

Modeling: Perspectives and Applications. United States of America by Information Science Reference, Hershey, PA, pp. 199–218.

Trisurat, Y., Shirakawa, H., Johnston, J.M., 2019. Land-use/land-cover change from socio-economic drivers and their impact on biodiversity in Nan Province, Thailand. Sustainability 11 (3), 1–22.

Tropek, R., et al., 2014. Comment on "high-resolution global maps of 21st-century forest cover change". Science 344 (6187), 981-d.

Tung, C., 2009. Carbon law and practice in China. In: Freestone, D., Streck, C. (Eds.), Legal Aspects of Carbon Trading. Kyoto, Copenhagen, and Beyond. Oxford University Press, Oxford, pp. 488–514.

Turner, J.L., Ellis, L., 2007. China's Growing Ecological Footprint. The China Monitor. Available from: <https://www.wilsoncenter.org/sites/default/files/china_monitor_article.pdf>.

Turvey, S.T., et al., 2007. First human-caused extinction of a cetacean species? Biol. Lett. 3, 537–540.

UNDP, 2004. Enhancing the Effectiveness and Catalyzing the Sustainability of the W-Arly-Pendjari (WAP) Protected Area System. UNDP Project Document PIMS 1617. United Nations Development Programme (UNDP), Nairobi.

UNDP, 2012a. Project Document: Biodiversity Conservation in Multiple-Use Forest Landscapes in Sabah, Malaysia. United Nations Development Programme (UNDP), New York.

UNDP, 2012b. Climate Change and Development in China. 3 Decades of UNDP Support. United Nations Development Programme (UNDP), China.

UNDP, 2017. A Revised Law Sparks New Hope for Restoring Paraguay's Biodiversity. United Nations Development Programme (UNDP), New York.

UNDP-GEF, 2009. Removing Barriers Hindering Protected Area Management Effectiveness in Viet Nam (PIMS 3965). UNDP Project Document. United Nations Development Programme (UNEP). Global Environment Facility (GEF).

UNDP-UN Environment, 2019. Reward and Renewal. UNDP-UN Environmental Poverty-Environmental Initiative. Phase 2 Final Progress Report 2014–2018. United Nations Development Programme (UNDP), New York; United Nations Environment Programme (UN Environment), Nairobi.

UNEP (United Nations Environment Programme), 2002. Global Environment Outlook 3: Past, Present and Future Perspectives. Earthscan, London Sterling, VA.

UNEP, 2008. Africa Atlas of Our Changing Environment. United Nations Environment Programme (UNEP), Nairobi.

UNEP, 2010a. State of Biodiversity in Africa. United Nations Environment Programme, Nairobi. Available from: <https://www.cbd.int/iyb/doc/celebrations/iyb-egypt-state-of-biodiversity-in-africa.pdf>.

UNEP, 2010b. Green Economy: Developing Countries Success Stories. United Nations Environment Programme, Nairobi.

UNEP, 2013. Africa Environment Outlook 3: Summary for Policy Makers. United Nations Environment Programme (UNEP), Nairobi, New York.

UNEP, 2016. The State of Biodiversity in Africa: A Mid-Term Review of Progress Towards the Aichi Biodiversity Targets. UNEP-WCMC, Cambridge.

UNEP, 2019a. Deforestation in Borneo Is Slowing, but Regulation Remains Key. United Nations Environmental Programme (UNEP), <https://www.unenvironment.org/news-and-stories/story/deforestation-borneo-slowing-regulation-remains-key> (accessed 19.12.19.).

UNEP, 2019b. Global environment outlook-6: healthy people, healthy planet. In: Ekins, P., Gupta, J., Boileau, P. (Eds.), United Nations Environmental Programme (UNEP), Nairobi. Cambridge University Press, United Kingdom.

UNEP-WCMC, 2014. Megadiverse Countries. Biodiversity A-Z, United Nations Environment Programme-World Conservation Monitoring Centre (UNEP-WCMC). Available from: <http://www.biodiversitya-z.org/content/megadiverse-countries>.

UNEP-WCMC, 2016. The State of Biodiversity in Africa: A Mid-Term Review of Progress Towards Aichi Biodiversity Targets. United Nations Environment Programme. World Conservation Monitoring Centre (UNEP-WCMC).

UNEP-WHRC, 2007. Reactive Nitrogen in the Environment. Too Much or Too Little of a Good Thing. United Nations Environment Programme (UNEP), Paris, Woods Hole Research Center (WHRC), Falmouth MA.

UNESCO, 2010. World Heritage in the Congo Basin. UNESCO World Heritage Centre, France.

UNESCO, 2017. Convention Concerning the Protection of the World Cultural and Natural Heritage. UNESCO World Heritage Centre, France.

UNFCCC, 2011. Fact Sheet: Reducing Emissions From Deforestation in Developing Countries: Approaches to Stimulate Action. United Nations Framework Convention on Climate Change (UNFCCC), Bonn. Available from: <https://unfccc.int/files/press/backgrounders/application/pdf/fact_sheet_reducing_emissions_from_deforestation.pdf>.

UNFCCC, 2015. Adoption of the Paris Agreement (Document FCCC/CP/2015/L.9/Rev.1). United Nations Framework Convention on Climate Change (UNFCCC), Bonn.

Union of Concerned Scientists, 2009. Why Does CO_2 Get Most of the Attention When There Are so Many Other Heat-Trapping Gases? Union of Concerned Scientists, <https://www.ucsusa.org/resources/why-does-co2-get-more-attention-other-gases> (accessed 12.12.19.).

Union of Concerned Scientists, 2013. Measuring the Role of Deforestation in Global Warming. Union of Concerned Scientists, Cambridge, MA.

United Nations, 1995. Report of the Opened-Ended Ad Hoc Intergovernmental Panel on Forests on Its First Session (Document E/CN.17/IPF/1995/3). United Nations Economic and Social Council, New York.

United Nations, 2000. Report of the Intergovernmental Forum on Forests on Its Fourth Session, (Document 2000/35). United Nations Economic and Social Council, New York.

United Nations, 2001. United Nations Forum on Forests Report on the Organizational and First Sessions (12 and 16 February and 11–22 June 2001) (Document E/2001/42/Rev.1E/CN.18/2001/3/Rev.1). United Nations Economic and Social Council, New York.

United Nations, 2002a. Collaborative Partnership on Forests (CPF). United Nations, New York.

United Nations, 2002b. Report on the Second Session (22 June 2001 and 4 to 15 March 2002) (Document E/2002/42, E/CN.18/2002/14). United Nations Economic and Social Council, New York.

United Nations, 2003. United Nations Forum on Forests. Report on the Third Session (15 March 2002 and 26 May to 6 June 2003) (Document E/2003/42 E/CN.18/2003/13). United Nations Economic and Social Council, New York.

United Nations, 2004. United Nations Forum on Forests. Report on the Fourth Session (6 June 2003 and 3 to 14 May 2004) (Document E/2004/42E/CN.18/2004/17). United Nations Economic and Social Council, New York.

United Nations, 2005. United Nations Forum on Forests. Report of the Fifth Session (14 May 2004 and 16 to 27 May 2005) (Document E/2005/42E/CN.18/2005/18). United Nations Economic and Social Council, New York.

United Nations, 2006a. Press Release. Sixth Session of UN Forum on Forests Opens at Headquarters. United Nations Economic and Social Council, New York.

United Nations, 2006b. United Nations Forum on Forests. Report of the Sixth Session (27 May 2005 and 13 to 24 February 2006) (Document E/2006/42E/CN.18/2006/18). United Nations Economic and Social Council, New York.

United Nations, 2007. Seventh Session of the United Nations Forum on Forests, 1627 April 2007. United Nations, New York.

United Nations, 2008. Resolution Adopted by the General Assembly on 17 December 2007: Non-Legally Binding Instrument on All Types of Forests (Document A/RES/62/98). United Nations Economic and Social Council, New York.

United Nations, 2009. United Nations Forum on Forests. Report of the Eighth Session (27 April 2007 and 20 April to 1 May 2009) (Document E/2009/42 E/CN.18/2009/20). United Nations Economic and Social Council, New York.

United Nations, 2011. United Nations Forum on Forests. Report on the Ninth Session (1 May 2009 and 24 January to 4 February 2011) (Document E/2011/42 E/CN.18/2011/20). United Nations Economic and Social Council, New York.

United Nations, 2013a. United Nations Forum on Forests. United Nations Economic and Social Council, New York, Report on the Tenth Session (4 February 2011 and 8–19 April 2013) (Document E/2013/42E/CN.18/2013/18).

United Nations, 2013b. Amazon Treaty Body Hailed as Model for Regional Conservation Efforts. United Nations Forum on Forests, New York.

United Nations, 2015a. United Nations Forum on Forests. United Nations Economic and Social Council, New York, Report on the Eleventh Session (19 April 2013 and 4 to 15 May 2015) (Document E/2015/42-E/CN.18/2015/14).

United Nations, 2015b. UNFF 11 Fact Sheet. United Nations Economic and Social Council, New York. Available from: Council, <http://www.un.org/esa/forests/wp-content/uploads/bsk-pdf-manager/280_UNFF11-FACTSHEET.PDF>.

United Nations, 2015c. Forests Pivotal to New Post-2015 Development Agenda. United Nations Economic and Social Council, New York.

United Nations, 2015d. National Reports UNFF11. United Nations Economic and Social Council, New York.

United Nations, 2017a. United Nations Forum on Forests. Report on the Twelfth Session (25 April 2016 and 1 to 5 May 2017) (Document E/2017/42-E/CN.18/2017/8). United Nations Economic and Social Council, New York.

United Nations, 2017b. United Nations Forum on Forests. Twelfth Session (Document E/CN.18/2017/1). United Nations Economic and Social Council, New York.

United Nations, 2017c. Resolution Adopted by the Economic and Social Council on 20 April 2017. United Nations Strategic Plan for Forests 2017–2030 and Quadrennial Programme of Work of the United Nations Forum on Forests for the period 2017–2020 (Document E/RES/2017/4). United Nations Economic and Social Council, New York.

United Nations, 2018. United Nations Forum on Forests. Report on the Thirteenth Session (5 May 2017 and 7 to 11 May 2018) (Document E/2018/42-E/CN.18/2018/9). Economic and Social Council, United Nations, New York.

United Nations, 2019. United Nations Forum on Forests. Report on the Fourteenth Session (11 May 2018 and 6–10 May 2019). (Document E/2019/42-E/CN.18/2019/9). Economic and Social Council, United Nations, New York.

Universitat Autònoma de Barcelona, 2018. Bolivian Amazon on Road to Deforestation. Universitat Autònoma de Barcelona, Spain. Available from: <https://www.uab.cat/web/newsroom/news-detail/bolivian-amazon-on-road-to-deforestation-1345668003610.html?noticiaid = 1345742134437> (accessed 09.12.19.).

University of Cambridge, 2015. Amazon Deforestation 'Threshold' Causes Species Loss to Accelerate. University of Cambridge, <http://www.cam.ac.uk/research/news/amazon-deforestation-threshold-causes-species-loss-to-accelerate> (accessed 10.12.19.).

UNODC, 1992. Wild Animal Reservation and Protection Act, B.E. 2535 (1992), United Nations Office on Drugs and Crime (UNODC). Available from: <https://sherloc.unodc.org/res/cld/document/wildlife-preservation-and-protection-act--b-e-2535_html/Wildlife_Preservation_and_Protection_Act_B.E._2535.pdf>.

Upadhyay, S., 2013. Community Based Forest and Livelihood Management in Nepal. In: Bollier, D., Helfrich, S., (Eds.), The Wealth of the Commons. A World Beyond Market & State. Levellers Press, Amherst, MA, pp. 265–270.

Uryu, Y., et al., 2008. Deforestation, Forest Degradation, Biodiversity Loss and CO_2 Emissions in Riau, Sumatra, Indonesia. World Wildlife Fund (WWF) Indonesia Technical Report. World Wildlife Fund (WWF), Jakarta.

USDA Forest Service, 2011. Final Environmental Impact Statement. Concow Hazardous Fuels Reduction Project. USDA Forest Service, Plumas National Forest, California.

US Department of Commerce, 2002. China Environmental Technologies Export Market Plan. International Trade Administration, Washington, DC.

USGS, 1952. Bureau of Mines minerals yearbook 1952. U.S. Geological Survey, United States.

USGS, 1976–2019. Cement: Statistics and Information. U.S. Geological Survey (USGS), United States. Available: <https://minerals.usgs.gov/minerals/pubs/commodity/cement/>.

van Noordwijk, M., Minang, P.A., 2009. If we cannot define it, we cannot save it: forest definitions and REDD. ASB Policy Brief 15, 1–4.

van Noordwijk, M., et al., 2009. Reducing emissions from all land uses (REALU): the case for a whole landscape approach. ASB Policy Brief 13, 1–4.

Vesilind, P.J., 2002. Hotspot: The Philippines. National Geographic Magazine. <http://ngm.nationalgeographic.com/print/features/world/asia/philippines/philippines-text> (accessed 09.12.19.).

Verheij, P.M., Foley, K.E., Engel, K., 2010. Reduced to Skin and Bones. An Analysis of Tiger Seizures From 11 Tiger Range Countries (2000–2010). TRAFFIC International, Cambridge.

Vidal, J., 2016. Elephants Could Vanish From One of Africa's Key Reserves Within Six Years. The Guardian, United Kingdom. <https://www.theguardian.com/environment/2016/jun/01/elephants-vanish-africas-key-reserves-six-years-tanzania-selous-national-park> (accessed 10.12.19.).

Vié, J.-C., Hilton-Taylor, C., Stuart, S.N. (Eds.), 2009. Wildlife in a Changing World—An Analysis of the 2008 IUCN Red List of Threatened Species. International Union for Conservation of Nature and Natural Resources (IUCN), Gland.

Vietnam Net, 2013. Vietnam to Have 41 Biodiversity Reserves by 2020. <http://www.vfej.vn/en/4647n/vietnam-to-have-41-biodiversity-reserves-by-2020.html> (accessed 11.12.19.).

Vietnam News, 2013. Central Highlands Suffers Deforestation, Vietnam News. <http://vietnamnews.vn/environment/247453/central-highlands-suffers-deforestation.html#Kw0xT83m3Cmd68hf.97> (accessed 09.12.19.).

Vietnam News, 2015. Deforestation Continues to Deplete Nation's Wealth. Vietnam News. <http://vietnamnews.vn/opinion/265119/deforestation-continues-to-deplete-nations-wealth.html#GhGLtjTzsXzmVjvZK.97> (accessed 09.12.19.).

Viet Nam News, 2018. Illegal Deforestation and Encroachment Still Rampant in Đắk Lắk. Viet Nam News. <https://vietnamnews.vn/environment/467020/illegal-deforestation-and-encroachment-still-rampant-in-dak-lak.html#TRDBUsrrXLYhjmKR.97> (accessed 09.12.19.).

Viet Nam News, 2019. Illegal Logging Threatens Old Forest. Viet Nam News. <https://vietnamnews.vn/environment/484440/illegal-logging-threatens-old-forest.html#1gwjTjuKckStAlyQ.97> (accessed 09.12.19.).

Vittert, L., 2019. Statistic of the Decade: The Massive Deforestation of the Amazon. The Conversation. <https://theconversation.com/statistic-of-the-decade-the-massive-deforestation-of-the-amazon-128307> (accessed 27.12.19.).

Voigt, M., et al., 2018. Global demand for natural resources eliminated more than 100,000 Bornean orangutans. Curr. Biol. 28, 761—769.

Waggener, T.R., 2001. Logging bans and the Asia-Pacific: an overview. In: Durst, P.B., Waggener, T.R., Enters, T., Cheng, T.L. (Eds.), Forests Out of Bounds: Impacts and Effectiveness of Logging Bans in Asia-Pacific. Asia-Pacific Forestry Commission, Food and Agriculture Organization of the United Nations, Regional Office for Asia and the Pacific, Bangkok.

Walston, J., Karanth, K.U., Stokes, E.J., 2010. Avoiding the Unthinkable: What Will It Cost to Prevent Tigers Becoming Extinct in the Wild? Wildlife Conservation Society, New York.

Walton, M., 2003. Study: Only 10 Percent of Big Ocean Fish Remain. CNN. <https://edition.cnn.com/2003/TECH/science/05/14/coolsc.disappearingfish/> (accessed 10.12.19.).

Wang, Z., 2000. Sustainable development and its legal situation in China note and commentary. Asia Pac. J. Environ. Law 5 (2), 175—196.

Wang, M., 2008. In: Zillman, D., et al., (Eds.), China's Plight in Moving Towards a Low-Carbon Future: Analysis From the Perspective of Energy Law. Oxford University Press, Oxford, pp. 379—398.

Wang, L., 2010. The changes of China's environmental policies in the latest 30 years. Procedia Environ. Sci. 2, 1206—1212.

Wang, K., 2011. The Role of Cities in Meeting China's Carbon Intensity Goal. World Resource Institute, <http://www.wri.org/blog/2011/09/role-cities-meeting-china%E2%80%99s-carbon-intensity-goal> (accessed 12.12.19.).

Wang, Z.H., He, H.L., Fan, M.J., 2014. The Ecological Civilization Debate in China. Monthly Review, New York, <https://monthlyreview.org/2014/11/01/the-ecological-civilization-debate-in-china/#fn5> (accessed 12.12.19.).

Watkinson, W., 2016. Congo Giraffes on Brink of Extinction Say Conservationists. International Business Times, United Kingdom, <http://www.ibtimes.co.uk/congo-giraffes-brink-extinction-say-conservationists-1538465> (accessed 10.12.19.).

Watts, J., 2014. Amazon Rainforest Losing Ability to Regulate Climate, Scientist Warns. The Guardian, United Kingdom, <https://www.theguardian.com/environment/2014/oct/31/amazon-rainforest-deforestation-weather-droughts-report> (accessed 07.12.19.).

WCS (Wildlife Conservation Society), 2016. Grauer's Gorilla at Extremely High Risk of Extinction in the Wild. National Geographic. <http://voices.nationalgeographic.com/2016/09/04/grauers-gorilla-at-extremely-high-risk-of-extinction-in-the-wild/> (accessed 10.12.19.).

Wearn, O.R., Reuman, D.C., Ewers, R.M., 2012. Extinction debt and windows of conservation opportunity in the Brazilian Amazon. Science 337, 228—232.

Weisse, M., 2016. When Tree Cover Loss Is Really Forest Loss: New Plantation Map Improve Forest Monitoring. Global Forest Watch, Washington, DC. Available from: <https://blog.globalforestwatch.org/data/when-tree-cover-loss-is-really-forest-loss-new-plantation-maps-improve-forest-monitoring.html>.

Weisse, M., Goldman, E.D., 2017a. Global Tree Cover Loss Rose 51 Percent in 2016. Global Forest Watch, Washington, DC. Available from: <http://www.wri.org/blog/2017/10/global-tree-cover-loss-rose-51-percent-2016>.

Weisse, M., Goldman, E.D., 2017b. Technical Blog: Caveats to the 2016 Tree Cover Loss Data, Explained. Global Forest Watch, Washington DC. Available from: <http://www.wri.org/blog/2017/10/technical-blog-caveats-2016-tree-cover-loss-data-explained>.

Weisse, M., Goldman, E.D., 2019. The World Lost a Belgium-Sized Area of Primary Rainforests Last Year. World Resource Institute, Washington, DC. Available from: <https://www.wri.org/blog/2019/04/world-lost-belgium-sized-area-primary-rainforests-last-year>.

Wells, P., Paoli, G., 2011. Preliminary Observations on the Indonesian Ministry of Forestry Decree SK .323/Menhut-II/2011and Indicative Maps Concerning the Suspension of New Licenses for Forest and Peatland Utilisation. Daemeter Consulting, Bogor.

Weng, X.X., et al., 2015. China's Path to a Green Economy. Decoding China's Green Economy Concepts and Policies. International Institute for Environment and Development, London, IIED Country Report.

Western, D., Russell, S., Cuthill, I., 2009. The status of wildlife in protected areas compared to non-protected areas of Kenya. PLoS One 4 (7), e6140. Available from: https://doi.org/10.1371/journal.pone.0006140.

Wetlands International, 2011. Impact of Oil Palm Plantations on Peatland Conversion in Sarawak 2005–2010. Wetland International, The Netherlands.

White House, The, 2015. U.S.-China Joint Presidential Statement on Climate Change. The White House. Office of the Press Secretary, Washington, DC. Available from: <https://obamawhitehouse.archives.gov/the-press-office/2015/09/25/us-china-joint-presidential-statement-climate-change>.

White House, The, 2017. Remarks by President Trump at the Unleashing American Energy Event. The White House, U.S. Department of Energy, Washington, DC. Available from: <https://www.whitehouse.gov/briefings-statements/remarks-president-trump-unleashing-american-energy-event/>.

Wijaya, A., et al., 2017. 6 Years After Moratorium, Satellite Data Shows Indonesia's Tropical Forests Remain Threatened. World Resource Institute, Washington DC. Available from: <https://www.wri.org/blog/2017/05/6-years-after-moratorium-satellite-data-shows-indonesia-s-tropical-forests-remain>.

Wittemyer, G., et al., 2014. Illegal killing for ivory drives global decline in African elephants. PNAS (Proc. Natl Acad. Sci. U.S.A.) 111 (36), 13117–13121.

Wong, C.M., et al., 2007. World's Top 10 Rivers at Risk. World Wildlife Fund (WWF), Washington, DC.

World Bank, 2005. Vietnam Environment Monitor 2005. Biodiversity. The World Bank, Washington, DC.

World Bank, 2010. Socialist Republic of Vietnam. Forest Law Enforcement and Governance. The World Bank. Sustainable Development Department, East Asia and Pacific Region.

World Bank, 2013. World Development Indicators: Rural Environment and Land Use. The World Bank, Washington, DC. Available from: <http://wdi.worldbank.org/table/3.1#>.

World Bank, 2016a. Countries and Economies: China. World Bank, Washington, DC. Available from: <http://data.worldbank.org/country>.

World Bank, 2016b. The Global Tiger Initiative. World Bank, Washington, DC, <https://www.worldbank.org/en/topic/environment/brief/the-global-tiger-initiative> (accessed 25.02.20.).

World Bank, 2017. Protecting the Atlantic Forest: Creating a Biodiversity Corridor in Eastern Paraguay. The World Bank, Washington, DC. Available from: <http://www.worldbank.org/en/results/2017/10/30/protecting-the-atlantic-forest-creating-a-biodiversity-corridor-in-eastern-paraguay>.

World Bank, 2018. Data. China. World Bank, Washington, DC. Available from: <https://data.worldbank.org/country/china>.

World Bank Inspection Panel, 2006. The Inspection Panel Report and Recommendation on Request for Inspection Democratic Republic of Congo: Transitional Support for Economic Recovery Credit Operation (TSERO) (IDA Grant No. H 192-DRC) and Emergency Economic and Social Reunification Support Project (EESRSP). World Bank, Washington, DC. Available from: <http://ewebapps.worldbank.org/apps/ip/PanelCases/37-Eligibility%20Report%20(English).pdf>.

World Steel Association, 2016. Steel Statistical Yearbook 2016. World Steel Association, Brussels. Available from: <https://www.worldsteel.org/en/dam/jcr:37ad1117-fefc-4df3-b84f-6295478ae460/Steel + Statistical + Yearbook + 2016.pdf>.

World Steel Association, 2018. World Crude Steel Output Increases by 5.3% in 2017. World Steel Association, Brussels, <https://www.worldsteel.org/media-centre/press-releases/2018/World-crude-steel-output-increases-by-5.3-in-2017.html> (accessed 12.12.19.).

World Steel Association, 2019. World Steel in Figures. World Steel Association, Brussels. Available from: <https://www.worldsteel.org/en/dam/jcr:96d7a585-e6b24d63-b943-4cd9ab621a91/World%2520Steel%2520in%2520Figures%25202019.pdf>.

WRI, 2015. Release: Orbital Insight and World Resources Institute Partner on Satellite Imagery to Curb Deforestation. World Forest Resources Institute, Washington, DC. Available from: <http://www.wri.org/news/2015/04/release-orbital-insight-and-world-resources-institute-partner-satellite-imagery-curb>.

WRI, 2016. Indonesia: Law and Regulation. Forest Legal Initiatives. World Resource Institute (WRI), Washington, DC. Available from: <http://www.forestlegality.org/risk-tool/country/indonesia>.

Wu, J.G., et al., 2003. Three-Gorges Dam—experiment in habitat fragmentation? Science 300 (5648), 1239–1240.

Wu, K.X., et al., 2014. Prediction of COD emission in Hubei Province based on the grey metabolizing model (Section 4: sustainability and economics). In: Lee, G. (Ed.), Environment and Sustainability. WIT Press, Southampton, pp. 559–568.

Wulffraat, S., et al., 2012. The Environmental Status of the Heart of Borneo. WWF's HoB Initiative, World Wildlife Fund, Switzerland. Available from: <http://www.hobgreeneconomy.org/downloads/wwf_environmental_status_heart_borneo.pdf>.

WWF, 1998. Root Causes of Biodiversity Loss in Vietnam: Summary. World Wildlife Fund (WWF), Washington, DC. Available from: <https://d2ouvy59p0dg6k.cloudfront.net/downloads/vietnam.pdf>.

WWF, 2004. Rivers at Risk. Dams and the Future of Freshwater Ecosystems. World Wildlife Fund (WWF), Washington, DC. Available from: <http://awsassets.panda.org/downloads/riversatriskfullreport.pdf>.

WWF, 2005. Borneo: Treasure Island at Risk. World Wildlife Fund (WWF), Germany.

WWF, 2007a. Forests of Borneo. Forest Area key Facts & Carbon Emissions From Deforestation. World Wildlife Fund (WWF), Indonesia.

WWF, 2007b. World's Top 10 Rivers at Risk. World Wildlife Fund (WWF), Washington, DC. Available from: <http://assets.wwf.org.uk/downloads/worldstop10riversatrisk.pdf>.

WWF, 2010. Feasibility Assessment for Financing the Heart of Borneo Landscape. Malaysia (Sabah and Sarawak). World Wildlife Fund (WWF), Washington, DC.

WWF, 2012. China Ecological Footprint Report 2012. World Wildlife Fund (WWF), Beijing.

WWF, 2013. Ecosystems in the Greater Mekong Past Trends, Current Status, Possible Futures. World Wildlife Fund (WWF), Greater Mekong.

WWF, 2014. In: McLellan, R., et al., (Eds.), Living Planet Report 2014: Species and Spaces, People and Places. World Wildlife Fund (WWF), Gland.

WWF, 2015a. The Brazilian Amazon: Challenges Facing an Effective Policy to Curb Deforestation. WWF Living Amazon Initiative and WWF Brazil, Brasilia.

WWF, 2015b. Magical Mekong: New Species Discoveries 2014. World Wildlife Fund (WWF), Greater Mekong, Washington, DC. Available from: <https://c402277.ssl.cf1.rackcdn.com/publications/800/files/original/magical_mekong_new_species_discoveries_2014_compressed.pdf?1432838557>.

WWF, 2016a. Brazil's New Forest Code: A Guide for Decision-Makers in Supply Chains and Governments. World Wildlife Fund (WWF), Brazil.

WWF, 2016b. Living Planet Report 2016. Risk and Resilience in a New Era. World Wildlife Fund (WWF), Gland.

WWF, 2016c. Atlantic Forests, South America. World Wildlife Fund, Washington, DC, <http://wwf.panda.org/what_we_do/where_we_work/atlantic_forests/> (accessed 10.12.19.).

WWF, 2016d. Elephants Could Disappear From Tanzania World Heritage Site Within Six Years. World Wildlife Fund (WWF), Washington, DC, <http://wwf.panda.org/?269211/Elephants-could-disappear-from-Tanzania-World-Heritage-site-within-six-years> (accessed 10.12.19.).

WWF, 2017a. How Would Mae Wong Dam Affect Forest and Wildlife? World Wildlife Fund (WWF), Thailand.

WWF, 2017b. New Species Discovered in the Greater Mekong. World Wildlife Fund (WWF), Greater Mekong. Available from: <https://www.worldwildlife.org/stories/new-species-discovered-in-the-greater-mekong>.

WWF, 2019. Sunda Tiger. World Wildlife Fund (WWF), Washington, DC, <https://www.worldwildlife.org/species/sunda-tiger> (accessed 11.12.19.).

Xie, P., 2003. Three-Gorges Dam: risk to ancient fish. Science 302 (5648), 1149–1151.

Xu, B., Lin, B.Q., 2016. Assessing CO_2 emissions in China iron and steel industry: a dynamic vector autoregression model. Appl. Energy 161, 375–386.

Xu, W.Q., et al., 2016. CO_2 emissions from China's iron and steel industry. J. Clean. Prod. 139, 1504–1511.

Yeager, C., 2008. Conservation of Tropical Forests and Biological Diversity in Indonesia. Report submitted to USAID (U.S. Agency for International Development), Indonesia.

Yew, F.K., 2012. Oil Palm Has Closest Resemblance to Forest Than Other Major Oil Crops. Malaysian Palm Oil Council (MPOC), Mumbai, <http://www.mpoc.org.in/2016/11/18/oil-palm-has-closest-resemblance-to-forest-than-other-major-oil-crops/> (accessed 07.12.19.).

Yousefi, A., Bellantonio, M., Hurowitz, G., 2018. The Avoidable Crisis. The European Meat Industry's Environmental Catastrophe. Mighty Earth, Washington, DC, Rainforest Foundation, Oslo; Fern, Belgium, United Kingdom, France.

Yue, P.Q., Chen, Y.Y., 1998. China Red Data Book of Endangered Animals−Pisces. Science Press, Beijing, Hong Kong, New York.

Zachos, E., 2018. Survival of Northern White Rhino Hinges on Last Sick Male. National Geographic Society. <https://news.nationalgeographic.com/2018/03/northern-white-rhino-endangered-species-spd/?utm_source = Facebook&utm_medium = Social&utm_content = link_fb20180302news-whiterhinodying&utm_campaign = Content&sf183517633 = 1> (accessed 10.12.19.).

Zhang, Z.H., 1999. Rural industrialisation in China: from backyard furnaces to township village enterprises. East. Asia 17 (3), 61−87.

Zhang, J., 2012. Delivering Environmentally Sustainable Economic Growth: The Case of China. Asia Society, New York; Hong Kong, Houston.

Zhang, Q.F., Lou, Z.P., 2011. The environmental changes and mitigation actions in the Three Gorges Reservoir region, China. Environ. Sci. Policy 14, 1132−1138.

Zheng, Y.B., 2019. China Steps Up Efforts to Protect Water Quality. China Global Television Network (CGTN). <https://news.cgtn.com/news/3d3d774e3545444d33457a6333566d54/index.html> (accessed 12.12.19.).

Zimdahl, R.L., 2015. Six Chemicals That Change Agriculture. Academic Press, Amsterdam, Boston, Heidelberg, London.

Zorthian, J., 2015. Sumatran Rhino Deemed Extinct in Malaysia. Time. <http://time.com/4006981/sumatran-rhino-extinct-malaysia/> (accessed 11.12.19.).

Chapter 4

Anderson, J.C., 1993. Species equality and the foundations of moral theory. Environ. Values 2 (4), 347−365.

Babor, E.R., 2007. Ethics. The Philosophical Discipline of Action. Rex Bookstore, Manila.

Barsam, P.A., 2008. Reverence for Life: Albert Schweitzer's Great Contribution to Ethical Thought. Oxford University Press, Oxford.

Bentham, J., 1823. An Introduction to the Principles of Morals and Legislation. Henry Frowe, Oxford at the Clarendon Press, London, New York, Toronto.

Bentham, J., 1907. An Introduction to the Principles of Morals and Legislation. Clarendon Press, Oxford.

Bessinger, D., 2000. Emerging From Chaos: Wholeness, Ethic, and New World Order. Orchard Park Press (Internet Version). Available from: <http://home.earthlink.net/~dbscr2/we2/hpefc.htm>.

Braybrooke, M., 2015. Peace in Our Hearts, Peace in Our World: A Practical Interfaith Daily Guide to a Spiritual Way of Life. Lulu.com., Morrisville, North Carolina.

Bruno, K., et al., 2010. Tar Sand Invasion. How Dirty and Expensive Oil From Canada Threatens America's New Energy Economy. Corporate Ethics International, Suisun City, CA; EARTHWORKS, Washington DC; Natural Resources Defense Council, New York; Sierra Club, San Francisco, CA. Available from: <https://www.nrdc.org/sites/default/files/TarSandsInvasion-full.pdf>.

Caldwell, L.K., Shrader-Frechette, K. (Eds.), 1993. A national policy for land? In: Policy for Land: Law and Ethics. Rowman & Littlefield Publishers, United States, pp. 245−260.

Callicott, J.B., 1989. In Defense of Land Ethic: Essays in Environmental Philosophy. State University of New York Press, Albany, NY.

Callicott, J.B., 2001. The land ethic. In: Jamieson, D. (Ed.), A Companion to Environmental Philosophy. Blackwell Publisher, Massachusetts, Oxford, pp. 204−217.

Callicott, J.B., 2004. Environmental ethics: I. Overview. In: third ed. Post, S.G. (Ed.), Encyclopedia of Bioethics, vol. 2. Macmillan Reference, New York, pp. 757—769.

Carlson, M., Wells, J., Roberts, D., 2009. Carbon the World Forgot. Conserving the Capacity of Canada's Boreal Forest Region to Mitigate and Adapt to Climate Change. Boreal Songbird Initiative, Seattle, WA; Canadian Boreal Initiative, Ottawa. Available from: <https://www.borealbirds.org/sites/default/files/publications/carbon%20report-full.pdf>.

Carson, R., 1962. Silent Spring. Fawcett Publications, Greenwich, CT.

Carson, R., 1963. Rachel Carson speaks on reverence of life. Animal Welfare Institute Information Report 12 (1), 221—224. New York.

Casselman, A., 2018. Environmental Disaster Is Canada's New Normal. Are We Ready? The Walrus. <https://thewalrus.ca/environmental-disaster-is-canadas-new-normal-are-we-ready/> (accessed 26.05.20.).

Chewinsk, M., 2016. Liberals Failing on Duty to Hold Canadian Mining Companies Accountable Abroad. National Observer, Vancouver, <https://www.nationalobserver.com/2016/11/04/opinion/liberals-failing-duty-hold-canadian-mining-companies-accountable-abroad> (accessed 14.12.19.).

Chow, L., 2016. Peru Declares State of Emergency as Mercury Contamination From Illegal Gold Mining Poisons People and Planet. EcoWatch. Available from: <http://www.ecowatch.com/peru-declares-state-of-emergency-as-mercury-contamination-from-illegal-1891145093.html>.

Choy, Y.K., 2005. Sustainable development—an institutional enclave (with special reference to the Bakun dam-induced development strategy in Malaysia). J. Econ. Issues 39 (4), 951—971.

Choy, Y.K., 2014. Land ethic from the Borneo tropical rainforests in Sarawak, Malaysia: an empirical and conceptual analysis. Environ. Ethics 36 (4), 421—441.

Choy, Y.K., 2016. Economic growth, sustainable development and ecological conservation in the Asian developing countries: the way forward. In: Indraneil Das, I., Tuen, A.A. (Eds.), Naturalists, Explorers and Field Scientists in South-East Asia and Australasia. Topics in Biodiversity and Conservation Series, vol. 15. Springer, Cham, Heidelberg, New York, Dordrecht, London, pp. 239—283.

Cochrane, A., 2017. Environmental Ethics. London School of Economics and Political Science. Internet Encyclopedia of Philosophy, United Kingdom, ISSN 2161-0002, Available from: <https://www.iep.utm.edu/envi-eth/>.

COHA, 2014. Canadian Mining in Latin America: Exploitation, Inconsistency, and Neglect. Council on Hemispheric Affairs (COHA), Washington, DC, <http://www.coha.org/canadian-mining-in-latin-america-exploitation-inconsistency-and-neglect/>.

Commission on Global Governance, 1995. Our Global Neighbourhood The Report of the Commission on Global Governance. Commission on Global Governance. Oxford University Press, Oxford, New York.

Cushman, J.H., 2017. Carbon Footprint of Canada's Oil Sands Is Larger Than Thought. Inside Climate News. <https://insideclimatenews.org/news/04042017/tar-sands-greenhouse-gas-emissions-climate-change-keystone-xl-pipeline-donald-trump-enbridge> (accessed 14.12.19.).

Delon, N., 2016. The replaceability argument in the ethics of animal husbandry. In: Thompson, P.B., Kaplan, D.M. (Eds.), Encyclopedia of Food and Agricultural Ethic. Springer, The Netherlands, pp. 1—7.

Doctorow, C., 2014. Canadian Libricide: Tories Torch and Dump Centuries of Priceless, Irreplaceable Environmental Archives. Boing Boing. <https://boingboing.net/2014/01/04/canadian-libraricide-tories-t.html> (accessed 14.12.19.).

Droitsch, D., Huot, M., Partington, P.J., 2010. Canadian Oil Sands and Greenhouse Gas Emissions: The Facts and Perspective. PEMBINA Institute. Calgary. Available: <https://www.pembina.org/reports/briefingnoteosghg.pdf>.

Dyer, S., 2010. Syncrude Found Guilty, but Has Justice Been Served? PEMBINA Institute, Calgary. <https://www.pembina.org/blog/syncrude-found-guilty-but-has-justice-been-served> (accessed 14.12.19.).

ERCB, 2010. ERCB Approves Fort Hills and Syncrude Pond Plans With Conditions. Energy Resources Conservation Board (ERCB), Government of Alberta, Canada. Available from: <https://www.alberta.ca/release.cfm?xID = 282012777C01C-9D59-9B31—78BF9F5F4EBE946B >.

Frampton, C., Redl, B., 2015. Stuck in the muck. The Harper tar sands legacy. In: Heal, T. (Ed.), The Harper Record. Canadian Centre for Policy Alternatives, Ottawa, pp. 257—273.

Francione, G.L., Steiner, G., 2016. Challenging Peter Singer's Paternity Claim. Rutgers University School of Law; Bucknell University. Available from: <https://www.abolitionistapproach.com/challenging-peter-singers-paternity-claim/>.

Fraser, B., 2011. Peruvian Gold Comes With Mercury Health Risks. Scientific American. Available: <https://www.scientificamerican.com/article/peruvian-gold-health-risks/>.

Fraser, B., 2016. Peru's gold rush prompts public-health emergency. Nature 534, 162. Available from: https://doi.org/10.1038/nature.2016.19999. Available from: <https://www.nature.com/news/polopoly_fs/1.19999!/menu/main/topColumns/topLeftColumn/pdf/nature.2016.19999.pdf>.

Free, A.C., 1992. Since Silent Spring: Our Debt to Albert Schweitzer & Rachel Carson. An Address by Ann Conttrell Free delivered August 13, 1992 at St. Bartholomew's Church, New York as part of an International Albert Schweitzer Symposium. Reverence for Life: Ethical Solutions to Environmental Problems. The Albert Schweitzer Institute for Humanities. The United Nations Environment Programme.

French, W.C., 1999. Against biospherical egalitarian. In: Witoszek, N., Brennan, A. (Eds.), Philosophical Dialogues: Arne Næss and the Progress of Ecophilosophy. Rowan & Littlefield Publishers, London, Boulder, New York, Oxford, pp. 127–145.

Friends of the Earth Europe, 2015. Tar Sands: Europe's Complicity in Canada's Climate Crime. Friend of the Earth Europe, Belgium. Available from: <https://www.foeeurope.org/sites/default/files/tar_sands/2015/foee-tar_sands-europes-complicity-011215.pdf>.

Germanos, A., 2013. Noam Chomsky: Canada on Fast-Speed Race "to Destroy the Environment. Common Dreams, Portland. Available: <https://www.commondreams.org/news/2013/11/01/noam-chomsky-canada-fast-speed-race-destroy-environment> (accessed 14.12.19.).

Gordon, J.S., 2017. Bioethics. Internet Encyclopedia of Philosophy. ISSN 2161-0002. Available from: <http://www.iep.utm.edu/bioethic/>.

Grant, J., Dyer, S., Woynillowicz, D., 2009. Oil Sand Myths. Clearing the Air. PEMBINA Institute, Drayton Valley, Alberta. Available from: <https://www.pembina.org/reports/clearing-the-air-report.pdf>.

Gray, J., Whyte, I., Curry, P., 2018. Ecocentrism: what it means and what it implies. Ecol. Citizen 1, 130–1.

Greenfield, N., 2015. 10 Threats From the Canadian Tar Sands Industry. Natural Resources Defense Council (NRDC), New York, <https://www.nrdc.org/stories/10-threats-canadian-tar-sands-industry> (accessed 14.12.19.).

Guardian, The, 2010. Tarnished Earth: The Destruction of Canada's Boreal Forest. The Guardian, United Kingdom, <https://www.theguardian.com/environment/gallery/2010/sep/07/tarnished-earth-oil-sands> (accessed 14.12.19.).

Guardian, The, 2011. Canada Pulls Out of Kyoto Protocol. The Guardian, United Kingdom, <https://www.theguardian.com/environment/2011/dec/13/canada-pulls-out-kyoto-protocol> (accessed 14.12.19.).

Harris, R., 2016. Schweitzer and Africa. Hist. J. 59 (4), 1107–1132.

Hatch, C., Price, M., 2008. Canada's Toxic Tar Sands. The Most Destructive Project on Earth. Environmental Defense, Toronto. Available from: <https://d36rd3gki5z3d3.cloudfront.net/wp-content/uploads/2016/01/TarSands_TheReport.pdf?x73948>.

Hayden, A., 2014. When Green Growth Is Not Enough: Climate Change, Ecological Modernization and Sufficiency. McGill Queen University Press, Canada.

Hill, D., 2014. Canadian Mining Doing Serious Environmental Harm, the IACHR Is Told. The Guardian, United Kingdom, <https://www.theguardian.com/environment/andes-to-the-amazon/2014/may/14/canadian-mining-serious-environmental-harm-iachr>.

Hill, D., 2016. Gold-Mining in Peru: Forests Razed, Millions Lost, Virgins Auctioned. The Guardian, United Kingdom. Available: <https://www.theguardian.com/environment/andes-to-the-amazon/2016/may/01/gold-mining-in-peru-forests-razed-millions-lost-virgins-auctioned> (accessed 14.12.19.).

Holden, M., 2019. Trump Opens Protected Alaskan Arctic Refuge to Oil Drillers. The Guardian, United Kingdom, <https://www.theguardian.com/us-news/2019/sep/12/trump-arctic-national-wildlife-refuge-oil-gas-drilling> (accessed 13.12.19.).

Ismi, A., 2009. Path of Destruction. Canadian Mining Companies on Rampage Around the World. Canadian Centre for Policy Alternatives, Ottawa, <https://www.policyalternatives.ca/publications/monitor/path-destruction> (accessed 14.12.19.).

Jamasmie, C., 2015. Latin America's Mining Industry Remains Optimistic—Report. MINING.com, Vancouver, BC, <https://www.mining.com/latin-americas-mining-industry-remains-optimistic-report/> (accessed 14.12.19.).

Kalman, J., 2015. Downstream from the tar sands, people are dying. In: Heal, T. (Ed.), The Harper Record. Canadian Centre for Policy Alternatives, Ottawa, pp. 275–280.

Kant, I., 1909–14. Fundamental Principles of the Metaphysic of Morals. Second Section: Transition from popular moral philosophy to the Metaphysic of Morals. In: Literary and Philosophical Essays. The Harvard Classics. 1909–14. Available from: <http://www.bartleby.com/32/603.html>.

Kant, I., 1963. Of duties to animals and spirits. Immanuel Kant Lectures on Ethics (Louis Infield, Trans.). Foreword to the Touchbook Edition by Lewis White Beck. Harper Torchbooks, New York, pp. 239–241.

Kemmerer, L., 2016. Speaking Up for Animals: An Anthology of Women's Voices. Routledge, Oxon, New York.

Kluckhohn, C., 1951. Values and value orientations in the theory of action: an exploration in definition and classification. In: Parsons, T., Edward Shils, E. (Eds.), Toward a General Theory of Action. Harvard University Press, Cambridge, pp. 388–433.

Klonoski, R.J., 1991. Callicott's holism: a clue for a classical realist contribution to the debate over the value of animals. Between Species 7 (2), 97–101.

Kravchenko, S., et al., 2013. Principles of international environmental law. In: Alam, S., et al., (Eds.), Routledge Handbook of International Environmental Law. Routledge, Oxon, New York, pp. 43–60.

Lamb, K.L., 1999. Ethical discourse. An exploration of theories in environmental ethics. In: Soden, D., Steel, B.R. (Eds.), Handbook of Global Environmental Policy and Administration. Marcel DeKker, Inc., New York, Basel, pp. 243–259.

Leopold, A., 1949. A Sand County Almanac. Oxford University Press, New York.

Leopold, A., 1999. In: Callicott, J.B., Freyfogle, E.T. (Eds.), For the Health of the Land: Previously Unpublished Essays and Other Writings by Aldo Leopold. Island Press, Washington, DC.

Lewis, S., 2019. Trump Administration Reauthorizes Use of "Cyanide Bombs" to Kill Wild Animals. CBS News. <https://www.cbsnews.com/news/trump-administration-reauthorizes-use-of-cyanide-bombs-to-kill-wild-animals/> (accessed 14.12.19.).

Linnitt, C., 2014. How Beaver Lake Cree Heats Up the Landmark Tar Sands Trial. The Huffington Post. <https://www.huffingtonpost.ca/carol-linnitt/tar-sands-trial_b_4352278.html?guccounter = 1> (accessed 14.12.19.).

MacDonald, H.P., 2004. John Dewey and Environmental Philosophy. State University of New York Press, Albany, NY.

Mackellar, C., Jones, D.A. (Eds.), 2012. Chimera's Children: Ethical, Philosophical and Religious Perspectives on Human-Nonhuman Experimentation. Continuum International Publishing Group, London, New York.

Martin, M.W., 1993. Rethinking reverence for life. Between Species 9 (4), 204–213.

Martin, A.M., 2006. How to argue for the value of humanity. Pac. Philos. Q. 87, 96–125.

Mathews, F., 2010. Environmental philosophy. In: Trakakis, N., Oppy, G. (Eds.), A Companion to Philosophy in Australia and New Zealand. Monash University Publishing, Melbourne, pp. 162–171.

McDonald, L., 2013. Crimes against ecology. Is the Harper government guilty? You be the judge. Alternatives J. <https://www.alternativesjournal.ca/policy-and-politics/crimes-against-ecology> (accessed 26.05.20.).

Mill, J.S., 1907. Utilitarianism. Longmans, Green and Co., London, New York, Bombay, Calcutta.

Mill, J.S., 1985. The collected works of John Stuart Mill, volume X—essays on ethics, religion, and society. In: Robson, J.M. (Ed.), Introduction by F.E.L. Priestley. University of Toronto Press, Toronto; Routledge and Kegan Paul, London.

Muradov, N., 2014. Liberating Energy From Carbon: Introduction to Decarbonization. Springer, New York, Heidelberg, Dordrecht, London.

Nadeau, C., 2010. Rogue in Power: Why Stephen Harper Is Remaking Canada by Stealth? James Lorimer & Company Ltd., Publishers, Toronto.

Najam, A., Papa, M., Taiyab, N., 2006. Global Environmental Governance. A Reform Agenda. International Institute for Sustainable Development.

National Post, 2012. Oil Sands Death of Hundreds of Ducks in 2010 Blamed on Weather, No Charges Laid. National Post, Canada. <http://news.nationalpost.com/news/canada/oil-sands-death-of-hundreds-of-ducks-in-2010-blamed-on-weather-no-charges-laid> (accessed 14.12.19.).

Nikiforuk, A., 2010. Tar Sands: Dirty Oil and the Future of a Continent. D & M Publisher, Vancouver, Toronto, Berkeley, CA.

NRDC, 2013. Climate Impacts of the Keystone XL Tar Sands Pipeline. Natural Resources Defense Council (NRDC), New York.

NRDC, 2014. Tar Sands Crude Oil: Health Effects of a Dirty and Destructive Fuel. Natural Resources Defense Council (NRDC), New York. Available from: <https://www.nrdc.org/sites/default/files/tar-sands-health-effects-IB.pdf>.

Owens, B., 2014. Canadian government accused of destroying environmental archives. Nature. Available from: https://doi.org/10.1038/nature.2014.14539. <https://www.nature.com/news/canadian-government-accused-of-destroying-environmental-archives-1.14539>.

Oxfam, 2014. Guatemala: Conflict Over Mining Fuels Violence as Companies Fail to Respect Community Rights. Oxfam America Inc, Boston, MA, <https://policy-practice.oxfamamerica.org/work/in-action/guatemala-conflict-over-mining-fuels-violence-as-companies-fail-to-respect-community-rights/> (accessed 14.12.19.).

Palmer, C., 1994. A bibliographical essay on environmental ethics. Stud. Christ. Ethics 7, 68–79.

Preece, R., Chamberlain, L., 1993. Animal Welfare & Human Values. Wilfrid Laurier University Press, Waterloo.

Price, M., 2008. The Tar Sands' Leaking Legacy. Environmental Defense, Toronto. Available from: <https://www.macleans.ca/wp-content/uploads/2008/12/tailingsreport_finaldec8.pdf>.

Raffensperger, C., Tickner, J.A. (Eds.), 1999. Protecting Public Health and the Environment: Implementing the Precautionary Principle. Foreword by Wes Jackson. Island Press, Washington, DC.

Regan, T., 1975. The moral basis of vegetarianism. Can. J. Philos. 5 (2), 181–214.

Regan, T., 2002. The case for animal rights. In: LaFollette, H. (Ed.), Ethics in Practice. Wiley-Blackwell, Malden, MA, pp. 140–148.

Regan, T., 2003. Animal Rights, Human Wrongs: An Introduction to Moral Philosophy. Rowan & Littlefield Publishers, Inc., Lanham, Boulder, New York, Toronto, Oxford.

Regan, T., 2004. The Case for Animal Rights. University of California, Berkeley, CA.

Regan, T., 2017. The radical egalitarian case for animal rights. In: Pojman, L.J., Pojman, P., McShane, K. (Eds.), Environmental Ethics: Readings in Theory and Application. Cengage Learning, Boston, MA, pp. 106–113.

Remele, K., 2013. Animal protection and environmentalism. In: Linzey, A. (Ed.), The Global Guide to Animal Protection. University of Illinois Press, Urbana, Chicago, IL, Springfield, pp. 68–69.

Robinson, R. (Ed.), 2010. Political politics of economic development. In: Developing the Third World: The Experience of the Nineteen-Sixties. Cambridge University Press, Cambridge, pp. 1–17.

Rolston III, H., 2003. Environmental ethics. In: Bunnin, N., Tsui-James, E.P. (Eds.), The Blackwell Companion to Philosophy, second ed. Blackwell Publishing, Oxford, pp. 517–530.

Ryder, R.D., 1975. Victims of Science: The Use of Animals in Research. Davis-Poynter Ltd, London.

Ryder, R.D., 2010. Speciesism again: the original leaflet. Crit. Soc. 2 (Spring), 1–2.

Sagoff, M., 1988. The Economy of the Earth. Cambridge University Press, Cambridge.

Sanger, T., Saul, G., 2015. The Harper government and climate change. Lost at sea? In: Heal, T. (Ed.), The Harper Record. Canadian Centre for Policy Alternatives, Ottawa, pp. 281–297.

Schmidtz, D., 1998. Are all species equal? J. Appl. Philos. 15 (1), 57–67.

Schneider, K., 2010. Tar Sands Oil Production, An Industrial Bonanza, Poses Major Water Use Challenges. Circle of Blue. <https://www.circleofblue.org/2010/world/tar-sands-oil-production-is-an-industrial-bonanza-poses-major-water-use-challenges/> (accessed 14.12.19.).

Schultz, C., 2013. Alberta's Oil Sands Account for 9 Percent of Canada's Carbon Dioxide Emissions. Smithsonian Institution, Washington, DC, <https://www.smithsonianmag.com/smart-news/albertas-oil-sands-account-for-9-percent-of-canadas-carbon-dioxide-emissions-180947747/> (accessed 14.12.19.).

Schwartz, S.H., 1992. Universals in the content and structure of values: theoretical, advances and empirical tests in 20 countries. In: Zanna, M. (Ed.), Advances in Experimental Social Psychology, vol. 25. Academic Press, New York, pp. 1–65.

Schweitzer, A., 1923a. The Philosophy of Civilization. The Decay and the Restoration of Civilization. A & C Black, Ltd., London.

Schweitzer, A., 1923b. Dale Lecturer at Mansfield College, Oxford, 1923. Oxford Centre for Animal Ethics. Available from: <http://www.oxfordanimalethics.com/2009/01/692/>.

Schweitzer, A., 1929. Civilization and Ethics: The Philosophy of Civilization Part II. A & C Black, London.

Schweitzer, A., 1948. My Life and Thought. George Allen & Unwin Ltd, London.

Schweitzer, A., 1949. The Philosophy of Civilization. Civilization and Ethics, third ed. Adam & Charles Black, London.

Schweitzer, A., 1965. The Teaching of Reverence for Life. Holt, Rinehart and Winston, New York.

Schweitzer, A., 1987. The Philosophy of Civilization. Prometheus, Amherst, NY.

Schweitzer, A., 1998. Out of My Life and Thought. An Autobiography. The Johns Hopkins University Press, Baltimore, MD, London.

Sebo, J., 2004. A critique of the Kantian theory of indirect moral duties to animals. Anim. Liberation Philos. Policy J. II (2), 1−19.

Singer, P., 1974. All animals are equal. Philos. Exch. 1 (5), 243−257.

Singer, P., 1975. Animal Liberation: A New Ethics for Our Treatment of Animals. Random House, New York.

Singer, P., 2009. Speciesism and moral status. Metaphilosophy 40 (3−4), 567−581. July.

Singer, P., 2011. Practical Ethics. Cambridge University Press, Cambridge, New York.

Smart, J.J.C., 1973. An outline of a system of utilitarian ethics. In: Smart, J.J.C., Williams, B. (Eds.), Utilitarianism: For and Against. Cambridge University Press, New York, pp. 1−74.

Sowunmi, J., 2014. The Harper Government Has Trashed and Destroyed Environmental Books and Documents. VICE News, Canada. <https://www.vice.com/en_ca/article/4w578d/the-harper-government-has-trashed-and-burned-environmental-books-and-documents> (accessed 14.12.19.).

Steward, G., 2015. Tailings Ponds a Toxic Legacy of Alberta's Oilsands. The Toronto Star. Available: Canada. <https://www.thestar.com/news/atkinsonseries/2015/09/04/tailings-ponds-a-toxic-legacy-of-albertas-oil-sands.html> (accessed 14.12.19.).

Taylor, P.W., 1981. The ethics of respect for nature. Environ. Ethics 3 (3), 197−218.

Taylor, P.W., 1983. In defense of biocentrism. Environ. Ethics 5, 237−243.

Taylor, P.W., 1986. Respect for Nature: A Theory of Environmental Ethics. Princeton University Press, Princeton, NJ.

Thompson, L., 2017. Anthropocentrism. Humanity as Peril and Promise. In: Gardiner, S.M., Thompson, L. (Eds.), The Oxford Handbook of Environmental Ethics. Oxford University Press, Oxford, pp. 77−90.

Walsh, B., 2011. Bienvenue au Canada: Welcome to Your Friendly Neighborhood Petro-State. Time. <http://science.time.com/2011/12/14/bienvenue-au-canada-welcome-to-your-friendly-neighborhood-petrostate/> (accessed 14.12.19.).

Warren, M.A., 1987. Difficulties with the strong rights position. Between Species 2 (4), 433−441.

WCED, 1987. Our Common Future: Report of the World Commission on Environment and Development: Our Common Future. World Commission on Environment and Development (WCED), United Nations. Oxford University Press, Oxford, New York.

Wells, J., et al., 2008. Danger in the Nursery: Impacts of Tar Sands Oil Development in Canada's Boreal Forest. The Natural Resources Defense Council (NRDC), New York. The Boreal Songbird Initiative (BSI), Seattle. Pembina Institute, Calgary. Available from: <https://www.nrdc.org/sites/default/files/borealbirds.pdf>.

Wilson, S.D., 2017. Bioethics. Wright State University. United States. Internet Encyclopedia of Philosophy (ISSN 2161-0002). Available from: <http://www.iep.utm.edu/bioethic/>.

Yancy, G., Singer, P., 2015. Peter Singer: On Racism, Animal Rights and Human Rights. The Stone. The New York Times. <https://opinionator.blogs.nytimes.com/2015/05/27/peter-singer-on-speciesism-and-racism/#more-157145> (accessed 14.12.19.).

Yoshida, F., 2012. Lecture on environmental economics. The Theory of Environment Governance. Hokkaido University Press, Sapporo, pp. 75−103.

Zhang, S., 2017. Looking Back at Canada's Political Fight Over Science. The Atlantic. <https://www.theatlantic.com/science/archive/2017/01/canada-war-on-science/514322/> (accessed 26.05.20.).

Chapter 5

Alexander, J., Diana Betz, D., Gonnerman, C., Waterman, J.P., 2017. Framing how we think about disagreement. Philos. Stud. 1−28.

Borneo Project, 2018. Villagers' Water Source Destroyed by Company, Police Brutally Dismantled Peaceful Blockade. Borneo Project, Earth Island Institute, Berkeley, CA. <https://borneoproject.org/updates/villagers-water-source-destroyed-by-company-police-brutally-dismantled-peaceful-blockade> (accessed 10.03.20.).

Callicott, J.B., 1989. In Defense of Land Ethic: Essays in Environmental Philosophy. State University of New York Press, Albany, NY.

Cannon, J.C., 2017. Leading US plywood firm linked to alleged destruction, rights violations in Malaysia. Mongabay. <https://www.news.mongabay.com/2017/10/leading-us-plywood-firm-linked-to-alleged-destruction-rights-violations-in-malaysia/#:~:text=Leading%20US%20plywood%20firm%20linked%20to%20alleged%20destruction%2C%20rights%20violations%20in%20Malaysia,-by%20John%20C&text=An%20investigation%20has%20found%20that,logging%20and%20indigenous%20rights%20violations> (accessed 15.01.20.).

Choy, Y.K., 2004. Sustainable development and the social and cultural impacts of dam-induced development strategy—the Bakun experience. Pac. Aff. 77 (1), 50–68.

Choy, Y.K., 2014. Land ethic from the Borneo tropical rainforests in Sarawak, Malaysia: an empirical and conceptual analysis. Environ. Ethics 36 (4), 421–441.

Choy, Y.K., 2018a. Cost-benefit analysis, values, wellbeing and ethics: an indigenous worldview analysis. Ecol. Econ. 145, 1–9.

Choy, Y.K., 2018b. Sustainable development and environmental stewardship: the Heart of Borneo paradox and its implications on green economic transformations in Asia. In: Hsu, S. (Ed.), Routledge Handbook of Sustainable Development in Asia. Routledge, Oxon, New York, pp. 532–549.

Choy, Y.K., Onuma, A., 2014. Sustainable biodiversity use and traditional knowledge of the indigenous society in the tropical rainforest in Sarawak. Rainforest. Environ. Econ. Policy Stud. 7 (1), 69–73 (in Japanese).

Colchester, M., 1993. Pirates, squatters and poachers: the political ecology of dispossession of the native peoples of Sarawak. Glob. Ecol. Biogeogr. Lett. 3, 158–179.

Durkheim, E., 1915. The Elementary Forms of the Religious Life. Translated From the French by Joseph Ward Swain. George Allen & Unwin Ltd., London.

Durkheim, E., 1953. Sociology and Philosophy. Translated by D.F. Pocock With an Introduction by J.G. Peristiany. Cohen & West, London.

Durkheim, E., 1972. Emile Durkheim: Selected Writings. Edited, Translated, and With a Introduction by Anthony Giddens. Cambridge University Press, London.

Erenberg, D., 2015. Enforcing human rights for people, animals, and the planet. In: Kemmere, L. (Ed.), Animals and the Environment: Advocacy, Activism, and the Quest for Common Ground. Routledge, Oxon, New York, pp. 239–248.

Fiske, A.P., et al., 1998. The cultural matrix of social psychology. In: Daniel Gilbert, T., Fiske, S.T., Lindzey, G. (Eds.), The Handbook of Social Psychology. McGraw-Hill, New York, pp. 915–981.

Gabungan (The Coalition of Concerned NGOs on Bakun, Malaysia), 1999. The Resettlement of Indigenous People affected by the Bakun Hydro-Electric Project, Sarawak. World Commission on Dams, Cape Town. Available from: <https://www.internationalrivers.org/sites/default/files/attached-files/resettlement_of_indigenous_people_at_bakun.pdf>.

Global Witness, 2012. In the Future, There Will Be No Forests Left. Global Witness, London.

Global Witness, 2014. Japan's Timber Imports Fuelling Rainforest Destruction in Sarawak and Violation of Indigenous Land Rights: New Findings From Recent Research and Field Investigations. Global Witness, London.

Hsu Francis, L.K., 1948. Under the Ancestor's Shadow: Chinese Culture and Personality. Columbia University Press, New York.

Kant, I., 2002. Metaphysis of Morals. Edited and translated by Allen W. Wood with Essays by Schneewind, J.B., Baron, M., Allen W. Wood. Yale University Press, New Haven, CT, London.

Lazier, J., 2010. Kantian ethics. In: Corrigan, R.H., Farrell, M.E. (Eds.), Ethics: A University Guide. Progressive Frontier Press, Gloucester, pp. 207–220.

Markus, H.R., Kitayama, S. (Eds.), 1994. The cultural construction of self and emotion: implications for social behavior. In: Emotion and Culture: Empirical Studies of Mutual Influence. American Psychological Association, Washington, DC, pp. 89–132.

McCormick, M., 2018. Immanuel Kant: Metaphysics. Internet Encyclopedia of Philosophy. Available from: <http://www.iep.utm.edu/kantmeta>.

Rachels, J., 1997. The best action is the one in accord with universal rule. In: Ermann, M.D., Williams, M.B., Shauf, M.S. (Eds.), Computers, Ethics, and Society, second ed. Oxford University Press, New York, Oxford, pp. 42–46.

SAM, 1996. The Social Impact of the Bakun Hydroelectric Dam Project on the Indigenous Peoples of the Balui Region. Sarawak, Malaysia, A Report by Sahabat Alam Malaysia (SAM). Penang, Malaysia: Sahabat Alam Malaysia, [Friends of the Earth Malaysia].

Semken, S., 2005. Sense of place and place-based introductory geosciences teaching for American Indian and Alaska native undergraduates. J. Geosci. Educ. 53 (2), 149–157.

Stern, P.C., Dietz, T., 1994. The value basis of environmental concern. J. Soc. Issues 50 (3), 65–84.

Stern, P.C., Dietz, T., Kalof, L., 1993. Value orientations, gender, and environmental concern. Environ. Behav. 25 (5), 322–348.

Venn, T.J., Quiggin, J., 2007. Accommodating indigenous cultural heritage values in resource assessment: Cape York Peninsula and the Murray–Darling Basin, Australia. Ecol. Econ. 61 (2), 334–344.

Chapter 6

Ariew, R., et al., 2010. The A to Z of Descartes and Cartesian Philosophy. The A to Z Guide Series, No. 155. The Scarecrow Press, Inc., Lanham, Toronto, Plymouth.

Ballantine, J., Hammack, F.M., 2012. The Sociology of Education: A Systematic Analysis, seventh ed. Routledge, London, New York.

Barton, B., 1981. The sea. In: Carman, B. (Ed.), Nature: The World's Best Poetry, Series 1, vol. V. Granger Book Co., Inc., Great Neck, NY, pp. 377–378.

Bernauer, O.M., Gaines-Day, H.R., Steffan, S.A., 2015. Colonies of bumble bees (Bombus impatiens) produce fewer workers, less bee biomass, and have smaller mother queens following fungicide exposure. Insects 6, 478–488.

Berryman, T., 1999. Relieving modern day atlas of an illusionary burden: abandoning the hypermodern fantasy of education to manage the globe. Can. J. Environ. Educ. 4, 50–68.

Bijay, P., Boodhan, M., Chmutina, K., 2012. Transnational Recommendation Report. Prepared for DIREKT—Small Developing Island Renewable Energy Knowledge and Technology Transfer Network. DIREKT Project Consortium.

Brandt, A., et al., 2016. The neonicotinoids thiacloprid, imidacloprid, and clothianidin affect the immunocompetence of honey bees (Apis mellifera L.). J. Insect Physiol. 86, 40–44.

Buchmann, S.L., Nabhan, G.P., 1996. The Forgotten Pollinators. Island Press/Shearwater Books, Washington, DC, Covelo, CA.

Byrne, J.M., 1996. Religion and the Enlightenment: From Descartes to Kant. Westminster John Knox Press, Louisville, KY.

Calhoun, C., et al., (Eds.), 2012. Classical Sociological Theory. third ed. Wiley-Blackwell, United Kingdom.

Carroll, L., 2010. Alice in Wonderland and Through the Looking Glass. William Collins.

CEE, 2007. Final Report. 4th International Conference on Environmental Education. Environmental Education Towards a Sustainable Future—Partners for the Decade of Education for Sustainable Development. Centre for Environment Education (CEE), Ahmedabad.

Čiegis, R., Gineitienė, D., 2006. The role of universities in promoting sustainability. Eng. Econ. 3 (48), 56–62.

Coffey, A., 2004. Reconceptualizing Social Policy: Sociological Perspectives on Contemporary Social Policy. Open University Press, United Kingdom.

Cooke, S.J., et al., 2017. Troubling issues at the frontier of animal tracking for conservation and management. Conserv. Biol. 31 (5), 1205–1207.

Dahnke, M.D., Dreher, H.M., 2016. Philosophy of Science for Nursing Practice, Concepts and Applications, second ed. Springer, New York.

de Ciurana, A.M.G., Leal Filho, W., 2006. Education for sustainability in university studies experiences from a project involving European and Latin American universities. Int. J. Sustain. High. Educ. 7 (1), 81–93.

Descartes, R., 1913. Descartes' Meditations and Selections From the Principles of Philosophy (J. Veitch, Trans.). Open University Press, Chicago, IL.

Descartes, R., 1985. The Philosophical Writings of Descartes, vol. I (J. Cottingham, R. Stoothoff, and D. Murdoch, Trans.). Cambridge University Press, Cambridge, New York.

De Trincheria, J., et al., 2015. Adapting agriculture to climate change by developing promising strategies using analogue locations in Eastern and Southern Africa: a systematic approach to develop practical solutions. Transforming rural livelihoods. In: Leal Filho, et al., (Eds.), Adapting African Agriculture to Climate Change. Springer, Cham, Heidelberg, New York, Dordrecht, London, pp. 1–23.

Dickson, A., 2010. The carbon dioxide system in seawater: equilibrium chemistry and measurements. In: Riebesell, U., Fabry, V.J., Hansson, L., Gattuso, J.-P. (Eds.), Guide to Best Practices for Ocean Acidification Research and Data Reporting. Publications Office of the European Union, Luxembourg, pp. 17–40.

Dodds, J., 2011. Psychoanalysis and Ecology at the Edge of Chaos. Routledge, East Sussex, New York.

Dodds, W.K., Whiles, M.R., 2010. Freshwater Ecology. Concepts and Environmental Applications of Limnology, second ed. Elsevier Ltd., United States, United Kingdom.

Durkheim, E., 1893. The Division of Labour in Society. English Translation by George Simpson 1933. Macmillan, New York.

Durkheim, E., 1915. The Elementary Forms of the Religious Life. Translated From the French by Joseph Ward Swain. George Allen & Unwin Ltd., London.

Durkheim, E., 1951. Suicide (J.A. Spaulding, G. George Simpson, Trans.). Free Press, New York.

Durkheim, E., 1958. Professional Ethics and Civic Morals (C. Brookfield, Trans.). The Free Press, Glencoe, IL.

Durkheim, E., 1961. Moral Education: A Study in the Theory and Application of the Sociology of Education (E.K. Wilson, H. Schnurer, Trans.). Free Press, New York.

Durkheim, E., 1973a. Moral Education. A Study in the Theory and Application of Sociology of Education. Edited with a new introduction by Everett K. Wilson. The Free Press, New York.

Durkheim, E., 1973b. Emile Durkheim on Morality and Society. Selected Writings. Edited and with an Introduction by Robert N. Bellah. The University of Chicago Press, Chicago, IL, London.

Durkheim, E., 1973c. The dualism of human nature and its social conditions. Emile Durkheim on Morality and Society. Selected Writings. Edited and with an Introduction by Robert N. Bellah. The University of Chicago Press, Chicago IL, London, pp. 149–163.

Durkheim, E., 1982. The Rules of Sociological Method. Lukes, S. (Ed.); W.D. Halls (Trans.). The Free Press, New York, London, Toronto, Sydney.

Durkheim, M., 2002. Moral Education. Émile Durkheim. Translated and with a Preface by Everett K. Wilson and Herman Schnurer. Foreword by Paul Fauconnet. Dover Publications, Inc., Mineola, NY.

Farinha, C., Caeiro, S., Azeiteiro, U., 2019. Sustainability strategies in Portuguese Higher Education Institutions: commitments and practices from internal insights. Sustainability 2019 (11), 1–25. Available from: https://doi.org/10.3390/su11113227.

Feely, R.A., Sabine, C.L., Takahashi, T., Wanninkhof, R., 2001. Uptake and storage of carbon dioxide in the ocean: the Global CO_2 Survey. Oceanography 14 (4), 18–32.

Fish, J.S., 2013. Homo duplex revisited: a defense of Émile Durkheim's theory of the moral self. J. Class. Sociol. 13 (3), 338–358.

Flynn, R., Bellaby, P., Ricci, M., 2009. The 'value-action gap' in public attitudes towards sustainable energy: the case of hydrogen energy. Sociol. Rev. 57 (2), 159–180.

Forrester, J.W., 1961. Industrial Dynamics. The MIT Press, Cambridge, MA.

Forrester, J.W., 1971a. World Dynamics. Wright-Allen Press, Inc., Cambridge, MA.

Forrester, J.W., 1971b. Counterintuitive behavior of social systems. Theory Decis. 2, 109–140.

FTS-ALS, 2015. CELA—Network of Climate Change Technology Transfer Centres in Europe and Latin America. Hamburg University of Applied Sciences Faculty of Life Sciences Research and Transfer Centre (FTZ-ALS).

Gadotti, M., 2009. Education for Sustainability: A Contribution to the Decade of Education for Sustainable Development. Editora e Livraria Instituto Freire, Sao Paulo.

Gough, A., 1999. Recognizing women in environmental education pedagogy and research: toward an ecofeminist poststructuralist perspective. Environ. Educ. Res. 5 (2), 143–161.

Goulson, D., et al., 2015. Bee declines driven by combined stress from parasites, pesticides, and lack of flowers. Science 1–16. Available from: https://doi.org/10.1126/science.1255957.

Greenpeace, 2013. A review of factors that put pollinators and agriculture in Europe at risk. Greenpeace Research Laboratories Technical Report (Review) 01/2013. Greenpeace International, The Netherlands.

Grossman, E., 2013. Declining Bee Populations Pose a Threat to Global Agriculture. e-360 Yale. Available from: <https://e360.yale.edu/features/declining_bee_populations_pose_a_threat_to_global_agriculture>.

Guidi, L., et al., 2016. Plankton networks driving carbon export in the oligotrophic ocean. Nature 532, 465–470. Available from: https://doi.org/10.1038/nature16942.

Gurbuz, M.E., 2008. Collective consciousness. In: Parrillo, V.N. (Ed.), Encyclopedia of Social Problems, vol. 1. Sage Publication, United Kingdom, pp. 142–143.

Honjo, S., et al., 2014. Understanding the role of the biological pump in the global carbon cycle: an imperative for ocean science. Oceanography 27 (3), 10–16.

IAU, 2007. Annual Report. International Association for Universities (IAU). UNESCO House, Paris.

ICRISAT, 2014. Adapting agriculture to climate change: developing promising strategies using analogue locations in Eastern and Southern Africa. Final Report 1 January 2011−30 June 2014. International Crops Research Institute for the Semi-Arid Tropics (ICRISAT), Addis Ababa.

IGES, 2002. Regional Strategy on Environmental Education in the Asia-Pacific. Institute for Global Environmental Strategies (IGES), Hayama, Kanagawa.

Interreg Baltic Sea Region, 2010. WATERPRAXIS—HAW Hamburg: Summer Course on Sustainability, River Basin Management and Climate Change. Baltic Sea Region, Rostock. <http://eu.baltic.net/ Project_Database.5308.html?contentid = 1&contentaction = single&bsr_usercontent = news&bsr_usercontent_ action = view&news_id = 63> (accessed 14.03.20.).

IPCC, 2005. IPCC Special Report on Carbon Dioxide Capture and Storage. Bert Metz, B. et al (Eds.). Cambridge University Press, Cambridge, New York.

IPCC, 2013. Climate change 2013: the physical science basis. In: Stocker, T.F., et al., (Eds.), Contribution of Working Group I to the Fifth Assessment Report of the Intergovernmental Panel on Climate Change (IPCC). Cambridge University Press, Cambridge, New York.

IUCN, 1970. International Working Meeting on Environmental Education in the School Curriculum. Final Report From IUCN-UNESCO Seminar. Foresta Institute for Ocean and Mountains Studies, Carson City, NV.

IUCN-UNEP-WWF, 1991. Caring for the Earth: A Strategy for Sustainable Living. The World Conservation Union (IUCN). United Nations Environment Programmes (UNEP). World Wildlife Fund (WWF), Gland.

Japanese National Commission for UNESCO, 2000. The Third UNESCO/Japan Seminar on Environmental Education in Asia-Pacific Region, November 30−December 2, Tokyo. Final Report. Japanese National Commission for UNESCO, Ministry of Education, Science, Sports and Culture, Japan.

JELARE partner consortium, 2011. JELARE Transnational Recommendation Report. Fostering Innovative Labour Market-Oriented Educational and Research Approaches in the Field of Renewable Energy at Latin American and European Institutes of Higher Education. JELARE/Alfa, Europe Aid Cooperation Office, Hamburg.

Kämpf, J., Chapman, P., 2016. Upwelling Systems of the World: A Scientific Journey to the Most Productive Marine Ecosystems. Springer, Switzerland.

Keep.eu, 2017. Project—RECO Baltic 21 Tech. Keep.eu, European Union. <https://www.keep.eu/project/15692/ reco-baltic-21-tech> (accessed 13.03.20.).

Kelsey, E., 2003. Constructing the public: implications of the discourse of international environmental agreements on conceptions of education and public participation. Environ. Educ. Res. 9 (4), 403−427.

Kifle, W.H., Teklemichael, M., Gesesew, H.A., 2014. Pesticide Use Knowledge, Attitudes, and Practice and Related Short-Term Health Problems Among Farmers Using Irrigation in Southwest Ethiopia, 2014. Global Disaster Preparedness Center, Ethiopia.

Kirk, J.A., 2007. The Future of Reason, Science and Faith. Following Modernity and Post-Modernity. Ashgate Publishing Limited, United Kingdom.

Ko, N., Osamu, A., 2015. Sustainability and Higher Education in Asia and the Pacific. The Global University Network for Innovation (GUNi), Barcelona.

Kollmuss, A., Agyeman, J., 2002. Mind the gap: why do people act environmentally and what are the barriers to pro-environmental behavior? Environ. Educ. Res. 8 (3), 239−260.

Krnel, D., Naglič, S., 2009. Environmental literacy comparison between eco-schools and ordinary schools in Slovenia. Sci. Educ. Int. 20 (1/2), 5−24.

Leal Filho, W. (Ed.), 2011. Applied sustainable development: a way forward in promoting sustainable development in higher education institutions. In: World Trends in Education for Sustainable Development. Peter Lang, Internationaler Verlag der Wissenschaften.

Leal Filho, W., Mannke, F., 2011. Adapting agriculture to climate change by developing promising strategies using analogue locations in Eastern and Southern Africa: introducing the Calesa Project. In: Leal Filho (Ed.), Experiences of Climate Change Adaptation in Africa. Springer-Verlag, Berlin, Heidelberg, pp. 247−253.

Leal Filho, W., Manolas, E., Pace, P., 2015. The future we want: key issues on sustainable development in higher education after Rio and the UN decade of education for sustainable development. Int. J. Sustain. High. Educ. 16, 112−129.

Leal Filho, et al., 2017. Identifying and overcoming obstacles to the implementation of sustainable development at universities. J. Integr. Environ. Sci. 14 (1), 93−108.

Lewinsohn, T.M., et al., 2015. Ecological literacy and beyond: problem-based learning for future professionals. Ambio 44 (2), 154−162.

Lewis, A., 2017. Is Tracking Technology Putting Creatures in Danger? The Irish Times. <http://www.irishtimes.com/news/science/is-tracking-technology-putting-creatures-in-danger-1.3035028> (accessed 15.12.19.).

Long, J., 2014. Fighting back against the neonicotinoid lobby. In: Harrison, R. (Ed.), The Product of Forced Labour? Friend of the Earth, p. 13. Available from: <https://cdn.friendsoftheearth.uk/sites/default/files/downloads/Ethical%20consumers%20guide%20to%20honey.pdf>.

Lozanno, R., 2016. Higher education for Sustainable Development. Universidad de Castilla La Mancha Albacete, Spain.

Marans, S., Dahl, K., Schowalter, J., 2002. Child and adolescent psychotherapy: psychoanalytic principles. In: Hersen, M., Sledge, W. (Eds.), Encyclopedia of Psychotherapy, vol. 1. Academic Press, San Diego, CA, pp. 381−400.

Mills, S.L., Soulé, M.E., Doak, D.F., 1993. The keystone-species concept in ecology and conservation. BioScience 43 (4), 219−224.

Ministry for the Environment Land and Sea of Italy, 2006. UNECE Steering Committee on Education for Sustainable Development. Fourth Meeting Geneva, 19−20 February. Ministry for the Environment Land and Sea, Italy.

Mostafa, K.T., et al., 1992. The World Environment 1972−1992. Two Decades of Challenge. Chapman & Hall, New York.

Myers, J., Noebel, D.A., 2015. Understanding the Times. A Survey of Competing Worldviews. David C. Cook, Colorado Springs.

NAAEE, 2010. Excellence in Environmental Education: Guidelines for Learning (K-12). North American Association for Environmental Education (NAAEE), Washington, DC. Available from: <http://resources.spaces3.com/89c197bf-e630-42b0-ad9a-91f0bc55c72d.pdf>.

NOAA, 2006. Learning Ocean Science Through Ocean Exploration, third ed. National Oceanic and Atmospheric Administration (NOAA), Washington, DC.

NRDC, 2011. Why We Need Bees: Nature's Tiny Workers Put Food on Our Tables. Natural Resources Defense Council (NRDC), United States. Available from: <https://www.nrdc.org/sites/default/files/bees.pdf>.

Ocean and Climate, 2015. Ocean and Climate Scientific Note. Ocean and Climate Platform. Ocean and Climate Organization, France.

OSSE, 2017. 2017 DC Environmental Literacy Pan. Integrating Environmental Education into the K-12 Curriculum. Office of the State Superintendent of Education (OSSE), Washington, DC. Available from: <https://osse.dc.gov/sites/default/files/dc/sites/osse/page_content/attachments/2017%20Environmental%20Literacy%20Plan.pdf>.

Paine, R.T., 1969. A note on trophic complexity and community stability. Am. Nat. 103, 91−93.

Revkin, A. 2019. Most Americans Now Worry About Climate Change—And Want to Fix It. National Geographic. <https://www.nationalgeographic.com/environment/2019/01/climate-change-awareness-polls-show-rising-concern-for-global-warming/> (accessed 16.03.20.).

Ritchie, E.J., 2017. Good Intentions: Why Environmental Awareness Doesn't Lead to Green Behavior. Forbes, <https://www.forbes.com/sites/uhenergy/2017/06/30/good-intentions-why-environmental-awareness-doesnt-lead-to-green-behavior/#a2872da57d98> (accessed 16.03.20.).

Sabine, C.L., et al., 2004. Current status and past trends of the global carbon cycle. In: Field, C.B., Raupach, M.R. (Eds.), The Global Carbon Cycle: Integrating Humans, Climate, and the Natural World. Island Press, Washington, Covelo, London, pp. 17−44.

Sánchez-Bayo, F., et al., 2016. Are bee diseases linked to pesticides? Environ. Int. 89−90, 7−11.

Santos, M.A., Leal Filho, W., 2005. An analysis of the relationship between sustainable development and the anthroposystem concept. Int. J. Environ. Sustain. Dev. 4 (1), 78−87.

Sauve, L., 1996. Environmental education and sustainable development: a further appraisal. Can. J. Environ. Educ. 1, 7−34.

Saylan, C., Blumstein, D.T., 2011. The Failure of Environmental Education (and how we can fix it). University of California Press., Berkeley, Los Angeles, London.

Schacker, M.A., 2008. Spring Without Bees. How Colony Collapse Has Endangered our Food Supply. The Lyons Press, Guilford, CT.

SDC, 2006. I Will If You Will: Towards Sustainable Consumption. Sustainable Development Commission (SDC).

Sen Gupta, A., McNeil, B., 2012. Variability and change in the ocean. In: Henderson-Sellers, A., McGuffie, K. (Eds.), The Future of the World's Climate. Elsevier, Amsterdam, United States, United Kingdom, pp. 141–165.

Sidak, F., Sidak, S., 2014. A study on environmental knowledge and attitudes of teacher. Procedia Soc. Behav. Sci. 116, 2379–2385.

Siegel, D.A., et al., 2016. Prediction of the export and fate of global ocean net primary production: the exports science plan. Front. Mar. Sci. 3 (22), 1–22.

Simon, S., 2002. Participatory online environmental education at the Open University UK. In: Leal Filho, W. (Ed.), Sustainability—Towards Curriculum Greening. Environmental Education, Communication and Sustainability, 11. Peter Lan Scientific Publishers, Frankfurt, pp. 121–150.

Sobel, D., 1996. Beyond Ecophobia: Reclaiming the Heart in Nature Education. The Orion Society and The Myrin Institute, Great Barrington, MA.

Sobel, D., 1998. Beyond Ecophobia. YES Magazine, Madrona Way, Bainbridge Island. <http://www.yesmagazine.org/issues/education-for-life/803> (accessed 16.12.19.).

Sobel, D., 2008. Childhood and Nature: Design Principles for Educators. Stenhouse Publishers, Portland, ME.

Sobel, D., 2012. Look, Don't Touch. Orion Magazine, Great Barrington, MA. <https://orionmagazine.org/article/look-dont-touch1/> (accessed 16.12.19.).

Soule, M., 1998. Mind in the biosphere; mind of the biosphere. In: Wilson, E.O. (Ed.), Biodiversity. National Academy Press, Washington, DC, pp. 465–469.

Stanley, D.A., Raine, N.E., 2016. Chronic exposure to a neonicotinoid pesticide alters the interactions between bumblebees and wild plants. Funct. Ecol. 30, 1132–1139.

Stanley, D.A., et al., 2016. Investigating the impacts of field-realistic exposure to a neonicotinoid pesticide on bumblebee foraging, homing ability and colony growth. J. Appl. Ecol. 2016 (53), 1440–1449.

Šubrt, J., 2017. Homo sociologicus and the society of individuals. Historicá Sociologie 9–22.

Sundquist, E.T., Visser, K., 2005. The geologic history of carbon cycle. In: Schlesinger, W.H. (Ed.), Biogeochemistry. Elsevier, The Netherlands, United States, United Kingdom, pp. 425–472.

The Pacific Regional Data Repository (PRDR), 2015. Renewable Energy Knowledge and Technology Transfer (DIREKT) Project.

UNEP, 2009. Frequently Asked Questions. The Marrakech Process. Towards a 10-Year Framework Programmes on Sustainable Consumption and Production. United Nations Environment Programme (UNEP), Nairobi.

UNEP, 2010. Here and Now. Education for Sustainable Consumption. Recommendations and Guidelines. United Nations Environment Programme (UNEP), Nairobi.

UNESCO, 1976. The Belgrade Charter. The United Nations Educational, Scientific and Cultural Organization (UNESCO), Paris, UNESCO-UNEP Environmental Education Newsletter, 1 (1), 1–2.

UNESCO, 1978a. Intergovernmental Conference on Environmental Education. Organized by UNESCO in Co-Operation With UNEP. Tbilisi (USSR) 74—26 October 1977. Final Report. The United Nations Educational, Scientific and Cultural Organization (UNESCO), Paris.

UNESCO, 1978b. The Tbilisi Declaration. The United Nations Educational, Scientific and Cultural Organization (UNESCO), Paris, UNESCO-UNEP Environmental Education Newsletter 3 (1), 1–8.

UNESCO, 1984. Activities of the UNESCO-UNEP International Environmental Education Programme (1975–1983). The United Nations Educational, Scientific and Cultural Organization (UNESCO), Paris.

UNESCO, 1987. Moscow' 87: UNESCO-UNEP International Congress on Environmental Education and Training (USSR. 17–21 August 1987). The United Nations Educational, Scientific and Cultural Organization (UNESCO), Paris, UNESCO-UNEP Environmental Education Newsletter, 12(3),1–8.

UNESCO, 1989. Environmental Literacy for All. The United Nations Educational, Scientific and Cultural Organization, Paris, UNESCO-UNEP Environmental Education Newsletter, XIV (2), 1–2.

UNESCO, 1990. UNESCO-UNEP International Environmental Education Programme (IEEP). Environmental Education: Selected Activities of UNESCO-UNEP International Environmental Education Programme 1975–1990. The United Nations Educational, Scientific and Cultural Organization, Paris.

UNESCO, 1992a. Address by Mr. Federico Mayor, Director-General of the United Nations Educational, Scientific and Cultural Organization (UNESCO) at the Opening of the World Congress for Education and Communication on Environment and Development (ECO-ED), Held in Toronto, Ontario, 17 October 1992. United Nations Educational, Scientific and Cultural Organization (UNESCO), Paris.

UNESCO, 1992b. The International Environmental Education Programme 1992–1993. The United Nations Educational, Scientific and Cultural Organization (UNESCO), Paris, UNESCO-UNEP Environmental Education Newsletter, XVII (1), 1–8.

UNESCO, 1994a. Environment and Population Education and Information for Human Development: UNESCO's Interdisciplinary and Inter-Agency Cooperation Project. The United Nations Educational, Scientific and Cultural Organization (UNESCO), Paris, UNESCO-UNEP Environmental Education Newsletter, XIX (1), 1–4.

UNESCO, 1994b. Environment and Population Education and Information for Human Development (EPD). UNESCO's Interdisciplinary and Inter-Agency Cooperation Project. The United Nations Educational, Scientific and Cultural Organization (UNESCO), Paris.

UNESCO, 1995. Two Decades of Successful Interagency Cooperation for the Worldwide Promotion of Environmental Education. The United Nations Educational, Scientific and Cultural Organization (UNESCO), Paris, UNESCO-UNEP Environmental Education Newsletter, XX (3), 1–16.

UNESCO, 1997a. Educating for a Sustainable Future: A Transdisciplinary Vision for Concerted Actions. The United Nations Educational, Scientific and Cultural Organization (UNESCO), Paris.

UNESCO, 1997b. International Conference on Environment and Society: Education and Public Awareness for Sustainability—Declaration of Thessaloniki. The United Nations Educational, Scientific and Cultural Organization (UNESCO), Paris.

UNESCO, 2000a. Report of the Fifth UNESCO-ACEID International Conference on Education. Reforming Learning, Curriculum and Pedagogy: Innovative Visions for the New Century. United Nations Educational, Scientific and Cultural Organization (UNESCO), Paris.

UNESCO, 2000b. World Education Forum. Dakar, Senegel 26–28 April 2000. Final Report. United Nations Educational, Scientific and Cultural Organization (UNESCO), Paris.

UNESCO, 2001. The Environmental Dimensions of Dialogue Among Civilizations. Address by Mr. Koichiro Matsuura, Director-General of the United Nations Educational, Scientific and Cultural Organization (UNESCO) at the Working Dinner on Dialogue Among Civilizations at the UNEP Global Ministerial Forum. United Nations Educational, Scientific and Cultural Organization (UNESCO), Paris.

UNESCO, 2002. Education for Sustainability. From Rio to Johannesburg: Lessons Learnt From a Decade of Commitment. United Nations Educational, Scientific and Cultural Organization (UNESCO), Paris.

UNESCO, 2003a. Fifth Ministerial Conference. Environment for Europe. Kiev, Ukraine 21–23 May 2003. Declaration by the Environmental Ministers of the Region of the United Nations Economic Commission for Europe (UNECE). United Nations Educational, Scientific and Cultural Organization (UNESCO), Paris.

UNESCO, 2003b. General Conference 32nd Session. Commission III. Resolution 32 C/COM.III/DR.1. 1 October 2003. United Nations Educational, Scientific and Cultural Organization (UNESCO), Paris.

UNESCO, 2004. Second Regional Meeting on Education for Sustainable Development. Rome, 15–16 July 2004. Report of the Meeting. United Nations Educational, Scientific and Cultural Organization (UNESCO), Paris.

UNESCO, 2005. United Nations Decade of Education for Sustainable Development (2005–2014). International Implementation Scheme. United Nations Educational, Scientific and Cultural Organization (UNESCO), Paris.

UNESCO, 2007. The UN Decade of Education for Sustainable Development (DESD 2005–2014). The First Two Years. United Nations Educational, Scientific and Cultural Organization (UNESCO), Paris.

UNESCO, 2013. Education for Sustainable Development in Biosphere Reserves and other Designated Areas. A Resource Book for Educators in South-Eastern Europe and the Mediterranean. The United Nations Educational, Scientific and Cultural Organization (UNESCO), Venice.

UNESCO, 2014a. UNESCO World Conference on Education for Sustainable Development Conference Report by the General Rapporteur Heila Lotz-Sistika, Professor, Rhodes University. United Nations Educational, Scientific and Cultural Organization (UNESCO), Paris.

UNESCO, 2014b. UNESCO Roadmap for Implementing the Global Action Programme on Education for Sustainable Development. United Nations Educational, Scientific and Cultural Organization (UNESCO), Paris.

UNESCO, 2014c. Shaping the Future We Want. UN Decade of Education for Sustainable Development (2005–2014). Final Report. United Nations Educational, Scientific and Cultural Organization (UNESCO), Paris.

UNESCO, 2014d. Sustainable Lifestyles and Education Programme. United Nations Educational, Scientific and Cultural Organization (UNESCO), Paris.

UNESCO, 2015. World Education Forum 2015. Final Report. United Nations Educational, Scientific and Cultural Organization (UNESCO), Paris.

UNESCO-BMBF-German Commission for UNESCO, 2009. UNESCO World Conference on Education for Sustainable Development. 31 March–2 April 2009. Bonn, Germany. Proceedings. United Nations Educational, Scientific and Cultural Organization (UNESCO), Paris. Federal Ministry of Education and Research (BMBF), Berlin. German Commission for UNESCO, Bonn.

UNESCO-UNEP, 1988. International Strategy for Action in the Field of Environmental Education and Training for the 1990s. The United Nations Educational, Scientific and Cultural Organization, Paris, The United Nations Environment Programme, Nairobi.

UNESCO/UNEP- MIO/ECSDE, 1995. Inter-Regional Workshop: Re-Orienting Environmental Education for Sustainable Development, Athens, June 26–30, 1995. United Nations Educational, Scientific and Cultural Organization (UNESCO), Paris; United Nations Environment Programme (UNEP), Nairobi; Mediterranean Information Office for Environment, Culture and Sustainable Development (MIO/ECSDE), Athens.

UNESCO-UNFPA, 1994. First International Congress Population Education and Development. Action Framework for Population Education on the Eve of the Twenty-First Century. Istanbul Declaration. The United Nations Educational, Scientific and Cultural Organization (UNESCO), Paris, United Nations Population Fund (UNFPA), New York.

UNESCO-UNU, 2006. Globalization and Education for Sustainable Development. Sustaining the Future. United Nations Educational, Scientific and Cultural Organization (UNESCO), Paris, United Nations University (UNU), Tokyo, Japan.

United Nations, 1972. Report of the United Nations Conference on the Human Environment, Stockholm, June 1972. (UN.Doc. A/CONF/48/14/REV.1). United Nations, New York.

United Nations, 1992. United Nations Conference on Environment & Development. Rio de Janeiro, Brazil, 3 to 14 June 1992. Agenda 21. United Nations: United Nations for Sustainable Development.

United Nations, 2005. UNECE Strategy for Education for Sustainable Development. United Nations Economic and Social Council, New York.

United Nations, 2010. Mid-Decade Review of the United Nations Decade of Education for Sustainable Development, 2005–2014. Report of the Director-General of the United Nations Educational, Scientific and Cultural Organisation. United Nations, New York.

United Nations, 2016. The 2030 Agenda and the Sustainable Development Goals. An opportunity for Latin America and the Caribbean. United Nations, New York.

UNPF, 2014. Programme of Action Adopted at the International Conference on Population and Development Cairo, 5–13 September 1994, 20th Anniversary Edition United Nations Population Fund (UNPF), New York.

Verhulst, E., Lambrechts, W., 2015. Fostering the incorporation of sustainable development in higher education. Lessons learned from a change management perspective. J. Clean. Prod. 106, 189–204.

Vujica, M.S. 2013. Myth Busters: Will Bees Become Extinct? How Will Food Be Affected? The Epoch Times. < https://www.theepochtimes.com/myth-busters-will-bees-become-extinct-how-will-food-be-affected_344973.html > (accessed 03.09.19.).

Wall, J., et al., 2014. Novel opportunities for wildlife conservation and research with real-time monitoring. Ecol. Appl. 24 (4), 593–601.

WCED (World Commission on Environment and Development), 1987. Our Common Future. Oxford University Press, Oxford, New York.

WCEFA Inter-Agency Commission, 1990. World Conference on Education for All: Meeting the Basic Learning Needs. Final Report. WCEFA Inter-Agency Commission (UNDP, UNESCO, UNICEF, World Bank).

Wells, N.M., Lekies, K.S., 2006. Nature and the life. Course: pathways from childhood nature experiences to adult. Environmentalism. Child. Youth Environ. 16 (1), 1–24.

Weyler, R., 2013. Worldwide Honey Bee Collapse: A Lesson in Ecology. EcoWatch, <https://www.ecowatch.com/worldwide-honey-bee-collapse-a-lesson-in-ecology-1881760601.html> (accessed 16.12.19.).

Wold, C., Hunter, D., Powers, M., 2013. Climate Change and the Law, second ed. Carolina Academic Press, Durham, NC.

World Ocean Review, 2010. World Ocean Review 1. Living With the Ocean. Maribus, Hamburg in Cooperation With the Future Ocean. Germany and The International Ocean Institute, Malta.

Zeebe, R.E., 2012. History of seawater carbonate chemistry, atmospheric CO_2, and ocean acidification. Annu. Rev. Earth Planet. Sci. 40, 141−165.

Index

Printed in the United States
By Bookmasters